Principles and Prevention of

CORROSION

Second Edition

Denny A. Jones

Department of Chemical and Metallurgical Engineering
University of Nevada, Reno

Prentice Hall, Upper Saddle River, NJ 07458

KOREA edition published by PEARSON EDUCATION KOREA LTD. Copyright © 2008

This edition is manufactured in KOREA, and is authorized for sale only in KOREA.

ISBN 978-89-450-3501-1
ISBN 89-450-3501-X

Principles and Prevention of
CORROSION

Second Edition

Denny A. Jones

Department of Chemical and Metallurgical Engineering
University of Nevada, Reno

Prentice Hall, Upper Saddle River, NJ 07458

Library of Congress Cataloging-in-Publication Data

Jones, Denny A.
    Principles and prevention of corrosion / Denny A. Jones. — 2nd
ed.
        p. cm.
    Includes bibliographical references and index.
    ISBN 0–13–359993–0
    1. Corrosion and anti-corrosives   I. Title.
TA462.J59   1996
620.1´1223—dc20

Acquisitions editor: Bill Stenquist
Production editor: Rose Kernan
Copy editor: Patricia Daly
Cover designer: Bruce Kenselaar
Buyer: Donna Sullivan
Editorial assistant: Meg Weist

© 1996, 1992 by Prentice-Hall, Inc.
Simon & Schuster /A Viacom Company
Upper Saddle River, NJ 07458

The author and publisher of this book have used their best efforts in preparing this book.
These efforts include the development, research, and testing of the theories and programs to
determine their effectiveness. The author and publisher make no warranty of any kind,
expressed or implied, with regard to these programs or the documentation contained in this
book. The author and publisher shall not be liable in any event for incidental or
consequential damages in connection with, or arising out of, the furnishing, performance,
or use of these programs.

Printed in the United States of America

10   9

ISBN 0-13-359993-0

Prentice-Hall International (UK) Limited, London
Prentice-Hall of Australia Pty. Limited, Sydney
Prentice-Hall Canada Inc., Toronto
Prentice-Hall Hispanoamericana, S.A., Mexico
Prentice-Hall of India Private Limited, New Delhi
Prentice-Hall of Japan, Inc., Tokyo
Simon & Schuster Asia Pte. Ltd., Singapore
Editora Prentice-Hall do Brasil, Ltda., Rio de Janeiro

Contents

2. Electrochemical Thermodynamics and Electrode Potential   40

3. Electrochemical Kinetics of Corrosion   75

4. Passivity   116

5. Polarization Methods to Measure Corrosion Rate   143

6. Galvanic and Concentration Cell Corrosion   168

7. Pitting and Crevice Corrosion 199

8. Environmentally Induced Cracking 235

9. Effects of Metallurgical Structure on Corrosion   291

10. Corrosion-Related Damage by Hydrogen, Erosion, and Wear 334

11. Corrosion in Selected Corrosive Environments 357

12. Atmospheric Corrosion and Elevated Temperature Oxidation   399

13. Cathodic Protection   439

Preface

The title of this book was chosen to reflect a blend of science (*principles*) and engineering (*prevention*) in the study of metal corrosion. The instructor or the student can pick and choose between principles and prevention to obtain a blend that satisfies his or her background and interests. The book is planned so that the instructor can easily design a course at any desired level—or so that an individual can pick the book up and easily read selected sections to get a quick introduction to the subject. In general, most chapters progress from qualititative, descriptive sections, including methods of prevention and testing, to more quantitative sections, involving metallurgy and electrochemistry, and finally to the later sections on current research and the latest knowledge of the chapter topic. Thus, the casual reader or serious student can progress in a given chapter to any desired level.

The essentials of a senior/graduate level course are contained in the book, which deliberately contains more information than can be included in a fifteen-week semester. Most chapters include material beyond that which need be treated in an introductory course. Thus, the book can provide at least the starting point for a more-advanced graduate-level course as well. The additional material also adds a valuable component for future reference by practicing engineers, whether or not they may have taken a formal academic course in corrosion. I have included material that I have wished during my career could have been included in my first course on corrosion or had been available for concise reference along the way.

An understanding of the principles of electrochemistry is recommended to provide students with the tools to deal with diverse problems of analysis and prevention. However, the essential elements of electrochemical corrosion are presented in Chapter 1, and the more engineering-oriented instructor could easily omit the added detail of Chapters 2, 3, and 4, concentrating subsequently on the engineering aspects of corrosion testing and prevention.

The book does not provide encyclopedic coverage for the subject of corrosion and corrosion prevention. The emphasis is on the principles and methods leading to reasonable mitigation of the great majority of corrosion problems in engineering practice. Thus, the book is designed to be useful as a text for formal university engineering courses at the senior/first year graduate level and as a reference for the neophyte, the practicing engineer, or the beginning researcher. A primary reference within this book is the recent *Metals Handbook*, Volume 13, *Corrosion*, 9th ed., published by ASM International, which is recommended as a supplement.

Chapter 1 is an overview of the field, including social and economic importance, the forms of corrosion, the electrochemical nature of corrosion, and methods of simple exposure testing. Photographic and metallographic examples of the various forms of corrosion are introduced briefly to familiarize the student or engineer with the appearance of corrosive attack in practice. These examples are referred to in later chapters with additional examples, when appropriate. The first chapter concludes with a description of exposure testing for general and atmospheric corrosion, to introduce the student to the practice of corrosion engineering, as well as to impart information for future reference.

Descriptions of testing methods have been integrated with discussions of the various forms of corrosion throughout the book, avoiding a separate chapter on corrosion testing. In my opinion, close association of testing methods with discussion of the corresponding forms of corrosion benefits both. Principles are better understood from the practical aspects of testing, and testing procedures are better designed and studied on the basis of scientific principle.

The book proceeds with chapters on the thermodynamic and kinetic aspects of electrochemistry, including potential-pH (Pourbaix) diagrams, mixed potential theory, and the theory and application of passivity, all of which are necessary to understand the underlying mechanisms of corrosion. These fundamental principles provide the foundation for subsequent discussions of anodic protection, galvanic corrosion, electrochemical methods to study and measure corrosion, and pitting and crevice corrosion. The order of treatment thereafter is somewhat arbitrary and can be varied to suit the preferences of the instructor. Environmentally induced cracking (including stress corrosion, corrosion fatigue, and hydrogen induced cracking) is treated next, because of its importance in many modern applications. Subsequent topics also have considerable practical importance and include intergranular corrosion, dealloying, and effects of welding, hydrogen, erosion, and wear on corrosion. This constitutes the first ten chapters, which cover the electrochemical and metallurgical fundamentals of the various forms of aqueous corrosion. Chapter 11 continues with a description of solution parameters affecting aqueous corrosion, including pH, dissolved gases, temperature, dissolved salts, sulfides, and biological microorganisms. The chapter concludes with corrosion in some environments of special engineering interest—soils, concrete, and acid and alkaline chemical process streams. The first eleven chapters could be considered the basic core of the book, and some parts of them would be included in most introductory courses. The remaining chapters are optional and could be omitted to suit the student or instructor.

Chapter 12 is devoted to corrosion in the vapor phase: ambient-temperature atmospheric corrosion and elevated-temperature oxidation and scaling. This chapter is designed to introduce students to oxidation behavior of metals and alloys and to satisfy interested instructors, who wish to include the subject in their treatments of corrosion. Oxide structure and growth kinetics are described, and the engineering behavior of commercial alloys in high-temperature service conditions is included.

The last three chapters are devoted to the major methods of corrosion prevention. Prevention is also discussed in conjunction with corresponding corrosion phenomena throughout the earlier chapters. However, the topics of these concluding chapters merit separate treatment. Chapter 13 extensively treats the electrochemical principles and engineering design of cathodic protection systems. This treatment of cathodic protection is probably more thorough than would normally be necessary for an introductory textbook, because of my personal interest in the subject, and because a similar treatment at the same level is apparently unavailable for reference elsewhere. Chapter 14 describes the formulation and application of organic polymer coatings, the application and function of galvanic and barrier metal coatings, and the principles and uses of dissolved inhibitors. Chapter 15 is considered the capstone of the book. It treats materials and alloy selection, the major method for corrosion prevention, as well as engineering-design measures for corrosion prevention. In so doing, it reviews considerable information covered previously in the book. Chapter 15 concludes with a discussion of economic calculations applied to corrosion prevention methods, with examples.

Alloy compositions are not included in the chapters but have been assembled together in the rear endpapers for convenient reference. I have not attempted to cite original references for many of the concepts discussed. The field of corrosion has matured to the point that recent advances are more instructive to the student than historical references published decades ago, however interesting and important they may be. Thus, I have often cited more recent and accessible textbooks, handbooks, and symposia proceedings, that give a more generalized account and include the original references. An instructor's manual, available on request from the publisher, contains solutions to the exercises that conclude each chapter (except Chapters 10 and 14), and a set of multiple choice questions for further discussion or examination.

Acknowledgments

I cannot adequately express gratitude to my wife, Wanda, whose patience and encouragement played the greatest role in sustaining me through the challenge of writing this book. Our children, Rena, Gillie, Mike, and Bryce also deserve my appreciation for providing the ballgames, dances, missions, boyfriends, girlfriends, teasing, laughter, and grandchildren that keep my professional strivings in the proper perspective. On reflection, I find that writing a book and raising a family have much in common; both respond to recurring and sometimes subliminal desires to leave behind something of value.

Many professional mentors, co-workers, and friends have taught me, labored with me, advised me, and supported me through my career as a student, researcher, and instructor of corrosion. The following deserve special appreciation in chronological order: Professor L. J. Demer at The University of Arizona; R. E. Westerman, A. B. Johnson, and R. L. Dillon at the Battelle-Northwest Laboratories; Professor N. D. Greene, Jr. at Renesselaer Polytechnic Institute; T. J. Summerson and T. A. Lowe at Kaiser Aluminum and Chemical Corp.; R. Bandy at the University of Hawaii; B. E. Wilde, C. D. Kim, and A. W. Loginow at the U. S. Steel Research Laboratory; and J. P. N. Paine at the Electric Power Research Institute.

The following individuals have generously supplied photographs: A. K. Agrawal, Battelle-Columbus; P. J. Andersen, Zimmer Inc.; W. G. Ashbaugh, Cortest; A. I. Asphahani, Haynes International; A. Cohen, Copper Development Association; T. A. DeBold, Carpenter Technology; R. J. Eiber, Battelle-Columbus; E. Escalante, National Institute of Standards and Technology; W. A. Glaeser, Battelle-Columbus;

B. M. Gordon, General Electric; J. Gutzeit, Amoco; R. H. Heidersbach, California State Polytechnic University, San Luis Obispo; R. L. Jones, Electric Power Research Institute; R. M. Kain, LaQue Center for Corrosion Technology; C. D. Kim, USX Corp.; W. W. Kirk, LaQue Center for Corrosion Technology; L. J. Korb, Rockwell International; K. F. Krysiak, Hercules; B. M. Lifka, Alcoa; G. A. Minick, A. R. Wilfley & Sons (retired), C. G. Munger, Consultant; R. J. Neville, Dofasco; S. L. Pohlman, Kennecott; R. W. Revie, Canada Centre for Mineral Energy and Technology; G. Schick, Bell Communications Research; D. O. Sprowls, Alcoa (retired), N. B. Tipton, Singleton Corp.; D. J. Wulpi, Consultant; and R. A. Yeske, Institute of Paper Science and Technology. The National Association of Corrosion Engineers and ASM International kindly provided several printing-plate negatives of photographic prints, which were no longer available.

Invaluable review comments on selected chapters of the final manuscript have come from N. Birks, University of Pittsburgh; W. K. Boyd, Battelle-Columbus Laboratories (retired); T. M. Devine, University of California-Berkeley; S. C. Dexter, University of Delaware; R. A. Dodd, University of Wisconsin, Madison; W. H. Hartt, Florida Atlantic University; R. H. Heidersbach, California State Polytechnic University, San Luis Obispo; M. Marek, Georgia Institute of Technology; P. J. Moran, Johns Hopkins University, D. L. Olson, Colorado School of Mines; R. A. Rapp, Ohio State University; J. M. Rigsbee, University of Illinois; and E. D. Verink, University of Florida. I hasten to add, however, that reviewers can only help a manuscript, not perfect it. Furthermore, I have not always taken their advice and, in any event, accept full responsibility for any shortcomings.

A Faculty Development Leave from the University of Nevada, Reno made time available for the bulk of the writing. Office space and support from the Desert Research Institute, Dr. J. V. Taranik, President, made the leave both pleasant and productive in Las Vegas, Nevada.

D.A.J.
August 1991
August 1995

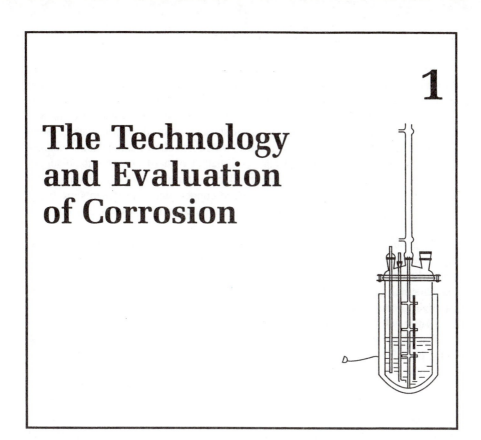

The Technology and Evaluation of Corrosion

1

1.1 Introduction

1.1.1 Significance of Corrosion

Corrosion intrudes itself into many parts of our lives. The great majority of us have a personal feeling for the definition and importance of corrosion. Far too many have cringed at the emergence of rust holes in the body panels of a relatively new automobile whose loan has not yet been amortized. The outdoor rusting of steel household and garden appliances is a common fact of life. Few homeowners have not borne the expense of a new water heater after the old one started leaking. Expensive repairs to home plumbing systems and the water and sewer utility lines serving our homes are all too common. All have seen the stains on cooking utensils from hot foods or experienced the metallic taste in acid foods stored too long in opened cans. That these effects are caused by "corrosion" is well known.

Furthermore, we are all familiar with many of the measures taken to combat corrosion. Many have spent hundreds of dollars on aftermarket corrosion protection systems for new automobiles to delay the dreaded rust appearance noted above. Automobile manufacturers have devoted millions of dollars to develop and test the materials and designs that allow guarantees against rust formation in current new-car models. Stainless steel cooking utensils are sold mainly to resist corrosion and maintain a bright, attractive, easy-to-clean surface. Glass-lined hot water heaters have the interior of their steel tanks covered with a porcelain enamel "glass" coating (Section 14.2.2) to forestall corrosion. The shelves of hardware and home supply stores are well stocked with paints and caulking materials to protect steel and other metals against corrosion. Older cast iron and copper plumbing systems are being replaced with plastic which is nominally corrosion resistant. Many readers could add still more examples to the list.

The public is, however, much less aware of many other burdens heaped on society by corrosion. Public and private water companies must not only treat their water routinely for potability and taste, but also for minimum corrosivity toward the company distribution lines and the home plumbing systems which they serve. Even comparatively unaggressive environments can have a devastating effect when combined with other deleterious conditions. The spectacular airline accident in which part of the fuselage tore away during an airline flight in Hawaii was in all probability due to the combined effects of cyclic mechanical stress and atmospheric corrosion in a semitropical environment.[1]

Electric power generation plants, whether nuclear or fossil fueled, require costly alloys as well as elaborate and expensive treatment systems for cooling water and steam, all to minimize corrosion. The petroleum industry is plagued by corrosion problems. Many wells now contain H_2S, which contributes not only acid, H^+, but also sulfide, S^{2-}, both of which accelerate corrosion in the high-temperature conditions in deep wells. Oil refineries have a myriad of complex chemicals, often at high temperature, and the constructional alloys and corrosion control systems are correspondingly complex and expensive. The alarming deterioration of concrete bridges, overpasses, and other public structures is accelerated by the corrosion of reinforcing steel within the concrete.

1.1.2 Purposes

Airplanes, electric power generating plants, chemical process and manufacturing plants, concrete structures, and many others—all are designed, operated, and maintained by or under the supervision of engineers. Engineers are consequently responsible for minimizing the costs and risks to the public from corrosion. Yet many engineers, past and present, are woefully ignorant of the causes of corrosion and methods to prevent and minimize its effects. Many university engineering departments have responded by incorporating into their curricula courses devoted to corrosion. *The purpose of this book is, first, to provide a textbook for upper-division and first-year graduate engineering courses in corrosion.* However, many graduate engineers now actively engaged in corrosion-related problem solving have little training in corrosion due to lack of availability or interest during their formal education. Thus, *a second purpose of this book is to provide an introduction and self-study reference for the practicing engineer.*

The chemical and metallurgical processes underlying corrosion are treated to assist engineers in innovation and evaluation of prevention programs. In the latter sections of many chapters, current advances in science and engineering are discussed for the benefit of the researcher seeking recent information. But, of course, all aspects of corrosion cannot be covered because of cost and size limitations on a university textbook. More encyclopedic works, already available, are liberally referenced, along with other key articles, to give readers access to more detailed information when needed.

The purpose of this chapter is to provide an overview for the remainder of the book. The reader can obtain a convenient introduction to the subject and follow up in subsequent chapters for more information as desired. The forms of corrosion are described briefly, but details are deferred to later chapters. Exposure testing for evaluation of corrosion resistance is generally applicable and is treated in this chapter to add further background as well as valuable information for reference. Additional testing procedures, including electrochemical methods, are distributed appropriately through the book, avoiding a separate chapter dedicated to corrosion testing and evaluation. The goal is to improve the correlation between the scientific principles of corrosion and the engineering evaluation of corrosion resistance.

1.2 Costs of Corrosion

1.2.1 Economic

The economic costs of corrosion in the United States of America alone have been estimated between $8 billion and $126 billion per year.[2] The most comprehensive study on the economic cost of corrosion in the United States, performed in 1976, estimated an annual cost of $70 billion,[3] while the Department of Commerce has stated that corrosion would cost the United States an estimated $126 billion in 1982.[4] These figures are somewhat misleading because zero corrosion assumed as the baseline is unattainable. Some measures, while reducing corrosion, will cost more than the parts or equipment protected. A perhaps more realistic annual $30 billion has been suggested[2] for the savings that could result if all economically useful measures were taken to prevent or minimize corrosion. *Nevertheless, the economic costs of corrosion are obviously enormous.*

The figures given above are only the *direct* economic costs of corrosion. The *indirect* costs resulting from actual or possible corrosion are more difficult to evaluate but are probably even greater. Some of the more important sources of indirect costs are summarized as follows.[5]

1. *Plant Downtime*: Parts and labor to replace corroded equipment are often minor compared to the loss of production while the plant is inoperable during repairs. For example, the cost to replace power from a shutdown nuclear power plant can run into millions of dollars per day.

2. *Loss of Product*: Leaking containers, tanks, and pipelines result in significant losses in product, which have a high cost. These leaks and spills have a correspond-

ing hazardous effect on the surrounding environment and safety of the populace. Thousands of leaking automotive service station fuel tanks must be repaired or replaced to prevent recurring losses of fuel and contamination of groundwater. The expense is enormous, with an impact on the livelihood of many independent station operators.[6]

3. *Loss of Efficiency*: Accumulated corrosion products on heat exchanger tubing and pipelines decrease the efficiency of heat transfer and reduce the pumping capacity, respectively.

4. *Contamination*: Soluble corrosion products can spoil chemical preparations of soap, dyes, and pharmaceuticals, among others. A major problem in many nuclear power generating plants is the transport of radioactive corrosion products from the reactor core followed by deposition in other parts of the cooling water system, necessitating dangerous and expensive shutdowns to decontaminate. In times past, lead pipes released toxic corrosion products into soft waters for public consumption. Fortunately, lead has long been prohibited from use in this application.

5. *Overdesign*: In the absence of adequate corrosion rate information, overdesign is required to ensure reasonable service life, resulting in wasted resources, and greater power requirements for moving parts. Engineers sometimes overdesign through ignorance of available information. The needed information may not be available, however. Geologic storage of high-level nuclear waste in deep repositories, mandated by the U.S. Congress, requires containers that will maintain their integrity for 300 to 1000 years.[7] Corrosion data in environments resembling the relevant groundwaters in radiation fields at elevated temperature are not available and are currently being generated in extensive test programs. However, extrapolations of such data, gathered in a few years at most, are highly uncertain, and considerable overdesign will almost certainly be included.

Were it not so insidious and ubiquitous, corrosion could be classified collectively as the greatest economic calamity known to humankind. As it is, corrosion is the root cause of many individual economic calamities.

1.2.2 Human Life and Safety

Corrosion takes an even more significant toll in human life and safety. Corrosion is never correctly reported as the cause of many fatal accidents, but a few examples can be cited readily. The airline accident in Hawaii has already been mentioned. The well-known bridge collapse at Pt. Pleasant, West Virginia, killed 46 in 1967 and has been attributed to stress corrosion cracking.[8] Selective corrosion of a welded liquid gasoline pipeline in Minnesota in 1986[9] resulted in massive fire damage to an entire town and the loss of two lives. Again in 1986, a steam pipe, suffering the combined effects of corrosion and erosion, burst at a Virginia nuclear power generating plant, severely burning eight nearby workers, four of whom died later.[10] These are only a few examples of the many such fatal accidents caused by corrosion.

1.3 Definitions

1.3.1 Corrosion

Corrosion is the destructive result of chemical reaction between a metal or metal alloy and its environment. Metal atoms in nature are present in chemical compounds (i.e., minerals). The same amounts of energy needed to extract metals from their minerals are emitted during the chemical reactions that produce corrosion. Corrosion returns the metal to its combined state in chemical compounds that are similar or even identical to the minerals from which the metals were extracted. Thus, corrosion has been called extractive metallurgy in reverse.[3]

Many nonmetallic materials, such as ceramics, consist of metals that have their chemical reactivity satisfied by the formation of bonds with other reactive ions, such as oxides and silicates. Thus, such materials are chemically unreactive, and they degrade by physical breakdown at high temperature or by mechanical wear or erosion. Similarly, organic polymers (plastics) are relatively unreactive because they have very stable covalent bonding, primarily between carbon atoms. The degradation of such materials is not defined here as corrosion, and the treatment of such processes is generally beyond the scope of this book. However, the degradation and failure of protective polymer coatings on metals is an important exception that is discussed in Section 14.1.4.

1.3.2 Corrosion Science and Engineering

Corrosion science is the study of the chemical and metallurgical processes that occur during corrosion. Corrosion engineering is the design and application of methods to prevent corrosion. Ideally, science should be married to engineering so as to invent new and better methods of prevention and apply existing methods more intelligently and effectively. However, scientists are sometimes devoted to the pursuit of pure knowledge, with little or no perspective on the possible applications of their work. On the other hand, engineers often apply time-honored methods with little or no understanding of the principles behind them. A goal of this book is to encourage further cooperation for the benefit of current and future students in both engineering and science. All are winners as the costs of corrosion to society are reduced.

1.4 Electrochemical Nature of Aqueous Corrosion

Nearly all metallic corrosion processes involve transfer of electronic charge in aqueous solutions. Thus, it is necessary to discuss the electrochemical nature of corrosion briefly before continuing with descriptions of the various forms of corrosion in Section 1.5.

1.4.1 Electrochemical Reactions

Consider an example of corrosion between zinc and hydrochloric acid represented by the following overall reaction:

$$Zn + 2HCl \rightarrow ZnCl_2 + H_2. \tag{1}$$

Zinc reacts with the acid solution forming soluble zinc chloride and liberating hydrogen bubbles on the surface. Reactions such as this are used for surface cleaning and pickling of many metals and alloys. In ionic form the reaction is

$$Zn + 2H^+ + 2Cl^- \rightarrow Zn^{2+} + 2Cl^- + H_2.$$

Eliminating Cl^- from both sides of the reaction gives

$$Zn + 2H^+ \rightarrow Zn^{2+} + H_2. \tag{2}$$

Thus, the same corrosion reaction would occur in sulfuric acid. Reaction (2) can be separated as follows:

$$Zn \rightarrow Zn^{2+} + 2e^- \quad \text{anodic reaction.} \tag{3}$$

$$2H^+ + 2e^- \rightarrow H_2 \quad \text{cathodic reaction.} \tag{4}$$

Reaction (3), defined as the anodic reaction, is an oxidation in which zinc valence increases from 0 to +2, liberating electrons, e, while (4), defined as the cathodic reaction, is a reduction in which the oxidation state of hydrogen decreases from +1 to 0, consuming electrons.

The composite reaction involving charge transfer or exchange of electrons is shown schematically in Figure 1.1. The metal dissolves by (3) liberating electrons into the bulk of the metal which migrate to the adjoining surface, where they react with H+ in solution to form H_2 by (4). The summation of (3) and (4), of course, gives

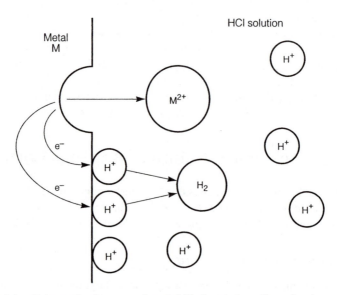

FIGURE 1.1 Schematic diagram of metal M dissolution, liberating into solution a metal ion M^{2+} and into the metal electrons, e^-, which are consumed by reduction of H^+ to H_2.

(2). Water is required as the carrier for ions, such as Zn^{2+} and H^+, and is called the electrolyte.

The separation of (2) to give (3) and (4) with exchange of electrons, e^-, is artificial unless there is some evidence that electrons are actually involved. When an excess of electrons is supplied to the metal in Figure 1.1, it is always observed that the rate of corrosion, expressed by the anodic reaction (3), is reduced, while the rate of hydrogen evolution reaction (4) is increased. All corrosion reactions in water involve an anodic reaction such as (3); application of a negative potential with attendant excess electrons always decreases the corrosion rate. This is the basis of cathodic protection for the mitigation of corrosion of pipelines, offshore oil drilling structures, and steel hot water tanks, as described in Chapter 13.

Thus, all aqueous corrosion reactions are considered to be electrochemical. Most corrosion reactions do involve water in either the liquid or condensed vapor phases. Even some "dry" corrosion reactions without water involve charge transfer in a solid state electrolyte (Chapter 12) and are considered still to be electrochemical.

For corroding metals, the anodic reaction invariably is of the form

$$M \rightarrow M^{n+} + ne^-. \tag{5}$$

Examples in addition to (3) are

$$Fe \rightarrow Fe^{2+} + 2e^-,$$

$$Ni \rightarrow Ni^{+2} + 2e^-,$$

$$Al \rightarrow Al^{+3} + 3e^-.$$

Cathodic reduction reactions significant to corrosion are few in number. The simplest and one of the most common is reduction of hydrogen ions (2) in acid solution. Another is reduction of an oxidized ion in solution by a so-called redox reaction; the most important example is reduction of ferric to ferrous ions,

$$Fe^{3+} + e^- \rightarrow Fe^{2+}.$$

Others, such as

$$Sn^{4+} + 2e^- \rightarrow Sn^{2+}$$

are present infrequently and are consequently less important. The reduction of dissolved oxygen is often observed in neutral and acid solutions exposed to ambient air. The respective reduction reactions are

$$O_2 + 2H_2O + 4e^- \rightarrow 4OH^-,$$

and

$$O_2 + 4H^+ + 4e^- \rightarrow 2H_2O.$$

In the absence of all other reduction reactions, water will be reduced by

$$2H_2O + 2e^- \rightarrow H_2 + 2OH^-,$$

which is equivalent to (4), assuming dissociation of water to H^+ and OH^- and subtracting OH^- from both sides of the reaction.

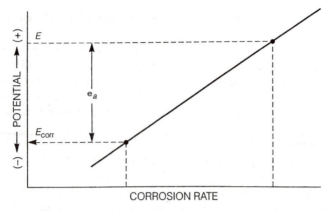

FIGURE 1.2   Schematic increase in corrosion rate with increasing potential E and anodic polarization ε_a.

1.4.2 Polarization

Electrochemical reactions such as (3) and (4) proceed only at finite rates. If electrons are made available to (4), the potential at the surface becomes more negative, suggesting that excess electrons with their negative charges accumulate at the metal/solution interface waiting for reaction. That is, the reaction is not fast enough to accommodate all the available electrons. This negative potential change is called cathodic polarization. Similarly, a deficiency of electrons in the metal liberated by (3) at the interface produces a positive potential change called anodic polarization. As the deficiency (polarization) becomes greater, the tendency for anodic dissolution becomes greater. Anodic polarization thus represents a driving force for corrosion by the anodic reaction (3). When surface potential measures more positive (Section 2.3), the oxidizing (or corrosive) power of the solution increases because the anodic polarization is greater.

In an aqueous electrolyte solution the surface will reach a steady state potential, E_{corr}, which depends on the ability and rate at which electrons can be exchanged by the available cathodic and anodic reactions (Section 3.3.1). As the surface potential increases above E_{corr} to E, the anodic reaction rate or corrosion rate generally increases, as shown schematically in Figure 1.2. Anodic polarization is defined as $\varepsilon_a = E - E_{corr}$. Without polarization, the slightest driving force would produce very high rates, and the line in Figure 1.2 would be horizontal with zero slope. Reaction rates and polarization of electrochemical reactions at metal surfaces are discussed in more detail in Chapter 3.

1.4.3 Passivity

In many metals, including iron, nickel, chromium, titanium, and cobalt, the corrosion rate decreases above some critical potential E_p, as shown in Figure 1.3. This corrosion resistance above E_p, despite a high driving force for corrosion (i.e., high anodic polarization), is defined as passivity. Below E_p the alloy corrodes at a rela-

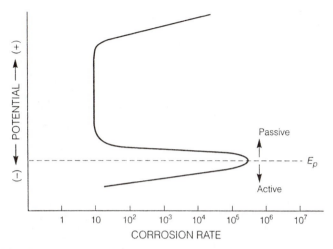

FIGURE 1.3   Passivity at oxidizing potentials above E_p.

tively high rate. Passive corrosion rates are very low; a reduction of 10^3 to 10^6 times below the corrosion rate in the active state is not unusual. Passivity is caused by formation of a thin, protective, hydrated oxide, corrosion-product surface film that acts as a barrier to the anodic dissolution reaction. Depending on the potential, or oxidizing power of the solution, an alloy may exist in the passive state above E_p or in the active state below it. For example, Type 304 stainless steel is passive in aerated but active in deaerated salt water.

Fortunately, the list of active-passive metals includes many used in our most common constructional alloys, which consequently rely on passivity for corrosion resistance. Chromium is a key alloying element forming resistant passive oxide films on the surface. Although it cannot be used alone because of brittleness, chromium enhances passivity when alloyed with other metals, especially iron and nickel in the stainless steels.

Passivity comes not without attendant problems. Because the passive film is thin and often fragile, its breakdown can result in unpredictable localized forms of corrosion, including pitting, corrosion in crevices, and embrittlement by stress corrosion cracking. In Chapter 4, we address passivity in more detail. Much of the rest of this book treats aspects of corrosion that result from losses or breakdown in passivity, beginning with the following sections describing the various forms of corrosion.

1.5 Forms of Corrosion

Various forms of corrosion are pictured schematically in Figure 1.4 and listed as follows:

- Uniform Corrosion
- Galvanic Corrosion
- Crevice Corrosion

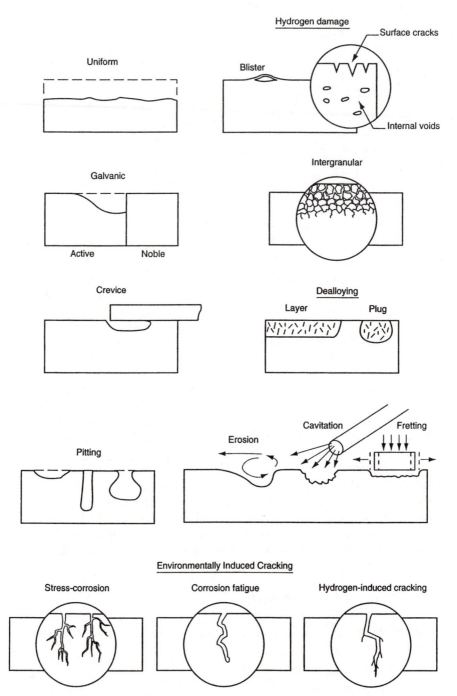

FIGURE 1.4   Schematic summary of the various forms of corrosion.

- Pitting Corrosion
- Environmentally Induced Cracking
- Hydrogen Damage
- Intergranular Corrosion
- Dealloying
- Erosion Corrosion

The order of listing is somewhat arbitrary, reflecting the personal preferences of the author in their order of presentation in this book.

Uniform corrosion accounts for the greatest tonnage of metal consumed. Yet the other localized forms of corrosion are more insidious and difficult to predict and control. While localized corrosion may not consume as much material, penetration and failure are often more rapid. Hydrogen damage, more a metallurgical result of, than a form of corrosion, is included because of its importance and frequency.

Each form of corrosion is described briefly in this chapter to introduce the domain of corrosion. The present descriptions are only preliminary; more details are given in later chapters and referenced as necessary. *The reader can obtain a convenient introduction to the various forms of corrosion here and pursue more detailed information, including methods of prevention, in subsequent chapters, as desired.*

1.5.1 Uniform Corrosion

A uniform, regular removal of metal from the surface is the usually expected mode of corrosion. For uniform corrosion, the corrosive environment must have the same access to all parts of the metal surface, and the metal itself must be metallurgically and compositionally uniform. These requirements are not prevalent in operating equipment, and some degree of nonuniformity is tolerated within the definition of uniform corrosion. Figure 1.5 shows uniform corrosion, thinning, and penetration of a carbon steel storage tank for sour (H_2S-containing) crude oil after only two years of service.[11]

Atmospheric corrosion is probably the most prevalent example of uniform corrosion at a visually apparent rate. The other frequently cited example is uniform corrosion of steel in an acid solution. A properly specified alloy must corrode uniformly at a low rate in service. The other localized forms of corrosion are much less predictable and are to be avoided whenever possible. Thus, uniform corrosion is preferred from a technical viewpoint because it is predictable and thus acceptable for design.

1.5.2 Galvanic Corrosion

When two dissimilar alloys are coupled in the presence of a corrosive electrolyte, one of them is preferentially corroded while the other is protected from corrosion. Corrosion prevention by galvanic coupling is described in Section 13.1.3. Any two alloys have differing corrosion potentials E_{corr}, as defined in Section 1.4.2. The var-

FIGURE 1.5   Crude oil storage tank showing uniform penetration from exterior and interior after two years service. *(From C. J. Munger,* Corrosion Prevention by Protective Coatings, *NACE, Houston, p. 176, 1984. Photograph by courtesy of C. G. Munger. Reprinted by permission, National Association of Corrosion Engineers.)*

ious useful alloys are listed in order of E_{corr} in the abbreviated Galvanic Series that appears in Table 1.1. Any alloy will be preferentially corroded when coupled to another alloy with a more positive or noble potential in the Galvanic Series. At the same time, the more noble alloy is protected from corrosion, as discussed in Section 13.1.3. The negative, preferentially corroded alloy in the couple is said to be active in the Galvanic Series.

Although the series in Table 1.1 is specific to aerated seawater, differences with other solutions of moderately high corrosivity are usually minor. It is notable that the stainless steels and nickel may exist in either the passive or active state. Potential reversals and unpredicted galvanic corrosion on either the stainless steel or the other coupled alloy may result if conditions are sufficiently unstable to allow the stainless steel to alternate between the passive and active states.

Figure 1.6 shows an example of galvanic corrosion between welded carbon steel and stainless steel. The carbon steel pipe flange is active to the stainless steel pipe, as shown in Table 1.1 and is corroded preferentially near the weld. Preferential cor-

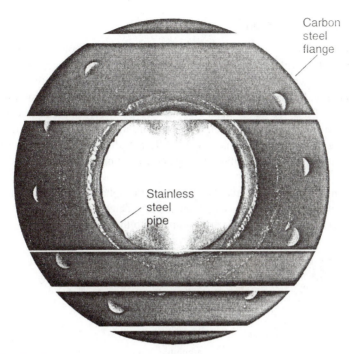

FIGURE 1.6   Galvanic corrosion of a carbon steel pipe flange welded to stainless steel pipe. *(Photograph by courtesy of R. A. Yeske, The Institute of Paper Science and Technology.)*

rosion near the junction between dissimilar alloys is characteristic of galvanic corrosion, as shown schematically in Figure 1.4. The corrosion decreases at points farther from the junction due to higher resistance through a longer electrolyte path. The electrolyte conductivity thus confines the current to a small surface area near the junction.

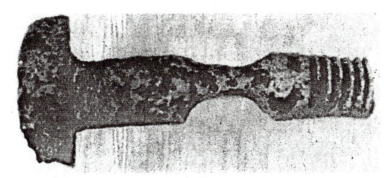

FIGURE 1.7   Corrosion of a steel bolt in an industrial environment resulting from continually wetted area within a crevice. *(From D. A. Jones, Forms of Corrosion, Recognition and Prevention, C. P. Dillon, ed., NACE, Houston, pp. 19–43, 1982. Reprinted by permission, National Association of Corrosion Engineers.)*

1.5.3 Crevice Corrosion

Corrosion of an alloy is often greater in the small sheltered volume of the crevice created by contact with another material. The second material may be part of a fastener (bolt, rivet, washer) of the same or a different alloy, a deposit of mud, sand, or other insoluble solid, or a nonmetallic gasket or packing. Crevice corrosion seems to be the preferred title for metal-metal crevices. Deposit corrosion and gasket corrosion are terms sometimes used when a nonmetallic material forms a crevice on the metal surface. If the crevice is made up of differing alloys or if the deposit is conductive (e.g., magnetite or graphite), crevice corrosion may be compounded by galvanic effects.

Corrosion within a crevice may be caused in atmospheric exposures by retention of water, while the outer surfaces can drain and dry. Thus, the bolt of Figure 1.7[12] was heavily corroded within the crevice as compared to the outer surfaces.

TABLE 1.1   Galvanic Series in Seawater

Cathodic (noble)
↑
platinum
gold
graphite
titanium
silver
zirconium
AISI Type 316, 317 stainless steels (passive)
AISI Type 304 stainless steel (passive)
AISI Type 430 stainless steel (passive)
nickel (passive)
copper-nickel (70-30)
bronzes
copper
brasses
nickel (active)
naval brass
tin
lead
AISI Type 316, 317 stainless steels (active)
AISI Type 304 stainless steel (active)
cast iron
steel or iron
aluminum alloy 2024
cadmium
aluminum alloy 1100
zinc
magnesium and magnesium alloys
↑
Anodic (active)

FIGURE 1.8   Crevice corrosion of drilled and tapped stainless steel test specimen in 3% NaCl solution at 90°C.

Crevice corrosion of stainless steels in aerated salt solutions is widely known. Corrosion products of Fe, Cr, and Ni, the main components of stainless steel, accumulate in the crevice and form very acid chloride solutions in which corrosion rates are very high. An example of crevice corrosion of stainless steel in hot salt solution is shown in Figure 1.8. The anode is isolated in the crevice with the surrounded passive surfaces acting as the cathode. The mechanisms of initiation and growth of crevice corrosion are discussed further in Section 7.3.

1.5.4 Pitting Corrosion

Localized attack in an otherwise resistant surface produces pitting corrosion. The pits may be deep, shallow, or undercut, as shown schematically in Figure 1.4. The stainless steels and nickel alloys with chromium depend on a passive film for corrosion resistance and are especially susceptible to pitting by local breakdown of the film at isolated sites. A deep pit causing wall penetration and leaking is shown in Figure 1.9.[13]

Pitting shares the same mechanism with crevice corrosion in stainless steels. The pit is a self-serving crevice that restricts transport between the bulk solution and the acid chloride pit anode. Pitting and crevice corrosion are discussed further in Chapter 7. Standard exposure and electrochemical testing methods are described, as well as the electrochemical theory behind crevice and pitting corrosion.

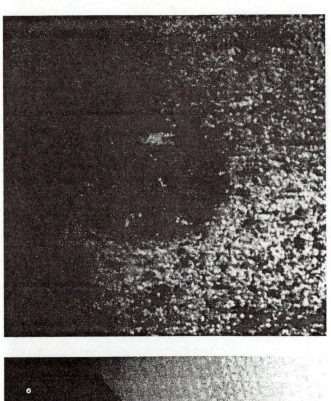

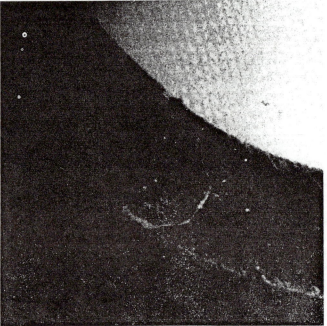

FIGURE 1.9 Pit formed in Type 410 stainless steel: (a) appearance of pit at the surface, about 50×; (b) cross section of pit, about 6×. *(From R. J. Franco*, Metals Handbook, *Vol. 11,* Failure Analysis and Prevention, *9th ed., ASM International, Metals Park, OH, p. 628, 1986. Reprinted by permission, ASM International.)*

1.5.5 Environmentally Induced Cracking

Brittle fracture of a normally ductile alloy in the presence of an environment that causes minimal uniform corrosion is defined as environmentally induced cracking (EIC). Three related but distinct types of failure are included in EIC: stress corrosion cracking (SCC), corrosion fatigue cracking (CFC), and hydrogen-induced cracking (HIC). The three are described briefly here but more extensively in Chapter 8. Alternative terms for HIC are hydrogen embrittlement, hydrogen-assisted cracking, and hydrogen stress cracking.

Stress corrosion cracking occurs in alloys with a static tensile stress in the presence of specific environmental conditions. Pure metals are comparatively resistant to stress corrosion cracking. A passive surface film under oxidizing conditions must be present, and corrosion rates are consequently quite low. A specific dissolved species is often required. For example, stainless steels are susceptible in hot chlorides, brass in ammonia solutions, and carbon steel in nitrates. Further discussion of critical environments for most of the important constructional alloys appears in Section 8.2 and is summarized in Table 8.1. Figure 1.10[14] shows an example of stress corrosion cracking in a U-bend test specimen.

Corrosion fatigue cracking occurs under cyclic stresses in a corrosive environment. Susceptibility to and rate of fatigue cracking without corrosion are usually increased in the presence of a corrosive environment. Both alloys and pure metals

FIGURE 1.10   Stress corrosion cracking at a U-bend. *(From B. J. Moniz,* Process Industries Corrosion, *B. J. Moniz and W. I. Pollock, eds., NACE, Houston, p. 67, 1986. Reprinted by permission, National Association of Corrosion Engineers.)*

are susceptible, and no specific environment is required. Macroscopic beach marks are often observed where tarnish or corrosion products accumulate during discontinuous propagation of the crack front. However, beach marks may also be caused by differences in microplastic deformation when crack propagation is interrupted, and beach marks may be present with little or no visible corrosion. A good example of beach marks is shown in Figure 1.11.[15]

Hydrogen induced cracking is caused by hydrogen diffusing into the alloy lattice when the hydrogen evolution reaction,

$$2H^+ + 2e^- \rightarrow H_2, \tag{4}$$

produces atomic hydrogen at the surface during corrosion, electroplating, cleaning and pickling, or cathodic protection. Although stress corrosion cracks often show

FIGURE 1.11 Beach marks on the fracture surface of a fatigue failure. *(From D. J. Wulpi,* Understanding How Components Fail, *ASM International, Metals Park, OH, p. 144, 1985. Reprinted by permission, ASM International. Photograph by courtesy of D. J. Wulpi, Consultant.)*

more branching, cracks have a very similar appearance otherwise. However, hydrogen induced cracking is accelerated by cathodic polarization, whereas stress corrosion cracking and corrosion fatigue cracking are suppressed.

Hydrogen induced cracking usually predominates over stress corrosion cracking in carbon and low alloy steels, stainless steels, aluminum alloys, and titanium alloys which have been alloyed, heat treated, or cold worked to near maximum strength. Hydrogen induced cracking also contributes to losses of fatigue life in such high strength alloys, which have been studied extensively by aircraft and aerospace manufacturers. Details of characteristics, identification, prevention and mechanism are given in Chapter 8 for all three: stress corrosion cracking, corrosion fatigue cracking, and hydrogen induced cracking.

1.5.6 Hydrogen Damage

Hydrogen induced cracking (HIC) and the impairment of ductility by relatively low levels of hydrogen are reversible to some extent if the hydrogen is permitted to escape by baking at elevated temperature. Other effects at higher hydrogen levels are usually irreversible. *Hydrogen attack* is the reaction of hydrogen with carbides in steel to form methane, resulting in decarburization, voids, and surface blisters. *Hydrogen blisters* (Figure 1.12[16]) or smaller *hydrogen cracks* become evident when internal hydrogen-filled voids erupt at the surface. Voids are formed when atomic hydrogen migrates from the surface to internal defects and inclusions, where molecular hydrogen gas can nucleate, generating sufficient internal pressure to deform and rupture the metal locally.

Hydride formation embrittles reactive metals such as titanium, zirconium, magnesium, tantalum, niobium, vanadium, uranium, and thorium. The other hydrogen damage mechanisms described are also possible in these metals.

1.5.7 Intergranular Corrosion

Reactive impurities may segregate, or passivating elements such as chromium may be depleted at the grain boundaries. As a result, the grain boundary or adjacent regions are often less corrosion resistant, and preferential corrosion at the grain boundary may be severe enough to drop grains out of the surface. Thus, intergranular corrosion (IGC), sometimes called intergranular attack, is a common problem in many alloy systems.

The best-known form of IGC occurs in austenitic stainless steels when heat treatments deplete the grain boundaries of chromium by metallurgical reaction with carbon. The resultant structure is susceptible or "sensitized" to IGC. In the temperature range 425 to 815°C (800 to 1500°F), chromium carbides (mainly $Cr_{23}C_6$) precipitate at the grain boundaries, depleting the grain boundary and nearby structure of chromium. Below about 10% Cr, these areas lose resistance and are corroded preferentially. Figure 1.13 shows an example of intergranular corrosion of sensitized austenitic stainless steel.

Sensitization to IGC is a common problem during welding of stainless steels and is discussed fully in Section 9.1. IGC in other alloy systems is also described in Chapter 9.

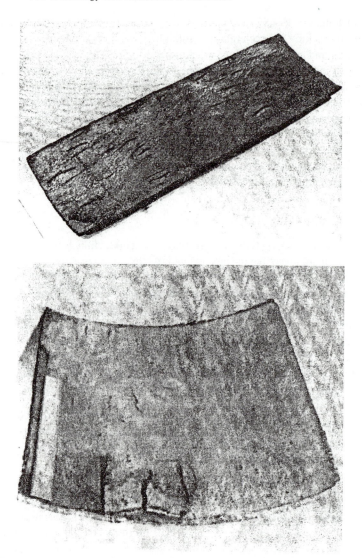

FIGURE 1.12   Hydrogen blisters in AISI 1020 carbon steel. *(From R. T. Jones, Process Industries Corrosion, B. J. Moniz and W. I. Pollock, eds., NACE, Houston, p. 387, 1986. Reprinted by permission, National Association of Corrosion Engineers.)*

1.5.8 Dealloying and Dezincification

An alloying element that is active (negative electrochemically) to the major solvent element is likely to be preferentially corroded by dealloying. Selective leaching and parting are alternative terms used occasionally for the same phenomenon. The dealloying of brass, known as dezincification, is a common and frequently cited example. Zinc is strongly active to copper (Table 1.1) and readily leaches out of brass, leaving behind relatively pure porous copper with poor mechanical properties. Uniform or

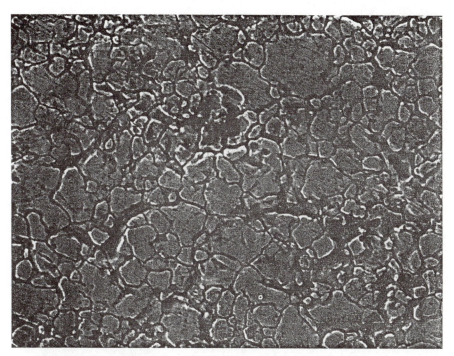

FIGURE 1.13   Intergranular corrosion of a sensitized austenitic stainless steel surface viewed by scanning electron microscopy.

layer dezincification is easily recognizable by the red, copper-colored material visible on the bolt surface in Figure 1.14. The deformed bolt, previously exposed to acidic cooling tower waters, shows a weak, brittle layer of uniform thickness. In plug dezincification the attack is localized and results in penetration and weakening of brass pipe and tubing.[17] Dezincification often occurs under deposits and in other nonvisible locations, and failures may be sudden and unpredictable. Copper is the most noble of the major constructional metals, and selective leaching of other more active alloying elements, such as nickel, silicon, and aluminum has also been reported.

The other major example of dealloying, known as graphitic corrosion, is the selective leaching of iron from gray cast iron, leaving behind a weak, porous network of inert graphite that can be scratched with a pen knife (Figure 1.15[18]). Graphitic corrosion has occurred primarily in buried cast iron pipe and becomes evident only after decades of service. When the old graphitized pipe is disturbed, escaping hazardous chemicals can contaminate the surrounding soil or cause fatal explosions and fires.[18]

1.5.9 Erosion-Corrosion and Fretting

The combination of a corrosive fluid and high flow velocity results in erosion-corrosion. The same stagnant or slow-flowing fluid will cause a low or modest corrosion rate, but rapid movement of the corrosive fluid physically erodes and removes

FIGURE 1.14 Brass bolt (30% Zn) which was bent to show uniform penetration of a dezincified layer. *(Sample provided by courtesy of S. L. Pohlman, Kennecott Corp.)*

the protective corrosion product film, exposes the reactive alloy beneath, and accelerates corrosion. Sand or suspended slurries enhance erosion and accelerate erosion-corrosion attack. Low-strength alloys that depend on a surface corrosion product layer for corrosion resistance are most susceptible. The attack generally follows the directions of localized flow and turbulence around surface irregularities.

Figure 1.16[19] shows characteristic horseshoe-shaped pits whose ends follow the downstream flow pattern in copper and brass condenser tubes. Harder alloys (brass and bronze) are more resistant, but the chemical component of the mechanism is evident by the dependence on composition of the flowing corrosive solution. Alloying elements such as aluminum and nickel, which form tighter, more adherent surface films, also improve resistance. Erosion-corrosion is a common problem in steel pipe with flowing steam carrying condensate water droplets. These and other examples are discussed further in Section 10.2.

FIGURE 1.15 Graphitic corrosion in a gray cast iron water pipe. *(From R. Steigerwald, Metals Handboook, Vol. 13, Corrosion, 9th ed., ASM International, Metals Park, OH, p. 132, 1987. Reprinted by permission, ASM International.)*

Cavitation is a special case of erosion-corrosion. It occurs where velocity is so high that pressure reductions in the flow are sufficient to nucleate water vapor bubbles, which then implode (collapse) on the surface. The implosions produce extreme pressure bursts that disrupt surface films and even dislodge particles from the metal itself. The attack takes the form of

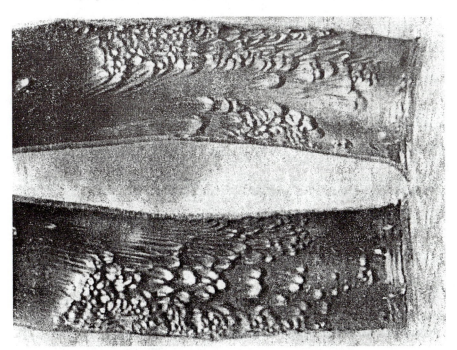

FIGURE 1.16 Brass condenser tube showing erosion corrosion in salt water. *(From C. P. Dillon, Process Industries Corrosion, B. J. Moniz and W. I. Pollock, eds., NACE, Houston, p. 6, 1986. Photograph by courtesy of A. Cohen, Copper Development Association. Reprinted by permission, National Association of Corrosion Engineers.)*

roughened pits, which may eventually result in penetration, as in Figure 1.17.[20] Cavitation occurs on turbine blades, pump impellers, ship propellers, and in tube and pipe where large pressure changes are present. Although largely mechanical, the chemical and electrochemical contributions are apparent in benefits from selection of corrosion-resistant alloys and control of solution composition.

Fretting is another type of erosion-corrosion, but in the vapor phase. The erosion is provided by repeated small movement, often from vibration, between the corroding metal and another contacting solid under load. The motion abrades surface oxide films off the metal surface, again exposing the reactive metal to increased oxide formation. The effect is compounded by the oxide debris, which acts as an additional abrasive between the contacting surfaces. A fine red rouge of iron oxide particles is produced on steel. Further discussion as well as micrographs of fretting attack and fretted surfaces appear in Section 10.4.

1.6 Corrosion Rate Determination

Methods of exposure testing for corrosion measurement are fundamental in corrosion engineering and are introduced in this section. The emphasis is on measure-

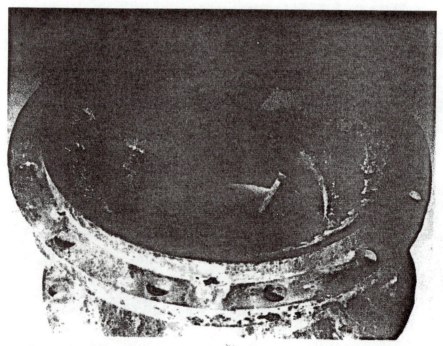

FIGURE 1.17 Cavitation of the cast iron suction bell from a low pressure water pump. *(From D. N. French*, Metals Handbook, *Vol. 11,* Failure Analysis and Prevention, *9th ed., ASM International, Metals Parks, OH, p. 624, 1986. Reprinted by permission, ASM International.)*

ment of uniform corrosion rates by weight loss of coupon specimens. However, exposure methods to determine localized attack are essentially the same, differing only in the form of the specimen and method of evaluation. It is necessary here to give only general descriptions of exposure procedures, which are detailed in appropriately referenced ASTM standards. Electrochemical methods are mentioned only in context with exposure tests and again are discussed more fully in subsequent chapters.

Corrosion testing can have one or more of the following objectives.[21]

- Screen available materials (alloys) to determine the best for a specific application.
- Determine probable service life of equipment or a product.
- Evaluate new commercial alloys or processes.
- Assist in the evaluation and development of new corrosion resistant alloys.
- Verify that an alloy lot (shipment) meets quality control specifications before either acceptance or release.
- Evaluate environmental controls and variations (e.g., inhibitors).
- Determine the most economical means for reducing corrosion.
- Study corrosion mechanisms.

Exposure tests may be conducted in the laboratory or in service. Laboratory tests are more flexible, less expensive, and can have any of the foregoing objectives because modifications or interruptions of plant processes are not required. However, it is nearly impossible to simulate plant conditions exactly in the laboratory. Time is usually at a premium, and accelerating factors, such as increased temperature, are often included. Thus, preliminary laboratory tests often require follow-up with plant qualification tests. Laboratory testing often seeks to determine mechanism, frequently using electrochemical methods, and offers the scientist and engineer significant challenges in relating the results to service and plant operation.

1.6.1 Specimen Preparation[22]

Specimen coupons are fabricated from sheet or plate stock, which is available for most common constructional alloys. Coupons are usually rectangular but may be round. The cheapest methods of fabrication are by punching, stamping, or shearing, which create cold-worked edges. Punched round coupons are produced very efficiently but create more waste material. Since cold work often has a significant effect on corrosion rate, the edges need to be ground or machined to remove an amount at least equal to the coupon thickness. Each coupon must be legibly, permanently, and uniquely identified, preferably by stamping. However, uncontrolled residual stresses are produced which may be undesirable in environments that cause environmentally induced cracking.

Surface finish of the alloy as placed in service should be reproduced in test coupons whenever possible. However, mill finish may include scale, which would cause nonreproducible effects on small coupons. Surface abrasion with carbide or diamond papers to 120 grit is common and can be produced without specialized equipment, although blasting with sand, steel shot, or glass beads may be less expensive if the necessary equipment is available. Surface milling and electrolytic polishing are also possible for research applications but are more expensive. Abrasion is usually conducted wet on successively finer papers to avoid heating, which could cause metallurgical changes in the coupons. Care must be taken to avoid transferring foreign materials to the surfaces. Clean abrasive paper must be used at each step. Blasting with steel shot leaves iron contamination behind and would not be appropriate for preparing nonferrous alloys.

Examples of round and rectangular coupons with typical surface finish and identification are shown in Figure 1.18.[23] Holes are included for mounting on racks during exposure.

1.6.2 Specimen Exposure

Methods and devices to mount coupons for corrosive exposures are limited only by the imagination of the investigator. In the laboratory a container must be selected for the corrosive environment and the racks to hold the coupons. The process piping or vessels serve as the container in plant exposures. Figure 1.19[24] shows a typical laboratory reactor with facilities to control temperature, aeration and reflux of vapor to prevent loss of solution by evaporation or boiling. A rack is shown to hang the specimens immersed, partially immersed, and in the vapor above the solution. Corrosion is often maximized at the liquid line or in the vapor. A minimum ratio of 100 to 200 liters per square meter of exposed surface area is usually recommended to avoid depletion of corrosive ingredients in solution. The vessel would be an autoclave when

FIGURE 1.18 Corrosion exposure test coupons. *(From B. J. Moniz,* Process Industries Corrosion, *B. J. Moniz and W. I. Pollock, eds., NACE, Houston, p. 69, 1986. Reprinted by permission, National Association of Corrosion Engineers.)*

temperatures and pressures above the atmospheric boiling point are required to simulate conditions in, for example, pressurized boilers and nuclear steam generators.

The coupons must be insulated from one another by the holding rack and should have uniform exposure to the environment. Simultaneous exposure of different alloy compositions may cause cross contamination of specimens, which can have significant effects on corrosion rate. Continuous circulation from a larger volume or continuous refreshment with exit to drain may be necessary if small changes in solution composition are critical.

Specimen racks for plant exposures are shown in Figure 1.20.[23] The racks must be fabricated of materials that are resistant to the process flow. Otherwise, holders

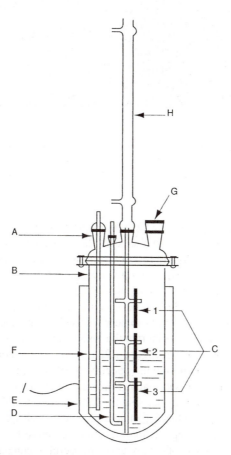

FIGURE 1.19   Flask to conduct immersion tests for coupons. A, entry for temperature sensor; B, resin flask; C, suspended specimens (C-1 vapor, C-2 partial immersion, C-3 complete immersion); D, gas bubbler; E, heating mantle; F, liquid surface; G, extra entry; H, reflux condenser. *(From D. O. Sprowls, Metals Handbook, Vol. 13, Corrosion, 9th ed., ASM International, Metals Park, OH, pp. 222–225, 1987. Reprinted by permission, ASM International.)*

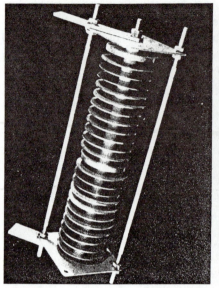 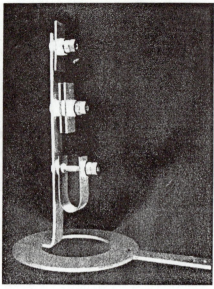

(a)   (b)

FIGURE 1.20 Racks for in-plant exposures of coupons: (a) spool rack, (b) pipeline rack. *(From B. J. Moniz,* Process Industries Corrosion, *B. J. Moniz and W. I. Pollock, eds., NACE, Houston, p. 69, 1986. Reprinted by permission, National Association of Corrosion Engineers.)*

and coupons can be (frequently are) lost in the process stream. A retractable coupon holder, shown in Figure 1.21[23], mounted on a flange behind a gate valve, overcomes the need to interrupt plant flow to install or remove coupons. It may also be convenient to install coupons in a bypass loop which can be isolated from the main flow during installation and removal of coupons.

Figure 1.22 shows racks to hold panel specimens during atmospheric exposures.[25] Specialized laboratory equipment is sometimes used to simulate or accelerate environmental conditions in order to shorten the long exposure times required by atmospheric-simulated service testing. Figure 1.23a[24] shows laboratory alternate immersion apparatus in which aluminum specimens of various types are typically immersed for 10 minutes in salt water, followed by 50 minutes out of solution to complete consecutive 1-hour cycles. Figure 1.23b shows a cabinet in which coated panel specimens are exposed to a fog of neutral or acidic salt spray. Specimen design and evaluation for coating degradation are discussed more fully in Section 14.1.3. The alternate immersion and salt spray tests are useful to compare the performance of various products under accelerated conditions. Correlation with actual service performance has been difficult because different mechanisms of corrosion are probable in testing and in service.

Uniform corrosion rates are normally calculated from coupon weight loss. Microscopic, visual, and metallographic examinations are usually needed instead of, or in addition to, weight-loss determinations for localized forms of corrosion.

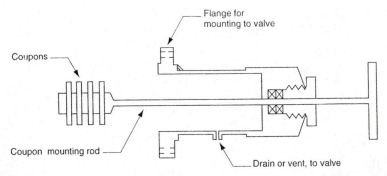

FIGURE 1.21   Retractable coupon holder for in-plant corrosion testing. *(From B. J. Moniz, Process Industries Corrosion, B. J. Moniz and W. I. Pollock, eds., NACE, Houston, p. 69, 1986. Reprinted by permission, National Association of Corrosion Engineers.)*

These are described in later chapters for each localized form of corrosion, when necessary.

After fabrication and surface preparation, coupons are weighed to the nearest 0.1-mg, exposed to the corrosive environment, cleaned of corrosion products, and weighed again to determine weight loss. The exposure time must be adequate to generate sufficient weight loss for accurate measurement. A rule of thumb for mod-

FIGURE 1.22   Test rack for atmospheric exposure of panel specimens. *(From K. L. Money, Metals Handbook, Vol. 13, Corrosion, 9th ed., ASM, Metals Park, OH, p. 205, 1987. Photograph by courtesy of LaQue Center for Corrosion Technology, Wrightsville Beach, NC. Test rack located at Kure Beach Atmospheric Test Lots, Kure Beach, NC. Reprinted by permission, ASM International.)*

(a)

(b)

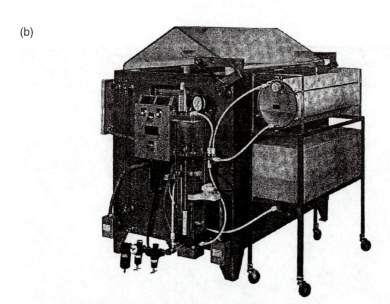

FIGURE 1.23 Accelerated atmospheric exposure apparatus: (a) alternate immersion tanks, (b) salt spray cabinets. *(From D. O. Sprowls*, Metals Handbook, *Vol 13, Corrosion, 9th ed., ASM International, Metals Park, OH, pp. 222-225, 1987. Photographs by courtesy of D. O. Sprowls, Aluminum Company of America, and N. B. Tipton, The Singleton Corp., respectively. Reprinted by permission, ASM International.)*

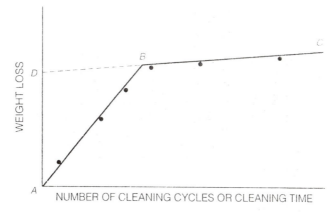

FIGURE 1.24   Schematic weight loss during coupon cleaning. *(From B. J. Moniz, Process Industries Corrosion, B. J. Moniz and W. I. Pollock, eds., NACE, Houston, p. 69, 1986. Reprinted by permission, National Association of Corrosion Engineers.)*

erate to low corrosion rates is: exposure hours = 2000/MPY, where MPY is mils (0.001-in.) penetration per year.

1.6.3 Specimen Cleaning[22]

Care must be exercised in cleaning and removal of all corrosion products and foreign matter from the surface of exposed coupons before the final weight can be measured. Cleaning may be mechanical, chemical, or often both. The ideal, in which the cleaning operation removes adherent corrosion products and leaves the underlying metal coupon unaffected, is seldom achieved. Instead, gradual loss of the base metal continues during the process after initial rapid removal of corrosion products (Figure 1.24[23]). Extrapolation of the metal loss period, *BC*, back to the beginning of the cleaning operation at *D* gives the most accurate final coupon weight. However, if the cleaning operation can be timed to stop at point *B*, the error will be uniform and small when extrapolation is omitted. Table 1.2 gives a selection of recommended[26] cleaning reagents.

1.6.4 Corrosion Rate Units and Calculations

The corrosion rate in mils (1 mil=0.001-in.) penetration per year (MPY) may be calculated from

$$MPY = \frac{534\ W}{DAT} \tag{6}$$

where *W* is weight loss in milligrams, *D* is density in grams per cubic centimeter, *A* is area in square inches, and *T* is time in hours. Equation (6) can be derived readily by dimensional analysis. Other units for *W*, *D*, *A*, and *T* simply change the value of the constant 534, as discussed later in this section. Units of penetration per unit time are most desirable from an engineering standpoint, but weight loss per unit area per

TABLE 1.2 Recommended Reagents for Cleaning and Removal of Corrosion Products from Corroded Coupons

Material	Chemical[a]	Time	Temperature	Remarks
Aluminum and aluminum alloys	70% HNO$_3$ or 2% CrO$_3$, 5% H$_3$PO$_4$ soln.	2–3 min 10 min	Room 175–185°F (79–85°C)	Follow by light scrub Used when oxide film resists HNO$_3$ treatment; follow by 70% HNO$_3$ treatment previously described
Copper and copper alloys	15–20% HCl or 5–10% H$_2$SO$_4$	2–3 min 2–3 min	Room Room	Follow by light scrub Follow by light scrub
Lead and lead alloys	1% acetic acid or 5% ammonium acetate	10 min 5 min	Boiling Hot	Follow by light scrub; removes PbO Follow by light scrub; Removes PbO and/or PbSO$_4$
	80 g/L NaOH, 50 g/L mannitol, 0.62 g/L Hydrazine sulfate	30 min or until clean	Boiling	Follow by light scrub
Iron and steel	20% NaOH, 200 g/L zinc dust[b] or conc. HCl, 50 g/L, SnCl$_2$+20 g/L SbCl$_3$	5 min Until clean	Boiling Cold	— —

32

Material	Solution	Time	Temperature	Remarks
Magnesium and magnesium alloys	15% CrO_3, 1% $AgCrO_4$ soln.	15 min	Boiling	—
Nickel and nickel alloys	15–20% HCl or	Until clean	Room	—
	10% H_2SO_4	Until clean	Room	—
Stainless steel	10% HNO_3	Until clean	140°F (60°C)	Avoid contamination with chlorides
Tin and tin alloys	15% Na_3PO_4	10 min	Boiling	Follow by scrub
Zinc	10% NH_4Cl	5 min	Room	Follow by light scrub
	followed by 5% CrO_3, 1% $AgNO_3$ soln.	20 s	Boiling	—
	or Saturated ammonium acetate	Until clean	Room	Follow by light scrub
	or 100 g/L NaCN	15 min	Room	—

[a]For a list of other effective chemicals, see *Materials Protection*, July 1967

[b]It is possible for zinc dust as well as the cleaning residue (containing zinc dust) to ignite spontaneously on exposure to air. Proper precautions should be taken in handling and disposing of these materials.

Source: "Laboratory Corrosion Testing of Metals for the Process Industries," NACE Standard TM-01-69 (1976 Revision). Reprinted by permission, National Association of Corrosion Engineers.

unit time, often milligrams per square decimeter per day (mdd), are sometimes used in research. For conversion, 1 mpy = 1.44(mdd)/specific gravity.

The unit MPY continues as the most popular for corrosion rate in the United States, despite increased use of metric units in recent years. The range of practical corrosion rates are expressed conveniently in terms of small whole integers from 0 to 200 mpy for ferrous alloys in a time period (one year) useful for engineering purposes. Conversions to equivalent metric penetration rates are: 1 mpy = 0.0254 mm/yr = 25.4 μm/yr = 2.90 nm/h = 0.805 pm/s, where 1 meter = 10^3 millimeter (mm), 10^6 micrometer or micron (μm), 10^9 nanometer (nm), and 10^{12} picometer (pm). Table 1.3[27] compares mpy with competing metric units. Equivalent mm/yr gives fractional numbers, and μm/yr gives large integers. Desirable small whole integers are produced in terms of nm/h and pm/s, but the time units, hours and seconds, are insignificant from an engineering standpoint.

Metric units of penetration rates will probably see still further use as future generations of engineers become more familiar with metric units and use them more frequently in design specifications. The most promising appear to be mm/yr and μm/yr for high and low corrosion rates, respectively. The proportionality constant, 534, in equation (6) varies depending on the units required for corrosion rate and used for the parameters in the equation.

$$\mu m / yr = \frac{87600 \ W}{DAT},$$

and

$$mm / yr = \frac{87.6 \ W}{DAT},$$

where W, D, and T have the same units as for equation (6) but area, A, is measured in square centimeters. Freeman and Silverman[28] estimate the errors in calculated corrosion rates resulting from uncertainties in measurements of W, A, and T in (6).

TABLE 1.3   Comparison of MPY with Equivalent Metric-Rate Expressions

Relative Corrosion Resistance[a]	mpy	mm/yr	μm/yr	nm/h	pm/s
Outstanding	< 1	< 0.02	< 25	< 2	< 1
Excellent	1–5	0.02–0.1	25–100	2–10	1–5
Good	5–20	0.1–0.5	100–500	10–50	20–50
Fair	20–50	0.5–1	500–1000	50–150	20–50
Poor	50–200	1–5	1000–5000	150–500	50–200
Unacceptable	200+	5+	5000+	500+	200+

[a]Based on typical ferrous and nickel alloys. Rates greater than 5 to 20 mpy are usually excessive for more expensive alloys, while rates above 200 mpy are sometimes acceptable for cheaper materials (e.g., cast iron) of thicker cross section.

Source: M. G. Fontana, Corrosion Engineering, McGraw-Hill, 3rd ed., p. 172, 1986. Reprinted by permission, McGraw-Hill Book Co.

1.6.5 Planned Interval Tests

Changes in the solution corrosivity and alloy corrodibility (corrosion rate) with time can be determined with a planned interval testing program. From a set of four weight-loss coupons, numbered 1 to 4, the first three are placed simultaneously in the same corrosive environment. Duplicate or more sets may be included as well for statistical validity. Number 1 is removed after exposure of "unit" time, which might be typically one day in a program of several days' total duration, or longer for lower corrosion rates. Number 2 is removed after exposure of time t. Number 3 is removed after exposure of time $t + 1$. Number 4 replaces number 2 after the initial interval t and is removed along with number 3 at time $t + 1$, giving an exposure time of unity for number 4.

The procedure and analysis are summarized in Table 1.4: A_1 is the corrosion (weight loss or calculated corrosion rate) obtained from number 1. A_t is the corrosion obtained from number 2. $A_t + 1$ is the corrosion obtained from number 3. A_2 is the difference in corrosion between numbers 2 and 3. B is the corrosion obtained from number 4. When $B = A_1$ the corrosiveness of the environment is unchanged during the time interval t. When $B < A_1$ corrosiveness has decreased, and when $B > A_1$ it has increased. When $A_2 = B$ the alloy corrodibility is unchanged. When $A_2 < B$ the corrodibility has decreased, and when $A_2 > B$ it has increased. Any combination of environment corrosivity and alloy corrodibility is possible and can be obtained from planned interval testing. Repeating the procedure with varying t will yield more detailed effects of time.

TABLE 1.4 Planned Interval Test

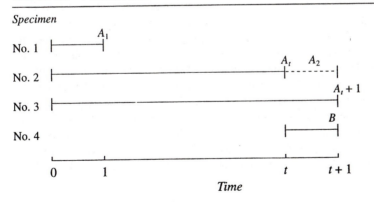

Criteria	Environment Corrosiveness	Criteria	Alloy Corrodibility
$B = A_1$	Unchanged	$A_2 = B$	Unchanged
$B < A_1$	Decreased	$A_2 < B$	Decreased
$B > A_1$	Increased	$A_2 > B$	Increased

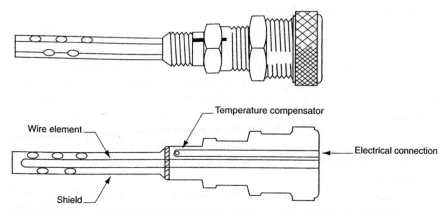

FIGURE 1.25 Electrical resistance probe for corrosion monitoring. *(From S. W. Dean,* ASTM STP 908, *G. C. Moran and P. Labine, eds., ASTM, Philadelphia, p. 197, 1986. Reprinted by permission, ASTM.)*

1.6.6 Electrical Resistance Probes

The electrical resistance of any conductor is given by $R=\rho L/A$, where ρ is the resistivity, L is the length, and A is the cross-sectional area of the conductor. As corrosion reduces A, the resultant increase in R of a wire sensing element can be used to monitor corrosion rate. Uniform loss of cross-section is measured by resistance increase instead of weight loss. Corrosion rate over a given interval can be calculated from the slope of resistance versus time. A commercial resistance probe design is shown in Figure 1.25.[29] A temperature sensor is usually included to compensate for resistance variations due to temperature changes.

Resistance probe results can be obtained from liquid, vapor, or mixed environments. The probes are small, easily installed, and can be wired into computerized control stations which give a running calculation of corrosion rate and remaining wall thickness. Alternatively, probe resistance can be measured manually with conventional portable bridge instruments. High conductivity environments such as liquid metals or molten salts allow parallel currents along the length of the sensing element and preclude use of resistance probes. Conductive deposits (e.g., magnetite) accumulating on the probe element during exposure have the same effect. Only uniform corrosion is measurable; pitting or otherwise localized corrosion does not have the necessary proportionate effect on cross-sectional area and probe resistance. As with any instrumental system, installation and maintenance costs are high, and leaks at wall entries are a hazard.

References

1. *Aviat. Week Space Technol.*, p. 29, July 4, 1988.
2. M. G. Fontana, *Corrosion Engineering*, 3rd ed., McGraw-Hill, New York, p. 1–5, 1986.

3. J. H. Payer, W. K. Boyd, D. B. Dippold, and W. H. Fisher, *Mater. Perform.*, May–Nov. 1980.
4. *Mater. Perform.*, p. 57, Feb. 1983.
5. H. H. Uhlig and R. W. Revie, *Corrosion and Corrosion Control*, 3rd ed., Wiley, New York, p. 3, 1985.
6. J. Millman, *Technol. Rev.*, Vol. 94, p. 74, Jan./Feb. 1986; *Bus. Week*, p. 28, Apr. 20, 1987.
7. L. J. Carter, *Issues Sci. Technol.*, p. 46, Winter 1987.
8. J. A. Bennet and H. Mindlin, *J. Test. Eval.*, p. 152, Mar. 1973.
9. R. J. Eiber and G. O. Davis, *Investigation of Williams Pipeline Company Mounds View, MN., Pipeline Rupture*, Final Report to Transportation Systems Center, U.S. Dept. of Transportation, Battelle-Columbus Lab., Oct. 14, 1987.
10. Pipe Break Causes Deaths at Surry, *Nucl. Eng. Int.*, Vol. 32, p. 4, Feb. 1987.
11. C. J. Munger, *Corrosion Prevention by Protective Coatings*, NACE, Houston, p. 176, 1984.
12. D. A. Jones, *Forms of Corrosion, Recognition and Prevention*, C. P. Dillon, ed., NACE, Houston, p. 19–43, 1982.
13. R. J. Franco, *Metals Handbook*, Vol. 11, *Failure Analysis and Prevention*, 9th ed., ASM International, Metals Park, OH, p. 628, 1986.
14. B. J. Moniz, *Process Industries Corrosion*, B. J. Moniz and W. I. Pollock, eds., NACE, Houston, p. 67, 1986.
15. D. J. Wulpi, *Understanding How Components Fail*, ASM International, Metals Park, OH, p. 144, 1985.
16. R. T. Jones, *Process Industries Corrosion*, B. J. Moniz and W. I. Pollock, eds., NACE, Houston, p. 387, 1986.
17. A. Cohen, *Process Industries Corrosion*, B. J. Moniz and W. I. Pollock, eds., NACE, Houston, p. 500, 1986.
18. R. Steigerwald, *Metals Handbook*, Vol. 13, *Corrosion*, 9th ed., ASM International, Metals Park, OH, p. 132, 1987.
19. C. P. Dillon, *Process Industries Corrosion*, B. J. Moniz and W. I. Pollock, eds., NACE, Houston, p. 6, 1986.
20. D. N. French, *Metals Handbook*, Vol. 11, *Failure Analysis and Prevention*, 9th ed., ASM International, Metals Park, OH., p. 624, 1986.
21. D. O. Sprowls, *Metals Handbook*, Vol. 13, *Corrosion*, 9th ed., ASM International, Metals Park, OH, p. 193, 1987.
22. ASTM Designation G1-88, *Annual Book of ASTM Standards*, Vol. 3.02, ASTM, Philadelphia, p. 61, 1988.
23. B. J. Moniz, *Process Industries Corrosion*, B. J. Moniz and W. I. Pollock, eds., NACE, Houston, p. 69, 1986.
24. D. O. Sprowls, *Metals Handbook*, Vol. 13, *Corrosion*, 9th ed., ASM International, Metals Park, OH, p. 222–5, 1987.
25. K. L. Money, *Metals Handbook*, Vol. 13, *Corrosion*, 9th ed., ASM International, Metals Park, OH, p. 205, 1987.
26. Laboratory Corrosion Testing of Metals for the Process Industries, NACE Standard TM-01-69 (1976 revision), NACE, Houston, 1976.
27. M. G. Fontana, *Corrosion Engineering*, 3rd ed., McGraw-Hill, New York, p. 172, 1986.
28. R. A. Freeman and D. C. Silverman, *Corrosion*, Vol. 48, p. 463, 1992.
29. S. W. Dean, *ASTM STP 908*, G. C. Moran and P. Labine, eds., ASTM, Philadelphia, p. 197, 1986.

Exercises

1-1. Radial steel belted automobile tires are commonly constructed of rubber reinforced with braided steel wire, each strand coated with copper. The copper coating is included to enhance adhesion of the rubber tire casing to the wire. How might corrosion contribute to tire failure? Which forms of corrosion might be involved?

1-2. What costs (Section 1.2) might be involved in tire failures in Problem 1-1 above?

1-3. In the steam return pipe failure described in Section 1.2.2, what are some of the costs which may result from this and similar failures?

1-4. For iron corroding in acidified, aerated seawater, write the probable anodic and cathodic reactions.

1-5. The corrosion rate of titanium was measured at 100 mpy in dilute sulfuric acid which was free of dissolved oxygen and other dissolved oxidizers. Iron was found to corrode at 250 mpy in the same conditions without titanium in separate tests. Contamination by Fe^{+3} ion produced a decrease in corrosion rate of titanium to 1.5 mpy and an increase in the corrosion rate of iron to 3500 mpy. Explain how this could be possible.

1-6. If polarization in Problem 1-5 were zero, what would be the corrosion rate in each case?

1-7. The material for the piping system in a new power plant must be specified to carry hot aerated seawater used to cool steam. Stresses both static and cyclic are present in the pipe due to welding, weight of the pipe, and vibrations from the pumps. Flow will vary from stagnant to very rapid. Austenitic stainless steel and 70Cu–30Zn brass are being considered for the pipe. What forms of corrosion might be possible for each material? Which alloy would be better?

1-8. (a) Derive the expression for corrosion rate in MPY given in Equation (9). (b) Show how the constant 534 in this equation changes to 87600 when units of corrosion rate change to micrometers per year with area measured in cm^2.

1-9. A carbon steel test specimen of dimensions 2-in. × 3-in. × 0.125-in. with a 0.25-in. hole for suspending in solution is exposed for 120 hours in an acid solution and loses 150 milligrams. Calculate the corrosion rate in mpy and mm/yr.

1-10. What experimental precautions must be taken to ensure maximum accuracy in the measured weight loss of Problem 1-9?

1-11. Three more identical test specimens are included with the one from Problem 1-9 in a planned interval testing program. One is exposed for 12 hours and loses 25 mg. A second is exposed for 108 hours and loses 130 mg. The third is inserted in the solution when the second is removed and shows a weight loss of 15 mg after removal with the specimen of Problem 1-9. What is the effect of time on the solution corrosiveness and the specimen corrodibility?

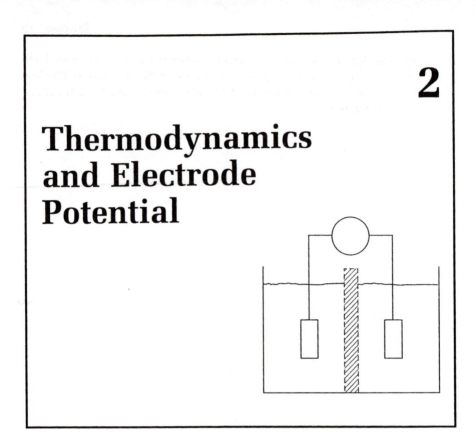

2

Thermodynamics and Electrode Potential

2.1 Electrode Potentials

Corrosion in aqueous solutions has been found to involve electron or charge transfer. A change in electrochemical potential or the electron activity or availability at a metal surface has a profound effect on the rates of corrosion reactions. Thus, corrosion reactions are said to be electrochemical, as discussed in Section 1.4.1. Thermodynamics gives an understanding of the energy changes involved in the electrochemical reactions of corrosion. These energy changes provide the driving force and control the spontaneous direction for a chemical reaction. Thus, thermodynamics shows how conditions may be adjusted to make corrosion impossible. When corrosion is possible, thermodynamics cannot predict the rate; corrosion may range from fast to very slow. The actual extent or rate of corrosion is governed by kinetic laws discussed in Chapter 3.

A conducting metal containing mobile electrons forms a complex interface in contact with an aqueous solution. Unsymmetrical, polar H_2O molecules (H-atoms positive, O-atoms negative in the molecule) are attracted to the conductive surface, forming an oriented solvent layer, which prevents close approach of charged species (ions) from the bulk solution. Charged ions also attract their own sheath of polar water-solvent molecules, which further insulate them from the conducting surface. The plane of closest approach of positively charged cations to the negatively charged metal surface is often referred to as the outer Helmholtz plane, as indicated in Figure 2.1. The result is an interfacial structure of separated charge commonly referred to as the electrical double layer, which behaves experimentally much like a charged capacitor (Figure 2.1).

The electrical field of the double layer structure prevents easy charge transfer, thereby limiting electrochemical reactions at the surface, as discussed in more detail in Section 3.2.1.

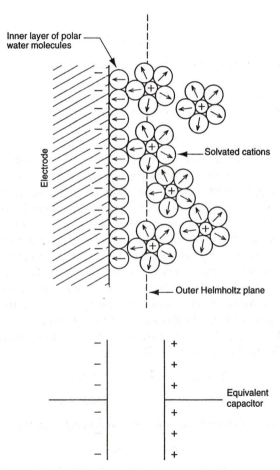

FIGURE 2.1   Schematic electrode surface structure with equivalent electric capacitor.

2.1.1 Free Energy and Electrode Potential

Consider again (see Section 1.4.1) the reaction for the corrosion of zinc in a solution of hydrochloric acid in water:

$$Zn + 2HCl \rightarrow ZnCl_2 + H_2. \tag{1}$$

There is a change of free energy, ΔG, associated with any such chemical reaction. When the reaction products have a lower energy than the reactants, ΔG is negative in the spontaneous reaction. As was demonstrated in Section 1.4.1, reaction (1) may be reduced to the simplest ionic reaction,

$$Zn + 2H^+ \rightarrow Zn^{2+} + H_2, \tag{2}$$

which may be separated into partial or *half-cell electrochemical reactions*,

$$Zn \rightarrow Zn^{2+} + 2e^-, \tag{3}$$

and

$$2H^+ + 2e^- \rightarrow H_2, \tag{4}$$

involving an exchange of electrons, e^-. A summation of reactions (3) and (4) results in reaction (2).

The free-energy change, ΔG, may be associated with an electrochemical potential, E, at equilibrium, by the fundamental relationship

$$\Delta G = -nFE, \tag{5}$$

where n is the number of electrons (or equivalents) exchanged in the reaction, and F is Faraday's constant, 96,500 coulombs per equivalent. Thus, we have the fundamental relationship, in which a charge, nF, taken reversibly at equilibrium through a potential, E, corresponds to an energy change, ΔG. From reactions (3) and (4), we see that n is 2 or the oxidation number change in reaction (1). The negative (−) in equation (5) is included to conform to the convention that a positive potential, E, results in a negative free-energy change, ΔG, for a spontaneous reaction.

The reaction (3) is an oxidation reaction with an increase of oxidation state of Zn from 0 to +2, defined in electrochemical terminology as an *anodic reaction*. Reaction (4) is a reduction reaction with a decrease in oxidation state for hydrogen from +1 to 0, defined electrochemically as a *cathodic reaction*.

The half-cell reactions (3) and (4) also have free-energy changes analogous to ΔG and corresponding potentials e_a and e_c. That is,

$$Zn = Zn^{2+} + 2e^- \qquad \text{(anodic).} \qquad\qquad e_a \tag{3}$$

$$2H^+ + 2e^- = H_2 \qquad \text{(cathodic).} \qquad\qquad e_c \tag{4}$$

The algebraic sum of these potentials is equal to E in equation (5). That is,

$$E = e_a + e_c. \tag{6}$$

The potentials e_a and e_c have been variously called half-cell, single electrode, or redox (reduction/oxidation) potentials for the corresponding half-cell reactions. The three terms are often used synonymously in the literature, leading to some confusion.

The usage is usually self-explanatory, but the term *half-cell electrode potential* seems unambiguous and easily recognizable and will be used in most instances throughout the remainder of this book.

When reactants and products in reactions (3) and (4) are all at some arbitrarily defined standard state, the half-cell electrode potentials have values designated e_a^o and e_c^o. The activity of each reactant and product is defined as unity for the standard state. For dilute or strongly dissociated solutes found in most instances of corrosion, activity may be reasonably approximated by concentration. For solid materials, the solid is taken as the standard state of unit activity, and for any gas, 1 atm pressure of the gas is taken as the standard state. Departures from standard half-cell electrode potentials at unit activity due to concentration changes are discussed in Section 2.1.3.

2.1.2 Electromotive Force or emf Series

A listing of standard half-cell electrode potentials constitutes the Electromotive Force (emf) Series. A partial emf series listing those half-cell reactions of particular interest for corrosion is given in Table 2.1. Notice that the reactions in Table 2.1 are all written as reduction reactions from left to right. Thus, the listed potentials are sometimes referred to as reduction potentials. Reduction potentials have the major advantage that they match the polarity of experimentally measured potentials (Section 2.3). In some older sources, all signs are reversed, and the potentials are referred to as oxidation potentials. However, the Stockholm Convention of 1953, used in Table 2.1, has been almost universally adopted, and exceptions are presently quite rare.

As indicated in Table 2.1, the noble end of the emf series derives its name from the presence of the well-known "noble" metals, Au and Pt. Similarly, the active end is characterized by the "active" metals, Na and K. This convention is intuitively obvious and is independent of any sign convention. It is apparent, however, in Table 2.1 that positive reduction potentials are noble, and negative are active.

Unfortunately, it is impossible to measure the absolute value of any half-cell electrode potential. Only cell potentials consisting of two such half-cell electrode potentials are measurable, and one must be selected as the primary reference. The zero or reference point in the emf series (Table 2.1) and all other electrochemical potential measurements are arbitrarily selected as the half-cell electrode potential, e_{H^+/H_2}^o, of the hydrogen half-cell reaction (4) at standard state. The actual absolute value of e_{H^+/H_2}^o is not zero; it is only assumed to be so for convenience. It would also be incorrect to infer from equation (5) that the free-energy change for reaction (4) is zero.

The reference half-cell electrode potential is established with the easily constructed *standard hydrogen electrode* (SHE). Additional details of construction are given in Section 2.3.1, but it is instructive at this point to schematically show how the potentials in Table 2.1 are measured with respect to SHE. The SHE consists of a platinum specimen immersed in unit activity acid solution, through which is bubbled H_2 gas at 1 atm pressure (standard state for H_2). Often, a porous "platinized" platinum coating is deposited on the surface to provide a larger area for the hydrogen reaction (4), which controls the half-cell electrode potential on the platinum surface.

TABLE 2.1 Standard Electromotive Force Potentials (Reduction Potentials)

	Reaction	Standard Potential, e° (volts vs. SHE)
Noble	$Au^{3+} + 3e^- = Au$	+1.498
	$Cl_2 + 2e^- = 2Cl^-$	+1.358
	$O_2 + 4H^+ + 4e^- = 2H_2O$ (pH 0)	+1.229
	$Pt^{2+} + 3e^- = Pt$	+1.118
	$NO_3^- + 4H^+ + 3e^- = NO + 2H_2O$	+0.957
	$O_2 + 2H_2O + 4e^- = 4OH^-$ (pH 7)a	+0.82
	$Ag^+ + e^- = Ag$	+0.799
	$Hg_2^{2+} + 2e^- = 2Hg$	+0.799
	$Fe^{3+} + e^- = Fe^{2+}$	+0.771
	$O_2 + 2H_2O + 4e^- = 4OH^-$ (pH 14)	+0.401
	$Cu^{2+} + 2e^- = Cu$	+0.342
	$Sn^{4+} + 2e^- = Sn^{2+}$	+0.15
	$2H^+ + 2e^- = H_2$	0.000
	$Pb^{2+} + 2e^- = Pb$	−0.126
	$Sn^{2+} + 2e^- = Sn$	−0.138
	$Ni^{2+} + 2e^- = Ni$	−0.250
	$Co^{2+} + 2e^- = Co$	−0.277
	$Cd^{2+} + 2e^- = Cd$	−0.403
	$2H_2O + 2e^- = H_2 + 2OH^-$ (pH 7)[a]	−0.413
	$Fe^{2+} + 2e^- = Fe$	−0.447
	$Cr^{3+} + 3e^- = Cr$	−0.744
	$Zn^{2+} + 2e^- = Zn$	−0.762
	$2H_2O + 2e^- = H_2 + 2OH^-$ (pH 14)	−0.828
	$Al^{3+} + 3e^- = Al$	−1.662
	$Mg^{2+} + 2e^- = Mg$	−2.372
	$Na^+ + e^- = Na$	−2.71
Active	$K^+ + e^- = K$	−2.931

[a]Not a standard state but included for reference.

Source: Handbook of Chemistry and Physics, *71st ed., CRC Press, 1991.*

Platinum itself is essentially inert in this solution, acting primarily as a catalyst for reaction (4). A cell potential may be measured when another half-cell is coupled with SHE. The other half-cell selected for Figure 2.2 is pure solid Zn metal in a solution of unit activity (approximately 1 N) Zn^{2+}. Thus, both Zn and Zn^{2+} are at the standard state in reaction (3). The potential experimentally measured in this cell with the voltmeter is 0.763 V, with Zn negative to the hydrogen reaction (4) on the Pt surface.

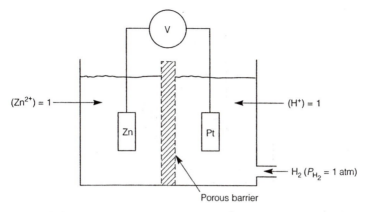

FIGURE 2.2   Schematic electrochemical cell showing measurement of cell potential between a zinc electrode at standard state and the standard hydrogen elec-

Again, it is impossible to measure the absolute value of potential on either the Pt or Zn surfaces. Only the potential difference or cell potential between the two half-cells is measurable, as shown in Figure 2.2. When the hydrogen electrode (SHE) is defined as zero, the measured potential of zinc with respect to the hydrogen electrode, -0.763 V, agrees with the half-cell electrode potential for the zinc reaction (3) listed in Table 2.1.

Of course, measurement of cell potential can include any pair of reactions selected from the emf series. Consider the cell shown schematically in Figure 2.3, in which the hydrogen electrode has been replaced by a pure copper electrode immersed in 1 N Cu^{2+} solution. The measured potential in this cell is 1.10 V, with the Cu positive with respect to the Zn. This cell potential may be predicted by the difference between the potential for the standard Cu half-cell, +0.337 V, and that of the Zn half-cell, −0.763 V. If the Cu half-cell were to replace the Zn half-cell in Figure 2.2, the potential of the Cu half-cell would be +0.337 V, with respect to SHE as listed in Table 2.1.

2.1.3 Concentration Effects on Electrode Potential

The cells shown in Figures 2.2 and 2.3 are carefully constructed to measure the standard half-cell electrode potentials in Table 2.1. Most corrosive conditions would only accidentally match the standard thermodynamic conditions shown for these half-cells. The standard state requires all reactants and products to be at unit activity. Thus, some means must be provided to calculate half-cell electrode potentials that deviate from the standard state. Taking a general equation for a half-cell reaction as,

$$aA + mH^+ + ne^- = bB + dH_2O, \tag{7}$$

departures from unit activity can be predicted from the Nernst equation, which is derived as follows.

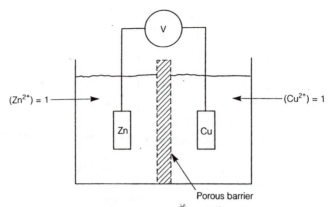

FIGURE 2.3   Schematic electrochemical cell incorporating standard zinc and copper half-cells.

Free-energy changes in the standard, $\Delta G°$, and nonstandard, ΔG, states are given for reaction (7) by

$$\Delta G° = (bG_B^o + dG_{H_2O}^o) - (aG_A^o + mG_{H+}^o)$$

and

$$\Delta G = (bG_B + dG_{H_2O}) - (aG_A + mG_{H+})$$

The departure of free-energy change in the nonstandard from the standard state is found from the difference,

$$\Delta G - \Delta G° = [b(G_B - G_B^o) + d(G_{H_2O} - G_{H_2O}^o)] \\ - [a(G_A - G_A^o) + m(G_{H^+} - G_H^o·)]. \tag{8}$$

Taking as an example the reactant A, the corrected concentration, (A), available for reaction, called the activity of A, is related to free-energy change from standard state, $(G_A - G_A^o)$, by

$$a(G_A - G_A^o) = aRT \ln (A) = RT \ln (A)^a,$$

where R is the gas constant and T is absolute temperature. Substituting this and equivalent expressions for other reactants and products into (8) gives

$$\Delta G - \Delta G° = RT \ln \frac{(B)^b (H_2O)^d}{(A)^a (H^+)^m}$$

which is equivalent to

$$e = e° - \frac{RT}{nF} \ln \frac{(B)^b (H_2O)^d}{(A)^a (H^+)^m} \tag{9}$$

when substituting $\Delta G = -nFe$ and $\Delta G° = -nFe°$ from (5). Equation (9) is the common form of the *Nernst equation* found in many physical chemistry textbooks. For convenience, we will use the equivalent form,

$$e = e^\circ + \frac{2.3RT}{nF} \log \frac{(A)^a(H^+)^m}{(B)^b(H_2O)^d},$$ (10)

which shows more readily that the half-cell electrode potential becomes more positive (noble) as activities of oxidized species, (A) and (H^+) increase, in agreement with experimental observation.

Activity, designated by (·), is defined as pressure in atmospheres for gases and is adequately approximated by concentration, C, in gram equivalents per liter for many relatively dilute corrosive solutions. However, quantitative calculations of activity require a correction by the activity coefficient, f, in which the activity of species A is (A) $= f C_A$. Activity coefficients are extensively tabulated in numerous chemical and electrochemical handbooks.

The quantity $2.3RT/F$ is equal to 0.059 V at 25°C when appropriate values for the constants are substituted. The activity of water is always defined as unity in aqueous solutions, and remembering the definition of pH $= -\log (H^+)$, equation (10) becomes

$$e = e^\circ + \frac{0.059}{n} \log \frac{(A)^a}{(B)^b} - \frac{m}{n} 0.059 \text{ pH}$$ (11)

for reaction (7).

In equations (10) and (11), any of the species A, B, H^+, or H_2O may be absent for any particular half-cell reaction. Taking as an example the hydrogen reaction,

$$2H^+ + 2e^- = H_2,$$ (4)

Reactants A and H_2O are absent, and product B is represented by H_2. The Nernst equation for reaction (4) becomes

$$e_{H^+/H_2} = e^\circ_{H^+/H_2} - 0.059 \text{ pH},$$ (12)

when assuming that (H_2) $= p_{H_2} = 1$ atm, standard state.

It is readily apparent that as the activity of the oxidizer, H^+, increases (as the pH decreases), the potential, e_{H^+/H_2}, also increases in agreement with experimental observation. This is also true of all other oxidizers, A, in the generalized half-cell reaction (7), according to the Nernst equation (10). As the activity of any dissolved oxidizer increases, more noble or positive potentials are routinely measured experimentally. Thus, *potential is commonly considered to be a measure of the oxidizing power of the solution.* Details of potential measurements are discussed in Section 2.3.

The Nernst equation correctly predicts that increased oxygen activity makes the half-cell electrode potential for reaction,

$$O_2 + 2H_2O + 4e^- = 4OH^-,$$

more noble, suggesting the basis for the *differential aeration cell,* which has important effects in many corrosion processes. However, simple differences in oxygen activity are not sufficient to explain corrosion by differential aeration. Nonuniform current distributions and/or local pH changes are also necessary, as discussed later in Section 6.5.1.

2.1.4 Reaction Direction by Free-Energy Determination

To predict the spontaneous direction of an electrochemical reaction, one must first separate the reaction into its composite half-cell reactions. This was done in Section 2.1.1 for

$$Zn + 2HCl \rightarrow ZnCl_2 + H_2, \tag{1}$$

in which the required half-cell reactions were found to be

$$Zn \rightarrow Zn^{2+} + 2e^-, \tag{3}$$

and

$$2H^+ + 2e^- \rightarrow H_2. \tag{4}$$

The spontaneous direction may be derived conventionally from the algebraic sum of the half-cell electrode potentials;

$$E = e_a + e_c. \tag{6}$$

This sum gives the cell potential, E, which is then substituted into equation (5) for the free-energy change, ΔG. A positive E corresponding to a negative ΔG indicates that the reaction is spontaneous in the written direction. *It must be emphasized, however, that a spontaneous reaction does not necessarily mean a rapid reaction.* Reaction rates are discussed in Chapter 3.

The Nernst equation (11) must be used usually to calculate e_a and e_c for the anodic and cathodic reactions, respectively, when determining the spontaneous direction of an electrochemical reaction. However, in the examples that follow, all half-cell reactions are assumed to be at standard state for simplicity, and the half-cell electrode potentials are taken directly from Table 2.1.

In the example above, reaction (3) is shown as an oxidation in order for reaction (1) to proceed as written. Therefore, when substituting into equation (6), the sign of the half-cell electrode potential must be reversed. Reaction (4) is written as a reduction and the sign can be used directly without change from Table 2.1. In this particular case, the sign for the half-cell electrode potential of reaction (4) is irrelevant since the value is zero. Therefore, the cell potential for reaction (1) is

$$E = +0.763 + 0 = +0.763 \text{ V}.$$

This will correspond to a negative ΔG when substituted into equation (5) and reaction (1) is spontaneous and proceeds from left to right as written.

Consider a second example,

$$3Pb + 2Al^{3+} \rightleftharpoons 3Pb^{2+} + 2Al. \tag{13}$$

The reaction consists of the two half-cells,

$$Pb^{2+} + 2e^- = Pb, \tag{14}$$

and

$$Al^{3+} + 3e^- = Al. \tag{15}$$

From Table 2.1, $e_{Pb/Pb^{2+}} = +0.126$ V for the half-cell reaction (14) written as an oxidation, assuming reaction (13) proceeds left to right as written. The sign must be reversed to account for the fact that the negative sign from Table 2.1 is for the reaction written as a reduction. Also, $eAl = -1.662$ V for reaction (15). The negative sign is correct in this case because reaction (13) is a reduction in reaction (15). The cell potential is then, $E = +0.126 + (-1.662) = -1.532$ V. Because the cell potential is negative, the reaction must proceed in the opposite direction from that written in reaction (13):

$$3Pb + 2Al^{3+} \leftarrow 3Pb^{2+} + 2Al. \tag{16}$$

The cell potential, 1.532 V, is superfluous because only the signs of ΔG and E are important, not the magnitudes.

Again it must be emphasized that the effects of solution activity or concentration have been ignored in these examples.

2.1.5 Reaction Direction by Inspection

The procedure described in the preceding section is widely used, but the following alternative[1] may be more intuitively convenient for some. A reasonably simple rule is apparent from the free energy calculations of the preceding section: *The half-cell reaction with the more active (negative) half-cell potential always proceeds as an oxidation, and the one with the more noble half-cell potential always proceeds as a reduction in the spontaneous reaction produced by the pair.* Thus, in reaction (1) the zinc reaction (3) is active to the hydrogen reaction (4). Therefore, Zn must be oxidized from a state 0 to +2, and H must be reduced from +1 to 0 in the spontaneous reaction between Zn and H^+. Both fulfill these requirements in reaction (1), which must then proceed spontaneously as written. Conventions for direction of half-cell reactions and corresponding potential signs in the emf series have no effect and need not be considered. Nor is it necessary to remember the appropriate signs for E and ΔG in equation (2).

For the second example in Section 2.1.4, the reaction between aluminum and lead, Table 2.1 shows that the aluminum reaction (15) is more active and would be an oxidation while the lead reaction (14) would be a reduction. These requirements are fulfilled only if the spontaneous direction is from right to left, as given in reaction (16). The reactions (14) and (15) are written as reductions, but the direction is of no consequence; they could be written either as oxidations or reductions.

Inspection of Table 2.1 explains why copper is etched (corroded) in a solution containing ferric ions. The half-cell reaction for ferric reduction, $Fe^{3+} + e^- = Fe^{2+}$ has the half-cell electrode potential noble to reaction for copper, $Cu^{2+} + 2e^- = Cu$. Therefore, copper is dissolved (oxidized) by reduction of ferric ions. On the other hand, copper is unattacked by most acid solutions because the half-cell electrode potential for H^+ reduction is active to the above copper dissolution reaction. Care must be used in making such predictions since the half-cell electrode potentials are dependent on activities (concentrations) of dissolved reactants and products, as discussed in Section 2.1.3.

Most half-cell electrode potentials occur only at values active to the very noble half-cell electrode potentials for gold and platinum. Thus, these two metals are relatively inert because only the reduced form of the metals is stable in most solution conditions. This allows platinum to be used in the presence of an acid solution with little corrosion when used for the hydrogen electrode of Figure 2.2 and elsewhere. Platinum is favored over gold in most electrode processes because most half-cell reactions including (4) are very rapid on platinum (Section 3.1.2).

2.2 Potential/pH (Pourbaix) Diagrams

2.2.1 Uses and Limitations

The potential/pH diagram may be thought of as a map showing conditions of solution oxidizing power (potential) and acidity or alkalinity (pH) for the various possible phases that are stable in an aqueous electrochemical system. Boundary lines on the diagram dividing areas of stability for different phases are derived from the Nernst equation (11). The diagrams have many applications, including fuel cells, batteries, electroplating, and extractive metallurgy. However, the discussion here is limited to corrosion of metals in aqueous electrolytes,[2] the first application considered by Professor Marcel Pourbaix, who pioneered the construction and use of the diagrams.

Pourbaix diagrams show the reactions and reaction products that will be present when equilibrium has been attained, assuming that all appropriate reactions have been included. The collection of such diagrams by Pourbaix[2] gives rather complete equilibria for most possible chemical reactions in pure water. Of special interest are conditions in which corrosion is thermodynamically impossible. Thus, potential and/or pH can in some cases be adjusted to prevent corrosion thermodynamically.

For the usual conditions in areas on the Pourbaix diagram where corrosion is possible, no predictions can be made as to corrosion rates, which may be fast or slow at the relatively low temperatures present in liquid aqueous solutions. Thermodynamics is generally more useful at high temperatures (Chapter 12), where rates are higher and equilibrium is reached relatively rapidly. Although the Pourbaix diagram gives the stable phases for given conditions of potential and pH, other thermodynamically unstable phases formed in the past may still be present because they are slow to decompose. As a general statement, thermodynamics, and the Pourbaix diagrams derived therefrom, give no information about the rates of the reactions they describe.

2.2.2 Water and Dissolved Oxygen

Consider again the equilibrium between hydrogen gas and an acid solution:

$$2H^+ + 2e^- \rightarrow H_2. \tag{4}$$

A reaction equivalent to (4) in neutral or alkaline solutions is

$$2H_2O + 2e^- \rightarrow H_2 + 2OH^-. \tag{a}$$

Adding sufficient OH⁻ to both sides of reaction (4) results in {a}. The braces { } are given for convenience in labeling this reaction and others for inclusion in subsequent Pourbaix diagrams. Thus, at higher pH, where OH⁻ is predominant over H⁺, {a} is the more appropriate reaction for consideration. However, because the two are equivalent,

$$e_{H^+/H_2} = e^{\circ}_{H^+/H_2} - 0.059 \text{ pH} \tag{12}$$

represents the pH dependence of the half-cell electrode potential for both. Equation (12) is plotted on a potential/pH (Pourbaix) diagram and labeled for the reaction {a} in Figure 2.4. As expected, $e^{\circ}_{H^+/H_2}$ is shown as 0 at pH = 0 for (H⁺) = 1, and the slope is −0.059 V.

Equation {a} demonstrates clearly that electrochemical evolution of hydrogen represents a decomposition of water. For a potential more active than the half-cell electrode potential, e_{H^+/H_2}, hydrogen is evolved and water is unstable thermodynamically. This should not be taken to mean that water will suddenly disappear when the potential is moved to such an active value. Instead, any water present will be steadily decomposed at an unknown rate to hydrogen gas. Even starting with acid solutions, the H⁺ will first be consumed by reaction (4) until the pH increases sufficiently that water is decomposed directly to hydrogen by {a}. Thus, in Figure 2.4, below line {a}, water is unstable and must decompose to H₂. Above the line, water is stable, and hydrogen gas, if present, is oxidized to H⁺ or H₂O.

As potential becomes more noble (positive), another reaction involving water becomes thermodynamically feasible. Similar to (4) and {a},

$$O_2 + 4H^+ + 4e^- \rightarrow 2H_2O \tag{17}$$

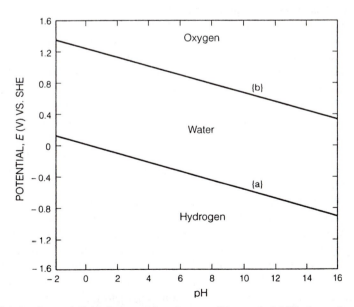

FIGURE 2.4 Potential/pH diagram showing conditions of stability for water and its decompostition products, oxygen and hydrogen.

and

$$O_2 + 2H_2O + 4e^- \rightarrow 4OH^- \qquad \{b\}$$

are equivalent reactions in acid and neutral or alkaline solutions, respectively. If $p_{O_2} = 1$ atm (unit oxygen activity), the Nernst equation for these becomes

$$e_{O_2/H_2O} = e^\circ_{O_2/H_2O} - 0.059 \text{ pH}, \qquad (18)$$

which is plotted again in Figure 2.4 and labeled for reaction $\{b\}$. For pH = 0, $e^\circ_{O_2/H_2O} = 1.229$ V, corresponding to the value for reaction (17) found in Table 2.1. For unit activity of OH^- at pH 14, $e^\circ_{O_2/H_2O} = 0.401$ V, corresponding to the standard value found in Table 2.1 for $\{b\}$. Again the slope is -0.059 V. At potentials noble (positive) to e_{O_2/H_2O} at any pH, water is unstable being oxidized to O_2. Below e_{O_2/H_2O} water is stable, and dissolved oxygen is reduced to water if present.

Figure 2.4 is divided into three regions. In the upper one, water can be electrolyzed anodically to form oxygen gas. In the lower one, water can be electrolyzed cathodically to form hydrogen gas. In the intermediate region, water is stable and cannot be electrolyzed. These two lines are frequently superimposed on the Pourbaix diagrams for corroding metals, showing conditions in which corrosion will cause hydrogen evolution or will reduce any dissolved oxygen present.

2.2.3 Corroding Metals

Two schematic examples of Pourbaix diagrams given in Figure 2.5 have sufficient detail for most purposes related to corrosion. Areas on the Pourbaix diagram are often labeled passive, corrosion, and immune, as indicated. However, these are only indications; actual rates cannot be derived from the diagrams. Corrosion is possible in areas of the Pourbaix diagram where soluble ions of the metal are stable. The metal is possibly resistant to corrosion or passive in areas where an oxide is stable. In areas where only the reduced form of the metal is stable, the metal is thermodynamically immune to corrosion. Potential can be located in the area of immunity by cathodic protection. More details on construction of the Pourbaix diagrams for these two metals, and additional examples of diagrams for other metals, are given in subsequent sections.

Aluminum is an amphoteric metal for which the protective oxide film dissolves at low and high pH. Thus, the Pourbaix diagram (Figure 2.5a) shows that the aluminum cation, Al^{3+}, is stable at low pH; the aluminate cation, AlO_2^-, at high pH; and the oxide, Al_2O_3, at intermediate pH. At sufficiently low, nonoxidizing potentials, the metal itself is stable and immune to corrosion. The boundaries between areas on the diagram represent the chemical reactions that define the equilibrium between the adjoining stable chemical species. The exact boundary positions depend on the activity of soluble aluminum bearing ions, as discussed in Section 2.2.4. Note for aluminum that there is a large potential difference (driving force) between the lines representing the cathodic half-cell reactions for hydrogen evolution and oxygen reduction and the half-cell reactions for anodic oxidation to dissolved ions or the oxide. However, aluminum is an excellent example of the fact that corrosion rate is

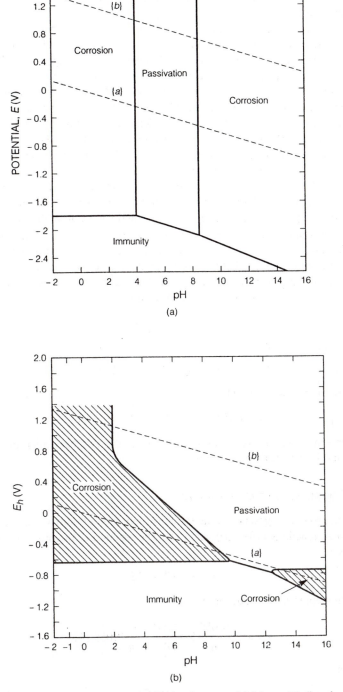

FIGURE 2.5 Pourbaix diagrams for (a) aluminum and (b) iron. All dissolved species at activities of 10^{-6} g-equiv/L.

relatively low due to kinetic limitations, despite the large driving force for the corrosion reactions.

In regions where an oxide film is stable, the metal may be passive. Passivity is the characteristic of corrosion resistance due to a protective surface film, although corrosion is favored thermodynamically (Chapter 4). The diagram cannot predict how rapidly the film may form or how low the resultant corrosion rate may be. Where a soluble ion is stable, corrosion is possible. In the areas of the diagram where the reduced form of the metal is stable, the metal is immune to corrosion. The boundaries between regions on the diagram depend on an arbitrarily defined activity (concentration) of dissolved species. The lower the activity, the more limited is the region of passivity, as discussed in Section 2.2.4. An activity of 10^{-6} is often selected[2] as a reasonably low value in Figure 2.5.

Iron is similar to aluminum in that a protective oxide forms in nearly neutral solutions. However, for iron the field of oxide stability is substantially greater at elevated pH, and iron is far more resistant to alkaline solutions. Contributing to the overall resistance of iron are the generally more noble half-cell electrode potentials for the anodic dissolution reactions which lower the driving force for corrosion reactions. It is apparent from Figure 2.5 that this resistance disappears in more acid solutions.

There are three general reactions by which a metal, M, may react anodically in the presence of water:[3] oxidation to aqueous cations,

$$M = M^{n+} + ne^-, \tag{19}$$

to metal hydroxide or oxide,

$$M + nH_2O = M(OH)_n + nH^+ + ne^-, \tag{20}$$

and to a soluble aqueous anions,

$$M + mH_2O = MO_m^{n-2m} + 2m\,H^+ + ne^- \tag{21}$$

These or similar reactions are used to construct the Pourbaix diagrams for aluminum and iron in following sections.

2.2.4 Construction for Aluminum

Aluminum provides a good example for construction of a Pourbaix diagram involving the three reactions (19), (20), and (21). The reaction analogous to (19) for aluminum is

$$Al = Al^{3+} + 3e^-. \tag{1}$$

Substituting appropriate values into

$$e = e^\circ + \frac{0.059}{n} \log \frac{(A)^a}{(B)^b} - \frac{m}{n} 0.059 \; \text{pH}, \tag{11}$$

we obtain

$$e_{Al/Al^{3+}} = -1.66 + \frac{0.059}{3} \log(Al^{3+}), \tag{22}$$

where $e^{\circ}_{Al/Al^{3+}} = -1.66$ V from Table 2.1. If (Al^{3+}) is unity, $e_{Al/Al^{3+}} = -1.66$ V - at all pH. Since this represents an unrealistically high concentration of Al^{3+} in a solution causing corrosion, a lower value, such as $(Al^{3+}) = 10^{-6}$, is often taken, giving $e_{Al/Al^{3+}} = -1.78$ V at all pH. Thus, the half-cell electrode potential for the formation of the aluminum cation is dependent on activity of aluminum in the corrosive solution and inde pendent of pH because no H^+ is involved in reaction {1}. Straight horizontal lines for (22) are plotted as potential versus pH and labeled {1} in Figure 2.6. These two lines are differentiated from one another by the corresponding values of (Al^{3+}).

For aluminum the corresponding equation for (20) is

$$2Al + 3H_2O = Al_2O_3 + 6H^+ + 6e^-. \tag{2}$$

The oxidized and reduced states are Al_2O_3 and Al, respectively, both of which are solids having activity of unity in equation (11), which reduces to

$$e_{Al/Al_2O_3} = -1.55 - 0.059 \text{ pH} \tag{23}$$

for the known value of $e^{\circ}_{Al/Al_2O_3} = -1.55$ V.[2] This standard value for e°_{Al/Al_2O_3} would be listed in a more complete emf series than one of Table 2.1. In Figure 2.7, equation (23) plots as a straight line of slope -0.059 labeled {2}, having a potential value of -1.55 V at pH $= 0$. The lines for reaction {1} from Figure 2.6 are superimposed in Figure 2.7 for comparison.

In Figure 2.7, the lines labeled {1} and {2} intersect at some pH that is dependent on (Al^{3+}). Above this pH, Al^{3+} reacts with water to form Al_2O_3 on the aluminum surface according to

$$2Al^{3+} + 3H_2O = Al_2O_3 + 6H^+. \tag{4}$$

Similarly, below this pH, the oxide Al_2O_3 dissolves to Al^{3+}. Oxidation states of the elements in this reaction are unchanged, and no electron (charge) transfer is required. Therefore, the equilibrium in the reaction depends on activities of products and reactants and is independent of potential. The dashed portions of {1} and {2} in

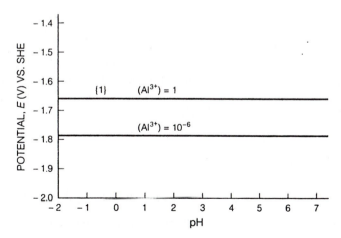

FIGURE 2.6   Potential/pH diagram showing equilibrium for the reaction Al = Al^{3+} + $3e^-$, {1} at Al^{3+} activities of 1 and 10^{-6}.

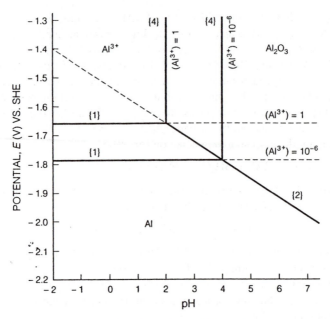

FIGURE 2.7   Potential/pH diagram showing equilibrium for the reaction $Al + 3H_2O = Al_2O_3 + 6H^+ + 6e^-$, {2}, superimposed on Figure 2.6.

Figure 2.7 indicate that the designated reactions have no significance in those potential-pH areas on the diagram.

The equilibrium pH where {1} and {2} intersect may be calculated from the known value of the equilibrium constant of $K = (H^+)^6/(Al^{3+})^2 = 10^{-11.4}$ for reaction {4}. For $(Al^{3+}) = 10^{-6}$,

$$\log K = 6 \log (H^+) - 2 \log (Al^{3+}) = -6 \, pH - 2 \log (Al^{3+});$$

$$pH = \frac{-\log K}{6} - \frac{\log (Al^{3+})}{3} = \frac{11.4}{6} + \frac{6}{3} = 3.9.$$

Thus, when potential is high enough, Al^{3+} is stable below and Al_2O_3 is stable above pH 3.9. When $(Al^{3+}) = 1$, the equilibrium pH between {1} and {2} is 1.9. Because these pH values are independent of potential, {4} plots as a vertical line in Figure 2.7.

The reaction equivalent to (21) for aluminum is

$$Al + 2H_2O = AlO_2^- + 4H^+ + 3e^-. \qquad \{3\}$$

Equation (11) becomes

$$e_{Al/AlO_2^-} = e^o_{Al/AlO_2^-} + \frac{0.059}{3} \log (AlO_2^-) - \frac{4}{3} 0.059 \, pH$$

$$= -1.262 + 0.020 \log (AlO_2^-) - 0.079 pH. \qquad (24)$$

Thus, the plot of equation (24), labeled {3} in Figure 2.8, is dependent on both pH and (AlO_2^-). Lines for (AlO_2^-) of both 1 and 10^{-6} are included in Figure 2.8. The

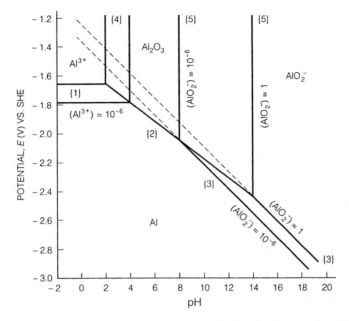

FIGURE 2.8 Potential/pH diagram showing equilibrium for the reaction Al + 2H₂0 = AlO_2^- + 4H⁺ + 3e⁻, {3}, superimposed on data from Figure 2.7.

slopes for both are −0.079. The intercept at pH = 0 for (AlO_2^-) = 1 is −1.262 V, which is the standard value of $e^°_{Al/AlO_2^-}$ for {3}. The plots for {1}, {2}, and {4} from Figure 2.7 are included for comparison in Figure 2.8.

At some high pH the oxide Al_2O_3 dissolves as the aluminate anion, AlO_2^-,

$$Al_2O_3 + H_2O = 2AlO_2^- + 2H^+. \tag{5}$$

Similar to {4}, there is no valence change or charge transfer, and the reaction is independent of potential. Again, from the equilibrium constant,

$$pH = 14.6 + \log(AlO_2^-).$$

When (AlO_2^-) = 10^{-6}, the equilibrium pH is calculated as 8.6, independent of potential. In Figure 2.8 this equilibrium is represented as a straight vertical line at pH 8.6. When potential is sufficiently high, Al_2O_3 is stable below pH 8.6 and AlO_2^- is stable above this pH. The similar vertical line for unit activity of the aluminate ion is at pH 14.6. The dashed portion of the line designated by {2} again indicates the area on the diagram where the reaction does not occur.

Figure 2.8 is a simplified Pourbaix diagram, showing the ranges of stability for all the major phases in the water-aluminum system, with some indication of the effects of dissolved ion activity. It is apparent that decreasing (Al^{3+}) results in a compressed field of stability for Al_2O_3. The similarity to Figure 2.5a is notable.

While it has been emphasized that the Pourbaix diagram cannot give the rates of corrosion, some qualitative predictions are possible nevertheless. Figure 2.9 shows corrosion rates for aluminum as a function of pH.[4] In the intermediate pH range between 4 and 8, where Al_2O_3 provides a protective film, the corrosion rate is low in all electrolytes. At high and low pH, where the oxide is soluble, the corrosion rate may be high depending on other ions present. Thus, areas of stability for the oxide film, soluble ions, and the metal itself are labeled "passivation," "corrosion," and "immunity," respectively, as in Figure 2.5a.

The complete Pourbaix diagram[2] for aluminum is given in Figure 2.5a, assuming an activity of 10^{-6} g-equiv/L for all dissolved ions. This activity level is frequently but arbitrarily selected[2] and depends on the preferences of the user. The stability boundaries for water are superimposed from Figure 2.4. It can be concluded that the cathodic reduction reaction during corrosion in the absence of dissolved oxygen will be evolution of hydrogen by reaction (4) in acid solutions or the equivalent reaction {a} in neutral or alkaline solutions. The anodic oxidation reaction will be {1} in acid solutions, {2} in nearly neutral solutions, and {3} in alkaline solutions. The actual surface potentials will fall somewhere between the half-cell electrode potentials for the anodic and cathodic reactions, where anodic oxidation of Al metal is possible (discussed more fully in Chapter 3). Thus, oxidation or corrosion is favored in most natural conditions, although reaction rates cannot be predicted from the Pourbaix diagram. By moving the potential into the range of immunity, corrosion can be made thermodynamically impossible with cathodic protection (Chapter 13).

2.2.5 Construction for Iron

Iron is the basis for the primary constructional alloy system, steel. It can exist in two oxidation states, +2 or +3, which complicate the Pourbaix diagram. The appropriate reactions and corresponding forms of the Nernst equation are:[2]

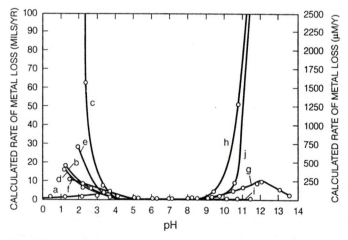

FIGURE 2.9 Effect of pH on corrosion rate of aluminum. a, acetic acid; b, hydrochloric acid; c, hydrofluoric acid; d, nitric acid; e, phosphoric acid; f, sulfuric acid; g, ammonium hydroxide; h, sodium carbonate; i, sodium disilicate; j, sodium hydroxide. *(From J. E. Hatch, ed.,* Aluminum: Properties and Physical Metallurgy, *ASM International, Metals Park, OH, p. 295, 1984. Reprinted by permission, ASM*

$$Fe = Fe^{2+} + 2e^- \qquad \{1\}$$
$$e_{Fe/Fe^{2+}} = -0.440 + 0.0295 \log (Fe^{2+})$$
$$Fe + 2H_2O = Fe(OH)_2 + 2H^+ + 2e^- \qquad \{2\}$$
$$e_{Fe/Fe(OH)_2} = -0.0470 - 0.0591 \text{ pH}$$
$$Fe + 2H_2O = HFeO_2^- + 3H+ + 2e^- \qquad \{3\}$$
$$e_{Fe/HFeO_2^-} = 0.493 - 0.0886 \text{ pH} + 0.0295 \log (HFeO_2^-)$$
$$Fe^{2+} + 2H_2O = Fe(OH)_2 + 2H^+ \qquad \{4\}$$
$$pH = 6.65 - 0.5 \log (Fe^{2+})$$
$$Fe(OH)_2 = HFeO_2^- + H^+ \qquad \{5\}$$
$$pH = 14.30 + \log (HFeO_2^-)$$
$$Fe^{2+} + 3H_2O = Fe(OH)_3 + 3H^+ + e^- \qquad \{6\}$$
$$e_{Fe^{2+}/Fe(OH)_3} = 1.057 - 0.1773 \text{ pH} - 0.0591 \log (Fe^{2+})$$
$$Fe^{3+} + 3H_2O = Fe(OH)_3 + 3H^+ \qquad \{7\}$$
$$pH = 1.613 - (1/3) \log (Fe^{3+})$$
$$HFeO_2^{2-} + H_2O = Fe(OH)_3 + 2e^- \qquad \{8\}$$
$$e_{HFeO_2^-/Fe(OH)_3} = -0.810 - 0.0591 \log (HFeO_2^-).$$
$$Fe(OH)_2 + H_2O = Fe(OH)_3 + H^+ + e^- \qquad \{9\}$$
$$e_{Fe(OH)_2/Fe(OH)_3} = 0.271 - 0.0591 \text{ pH}$$

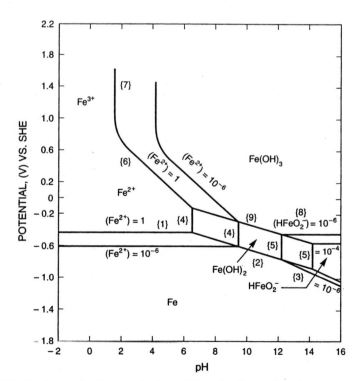

FIGURE 2.10   Pourbaix diagram for iron. *(From M. Pourbaix,* Atlas of Electrochemical Equilibria in Aqueous Solutions, *NACE, Houston, 1974. Reprinted by permission, National Association of Corrosion Engineers.)*

The Nernst equations for the various reactions are plotted in Figure 2.10 and labeled in the same numerical sequence as above. Lines are included for dissolved ion activities of both 1 and 10^{-6} g-equiv/L. Again, it is apparent that the lower dissolved ion activities result in compressed fields of stability for the precipitated oxide phases. However, the dissolved ion, $HFeO_2^-$, is stable only in a narrow potential range at very high pH, and iron is relatively resistant to alkaline solutions, as compared to aluminum.

The simplified Pourbaix diagram for iron with predictions of corrosion susceptibility is shown in Figure 2.5b, assuming dissolved ion activities of 10^{-6}. It is notable that the difference between the stability line for water and iron decreases significantly as pH increases. Thus, the corrosion rate drops also, although the actual rate cannot be predicted from the diagram. In nearly neutral solutions with dissolved oxygen, the potential is lifted into the passive region where a protective surface film is formed. However, below about pH 7, the film is not stable, and dissolved oxygen increases the corrosion rate. The complete and more complex Pourbaix diagrams for both aluminum and iron can be found in the collection by Pourbaix[2] and are reprinted in the next section.

2.2.6 Pourbaix Diagrams for Other Metals

Representative Pourbaix diagrams are assembled for some important pure metals in Figures 2.11a to 2.11h for easy reference. The shaded areas of Figure 2.11 show stability of soluble dissolution products where the metal is subject to corrosion. Unshaded areas show either thermodynamic immunity or corrosion resistance due to the stability of an oxide film. The labels 0, −2, −4, and −6 give the log of soluble ion activity for each indicated line.

Diagrams in Figure 2.11 for (a) aluminum and (b) iron are repeated for more detail and for easy comparison with others in the figure. Zirconium (c) is similar to aluminum with areas of corrosion in alkaline and acid solutions, but the area of oxide stability is much wider for zirconium, which is known for its high corrosion resistance. Nickel (d) with multiple oxidation states is similar to iron. However, the horizontal lines for anodic dissolution in acid solutions are significantly more noble for nickel, expanding the area of nickel stability. The kinetics of anodic dissolution are also intrinsically slower for nickel than iron (Section 15.1.3), contributing to the generally better corrosion resistance of nickel.

Copper (Figure 2.11e) is an example of a so-called noble metal in which the oxidation reactions occur at potentials noble to the line for hydrogen reduction {a}. Thus, copper cannot be oxidized by H^+ or water; it is thermodynamically resistant to corrosion. Chromium, on the other hand, shows corrosion resistance by passivity (Chapter 4) due to the presence of a resistant, protective oxidized surface film that is stable into areas of lower pH. Chromium is unsuitable as the basis for alloys because of brittleness, but it confers passivity and corrosion resistance as an alloying element in other metals, especially iron and nickel.

Titanium (g) is thermodynamically reactive, but is nevertheless very corrosion resistant because of a highly resistant passive film that is stable at all pHs in oxidizing potentials. It corrodes only at low pH in solutions without oxidizers. Tantalum (h) is even more resistant; the passive oxide film is stable at all pHs and potentials.

Engineering applications of alloys of these and other metals are discussed in more detail in Section 15.1.

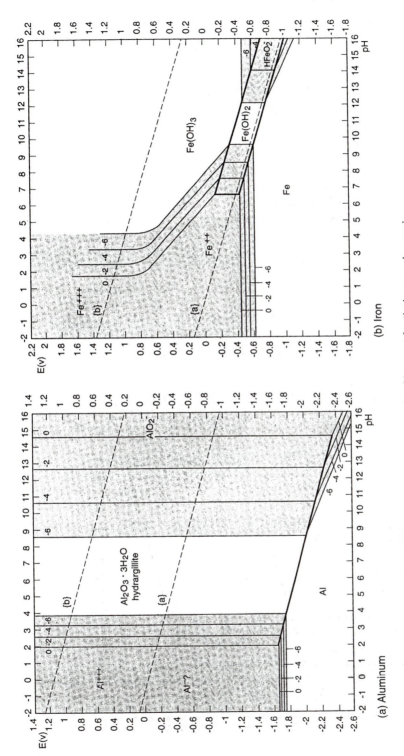

FIGURE 2.11 Representative Pourbaix diagrams for pure metals of interest for their corrosion resistance. Shaded areas indicate corrosion susceptibility. Labels 0, −2, −4, and −6 are the log of soluble ion activity for the indicated lines. *(From M. Pourbaix, Atlas of Electrochemical Equilibria in Aqueous Solutions, NACE, Houston, 1974. Reprinted by permission, National Association of Corrosion Engineers.)*

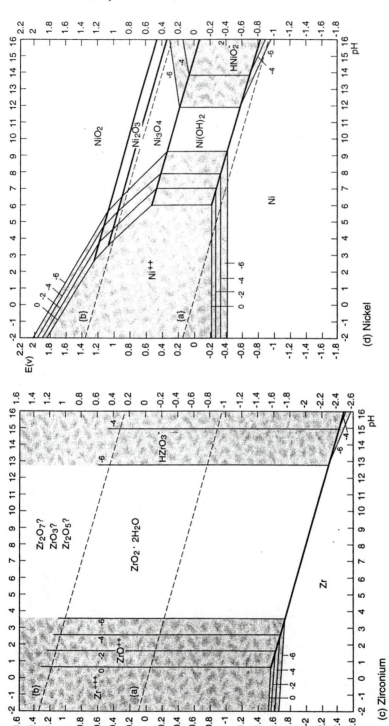

(d) Nickel

(c) Zirconium

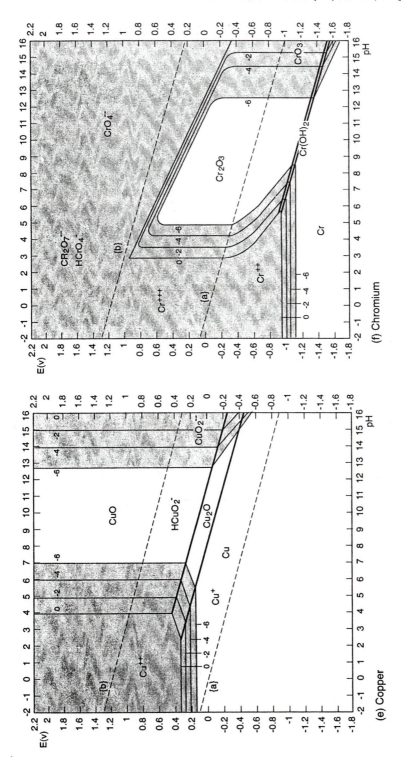

(f) Chromium

(e) Copper

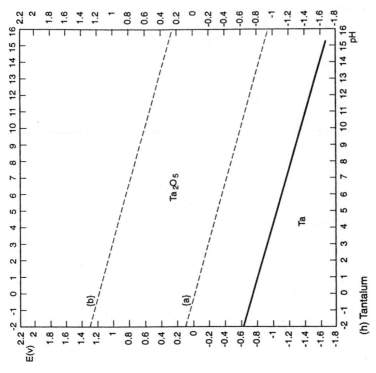

(h) Tantalum

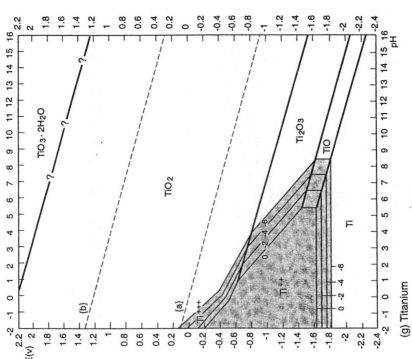

(g) Titanium

2.3 Experimental Measurements

Measurement of cell potential, E, is necessary to determine the driving force or free-energy change, ΔG, in an electrochemical cell from equation (1). Any electrochemical cell consists of two half-cells, as shown schematically in Figures 2.2 and 2.3. By making one of the half-cells a known or reference half-cell, we are able to isolate the second for measurement and study. The reference half-cell is commonly called the reference electrode, and the potential of the second, or unknown, half-cell is called the electrode potential, measured with respect to the particular reference electrode in the cell. Several reference electrodes in common use are described in the following section. Some physical features are included, but more details of construction may be found elsewhere.[5]

2.3.1 Reference Electrodes

The schematic construction of the hydrogen reference electrode is shown in Figure 2.12. A platinum foil specimen is suspended in a sulfuric acid solution of unit activity H^+, which is bubbled with purified hydrogen to remove dissolved oxygen and establish the standard state for H_2 gas at 1 atm pressure. The hydrogen electrode is

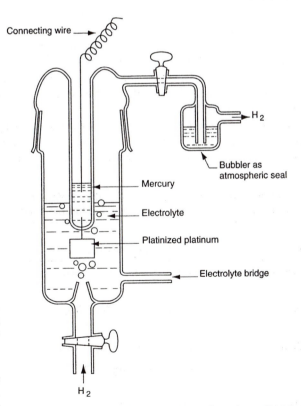

FIGURE 2.12 Standard hydrogen reference electrode, schematic.

connected to another half-cell through a solution bridge which contains a porous glass barrier to permit charge transfer and potential measurement but not mass transfer of the acid solution in the electrode.

The half-cell electrode potential, $e_{Pt^{3+}/Pt}$, for platinum dissolution from Table 2.1 is +1.2 V noble to $e^{\circ}_{H^+/H_2}$. Platinum can be dissolved only at very high potential, far above the redox potential for reaction (4) at 0 V. Thus, platinum will not dissolve to contaminate the solution and serves only as a catalyst for reaction (4), which establishes the potential on the surface. The ease and speed at which the half-cell electrode potential is assumed on platinum is related to the kinetics of the reaction (4) on the surface, as discussed in Section 3.1. The kinetics are improved by a porous "platinized" layer of platinum electroplated on the surface (Section 3.1.2).

The hydrogen half-cell reaction has been defined as having an electrode potential, $e_{H^+/H_2} = 0$, for all reactants and products at standard state. This standard hydrogen electrode (SHE) has been called the primary reference because it establishes the reference or zero point on the electrochemical scale by definition. All others are secondary reference electrodes; some of the more popular are listed in Table 2.2[6] with the corresponding half-cell reactions and half-cell electrode potentials on the SHE scale. Although SHE provides the zero point for the electrochemical potential scale, one of the listed secondary reference electrodes is used in the great majority of experimental measurements.

Electrode potentials may be reported directly in terms of the secondary reference, or a value may be calculated in terms of the SHE. Conversion of potential from one reference electrode scale to another is a simple arithmetic problem, but produces some confusion nevertheless. The difficulties may be minimized by examination of the potential scale in Table 2.2. For example, an electrode which measures −0.341 V vs. SCE would measure only −0.1 V vs. SHE; another, which measures −0.300 V vs. $Cu/CuSO_4$, would measure +0.018 vs. SHE.

The *saturated calomel electrode* (SCE) is the most popular secondary electrode for common laboratory use. It consists of mercurous chloride, Hg_2Cl_2, mixed in a mercury paste on a pool of liquid mercury in contact with a saturated KCl solution. Electrical con-

TABLE 2.2 Potential Values for Common Secondary Reference Electrodes. Standard Hydrogen Electrode included for reference. .

Name	Half-Cell Reaction	Potential V vs. SHE
Mercury-Mercurous Sulfate	$HgSO_4 + 2e^- = Hg + SO_4^{2-}$	+0.615
Copper-Copper Sulfate	$CuSO_4 + 2e^- = Cu + SO_4^{2-}$	+0.318
Saturated Calomel	$Hg_2Cl_2 + 2e^- = 2Hg + 2Cl^-$	+0.241
Silver-Silver Chloride	$AgCl + e^- = Ag + Cl^-$	+0.222
Standard Hydrogen	$2H^+ + 2e^- = H_2$	+0.000

tact is made via an inert platinum wire immersed in the mercury. Two types of construction are shown in Figure 2.13.[7] Design (a) is usually fabricated in the laboratory. However, it is more common to purchase commercial electrodes patterned after design (b).

For the calomel half-cell reaction,

$$Hg_2Cl_2 + 2e^- = 2Hg + 2Cl^-,$$

the Nernst equation is

$$e_{cal} = 0.268 - 0.059 \log (Cl^-).$$

The electrode potential, $e_{sce} = 0.241$ V, for SCE is somewhat lower than the standard potential for the half-cell reaction ($e_{cal}^\circ = 0.268$ V) because at saturation the chloride activity (Cl^-) is greater than unity. SCE is most convenient for corrosion purposes because the chloride activity can readily be controlled at a constant level by maintaining saturation in KCl. In more precise electrochemical studies, calomel electrodes with lower Cl^- activity have a lower temperature coefficient of variation.

The *silver-silver chloride electrode* is prepared by electrolytically oxidizing a silver surface to AgCl in hydrochloric acid. The silver surface may be provided by silver wire or an electroplated silver coating on platinum wire. The design may be sim-

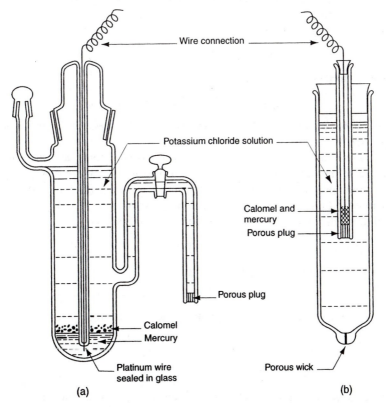

FIGURE 2.13 Two designs for the calomel reference electrode: (a) laboratory construction; (b) commercial construction.

ilar to either of those shown in Figure 2.13 with the wire sealed into Pyrex tubing and the exposed chloridized silver surface immersed in the required chloride solution. The half-cell reaction is

$$AgCl + e^- = Ag + Cl^-,$$

and the Nernst equation gives the electrode potential

$$e_{Ag/AgCl} = 0.222 - 0.059 \log (Cl^-).$$

It is apparent that the electrode potential for the Ag/AgCl electrode is dependent on (Cl^-) as for the calomel electrode. When Cl^- is provided by hydrochloric acid, $(Cl^-) = (H^+)$, and $-\log(Cl^-)$ in equation (22) can be replaced by pH:

$$e_{Ag/AgCl} = 0.222 + 0.059 \text{ pH}.$$

Thus, the half-cell electrode potential can be monitored by pH of the electrode solution. Alternatively, the electrode solution may be of KCl, as in the calomel electrode. The temperature coefficient has been well characterized for the Ag/AgCl electrode, and it finds frequent use in elevated temperature applications.

The *mercury-mercurous sulfate electrode* is a chemical analog of the calomel (mercury-mercurous chloride) electrode, finding use when possible contamination of the cell by chloride is undesirable. Electrode construction is identical to that in Figure 2.13, with $HgSO_4$ substituted for Hg_2Cl_2. The half-cell reaction is

$$HgSO_4 + 2e^- = Hg + SO_4^{2-},$$

and the electrode potential is predicted by

$$e_{Hg/HgSO_4} = 0.615 - 0.0295 \log (SO_4^{2-}).$$

The electrode potential is 0.615 V versus SHE when immersed in a solution of unit SO_4^{2-} activity.

The *copper-saturated copper sulfate electrode* is commonly used in field measurements to measure potentials on buried metal structures where rugged and simple construction is favored over higher precision. One design (Figure 2.14^8) consists of a copper rod immersed in a saturated $CuSO_4$ solution with electrolytic contact to the cell through a porous wooden plug. The half-cell reaction is

$$Cu^{2+} + 2e^- = Cu,$$

and the electrode potential is given by

$$e_{Cu^{2+}/Cu} = 0.342 + 0.0295 \log (Cu^{2+}).$$

Corrections for activity of Cu^{2+} at saturation place the electrode potential at 0.318 V versus SHE, or approximately 0.30 V for practical field purposes.

2.3.2 Reversible Cell Potential

Figure 2.15^9 shows a typical cell composed of reversible half-cells of zinc in zinc chloride solution of $(Zn^{2+}) = 1$ and platinum in a hydrogen-saturated sulfuric acid solution of unit activity. The experimentally measured cell potential is given by

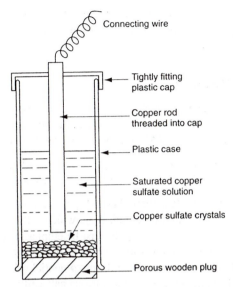

FIGURE 2.14   Copper-copper sulfate reference electrode.

$$E_{cell} = e°_{Zn^{2+}/Zn} + e°_{H^+/H_2} + e_{Zn/Cu} + e_{Cu/Pt} + e_j.$$

The standard half-cell electrode potentials, $e°_{Zn^{2+}/Zn}$ and $e°_{H^+/H_2}$, have been discussed in previous sections of this chapter. The last three terms are small contributions included to some extent in all cell potential measurements. Contact potentials, $e_{Zn/Cu}$ and $e_{Cu/Pt}$, arise from the energy required to move an electron from one metal lattice to another. They are universally quite small and often opposite in sign, canceling one another in the cell potential measurement. Nevertheless, half-cell electrode potentials, as listed in Table 2.1, invariably include some arguably small component of contact potential.

A liquid junction potential, e_j, is generated across the interface between two solutions of differing composition and/or concentration. The source of e_j is the ion mobility, which differs across a solution interface depending on ion type and concentration. For example, e_j between 0.001 N and 0.01 N HCl solutions is about 39 mV[10] because the H^+ ion migrates into the dilute solution more rapidly than the Cl^-. However, e_j between 0.001 N and 0.01 N KCl solutions is only about 1 mV because the mobilities of K^+ and Cl^- are nearly equal. The advantage of using KCl solutions in calomel and Ag/AgCl secondary reference electrodes becomes apparent. The e_j across the liquid junction of a calomel or Ag/AgCl reference electrode is composed of contributions from the internal dissolved KCl and all dissolved species in the external electrolyte solution. However, the e_j is minimized because the charge transfer across the junction is dominated by the relatively mobile KCl, especially when the reference electrode solution is saturated in KCl.[10]

Liquid junction potentials must be carefully removed from reversible cell potentials which are used to derive thermodynamic quantities. However, in polarization measurements for corrosion, the reference potential only provides a baseline for potential changes (polarization), and small constant contributions to the reference potential are of no consequence.

2.3.3 Corrosion Potential

The surface potential of a corroding metal, M, is measured with a voltmeter, V, with respect to a secondary reference electrode REF in the simple cell shown in Figure 2.16. Using zinc as an example of M in an acid solution, both reactions

$$Zn = Zn^{2+} + 2e^- \qquad e_{Zn/Zn^{2+}} \qquad \text{(anodic)} \qquad (3)$$
$$2H^+ + 2e^- = H_2 \qquad e_{H^+/H_2} \qquad \text{(cathodic)} \qquad (4)$$

occur on the Zn surface simultaneously. The cell potential measured in Figure 2.16 is a so-called mixed or corrosion potential, E_{corr}, which falls between the half-cell electrode potentials, for $e_{Zn/Zn^{2+}}$ and e_{H^+/H_2}. It must be emphasized that E_{corr} is not for a half-cell reaction but for a combination of two or more half-cell reactions which are drawn away (polarized) from their equilibrium potential values to E_{corr}. The exact value of E_{corr} depends on the kinetics of the various half-cell reactions, as discussed in Chapter 3.

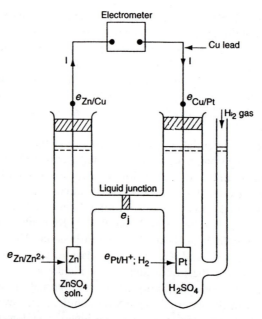

FIGURE 2.15 Typical schematic cell for measurement of reversible cell potential, showing contact potentials, $e_{Zn/Cu}$ and $e_{Pt/Cu}$, liquid junction potential, e_j, and half-cell potentials, $e_{Zn/Zn^{2+}}$ and e_{H^+/H_2}. *(From L. L. Shreir, Corrosion, Vol. I, Metal/Environment Reactions, Vol. II, Corrosion Control, 2nd ed., Newnes-Butterworths, Sevenoaks, Kent, England, p. 9:72, 1976. Reprinted by permission, Butterworths Ltd.)*

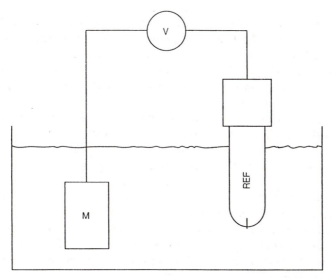

FIGURE 2.16   Simple cell for measurement of corrosion potential.

2.3.4 Instrumentation

Measurement of cell potential requires instrumentation of high internal resistance to prevent passage of currents which can change the potential of the electrodes. For example, according to Ohm's law ($E = IR$), a voltmeter of 10,000Ω resistance would allow passage of 100 μA when measuring a cell potential of 1 V in Figure

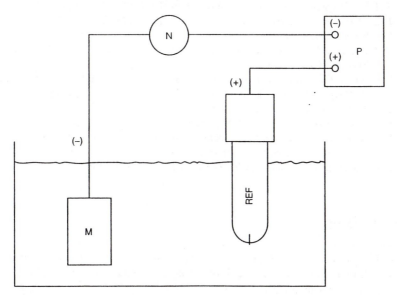

FIGURE 2.17   Schematic cell showing a potentiometer and electrometer to measure corrosion potential.

2.16. Currents of this magnitude will easily polarize (change the potential of) both corroding electrodes and reference electrodes. Electrometers with internal resistances up to 10^{14} Ω are used effectively in electrochemical potential measurements. Digital electrometers have sufficient sensitivity to < 0.1 mV, but analog electrometers with conventional analog meters give visual indications of potential change more conveniently with time.

A useful combination is a potentiometer connected in series with an analog electrometer, as shown schematically in Figure 2.17. A potentiometer consists of batteries with a voltage divider network to provide constant potential adjustable within < 0.1 mV from 0 to 1.6 V. The potentiometer is set up so that its potential output opposes that of the cell in Figure 2.16. The electrometer then serves as a high-resistance null detector which can be conveniently used to follow small potential changes in polarization studies (Chapters 3 and 5).

As a final precautionary note, all potential measurements—whether of half-cell, e, or corrosion potential, E_{corr}, values—tell us little about corrosion rates. Chapter 3 discusses mixed potential theory, which allows prediction of corrosion rates from fundamental principles.

References

1. M. G. Fontana, *Corrosion Engineering*, 3rd ed., McGraw-Hill, New York, p. 459, 1978.
2. M. Pourbaix, *Atlas of Electrochemical Equilibria in Aqueous Solutions*, NACE, Houston, 1974.
3. L. L. Shreir, *Corrosion*, Vol. I, *Metal/Environment Reactions*, Vol. II, *Corrosion Control*, 2nd ed., Newnes-Butterworths, Sevenoaks, Kent, England, p. 1:57, 1976.
4. J. E. Hatch, ed., *Aluminum: Properties and Physical Metallurgy*, ASM-International, Metals Park, OH, p. 295, 1984.
5. D. J. G. Ives and G. J. Janz, *Reference Electrodes: Theory and Practice*, Academic Press, New York, p. 198–213, 1961.
6. L. L. Shrier, *Corrosion*, Vol. I, *Metal/Environment Reactions*, Vol. II, *Corrosion Control*, 2nd ed., Newnes-Butterworths, Sevenoaks, Kent, England, p. 21:29, 1976.
7. E. C. Potter, *Electrochemistry*, Cleaver-Hume, London, p. 111, 1961.
8. E. C. Potter, *Electrochemistry*, Cleaver-Hume, London, p. 122, 1961.
9. L. L. Shreir, *Corrosion*, Vol. I, *Metal/Environment Reactions*, Vol. II, *Corrosion Control*, 2nd ed., Newnes-Butterworths, Sevenoaks, Kent, England, p. 9:72, 1976.
10. E. C. Potter, *Electrochemistry*, Cleaver-Hume, London, p. 97, 1961.

Exercises

2-1. Calculate the free energy change from cell potential for the net reaction in the cell of Figure 2.2. How does your answer compare with free energy determined calorimetrically?

2-2. Calculate the value of 2.303RT/F at 40C.

2-3. Assuming standard states for all reactants and products, determine the spontaneous direction of the following reactions by calculating the cell potential:

a. $Cu + 2HCl = CuCl_2 + H_2$
b. $Fe + 2HCl = FeCl_2 + H_2$
c. $2AgNO_3 + Fe = Fe(NO_3)_2 + 2Ag$
d. $Ag + FeCl_3 = FeCl_2 + AgCl$
e. $2Al + 3ZnSO_4 = Al_2(SO_4)_3 + 3Zn$

2-4. Repeat Problem 2-3 using the method of inspection described in Section 2.1.5.

2-5. Write the electrochemical half-cell reactions for oxidation and reduction during uniform corrosion in the following:

a. aluminum in air-free sulfuric acid
b. iron in air-free acid ferric sulfate solution
c. carbon steel in aerated seawater
d. zinc-tin alloy in an oxygen saturated solution of cupric chloride, stannic chloride, and hydrochloric acid
e. copper in deaerated seawater.

2-6. Copper immersed in concentrated pure hydrochloric acid is observed to corrode rapidly with evolution of gas bubbles. Is this contrary to the results of Problem 2-3a? Calculate a cell potential to prove your answer, making any necessary assumptions.

2-7. Iron in an NaCl solution, pH=1, shows a potential of +0.2 volts vs. SHE. What are the possible anodic and cathodic reactions, assuming that the Pourbaix diagram of Figure 2.4 is applicable?

2-8. A student suggests that two possible reactions for Problem 2.7 are

$$Cl_2 + 2e^- = 2Cl^- \text{ and}$$
$$Na = Na^+ + e^-.$$

Do you agree with either or both? If so, what assumptions must you make?

2-9. Construct the potential-pH diagram for zinc using the following reactions and assuming the activity of all dissolved substances is 10^{-6}. Assume any other reactions necessary to complete the diagram using the diagram for aluminum, Figure 2.11a, as a guide.

(1) $Zn = Zn^{2+} + 2e^-$, $e° = -0.763$ V
(2) $Zn + H_2O = ZnO + 2H^+ + 2e^-$, $e° = -0.439$ V
(3) $Zn + 2H_2O = ZnO_2^{2-} + 4H^+ + 2e^-$, $e° = +0.441$ V

2-10. Using the Pourbaix diagram for nickel, Figure 2.11d, give the anodic and cathodic reactions on Ni in water for the following conditions, assuming activity of 10^{-6} for all soluble species:

a. deaerated pH 2
b. deaerated pH 10
c. aerated pH 2
d. aerated pH 10

2-11. Write the electrochemical half-cell reaction and the corresponding Nernst equation for each of the lines separating the following phases in Figure 2.11(g):

    a. Ti and TiO

    b. TiO and Ti_2O_3

    c. Ti^{2+} and Ti_2O_3

2-12. A buried stainless steel nuclear waste storage tank contains significant amounts of dissolved radioactive Cr^{3+}, Al^{3+}, Fe^{3+}, and Ni^{2+} at pH 2. It is proposed to neutralize this acid tank solution to pH 7 in the tank with concentrated NaOH. The neutralized solution will then be pumped out of the tank for transport to other facilities for further processing. Explain the advantages and disadvantages of the neutralization step using the appropriate Pourbaix diagrams. On the basis of your evaluation, would you recommend neutralization?

2-13. **(a)** From solubility data for KCl, calculate the potential of SCE vs. SHE. Compare your result with that given in Table 2.2.

    (b) Calculate the potential of the calomel electrode in a 0.1 N KCl electrolyte and compare with SCE in (a).

2-14. A technician measured the potential of steel pipe with an analog multimeter in soil vs. a $Cu/CuSO_4$ electrode as -0.5615 volts. Do you think his/her measurements are accurate? Explain briefly.

2-15. A corrosion potential of -0.229 V versus SCE was measured for a corroding alloy. What is the potential versus SHE? Ag/AgCl (saturated)? Cu/saturated $CuSO_4$?

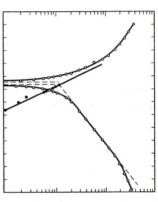

Electrochemical
Kinetics
of Corrosion

3

3.1 Introduction

Corrosion is thermodynamically possible for most environmental conditions. Thus, it is of primary importance to know how fast corrosion occurs. Fortunately, most alloys corrode only slowly in many environments. As new technology demands higher temperature, pressure, velocity, and concentration in aircraft, automobiles, energy generation, and manufacturing, corrosion rates are forced to higher levels. Chemical kinetics is a study of the rates of such reactions. Corrosion in aqueous systems is governed primarily by electrochemical reactions, as discussed in Section 2.1.1. An understanding of the fundamental laws of electrochemical reaction kinetics is thus essential to develop more corrosion-resistant alloys and to improve methods of protection against corrosion.

This chapter generally follows the treatment of Stern and Geary[1] and later Fontana and Greene,[2] who organized the principles into a sound pedagogical framework. Only general corrosion is of concern in this chapter. The various forms of localized nonuniform corrosion, described briefly in Chapter 1 and summarized in

Figure 1.4, will be described more thoroughly in subsequent chapters, using the electrochemical principles outlined in this chapter.

3.1.1 Faraday's Law

Electrochemical reactions either produce or consume electrons. Thus, the rate of electron flow to or from a reacting interface is a measure of reaction rate. Electron flow is conveniently measured as current, I, in amperes, where 1-ampere is equal to 1-coulomb of charge (6.2×10^{18} electrons) per second. The proportionality between I and mass reacted, m, in an electrochemical reaction is given by Faraday's Law:

$$m = \frac{Ita}{nF},$$

(1)

where F is Faraday's constant (96,500 coulombs/equivalent), n the number of equivalents exchanged, a the atomic weight, and t the time. Using the anodic reaction for zinc from Section 2.1.1 as an example:

$$Zn = Zn^{2+} + 2e^-.$$

(2)

Two equivalents are transferred for each atomic weight reacted. Thus, $n = 2$, and n is, in effect, the number of electrons transferred or the oxidation number changed during the reaction. Dividing equation (1) through by t and the surface area, A, yields the corrosion rate, r:

$$r = \frac{m}{tA} = \frac{ia}{nF},$$

(3)

where i, defined as *current density*, equals I/A. Equation (3) shows a proportionality between mass loss per unit area per unit time (e.g., mg/dm^2/day) and current density (e.g., $\mu A/cm^2$). The proportionality constant includes a/nF and any conversion factors for units. Current density rather than current is proportional to corrosion rate because the same current concentrated into a smaller surface area results in a larger corrosion rate. Corrosion rate is inversely proportional to area for the same dissolving current.

Current density can be routinely and precisely measured to values as low as 10^{-9} A/cm^2 up to several A/cm^2. Thus, electrochemical measurements are very sensitive and convenient tools for the study of corrosion in the laboratory and the field.

Units of penetration per unit time result from dividing equation (3) by the density, D, of the alloy. For corrosion rate in mils (0.001 in.) per year (mpy), equation (3) becomes

$$r = 0.129 \frac{ai}{nD} \text{(in mpy)}$$

(4)

for units of i, $\mu A/cm^2$, and D, g/cm^3. The proportionality constant, 0.129, becomes 0.00327 and 3.27 for mm/yr and $\mu M/yr$, respectively. The equivalence for iron between a current density of 1 $\mu A/cm^2$ and mpy is

$$1 \ \mu A \ / \ cm^2 = 0.129 \frac{(55.8)(1)}{(2)(7.68)} = 0.46 \ \text{mpy.}$$

A list of penetration rates equivalent to 1 $\mu A/cm^2$ appears in Table 3.1 for a number of pure metals and alloys.

Calculation of correspondence between penetration rate and current density for an alloy requires a determination of the equivalent weight, a/n, in equations (3) and (4) for the alloy. This alloy equivalent weight is a weighted average of a/n for the major alloying elements in any given alloy. The recommended procedure[3] for calculation of equivalent weight sums the fractional number of equivalents of all alloying elements to determine the total number of equivalents, N_{EQ}, which result from dissolving unit mass of the alloy. That is,

$$N_{EQ} = \Sigma \left(\frac{f_i}{a_i / n_i} \right) = \Sigma \left(\frac{f_i n_i}{a_i} \right) \tag{5}$$

where f_i, n_i and a_i are mass fraction, electrons exchanged, and atomic weight, respectively, of the ith alloying element. Equivalent weight, EW, is then the reciprocal of N_{EQ}; i.e., EW $= N_{EQ}^{-1}$.

Calculation of equivalent weight for Type 304 stainless steel[3] is given in the following example, assuming the following parameters for the alloy:

Cr: 19%, $n = 3$;
Ni: 9.25%, $n = 2$;
Fe: 71.75%, $n = 2$.

All other minor elements below 1% are neglected. From equation (5),

$$N_{EQ} = \frac{(0.19)(3)}{52.00} + \frac{(0.0925)(2)}{58.71} + \frac{(0.7175)(2)}{55.85} = 0.03981.$$

Equivalent weight for Type 304 stainless steel is then the reciprocal of 0.03981 or 25.12, which is entered in Table 3.1 along with values for a number of other commercial alloys.

3.1.2 Exchange Current Density

Consider the reaction for oxidation/reduction of hydrogen:

$$2H^+ + 2e^- \rightarrow H_2. \tag{6}$$

At the standard half-cell or redox potential, e_H^{o+}/H_2, this reaction is in the equilibrium state, where the forward rate (left to right), r_f, is equal to the reverse rate (right to left), r_r. From equation (3),

$$r_f = r_r = \frac{i_o a}{nF}, \tag{7}$$

where i_o is the *exchange current density* equivalent to the reversible rate at equilibrium. Whereas free-energy or half-cell electrode potential is the fundamental ther-

TABLE 3.1 Electrochemical and Current Density Equivalence with Corrosion Rate

Metal/Alloy	Element/ Oxidation State	Density (g/cm^3)	Equivalent Weight	Penetration Rate Equivalent to 1 μA/cm^2 (mpy)
Pure metals				
Iron	Fe/2	7.87	27.92	0.46
Nickel	Ni/2	8.90	29.36	0.43
Copper	Cu/2	8.96	31.77	0.46
Aluminum	Al/3	2.70	8.99	0.43
Lead	Pb/2	11.34	103.59	1.12
Zinc	Zn/2	7.13	32.69	0.59
Tin	Sn/2	7.3	59.34	1.05
Titanium	Ti/2	4.51	23.95	0.69
Zirconium	Zr/4	6.5	22.80	0.75
Aluminum alloys				
AA1100	Al/3	2.71	8.99	0.43
AA2024	Al/3,Mg/2, Cu/2	2.77	9.42	0.44
AA3004	Al/3,Mg/2	2.72	9.07	0.43
AA5052	Al/3,Mg/2	2.68	9.05	0.44
AA6070	Al/3,Mg/2	2.71	8.98	0.43
AA6061	Al/3,Mg/2	2.70	9.01	0.43
AA7072	Al/3,Zn/2	2.72	9.06	0.43
AA7075	Al/3,Mg/2 Zn/2,Cu/2	2.80	9.55	0.44
Copper alloys				
CDA110	Cu/2	8.96	31.77	0.46
CDA260	Cu/2,Zn/2	8.39	32.04	0.49
CDA280	Cu/2,Zn/2	8.39	32.11	0.49
CDA444	Cu/2,Sn/4	8.52	32.00	0.48
CDA687	Cu/2,Zn/2 Al/3	8.33	30.29	0.47
CDA608	Cu/2,Al/3	8.16	27.76	0.44
CDA510	Cu/2,Sn/4	8.86	31.66	0.46
CDA524	Cu/2,Sn/4	8.86	31.55	0.46
CDA655	Cu/2,Si/4	8.52	28.51	0.43
CDA706	Cu/2,Ni/2	8.94	31.55	0.46
CDA715	Cu/2,Ni/2	8.94	30.98	0.45
CDA752	Cu/2,Ni/2 Zn/2	8.94	31.46	0.45
Stainless steels				
304	Fe/2,Cr/3 Ni/2	7.9	25.12	0.41
321	Fe/2,Cr/3 Ni/2	7.9	25.13	0.41

TABLE 3.1 (Cont.)

Metal/Alloy	Element/ Oxidation State	Density (g/cm^3)	Equivalent Weight	Penetration Rate Equivalent to 1 $\mu A/cm^2$ (mpy)
309	Fe/2,Cr/3 Ni/2	7.9	24.62	0.41
316	Fe/2,Cr/3 Ni/2,Mo/3	8.0	25.50	0.41
430	Fe/2,Cr/3	7.7	25.30	0.42
446	Fe/2,Cr/3	7.6	24.22	0.41
20Cb3	Fe/2,Cr/3 Mo/3,Cu/1	7.97	23.98	0.39
Nickel alloys				
200	Ni/2	8.89	29.36	0.43
400	Ni/2,Cu/2	8.84	30.12	0.44
600	Ni/2,Fe/2 Cr/3	8.51	26.41	0.40
825	Ni/2,Fe/2 Cr/3,Mo/3 Cu/1	8.14	25.52	0.40
B	Ni/2,Mo/3 Fe/2	9.22	30.05	0.42
C-276	Ni/2,Fe/2 Cr/3,Mo/3 W/4	8.89	27.09	0.39
G	Ni/2,Fe/2 Cr/3,Mo/3 Cu/1,Nb/4 Mn/2	8.27	25.46	0.40

Source: Adapted from Proposed Standard, ASTM G01.11, *with permission, ASTM, Philadelphia.*

modynamic parameter characteristic of an electrochemical reaction (Section 2.1.1), i_o is the analogous fundamental kinetic parameter. Neither free energy (electrode potential) nor i_o can be calculated from first principles; both must be determined experimentally. Exchange current density is affected foremost by the nature of the surface on which it occurs. Figure 3.1[4] shows the half-cell electrode potential for the hydrogen reaction (6) plotted versus i_o. Two facts must be emphasized from this figure. First, the surface on which the reaction occurs has no effect on the electrode potential. Second, the i_o, in contrast, is strongly affected by the surface. That is, ΔG, the thermodynamic energy change is not affected by surface, but the reaction kinetics, as measured by i_o, are highly sensitive to surface properties.

The exchange current density for reaction (6) on mercury is only 10^{-12} A/cm^2 but is *nine orders of magnitude higher* on platinum at 10^{-3} A/cm^2. In fact, platinum is widely known as a catalyst because most reactions occur very rapidly on platinum surfaces, while the platinum itself is unreactive (Section 2.1.2). Exchange current

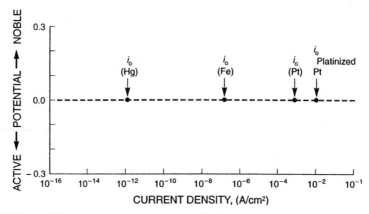

FIGURE 3.1 Effect of reaction surface on electrode potential and exchange current density for the hydrogen reaction. *(From M. G. Fontana,* Corrosion Engineering, *3rd ed., McGraw-Hill, New York, p. 457, 1986. Reprinted by permission, McGraw-Hill Book Company.)*

density is calculated on the basis of apparent geometric surface area. The "platinized" platinum electrode, with a porous electroplated platinum coating, has a greater actual surface area for the same apparent geometric surface area and, consequently, has a somewhat higher measured i_o. Exchange current density for the hydrogen reaction (6) on iron and all other common conducting surfaces falls between the extremes for mercury and platinum. Table 3.2 in Section 3.4.2 lists i_o's for various half-cell reactions on a number of surfaces, together with other polarization parameters described in the following section.

3.2 Electrochemical Polarization

Polarization, η, is the potential change, $E - e$, from the equilibrium half-cell electrode potential, e, caused by a net surface reaction rate for the half-cell reaction. For cathodic polarization, η_c, electrons are supplied to the surface, and a buildup in the metal due to the slow reaction rate causes the surface potential, E, to become negative to e. Hence, η_c is negative by definition. For anodic polarization, electrons are removed from the metal, a deficiency results in a positive potential change due to the slow liberation of electrons by the surface reaction, and η_a must be positive. Polarization is classified into two types—activation and concentration; both are discussed in the sections that follow.

3.2.1 Activation Polarization

When some step in the half-cell reaction controls the rate of charge (electron) flow, the reaction is said to be under activation or charge-transfer control, and activation polarization results. Hydrogen evolution by

$$2H^+ + 2e^- \rightarrow H_2. \tag{6}$$

at a metal surface occurs, for example, in three major steps. First, H^+ reacts with an electron from the metal,

$$H^+ + e^- \rightarrow H_{ads}, \tag{8}$$

to form an adsorbed hydrogen atom, H_{ads}, at the surface. Two of these adsorbed atoms must react in the second step to form the hydrogen molecule,

$$H_{ads} + H_{ads} \rightarrow H_2. \tag{9}$$

A third step requires sufficient molecules to combine and nucleate a hydrogen bubble on the surface. Any one of the steps can control the rate of reaction (6) and cause activation polarization.

The relationship between activation polarization or overpotential, η, and the rate of the reaction represented by current density, i_a or i_c, is

$$\eta_a = \beta_a \log \frac{i_a}{i_o} \tag{10a}$$

for anodic polarization, and

$$\eta_c = \beta_c \log \frac{i_c}{i_o} \tag{10b}$$

for cathodic polarization. *Overpotential* is a term used frequently for polarization. For anodic overpotential, η_a is positive, and β_a must also be positive, as a consequence. Similarly, for cathodic polarization, β_c is negative because η_c is negative. β_a and β_c are known as the Tafel constants for the half-cell reaction. The anodic, i_a, and cathodic, i_c, current densities flow in opposite directions. The Tafel relationships described by equations (10) have universally been observed by experiment for activation polarization. The experimental work leading to equations (10) is discussed in Section 3.5.

Inspection of equations (10) indicates that a plot of overpotential, η_{act}, versus log i is linear for both anodic and cathodic polarization, as shown in Figure 3.2. The slopes are given by the Tafel constants, which are assumed as ± 0.1 V per decade of current.. For zero η, either of equations (10) reduces to $i = i_o$. Below the reversible half-cell electrode potential, the reduction or forward reaction,

$$2H^+ + 2e^- \rightarrow H_2, \tag{6}$$

is favored, while above the same potential, the oxidation or reverse reaction,

$$H_2 \rightarrow 2H^+ + 2e^-,$$

is favored. The rate, as measured by i_a or i_c, increases by one order of magnitude for an overpotential change of $+0.1$ V for anodic polarization and -0.1 V for cathodic polarization, respectively, using the assumed values of β. The absolute values of the β Tafel constants usually range from 0.03 to 0.2 V and may not be equal for anodic and cathodic reactions, as assumed in Figure 3.2. However, $+0.1$ and -0.1 V are reasonable estimates for β_a and β_c, respectively, for many purposes.

The theoretical derivation of the Tafel relationships (10), expressed graphically in Figure 3.2, is as follows. Taking reaction (6) as an example, the half-cell electrode

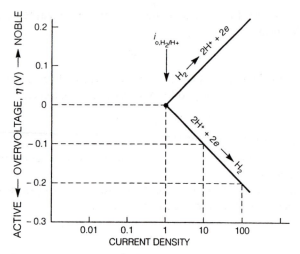

FIGURE 3.2 Activation overpotential showing Tafel behavior.

potential, e_{H^+/H_2}, is established when the reaction is at equilibrium. The rate of discharge of H^+ (forward) exactly balances the rate of ionization of H_2 (reverse). The presence of overpotential suggests the presence of energy barriers (activation energies), ΔG_f^* and ΔG_r^*, corresponding to the forward and reverse reactions, respectively, as shown schematically in Figure 3.3. The activation energy difference is related to half-cell electrode potential by the expression introduced in Section 2.1.1:

$$\Delta G_f^* - \Delta G_r^* = \Delta G_{H^+/H_2} = -nFe_{H^+/H_2}.$$

The Maxwell distribution law gives the energy distribution of reacting species and leads to expressions for forward, r_f, and reverse, r_r, reaction rates as a function of the respective activiation energies:

$$r_f = K_f \exp\left[-\frac{\Delta G_f^*}{RT}\right]$$

and

$$r_r = K_r \exp\left[-\frac{\Delta G_r^*}{RT}\right]$$

where K_f and K_r are the reaction rate constants for the forward and reverse reactions, respectively. At equilibrium,

$$r_f = r_r = \frac{i_o a}{nF}, \tag{7}$$

Thus

$$i_o = K_f' \exp\left(-\Delta G_f^*/RT\right) = K_r' \exp\left(-\Delta G_r^*/RT\right),$$

which clearly demonstrates that exchange current density is a function of the activation energies.

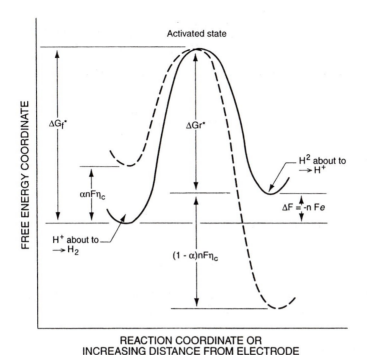

FIGURE 3.3   Activation energy model for activation overpotential. Equilibrium state (———); polarized state (- - - - -)

When a cathodic overpotential, η_c, is applied to the electrode, the discharge reaction rate is reduces and that of the ionization is increased. This is accomplished by decreasing the activation energy for the discharge reaction by an amount, $\alpha nF\eta_c$, and increasing that for the ionization reaction by an amount, $(1-\alpha)nF\eta_c$, as indicated by the dashed line in Figure 3.3. The factors α and $(1-\alpha)$ are the fractions of η_c taken by the discharge and ionization (forward and reverse) reactions, respectively. The cathodic discharge reaction rate in terms of current density becomes

$$i_c = K_f' \exp\left[-\frac{\Delta G_f{}^* - \alpha nF\eta_c}{RT}\right],$$

and the anodic ionization reaction rate becomes

$$i_a = K_r' \exp\left[-\frac{\Delta G_r{}^* + (1-\alpha)nF\eta_c}{RT}\right].$$

The net applied current is then

$$i_{app,c} = i_c - i_a = i_o \exp\left[\frac{\alpha nF\eta_c}{RT}\right] - i_o \exp\left[\frac{-(1-\alpha)nF\eta_c}{RT}\right], \tag{11}$$

For an applied anodic current favoring ionization to H^+,

$$i_{app,a} = i_a - i_c = i_o \exp\left[\frac{\alpha nF\eta_a}{RT}\right] - i_o \exp\left[\frac{-(1-\alpha)nF\eta_a}{RT}\right], \tag{12}$$

where α is now the fraction of η_a taken by the anodic ionization reaction. For high values of η_c, (11) simplifies to

$$i_{app,c} = i_c - i_c = i_o \exp\left[\frac{\alpha nF\eta_c}{RT}\right]$$

which is identical to

$$\eta_{act,c} = \beta_c \log\frac{i_c}{i_o} \tag{10b}$$

with $\beta_c = 2.3RT/\alpha nF$; and (12) becomes

$$\eta_{act,a} = \beta_a \log\frac{i_a}{i_o} \tag{10a}$$

for high values of η_a. For $\alpha = 0.5$, β_c or β_a is 0.12 V, near the assumed value of 0.1 V used for the approximate calculations in this book.

3.2.2 Concentration Polarization

At high rates, cathodic reduction reactions deplete the adjacent solution of the dissolved species being reduced. The concentration profile of H^+, for example, is shown schematically in Figure 3.4. C_B is the H^+ concentration of the uniform bulk solution, and δ is the thickness of the concentration gradient in solution. The half-cell electrode potential e_{H^+/H_2} of the depleted surface is given by the Nernst equation (Section 2.1.3) as a function of H^+ concentration or activity (H^+),

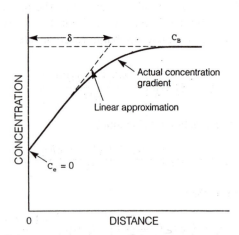

FIGURE 3.4   Concentration of H^+ in solution near a surface controlled by concentration polarization.

$$e_{H^+/H_2} = e^o_{H^+/H_2} + \frac{2.3RT}{nF} \log \frac{(H^+)^2}{P_{H_2}}.$$

It is apparent that the half-cell electrode potential, e, decreases as (H^+) is depleted at the surface. This decrease (potential change) is concentration polarization, η_{conc}, which is given as a function of current density by

$$\eta_{conc} = \frac{2.3RT}{nF} \log\left[1 - \frac{i_c}{i_L}\right]$$

A plot of this equation in Figure 3.5a shows that η_{conc} is low until a limiting current density, i_L, is approached. i_L is the measure of a maximum reaction rate that cannot be exceeded because of a limited diffusion rate of H^+ in solution.

The limiting current density, i_L, can be calculated from

$$i_L = \frac{D_z n F C_B}{\delta} \tag{13}$$

where n and F were defined in (1), and D_z is the diffusivity of the reacting species, z (H^+ in this example). Thus, i_L is increased by higher solution concentration, C_B; higher temperature, which increases D_z; and higher solution agitation, which decreases δ; as shown in Figure 3.5b. For corrosion, concentration polarization is significant primarily for cathodic reduction processes. Concentration polarization for anodic oxidation during corrosion can usually be ignored because an unlimited supply of metal atoms is available at the interface. Some concentration polarization of the anodic reaction is possible at very high corrosion rates or during intentional anodic dissolution by impressed currents (e.g., electrochemical machining or electrorefining), when rates are limited by transport of soluble oxidation products away from the surface.

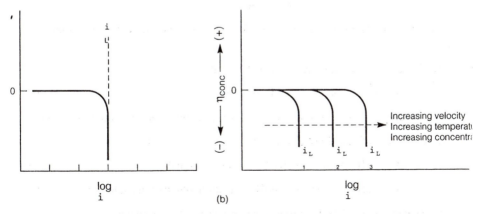

FIGURE 3.5 Cathodic concentration polarization: (a) plotted versus reaction rate or current density; (b) effect of solution conditions.

3.2.3 Combined Polarization

Total cathodic polarization, $\eta_{T,c}$, is the sum of activation and concentration polarization:

$$\eta_{T,c} = \eta_{act,c} + \eta_{conc},$$

which can be expanded to

$$\eta_{T,c} = \beta_c \log \frac{i_c}{i_o} + \frac{2.3RT}{nF} \log\left[1 - \frac{i_c}{i_L}\right] \tag{14}$$

Because concentration polarization is usually absent for anodic polarization of metal dissolution reactions, there remains

$$\eta_a = \beta_a \log \frac{i_a}{i_b}. \tag{15}$$

Equation (15) is essentially identical to (10a) but is renumbered here for easy reference together with (14).

3.3 Mixed Potential Theory

Four parameters, β_c, β_a, i_o, and i_L, can be used in equations (14) and (15) to describe virtually all electrochemical corrosion systems. Measurements of their values are desirable, but even estimates for each feasible half-cell reaction in various combinations allow predictions and explanations of many effects observed experimentally.

The principle of charge conservation is required to apply equations (14) and (15) to any number of half-cell reactions occurring simultaneously on a conducting surface. The total rate of oxidation must equal the total rate of reduction. That is, *the sum of anodic oxidation currents must equal the sum of cathodic reduction currents.* This must be true to avoid accumulating charge in the electrode.

The anodic reaction for metal corrosion is of the general form

$$M \rightarrow M^{n+} + ne^-.$$

Cathodic reactions are few in number (Section 1.4.1) and may be summarized as follows:

1. Evolution of H_2 from acid or neutral solutions,

$$2H^+ + 2e^- \rightarrow H_2 \quad \text{(acid solutions)} \tag{6}$$

$$2H_2O + 2e^- \rightarrow H_2 + 2OH^- \quad \text{(neutral and alkaline solutions)} \tag{17}$$

2. Reduction of dissolved oxygen in acid or neutral solutions,

$$O_2 + 4H^+ + 4e^- \rightarrow 2H_2O \quad \text{(acid solutions)} \tag{18}$$

$$O_2 + 2H_2O + 4e^- \rightarrow 4OH^- \quad \text{(neutral and alkaline solutions)} \tag{19}$$

3. Reduction of a dissolved oxidizer in a redox reaction such as,

$$Fe^{3+} + e^- \rightarrow Fe^{2+}. \tag{20}$$

The application of (14) and (15) to some of these reactions is illustrated by the examples that follow.

3.3.1 Corrosion Potential and Current Density

When a metal such as zinc is corroding in an acid solution, both the anodic,

$$Zn \rightarrow Zn^{2+} + 2e^-, \tag{2}$$

and cathodic,

$$2H^+ + 2e^- \rightarrow H_2, \tag{6}$$

half-cell reactions occur simultaneously on the surface. Each has its own half-cell electrode potential and exchange current density, as shown in Figure 3.6. However, the two half-cell electrode potentials $e_{Zn/Zn^{2+}}$ and e_{H^+/H_2} cannot coexist separately on an electrically conductive surface. Each must polarize or change potential to a common intermediate value, E_{corr}, which is called the *corrosion potential*. E_{corr} is referred to as a *mixed potential* since it is a combination or mixture of the half-cell electrode potentials for reactions (2) and (6). Hydrogen is sometimes referred to as an *oxidizer* because it serves to "oxidize" the metal to dissolved cations by its reduction in the cathodic reaction (6).

As reactions (2) and (6) polarize on the same surface, the half-cell electrode potentials change respectively, according to

$$\eta_a = \beta_a \log \frac{i_a}{i_o} \tag{15}$$

and

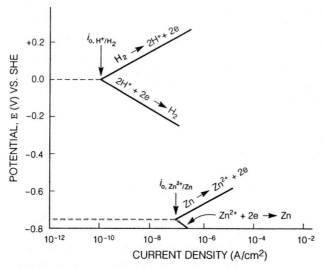

FIGURE 3.6   Anodic and cathodic half-cell reactions present simultaneously on a corroding zinc surface.

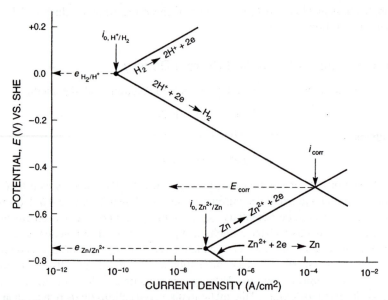

FIGURE 3.7   Polarization of anodic and cathodic half-cell reactions for zinc in acid solution to give a mixed potential, E_{corr}, and a corrosion rate (current density), i_{corr}. *(From M. G. Fontana, Corrosion Engineering, 3rd ed., McGraw-Hill, New York, p. 457, 1986. Reprinted by permission, McGraw-Hill Book Company.)*

$$\eta_c = \beta_c \log \frac{i_c}{i_o},$$

$$(10b)$$

until they become equal at E_{corr}, as shown in Figure 3.7. Equation (10b) is identical to equation (14), assuming for the moment that concentration polarization is absent. The relationships (15) and (10b) for the activation polarization of reactions (2) and (6) are again (Figure 3.2) linear on the semilog plot of Figure 3.7. Uniform average values of β_a and β_c are estimated at ± 0.1 V in this example and others to follow. At E_{corr} the rates of the anodic (2) and cathodic (6) reactions are equal. The rate of anodic dissolution, i_a, is identical to the corrosion rate, i_{corr}, in terms of current density, and

$$i_c = i_a = i_{corr}$$

at E_{corr}, as indicated in Figure 3.7.

3.3.2 Effect of Exchange Current Density

The exchange current density of each half-cell reaction often outweighs the thermodynamic driving force in determining the rate of the reaction. For example, the mixed potential description of iron corrosion in an acid (Figure 3.8) is similar in appearance to that of zinc (Figure 3.7). Because the half-cell electrode potential, $e_{Fe/Fe^{2+}}$, for the anodic reaction for iron, Fe $\rightarrow$ Fe^{2+} + 2e^{-}, is equal to about -0.44 V versus the standard hydrogen electrode, SHE, the corrosion rate would be expected

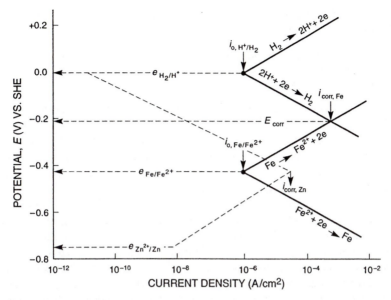

FIGURE 3.8 Comparison of electrochemical parameters for iron and zinc in acid solution, demonstrating the importance of i_o on determination of corrosion rates. Dashed lines represent lines from Figure 3.7 superimposed for comparison.

to be much lower than that of zinc, whose half-cell electrode potential is much more active, at −0.76 V (Table 2.1). The polarization diagram for zinc is shown by dashed lines in Figure 3.8, for comparison. The half-cell electrode potential for hydrogen reduction is the same for both at 0.00 versus SHE, and the driving force or difference between the half-cell reaction potentials is much larger for zinc than for iron. However, the corrosion rate of zinc, $i_{corr,Zn}$, is actually somewhat lower than that of iron, $i_{corr,Fe}$, because of the low exchange current density for hydrogen reduction on zinc compared to iron and the comparatively low exchange current density for zinc dissolution, as shown in Figure 3.8.

This example illustrates the importance of half-cell reaction kinetics on corrosion rates and demonstrates the error that can result from drawing uninformed conclusions from thermodynamic data (i.e., half-cell electrode potentials).

3.3.3 Effect of Added Oxidizer

The driving force for corrosion is increased by the addition of a stronger oxidizer, that is, a redox system with a half-cell electrode potential much more noble than that of any others present. Consider the addition of mixed ferric-ferrous salts to a hypothetical metal, M, corroding in an acid solution. This is a more realistic example of conditions in service where a pure system with only a single redox reaction would be a rarity. Industrial acids are frequently contaminated with ferric-ferrous salts and other cationic impurities added by corrosion or unremoved during processing.

Three experimental observations on the corrosion of iron result from adding an oxidizer to an acid solution:

- The corrosion potential, E_{corr}, is shifted to more noble (positive) values.
- The corrosion rate is increased.
- The hydrogen evolution rate is reduced.

Analysis by mixed potential theory is complicated by the presence of two oxidizers simultaneously: $2H^+ + 2e^- \rightarrow H_2$, and the added redox reaction, $Fe^{3+} + e^- \rightarrow Fe^{2+}$, which has a noble half-cell electrode potential of +0.77 V. The system is analyzed again in a similar manner to the examples of Sections 3.3.1 and 3.3.2. The i_o values for each reaction are located at the corresponding half-cell reaction potentials in Figure 3.9, and the Tafel constants (β's) are estimated at ± 0.1 V.

Using the charge conservation principle, the total oxidation and total reduction current densities are determined as a function of potential, and E_{corr} is determined where the two are equal. Beginning at the most noble half-cell electrode potential, $e_{Fe^{3+}/Fe^{2+}}$, and proceeding to more negative (active) potentials, the total reduction current density is equal to that for reduction of Fe^{3+} to Fe^{2+} until the half-cell electrode potential for hydrogen reduction, e_{H_2/H^+}, is reached, where the reduction current density for hydrogen must be added. The total reduction current density then follows the parallel dashed line marked "total reduction" until the half-cell electrode potential for metal oxidation is reached, when another increase is present due to reduction of M^+ to M.

Total oxidation is determined in a similar manner, starting with the most active half-cell electrode potential in the system, e_{M/M^+}. Total oxidation follows the line to more positive potentials for oxidation of M to M^+ until e_{H_2/H^+} is reached, where a current density for oxidation of H_2 to H^+ must be included. The total current density for

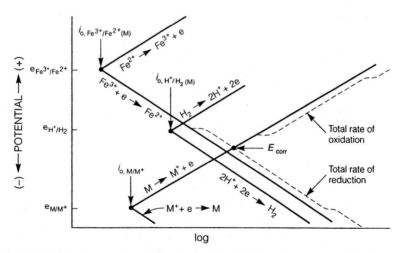

FIGURE 3.9 Determination of the mixed potential E_{corr} for a corroding metal M exposed to acid solution with a second oxidizer, Fe^{3+}/Fe^{2+}, present. *(From M. G. Fontana,* Corrosion Engineering, *3rd ed., McGraw-Hill, New York, p. 466, 1986. Reprinted by permission, McGraw-Hill Book Company.)*

oxidation follows the parallel line marked "total oxidation" until $e_{Fe^{3+}/Fe^{2+}}$ is reached and another addition is included for oxidation of Fe^{2+} to Fe^{3+}.

The corrosion potential, E_{corr}, is defined in Figure 3.9 by the intersection of the total oxidation and total reduction lines where the two are equal, fulfilling the charge conservation principle. Because $M \rightarrow M^+ + e^-$ is the only oxidation reaction present, the total oxidation current density is also the corrosion rate, i_{corr}. However, both Fe^{3+} and H^+ are being reduced in this system, and the sum of their rates, $i_{Fe^{3+} \rightarrow Fe^{2+}} + i_{H^+ \rightarrow H_2}$, is equal to total reduction, which is in turn equal to total oxidation or i_{corr}. The two reduction rates are defined in Figure 3.9 by the intersection of the horizontal equipotential line at E_{corr} with the polarization curves for the reduction rates of each half-cell reaction.

Examination of Figure 3.10 shows that the theory explains the experimental observations listed earlier in this section. Indeed, the corrosion rate of the metal is increased, the corrosion potential is made more noble, and the hydrogen evolution is decreased by the addition of the oxidizing agent to an acid solution. The reduction of hydrogen evolution has been interpreted sometimes as the result of reaction with the oxidizer. Hydrogen accumulation was thought to act as a barrier to corrosion. The removal of hydrogen would unblock, or *depolarize,* surface hydrogen and increase the corrosion rate. However, the mixed potential analysis of Figures 3.9 and 3.10 shows that a chemical reaction is unnecessary to explain the experimental observations, which are simply the result of the electrochemical behavior of the various half-cell reactions combined on the surface.

Addition of a strong oxidizer such as Fe^{3+} is not a guarantee that the corrosion rate is increased. The exchange current density for the oxidizer on the corroding surface must be high enough to justify an increase in corrosion rate. The assumption of a high exchange current density is implicit in Figure 3.9. Consider the alternative in Figure 3.11, where the i_o for the oxidizer is assumed to be very low. The

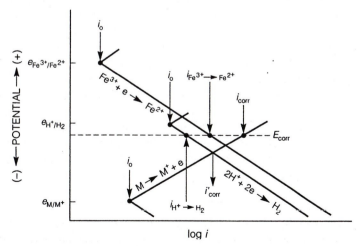

FIGURE 3.10 Rates of reduction ($i_{Fe^{3+} \rightarrow Fe^{2+}}$ and $i_{H^+ \rightarrow H_2}$) and oxidation (i_{corr}) from Figure 3.8 *(From M. G. Fontana,* Corrosion Engineering, *3rd ed., McGraw-Hill, New York, p. 467, 1986. Reprinted by permission, McGraw-Hill Book Company.)*

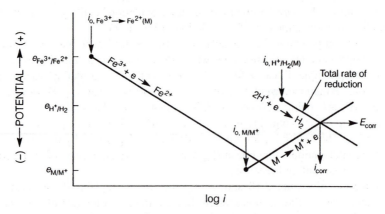

FIGURE 3.11   No effect on corrosion when oxidizer of low i_o is added to an acid solution. *(From M. G. Fontana, Corrosion Engineering, 3rd ed., McGraw-Hill, New York, p. 467, 1986. Reprinted by permission, McGraw-Hill Book Company.)*

rate (current density) for reduction is not high enough to have a significant effect on the corrosion of the metal. The corrosion potential is unchanged after the oxidizer addition because the current density for reduction of the oxidizer is not sufficiently high to have a measurable effect on the total reduction current and in turn on the corrosion rate.

3.3.4 Effect of Concentration Polarization

Concentration polarization is assumed absent in the preceding examples. This is equivalent to assuming that the limiting diffusion currents for all cathodic reduction processes are substantially higher than the graphically derived corrosion current densities. The assumption is invalid in many real situations, especially when the concentration of the oxidizer is relatively low, as shown by a decrease in C_B in equation (13). Figure 3.12 shows a plot of equation (14), which is the sum of activation and concentration polarization contributions in the total cathodic polarization. When the reduction current density approaches i_L, concentration polarization takes over from activation polarization, and the corrosion rate becomes limited by the diffusion of the oxidizer from the bulk solution. At low cathodic polarization the reduction process is activation controlled, but at high polarization it is diffusion or concentration controlled.

A common example of corrosion controlled by concentration polarization is iron or steel in dilute aerated salt solution (e.g., seawater). The cathodic process is reduction of dissolved oxygen, according to

$$O_2 + 2H_2O + 4e^- \rightarrow 4OH^-. \tag{19}$$

The maximum solubility of dissolved oxygen in water is relatively low, about 8 ppm at ambient temperature. In quiescent conditions, corrosion is controlled by diffusion of dissolved oxygen to the steel surface. However, if the solution is stirred, increasing i_L, a crossover is possible, as shown in Figure 3.13a, when i_L becomes greater than the anodic oxidation rate or current density, i_a. Corrosion rate, i_{corr}, increases

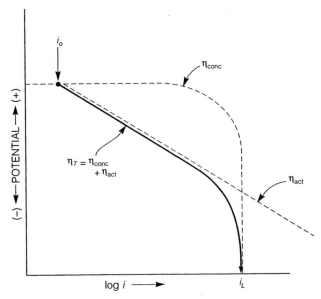

FIGURE 3.12 Combined polarization: sum of activation, η_{act}, and concentration, η_{conc}, polarization.

with stirring rate (Figure 3.13b) but levels off and becomes independent of stirring rate, when $i_L > i_a$, and the reduction reaction becomes activation controlled.

3.4 Experimental Polarization Curves

Mixed potential theory was derived originally[5] to explain experimental electro-chemical laboratory measurements. For purposes of instruction, the theory is usual-ly presented ahead of the experimental foundation which is given in this section. The student should not come to an incorrect conclusion that experimental results confirm a previously developed theory, when such is not the case.

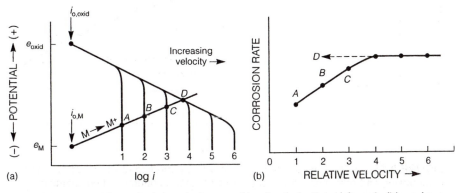

FIGURE 3.13 Effect of stirring during combined polarization: (a) on i_L, (b) on i_{corr}.

3.4.1 Cathodic Polarization

Assume a system in which anodic and cathodic half-cell reactions are given by reactions (2) and (6), respectively. Let us suppose that an excess of electron flow (a current density, $i_{app,c}$) is applied to this corroding zinc electrode in acid with E_{corr} and i_{corr} defined by mixed potential theory (Figure 3.14a). The excess of electrons

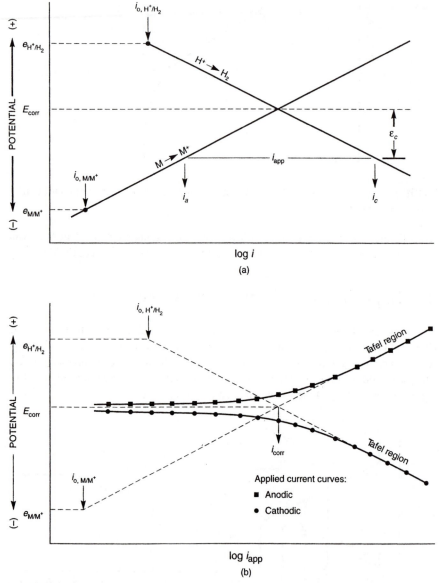

FIGURE 3.14   (a) Current density, i_{app}, applied to corroding electrode of E_{corr} and i_{corr}, causing cathodic overvoltage of ε_c; (b) simulated experimental polarization curves derived from (a).

causes the electrode potential to shift negatively from E_{corr} to E. The negative potential shift, $\varepsilon_c = E - E_{corr}$, is defined as cathodic overpotential. The term ε is used to distinguish between overpotential from a mixed potential, E_{corr}, and similar overpotential, η, from a half-cell electrode potential, e. The excess of electrons suppresses the rate of the anodic reaction (2) from i_{corr} to i_a and similarly increases the cathodic reduction reaction (6) from i_{corr} to i_c. To fulfill the principle of charge conservation (i.e., avoid buildup of charge in the electrode), the difference between the increase in the cathodic reduction rate and the decrease in the anodic oxidation rate, prompted by cathodic overpotential, ε_c, must be equal to the applied current:

$$i_{app,c} = i_c - i_a.$$

Figure 3.14b shows $i_{app,c}$ represented by simulated data points (●) for various values of ε_c. At low ε_c, i_c is only slightly higher than i_a, and $i_{app,c}$ is very low. As ε_c increases, i_c increases while i_a decreases, both quite rapidly, until i_a becomes insignificant compared to i_c, and the simulated experimental cathodic polarization curve coincides with the dashed line for the cathodic half-cell overpotential of reaction (6). Thus, the simulated cathodic polarization curve of potential versus log i_{app} is curved at low overpotential but becomes linear at higher overpotential. The linearity on a semilog plot is termed *Tafel behavior,* for the German investigator who first reported it.[6] The simulated anodic polarization curve defined by (■) in Figure 3.14b is discussed in Section 3.4.2.

Experimental Tafel behavior defines the cathodic branch of the half-cell reduction reaction. Figure 3.14b shows that extrapolation of the region of Tafel behavior gives the corrosion rate, i_{corr}, at E_{corr}. Thus, corrosion rates can be measured from polarization data, as discussed further in Section 5.1. If the Tafel line is further extrapolated to e_{H^+/H_2}, the exchange current density for the reduction process, $i_{0,H^+/H_2}$, can also be obtained.

An experimental cathodic polarization curve for steel in acid solution is shown in Figure 3.15.[7] The linearity, or Tafel behavior, is limited to only about one decade of current density before interferences become apparent at higher current density. These interferences may be caused by near surface depletion of the oxidizer, H^+ (concentration polarization), or ohmic gradients in the solution (Section 3.5.4). Higher concentration solutions reduce concentration polarization, increase conductivity, and thereby extend Tafel behavior.

The effects of solution pH in 4% NaCl on cathodic polarization of pure iron are shown in Figure 3.16.[8] At the lowest pH, 1.42, there is about 1.5 decades of Tafel behavior for the hydrogen reduction reaction (6) before concentration polarization begins with an eventual limiting current density of about 7000 $\mu A/cm^2$. As pH increases (e.g., to 2.91), hydrogen ion activity, (H^+), becomes much lower, concentration polarization of reaction (6) begins much sooner, and the limiting current density decreases to about 100 $\mu A/cm^2$. In the higher pH solutions, under high cathodic polarization, hydrogen evolution by direct reduction of water

$$2H_2O + 2e^- \rightarrow H_2 + 2OH^- \tag{17}$$

occurs above the limiting current density for reaction (6). At the highest pH, 5.26, reduction of H^+ is still more limited, and direct reduction of water by reaction (17) predominates, except for the lowest cathodic currents.

3.4.2 Anodic Polarization

Polarization in the anodic (positive) direction is analogous to cathodic polarization. Electrons are drawn out of the metal, and the current flows in the opposite direction from cathodic currents above. The deficiency of electrons makes the potential change positive with respect to E_{corr}. The anodic oxidation rate, i_a, is increased, while the cathodic reduction rate, i_c, is decreased, so that the applied anodic current density is

$$i_{app,a} = i_a - i_c. \tag{21}$$

The simulated anodic polarization curve, derived from Figure 3.14a and shown in Figure 3.14b, is also linear on the semilog plot, in a manner analogous to the cathodic data.

The experimental anodic polarization curves often do not correspond to the idealization of Figure 3.14. The anodic data of Figure 3.15 are generally curved on the semilog plot, and no sensible linearity or value of β_a can be found. The reasons for

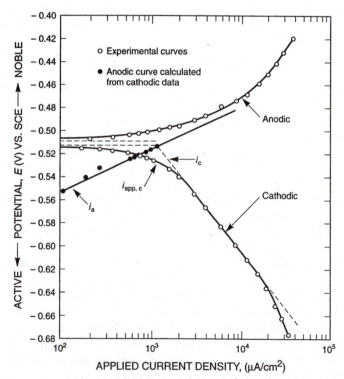

FIGURE 3.15 Experimental polarization curves for 1080 steel in deaerated 1 N H_2SO_4. β_c = −98 mV, β_a=38 mV (derived from cathodic data), i_{corr} = 1180 $\mu A/cm^2$. *(From R. Bandy and D. A. Jones, Corrosion, Vol. 32, p. 126, 1976. Reprinted by permission National Association of Corrosion Engineers.)*

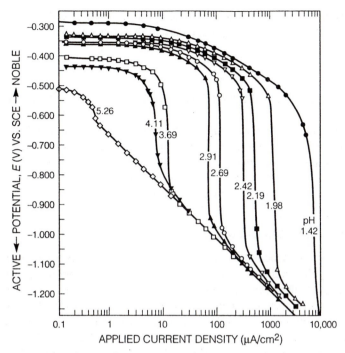

FIGURE 3.16 Cathodic polarization of pure iron in deaerated 4% NaCl. *(From M. Stern, J. Electrochem. Soc., Vol. 102, p. 609, 1955. Reprinted by permission, The Electrochemical Society.)*

the nonlinearity are not well understood, but some speculations seem reasonable. Anodic dissolution of metal is irreversible in dilute corroding solutions. Rapid anodic dissolution can cause unacceptable solution contamination before the anodic polarization curve is complete. The surface may be roughened or otherwise changed as liberated corrosion products accumulate and precipitate or form oxide/hydroxide films on the surface. In weakly or moderately corrosive solutions, the anodic overpotential is often higher than would be expected from cathodic data, probably due to formation of inhibiting surface films.

Anodic polarization data derived from cathodic data[9] have sometimes provided a reasonable alternative. The anodic current density, i_a, can be calculated from equation (21) in the potential region near E_{corr}, where $i_{app} \neq i_c$. The extrapolated Tafel line gives i_c, and the data points give i_{app}. Substituting these values into equation (21) gives corresponding values of i_a at a number of potentials. The solid data points in Figure 3.15 were derived in this way and give reasonable anodic Tafel behavior.

Anodic and cathodic polarization curves are symmetrical about E_{corr} when $|\beta_a| = |\beta_c|$, as assumed in Figure 3.14. However, for a corroding metal $|\beta_a|$ and $|\beta_c|$ are seldom equal for the separate and distinct anodic and cathodic half-cell reactions which constitute the mixed potential E_{corr}. The inequality is demonstrated typically in Figure 3.15, where $\beta_a = 38$ mV and $\beta_c = -98$ mV. Table 3.2[10] shows that in general $|\beta_a|$ for anodic dissolution reactions are usually half or less than $|\beta_c|$

for the cathodic reduction reactions. Table 3.2 lists useful cathodic and anodic polarization data as well as exchange current densities on a number of corroding metals.

3.5 Instrumentation and Experimental Procedures

Electrochemical polarization methods may be classified as controlled current (galvanostatic) and controlled potential (potentiostatic). Galvanostatic procedures are relatively simple instrumentally, were the first used historically, and are described initially in this section. Potentiostatic techniques are described in the next chapter on passivity, where they have particular application.

3.5.1 Galvanostatic Circuits

Figure 3.17 shows schematic circuitry for galvanostatic or constant-current electrochemical polarization measurements on a specimen or working electrode, WE. A filtered direct current (dc) power supply, PS, supplies current (I) to the working electrode, through a second counter or auxiliary electrode, AUX. The potential of the working electrode is measured with respect to a reference electrode, REF, with a series-connected potentiometer, P, and electrometer null detector, N, as described in Section 2.3.4. The simplest experimental arrangement would place the reference electrode directly in the experimental cell, which is the vessel containing the solu-

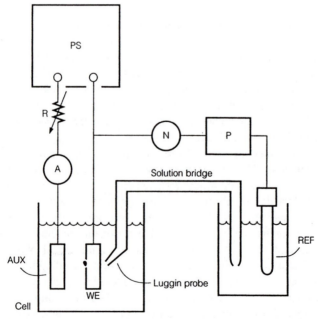

FIGURE 3.17   Schematic circuitry for galvanostatic polarization measurements.

tion electrolyte and electrodes. However, the salt bridge and Luggin probe are usually included to minimize ohmic resistance interferences in the electrolyte, as discussed further in Section 3.5.4.

The large variable resistor, R, is much greater than the sum of resistances, r_c, in the cell. Any small changes of r_c during polarization have no effect on the circuit resistance, which is dominated by R. Thus, a constant voltage applied from PS to the constant circuit resistance results in constant current, I, measured by the ammeter, A, in the circuit.

The working electrode is polarized as a cathode when connected to the negative terminal of the power supply, as shown in Figure 3.17. The auxiliary electrode, connected to the positive terminal, is simultaneously polarized as an anode. The working electrode becomes the anode and the auxiliary electrode the cathode when the terminals are reversed. We are usually interested only in the behavior of the working electrode, and the auxiliary electrode serves only to supply polarizing current.

TABLE 3.2 Electrode Kinetic Parameters, 20–25°C

Electrode	Solution	i_o (A/cm^2)	β_c (V)	β_a (V)	Reference
Reaction:	**$2H^+ + 2e^- \rightarrow H_2$**				
Pt	1 N HCl	10^{-2}	0.03		11
	0.1 N NaOH	0.7×10^{-3}	0.11		11
Pd	0.6 N HCl	2×10^{-3}	0.03		11
Mo	1 N HCl	10^{-5}	0.04		11
Au	1 N HCl	10^{-5}	0.05		11
Ta	1 N HCl	10^{-4}	0.08		11
W	5 N HCl	10^{-4}	0.11		11
Ag	0.1 N HCl	5×10^{-6}	0.09		11
Ni	0.1 N HCl	8×10^{-6}	0.31		11
	0.12 N NaOH	4×10^{-6}	0.10		11
Bi	1 N HCl	10^{-6}	0.10		11
Nb	1 N HCl	10^{-6}	0.10		11
Fe	1 N HCl	10^{-5}	0.15		11
	0.52 N H$_2$SO$_4$	2×10^{-5}	0.11		12
	4% NaCl, pH 1–4	10^{-6}	0.10		8
Cu	0.1 N HCl	2×10^{-6}	0.12		11
	0.15 N NaOH	10^{-5}	0.12		11
Sb	2 N H$_2$SO$_4$	10^{-8}	0.10		11
Al	2 N H$_2$SO$_4$	10^{-9}	0.10		11
Be	1 N HCl	10^{-8}	0.12		11
Sn	1 N HCl	10^{-7}	0.15		11
Cd	1 N HCl	10^{-6}	0.20		11
Zn	1 N H$_2$SO$_4$	2×10^{-10}	0.12		11
Hg	0.1 N HCl	7×10^{-12}	0.12		11
	0.1 N H$_2$SO$_4$	2×10^{-12}	0.12		11
	0.1 N NaOH	3×10^{-14}	0.10		11
Pb	0.01–8 N HCl	2×10^{-12}	0.12		11

TABLE 3.2   (Cont.)

Electrode	Solution	i_o (A/cm²)	β_c (V)	β_a (V)	Reference
Reaction:	**$O_2 + 4H^+ + 4e^- \rightarrow 2H_2O$**				
Pt	0.1 N H₂SO₄	9×10^{-11}	0.10		11
Reaction:	**$O_2 + 2H_2O + 4e^- \rightarrow 4OH^-$**				
Pt	0.1 N NaOH	4×10^{-12}	0.05		11
Au	0.1 N NaOH	5×10^{-12}		0.05	11
Reaction:	**$Cl_2 + 2e^- \rightarrow 2Cl^-$**				
Pt	1 N HCl	5×10^{-3}	0.11	0.13	11
Reaction:	**$M \rightarrow M^{n+} + ne^-$**				
Fe	0.52 N H₂SO₄	10^{-7}		0.060	12
	0.52 N H₂SO₄	10^{-10}		0.039	12
	0.63 N FeSO₄	3×10^{-9}		0.060–0.075	13
	4% NaCl, pH 1.5			0.068	9
	0.3 N H₂SO₄			0.10	14
	0.5 M FeSO₄, 0.1 M NaHSO₄	3×10^{-10}		0.030	15
	Perchlorate			0.030	16
Cu	0.001 N Cu(NO₃)₂	10^{-9}			11

Source: Adapted from H. H. Uhlig and R. W. Revie, Corrosion and Corrosion Control, 3rd ed., Wiley, p. 44, 1985. Reprinted by permission John Wiley and Sons, New York.

However, both electrodes become important in measurements on galvanic corrosion described in Section 6.3.2.

3.5.2 Charge Transport in Polarization Cells

Current is carried in an electrolyte solution by charged ions that move by three possible processes: (a) diffusion driven by concentration gradients, (b) electrostatic migration driven by potential gradients, (c) convection produced by physical mixing. In the absence of concentration differences and mixing, negative ions (e.g., Cl^-, SO_4^{2-}, OH^-) are attracted by (b) to the positive electrode where anodic reactions such as (16) liberate electrons into the electrode. The positive electrode by (b)where oxidation occurs is called the *anode*, and the negative ions reacting there are called *anions*.

The electrons generated at the anode pass through the electrical circuitry by normal electronic conduction. The power supply in Figure 3.17 can be thought of as an electron "pump" that pulls electrons from the anode and pushes them to the negative electrode. Positive ions (e.g., Fe^{2+}, Na^+, Ni^{2+}, H^+) also are attracted in the electrolyte to the negative electrode, where they react by cathodic reduction reactions which consume the excess electrons. The negative electrode, where cathodic reductions are favored, is thus called the *cathode*. The positive ions attracted to the cathode are called *cations*.

Only those ions that have favorable half-cell electrode potentials can react electrochemically after being attracted to their respective electrodes. For example, Cl^- attracted to the anode usually will not be oxidized to chlorine gas because the half-cell reaction occurs at very noble potentials below which only Cl^- is stable. Similarly, Na metal is unstable below a very active half-cell electrode potential in water. Thus, cathodic reduction of Na^+ is thermodynamically impossible in the usual electrolyte conditions. However, Cl^- can be oxidized to Cl_2 gas at impressed current anodes in seawater cathodic protection systems (Section 13.2.7) where anodic polarization is extremely high.

For the working electrode as an anode, metal dissolution reactions of the type

$$M \rightarrow M^{n+} + ne^- \tag{16}$$

are of interest in corrosion. When the auxiliary electrode is polarized as an anode (working electrode as cathode), M must be selected for the auxiliary electrode with a very noble $e_{M/M^{n+}}$ to prevent anodic dissolution, which would contaminate the electrolyte. Either platinum or carbon is the usual choice, as discussed in the next section. In the absence of anodic dissolution at the auxiliary electrode by reaction (16), other anodic oxidation reactions are possible to liberate electrons. These include oxidation in a redox reaction such as

$$Fe^{2+} \rightarrow Fe^{3+} + e^-, \tag{22}$$

and oxygen evolution by

$$4OH^- \rightarrow 2H_2O + O_2 + 4e^-. \tag{23}$$

Reactions (22) and (23) both must operate at potentials below $e_{M/M^{n+}}$ (e.g., $e_{Pt/Pt^{3+}}$) so that the noble-metal auxiliary electrode is not dissolved.

3.5.3 Design of Polarization Cells

Two cell designs are shown in Figure 3.18. The first, shown in (a), is the standardized design approved by ASTM,[17] which is available in slight variations from a number of commercial sources. The second, (b), made up from off-the-shelf components,[18] requires a minimum of glassblowing and is typical of numerous custom designs that still meet ASTM specifications. Both cells share most of the features described below.

The working electrode is centrally located in the cell with a pair of auxiliary electrodes on either side for better current distribution. If the specimen for the working electrode has minimum dimensions greater than about 6 mm ({uc,1/4} in.), it can be mounted as shown in Figure 3.19. A threaded stainless steel rod is screwed into a drilled and tapped hole in the specimen electrode, and a stainless steel nut and washer combination on the other end of the rod compresses a tapered Teflon gasket between the specimen and the lower end of heavy-wall Pyrex tubing.[19] The gasket thereby forms watertight seals with the specimen electrode and the Pyrex tubing, which shields the inner rod from exposure to the corrosive electrolyte in the cell. The assembly exposes only glass, Teflon, and a controlled working electrode surface area with a tight seal resistant to crevice attack.

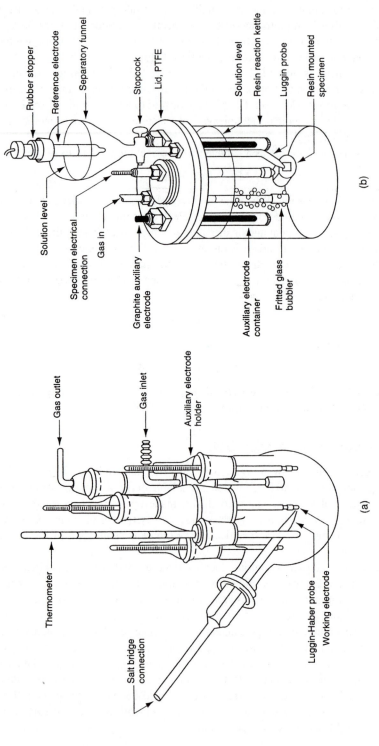

FIGURE 3.18 Electrochemical polarization cell designs: (a) commercially available model, (b) custom design using off-the-shelf components and minimal glass blowing. *(From D. A. Jones and A. J. P. Paul, Hydrometallurgical Reactor Design and Kinetics, R. G. Bautista, R. J. Wesley, and G. W. Warren, eds., THS-AIME, Warrendale, PA, p. 293, 1987. Reprinted by permission The Metallurgical Society.)*

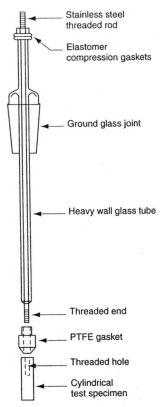

Stainless steel
threaded rod

Elastomer
compression gaskets

Ground glass joint

Heavy wall glass tube

Threaded end

PTFE gasket

Threaded hole

Cylindrical
test specimen

FIGURE 3.19 Stern-Makrides assembly for specimen electrode mounting. *(From M. Stern and A. C. Makrides, J.* Electrochem. Soc., *Vol. 107, p. 782, 1960. Reprinted by permission The Electrochemical Society.)*

The reference electrode, REF, is placed outside the cell, and the potential of the working electrode is measured through the Luggin probe and solution bridge with respect to the reference electrode. In both cell designs of Figure 3.18, the Luggin probe is maneuverable and the working electrode is eccentric in the entrance adapter to allow probe tip placement near the working electrode surface. This minimizes ohmic resistance interferences, which are discussed in Section 3.5.4.

Electrode materials that are too thin or brittle have been embedded in epoxy resin for drilling, tapping, and mounting, as in Figure 3.20.[18] Special assemblies have also been designed to accept thin sheet samples.[20,21] Rod and plate samples have been masked frequently with heat-shrink tubing, tape, and paint, to leave a measured surface area for exposure. Any of these methods may be satisfactory, but care must be used to ensure that masking materials are inert and that crevice corrosion is absent at the masked edges in aggressive environments.

Auxiliary electrodes must be made of materials that are inert to the electrolyte, even under strong anodic polarization. Platinum is often the choice and may be mounted as above if thick rod or plate stock is available. However, the cost has become detrimental in recent years of rising precious metal prices. A less expensive, lower mass/area alternative is platinum foil spot-welded to platinum wire which is sealed into Pyrex, as shown in Figure 3.21. Inexpensive graphite rod (available as carbon-arc lamp electrodes) is electrochemically inert, but organic binders often leach out to contaminate the electrolyte. Contamination from this source and other electrochemical reaction products at the auxiliary electrode can be minimized by isolation in a separate compartment, as shown in Figure 3.18b. The graphite electrodes are placed in glass tubes with fritted glass ends that allow current passage but restrict solution transport between the auxiliary electrode compartment and the main cell electrolyte.

3.5.4 Ohmic Electrolyte Resistance

An ohmic resistance gradient occurs through the electrolyte between working electrode and auxiliary electrode when current, I_{app}, is passed in any polarization cell. The magnitude depends upon the concentration and mobility of dissolved ionic species. Any polarization potential measured between working electrode and reference electrode includes some part, $\varepsilon_\Omega = I_{app}R_\Omega$, of the gradient, where R_Ω is the effective ohmic resistance between the two. ε_Ω, the ohmic resistance polarization, is

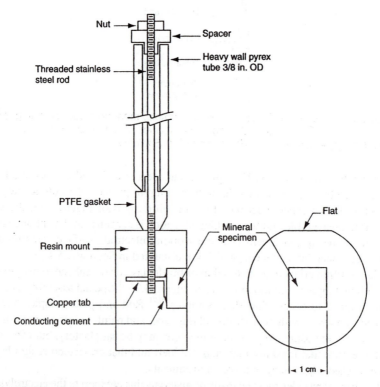

FIGURE 3.20 Epoxy mount for brittle or thin sheet specimens. *(From D. A. Jones and A. J. P. Paul, Hydrometallurgical Reactor Design and Kinetics, R. G. Bautista, R. J. Wesley, and G. W. Warren, eds., TMS-AIME, Warrendale, PA, p. 293, 1987. Reprinted by permission The Metallurgical Society.)*

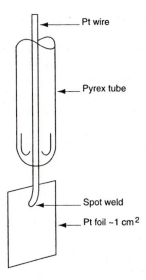

FIGURE 3.21   Platinum foil auxiliary electrode sealed into Pyrex tubing.

undesirable because it masks the other components of overpotential, which are of interest to determine corrosion rate and mechanism. Several methods to compensate or reduce ε_Ω are described in the remainder of this section.

The schematic cell in Figure 3.17 and the two cell designs in Figure 3.18 include the Luggin probe and salt bridge to the reference electrode containing an electrolytic path that does not carry any of the polarizing current, I_{app}. The length of the current-carrying electrolyte path between working electrode and reference electrode is reduced to the distance between the working electrode surface and the probe tip. When the probe tip is placed near the working electrode surface, both R_Ω and ε_Ω are reduced substantially as a result. The probe tip cannot be placed too close without shielding the surface from the applied polarizing current. A distance of about 1 mm is often recommended for usual conductive electrolyte solutions.

A "supporting" electrolyte (i.e., a soluble ionic salt), is sometimes added to increase the solution conductivity and minimize R_Ω. However, salt additions often have an indeterminate effect on the electrode reactions being studied and may be inadvisable in solutions designed to simulate service conditions. Polarization studies in actual service conditions, such as soil or concrete, also preclude addition of a supporting electrolyte.

ε_Ω disappears almost instantaneously when current is interrupted, whereas activation and concentration polarization require substantial time to decay, as shown schematically in Figure 3.22. The measurement of ε_Ω from the time trace of potential is simple in principle but more complex experimentally. Oscilloscope measurements are often required to monitor the decay of potential. Rather complex circuitry is necessary to ensure that switching speed is greater than decay time of the ε_Ω. High-impedance reference electrodes of the saturated calomel (SCE) type (Section 2.3.1) lead to slow response times or damped oscillations during current interruption measurements. Moran[22] reviews literature utilizing current interruption techniques and describes a low-impedance platinum wire electrode to replace SCE temporarily during measurement of ε_Ω. The exposed wire tip was placed adjacent to the solution junction of the SCE so that ε_Ω would be nearly the same whether measured versus Pt or SCE. ε_Ω was measured readily between the low-impedance Pt wire and the working electrode and served as a good measure of ε_Ω between SCE and the working electrode.

High-frequency signals are used commonly in bridge circuits to measure conductivity of solutions. The same instrumentation can in principle be utilized to measure the resistance, R_Ω, between working and reference electrodes. At high frequency, activation and concentration polarization do not have time to develop, but ε_Ω is readily measured. It may be necessary to use a low-impedance wire electrode as a temporary substitute, as described above,[22] to obtain valid high-frequency measurements. High-frequency measurements are a natural part of impedance spectroscopy methods discussed in Section 3.5.6.

A modified bridge circuit[23] shown in Figure 3.23 has been used to compensate ε_Ω during galvanostatic polarization measurements. When X is carefully adjusted to match R_Ω, the potential across the bridge measured by the potentiometer-electrometer null detector P-N is equal to $\frac{1}{2}E_{WE}$, where E_{WE} is the polarized potential of the working electrode. S and T are matched 100 MΩ resistors which are high enough to prevent polarization of the reference electrode. To stabilize the potential measurements, S and T must be shielded within a metal box, connected to the grounded case of the electrometer through shielded cable. X is adjusted at an intermediate polarizing current by momentary current interruption. When $X > R_\Omega$, the measured cathodic overpotential is too low, and potential momentarily jumps to a higher cathodic overpotential which is free of any ohmic $I_{app}R_\Omega$ contribution. Activation and concentration polarization thereafter decay more slowly back to zero. By decreasing X, the jump at the time of current interruption is eliminated when $X = R_\Omega$. A polarization curve run with the adjusted value of X is essentially free of ε_Ω. Figure 3.24 shows compensated and uncompensated cathodic polarization curves on aluminum.[23] The compensated curve shows expected Tafel behavior free of IR$_\Omega$ interferences. The bridge circuit has been used successfully in high temperature autoclaves,[24] simulated automotive cooling systems,[25] and soils in the field[26] and laboratory.[27]

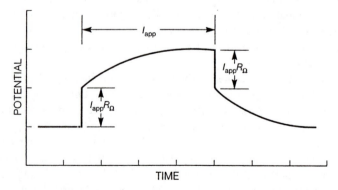

FIGURE 3.22 Current interruption method to determine ohmic resistance polarization, $\eta_\Omega = I_{app}R_\Omega$.

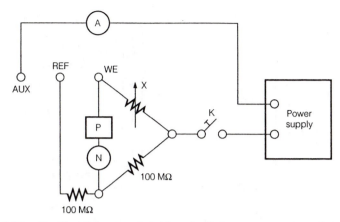

FIGURE 3.23 Modified Wheatstone bridge circuit for compensation of ohmic resistance polarization during galvanostatic polarization. AUX, auxiliary electrode; REF, reference electrode; WE, working electrode; P, potentiometer; N, null detector; X, variable resistor; K, key switch; A, ammeter. *(From D. A. Jones, Corros. Sci., Vol. 8, p. 19, 1968. Reprinted by permission, Pergamon Press.)*

Further instrumental methods of ε_Ω compensation utilizing potentiostat feedback circuits and electrochemical impedance spectroscopy are discussed in Sections 4.2.3 and 5.6.3, respectively.

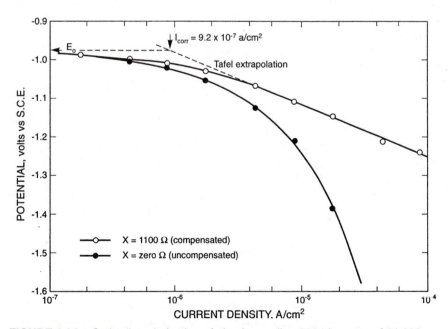

FIGURE 3.24 Cathodic polarization of aluminum alloy 3004 in water of 21,000 -cm resistivity, showing uncompensated and compensated potentials using the bridge circuit of Figure 3.23. *(From D. A. Jones, Corros. Sci., Vol. 8, p. 19, 1968, Reprinted by permission, Pergamon Press.)*

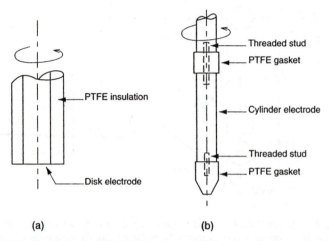

(a)   (b)

FIGURE 3.25 Electrode configurations for study of concentration polarization: (a) rotating disk; (b) rotating cylinder.

3.5.5 Rotating Electrodes: Concentration Polarization

Mass transport is improved and concentration polarization is reduced at a corroding electrode by increased mixing (convection). Reproducible control of convection is achieved under laminar flow, which is most easily produced experimentally by electrode rotation. Two configurations have been used in electrochemical studies: rotating disk (RDE) and rotating cylinder (RCE) electrodes (Figure 3.25). Both yield substantial decreases in concentration polarization and increases in i_L.

The rotating disc electrode has been well characterized mathematically with confirmation by experiment.[28] The limiting current at the RDE is

$$i_L = 0.620\, nFD_z^{2/3}\omega^{1/2}v^{-1/6}C_B,$$

where n and F were defined in equation (1), ω is the angular velocity in radians/time, v is kinematic viscosity, and C_B and D_z are the bulk solution concentration and diffusivity, respectively, of the reacting species. Mass transport (current density) is uniform over the surface of the RDE in the usual conditions where the solution is highly conductive and the reduction rate is at i_L.[28]

The rotating cylinder electrode is less amenable to mathematical analysis but has been used successfully in corrosion studies of controlled concentration polarization.[29] The limiting current is given[30] by

$$i_L = 0.0791\, nFC_B U(Ud/v)^{-0.30}(v/D_z)^{-0.644},$$

where d is the diameter and U the peripheral (linear) velocity of the cylinder electrode. The rotating cylinder electrode also gives uniform mass transport over the length of its surface but initiates turbulent flow at a much lower rotation rate than the rotating disk electrode. Thus, it has been used to study mass transport under turbulent flow conditions.[31,32] Turbulence is to be expected in plant pipe flow, and the rotating cylinder electrode has been used to simulate erosion corrosion (Section 10.2) in turbulent flow.[33]

3.5.6 Electrochemical Impedance Spectroscopy

The response of corroding electrodes to small-amplitude alternating potential signals of widely varying frequency has been analyzed by electrochemical impedance spectroscopy (EIS). EIS can determine in principle a number of fundamental parameters relating to electrochemical kinetics and has been the subject of vigorous research.

The time-dependent current response $I(t)$ of an electrode surface to a sinusoidal alternating potential signal $V(t)$ has been expressed[34] as an angular frequency (ω) dependent impedance $Z(\omega)$, where

$$Z(\omega) = V(t)/I(t),$$
$$t \equiv \text{time},$$
$$V(t) = V_o \sin \omega t,$$
$$I(t) = I_o \sin (\omega t + \theta), \text{ and}$$
$$\theta \equiv \text{phase angle between } V(t) \text{ and } I(t).$$

Various processes at the surface absorb electrical energy at discrete frequencies, causing a time lag and a measurable phase angle, θ, between the time-dependent excitation and response signals. These processes have been simulated by resistive-capacitive electrical networks. For example, the current response to a sinusoidal potential signal across a capacitor, shown in Figure 3.26, illustrates current lag behind potential.

The impedance, $Z(\omega)$, may be expressed in terms of real, $Z'(\omega)$, and imaginary, $Z''(\omega)$, components.

$$Z(\omega) = Z'(\omega) + Z''(\omega).$$

The impedance behavior of an electrode may be expressed in Nyquist plots of $Z''(\omega)$ as a function of $Z'(\omega)$ or in Bode plots of log $|Z|$ and log θ versus frequency f in cycles per sec (hertz), where $\omega = 2\pi f$. These plots for a simple parallel-connected resistance-capacitance circuit are shown schematically in Figure 3.27.[35] This circuit is often an adequate representation of a simple corroding surface under activation control. The Nyquist plot shows a semicircle, with increasing frequency in a counterclockwise direction. At very high frequency, the imaginary component, Z'', disappears, leaving only the solution resistance, R_Ω. At very low frequency, Z'' again

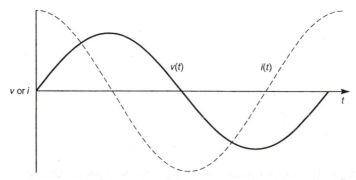

FIGURE 3.26   Current response, I, to a sinusoidal potential signal, V, for a capacitor.

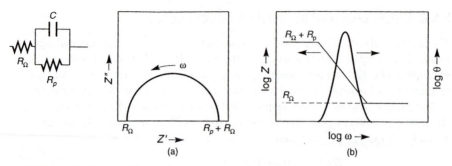

FIGURE 3.27   Data display for electrochemical impedance spectroscopy for a corroding electrode simulated by parallel-connected resistance R_p and capacitance C: (a) Nyquist plot; (b) Bode plot; both schematic.

disappears, leaving a sum of R_Ω and the faradaic reaction resistance or polarization resistance, R_p. The Bode plot gives analogous results. At intermediate frequencies, the capacitance plots linear with a slope of -1 and maximum phase angle, θ.

The faradaic reaction resistance, or polarization resistance, R_p, in Figure 3.27, is inversely proportional to the corrosion rate, as further discussed in Section 5.6.3. It is evident from Figure 3.27 that the R_Ω measured at high frequency can be subtracted from the sum of $R_p + R_\Omega$ at low frequency to give a compensated value of R_p free of ohmic interferences.

In the frequent case of control by diffusion in the electrolyte (concentration polarization) or in a surface film or coating, an additional resistive element called the Warburg impedance, W, must be included in the circuit. W is evidenced at low frequencies on the Nyquist plot by a straight line superimposed at 45° to both axes, as shown schematically in Figure 3.28.

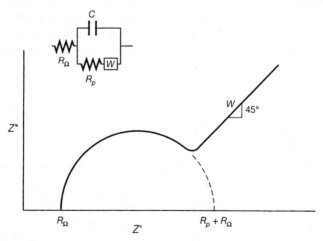

FIGURE 3.28   Schematic Nyquist plot showing effects of partial diffusion control with Warburg impedance W.

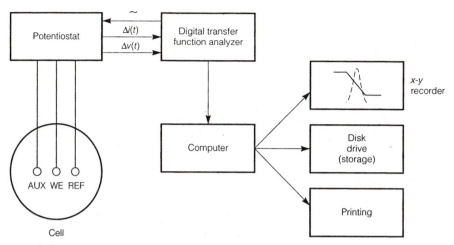

FIGURE 3.29 Schematic instrumentation for conducting electrochemical impedance spectroscopy.

Instrumentation (Figure 3.29) to conduct and interpret electrochemical impedance spectroscopy utilizes a function generator to apply a wide range of sinusoidal potential frequencies to a precision fast-acting potentiostat, which in turn applies the signal to the corroding electrode. The electrode response is fed back to a digital function analyzer that digitally displays the impedance response and phase angle at each frequency. Data storage, manipulation, and display are conducted separately through a dedicated microcomputer.

Recent research relating to electrochemical impedance spectroscopy has focused on applying the technique to real corroding systems. Figure 3.30[36] shows impedance spectra for carbon steel exposed to an aerated aggressive water for a minimum of 48 hours. Nyquist and Bode formats correspond reasonably well to the schematic data presentations in Figures 3.26 and 3.27 for a simple circuit with a Warburg impedance, which is evidenced by the beginnings of a 45° line in the low-frequency range on the Nyquist plot. The slightly nonhorizontal line at low frequency in the Bode plot is also evidence of diffusion control. The data points were plotted automatically in Figure 3.30, and the solid lines calculated by computerized regression analysis, using a system much like the one shown in Figure 3.28.

Impedance spectra often become more complex than the example cited in Figure 3.30. The semicircular plot may be distorted, and the center is often depressed below the horizontal axis. Some examples of such data are described briefly in Section 5.6.3 on polarization resistance measurements and Section 14.1.5 on coating evaluations from impedance spectra.

Electrochemical impedance spectroscopy retains all the advantages of traditional direct-current (dc) methods. It is sensitive, can be conducted *in situ*, and often does not require artificial accelerating factors for testing, such as increased temperature and concentration. However, the instrumentation is expensive and difficult to operate and maintain. Furthermore, data interpretation may be ambiguous and

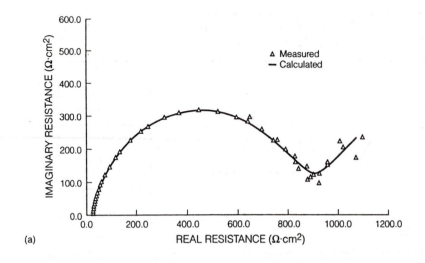

(a)

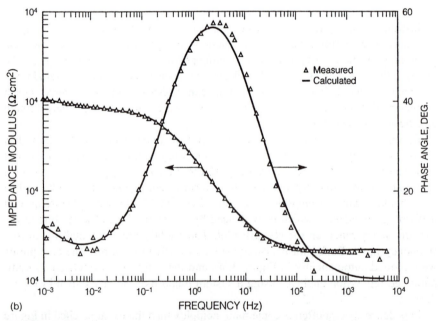

(b)

FIGURE 3.30 Impedance spectra for carbon steel in aerated water at 32°C containing 1000 ppm chloride with sulfate and bicarbonate: (a) Nyquist format; (b) Bode format. *(From D. C. Silverman and J. E. Carrico, Corrosion, Vol. 44, p. 280, 1988. Reprinted by permission, National Association of Corrosion Engineers.)*

difficult. Corrosion reactions are slow, requiring low frequency measurements which approach more simple interrupted dc measurements with much less expensive instrumentation. Nevertheless, the added dimension of frequency can provide essential mechanistic information which would be otherwise unavailable from dc measurements.

References

1. M. Stern and A. L. Geary, *J. Electrochem. Soc.*, Vol. 104, p. 56, 1957.
2. M. G. Fontana and N. D. Greene, *Corrosion Engineering*, 2nd ed., McGraw-Hill, New York, p. 297–346, 1978.
3. S. W. Dean, *Mater. Perf.*, Vol. 26, p. 51, Dec. 1987.
4. M. G. Fontana, *Corrosion Engineering*, 3rd ed., McGraw-Hill, New York, p. 457, 1986.
5. C. Wagner and W. Traud, *Z. Elektrochem.*, Vol. 44, p. 391, 1938.
6. J. Tafel, *Z. Phys. Chem.*, Vol. 50, p. 641, 1904.
7. R. Bandy and D. A. Jones, *Corrosion*, Vol. 32, p. 126, 1976.
8. M. Stern, *J. Electrochem. Soc.*, Vol. 102, p. 609, 1955.
9. M. Stern and R. M. Roth, *J. Electrochem. Soc.*, Vol. 104, p. 390, 1957.
10. H. H. Uhlig and R. W. Revie, *Corrosion and Corrosion Control*, 3rd ed., Wiley, New York, p. 44, 1985.
11. B. E. Conway, *Electrochemical Data*, Elsevier, New York, 1952.
12. A. C. Makrides, *J. Electrochem. Soc.*, Vol. 107, p. 869, 1960.
13. W. Roitar, W. Jura, and E. Polujan, *Acta Physichochim.*, Vol. 10, p. 389, 1939.
14. G. Okamoto, M. Nagayama, and N. Sato, *International Committee of Electrochemical Thermodynamics and Kinetics, 8th Meeting, Madrid, 1956*, Butterworths, Sevenoaks, Kent, England, p. 72, 1958.
15. T. P. Hoar and T. Hurlen, *International Committee of Electrochemical Thermodynamics and Kinetics, 8th Meeting, Madrid, 1956*, Butterworths, Sevenoaks, Kent, England, p. 445, 1958.
16. K. Bonhoeffer and K. Heusler, *Z. Phys. Chem.*, Vol. 8, p. 390, 1956.
17. Standard Method for Making Potentiostatic and Potentiodynamic Anodic Polarization Measurements, Standard Method G5-82, *Annual Book of ASTM Standards*.
18. D. A. Jones and A. J. P. Paul, *Hydrometallurgical Reactor Design and Kinetics*, R. G. Bautista, R. J. Wesley, and G. W. Warren, eds., TMS-AIME, Warrendale, PA, p. 293, 1987.
19. M. Stern and A. C. Makrides, *J. Electrochem. Soc.*, Vol. 107, p. 782, 1960.
20. W. D. France, Jr., *J. Electrochem. Soc.*, Vol. 114, p. 818, 1967.
21. D. A. Jones, *Corrosion*, Vol. 25, p. 187, 1969.
22. P. J. Moran, *Corrosion*, Vol. 42, p. 432, 1986.
23. D. A. Jones, *Corros. Sci.*, Vol. 8, p. 19, 1968.
24. B. E. Wilde, *Corrosion*, Vol. 23, p. 379, 1967.
25. M. S. Walker and W. D. France, *Mater. Prot.*, Vol. 8, No. 9, p. 47, 1969.
26. D. A. Jones and T. A. Lowe, *J. Mater.*, Vol. 4, p. 600, 1969.
27. D. Abraham, D. A. Jones, M. R. Whitbeck, and C. M. Case, ASTM STP 1056, ASTM, Philadelphia, p. 157, 1989.
28. A. J. Bard and L. R. Faulkner, *Electrochemical Methods*, Wiley, New York, p. 283, 1980.
29. A. C. Makrides, *Corrosion*, Vol. 18, p. 338t, Sept., 1962.
30. A. C. Makrides, *J. Electrochem. Soc.*, Vol. 107, p. 869, 1960.
31. D. R. Gabe, *J. Appl. Electrochem.*, Vol. 4, p. 91–108, 1974.
32. J. Pang and I. R. Ritchie, *Electrochim. Acta*, Vol. 26, p. 1345, 1981.

33. D. C. Silverman, *Corrosion*, Vol. 40, p. 220, 1984.
34. A. J. Bard and L. R. Faulkner, *Electrochemical Methods*, Wiley, New York, p. 316, 1980.
35. J. R. Scully, *DTNSRDC/SME-86/006*, David W. Taylor Naval Ship Research and Development Center, Bethesda, MD. 20084, Sept. 1986.
36. D. C. Silverman and J. E. Carrico, *Corrosion*, Vol. 44, p. 280, 1988.

Exercises

3-1. A zinc specimen exposed to an acid solution loses 25 milligrams during a 12 hour exposure. **(a)** What is the equivalent current flowing due to corrosion? **(b)** If the specimen area is 200-cm^2, what is the corrosion rate in mg per dm^2 per day due to this current? **(c)** what is the corrosion rate in mpy? μm/yr?

3-2. A single crystal (100) surface of nickel is corroding at a constant rate of 5 μA/cm^2. How many monolayers of nickel are dissolving in each minute?

3-3. An electrochemical measurement shows a uniform corrosion current density for nickel base Alloy 400 of 23.2 μA/cm^2. Calculate the equivalent corrosion rate in mpy for this alloy.

3-4. Calculate the penetration rate in mpy equivalent to 1 μA/cm^2 for a titanium alloy containing 5% aluminum and 2.5% tin.

3-5. Mercury is sometimes referred to as a "high overvoltage" surface and platinum as a "low overvoltage" surface. Justify this terminology, using the hydrogen reduction reaction, $2H^+ + 2e^- = H_2$, as an example.

3-6. Using appropriate polarization diagrams, determine the effect of the following parameters on the corrosion potential and corrosion rate of a metal M corroding to dissolved M^+ in an acid solution under activation control with all other parameters constant: **(a)** increasing i_o of the anodic reaction **(b)** increasing i_o of the cathodic reaction **(c)** increasing the concentration of dissolved H^+ **(d)** increasing the Tafel constant of the anodic reaction.

3-7. Assume that the corrosion of a metal M is controlled completely by activation polarization for both anodic and cathodic reactions. What is the effect of solution stirring on the corrosion rate and corrosion potential?

3-8. a. Plot the appropriate polarization curves for the following half cell reactions and determine the corrosion potential and corrosion rate (current density) assuming activation control of both the anodic and cathodic processes.

$$M = M^+ + e^-, \quad e = -0.7 \text{ V}, \quad i_o = 10^{-8} \text{ A/cm}^2, \quad \beta_a = +0.1 \text{ V}$$
$$2H^+ + 2e^- = H_2, \quad e = +0.1 \text{ V}, \quad i_o = 10^{-6} \text{ A/cm}^2, \quad \beta_c = -0.1 \text{ V}$$

(b) Same as **(a)**, but assume the limiting current density for the reduction reaction is 10^{-5} A/cm^2. Again determine the corrosion potential and corrosion rate from your plot.

3-9. A copper valve is used in the waste exit line from a pharmaceutical process reactor which is producing iodine-containing organics. The waste stream is composed of air saturated 15% sulfuric acid with the following impurities: ferric sulfate, ferrous sulfate, potassium iodide, and iodine.

From the electrochemical data listed below, determine the corrosion potential and corrosion rate in this brew. Assume all Tafel constants are 0.1 V.

$Cu = Cu^{+2} + 2e^-$: $e = 0.350$ V, $i_o = 10^{-4}$ A/cm^2

$H_2 = 2H^+ + 2e^-$: $e = 0.020$ V, $i_o = 2 \times 10^{-7}$ A/cm^2 (on Cu),

$\quad i_{Lim}(H^+ \rightarrow H_2) = 2 \times 10^{-1}$ A/cm^2, $i_{Lim}(H_2 \rightarrow H^+) = 10^{-4}$ A/cm^2.

$Fe^{+2} = Fe^{+3} + e^-$: $e = 0.850$ V, $i_o = 8 \times 10^{-5}$ A/cm^2 (on Cu).

$\quad i_{Lim}(Fe^{+3} \rightarrow Fe^{+2}) = 6 \times 10^{-3}$ A/cm^2,

$\quad i_{Lim}(Fe^{+2} \rightarrow Fe^{+3}) = 2 \times 10^{-4}$ A/cm^2.

$I_2 + 2e^- = 2I^-$, $e = 0.430$ V, $i_o = 10^{-3}$ A/cm^2 (on Cu),

$\quad i_{Lim}(I_2 \rightarrow I^-) = 2 \times 10^{-3}$ A/cm^2

$\quad i_{Lim}(I^- \rightarrow I_2) = 2 \times 10^{-2}$ A/cm^2.

3-10. Plot the polarization data shown below for an electrode of 1.13 cm^2 of the metal M in an acid solution of unit activity bubbled with pure hydrogen. How might the potentials in this data have been measured experimentally? From your plot, determine the corrosion rate (current density) and the exchange current density for the hydrogen reduction reaction on the surface of M. Comment on the accuracy of your determinations.

Current, A	Potential V vs. SHE	Current, A	Potential V vs. SHE
2×10^{-4}	−0.4843	1×10^{-2}	−0.6067
5×10^{-4}	−0.4939	2×10^{-2}	−0.6425
1×10^{-3}	−0.5075	5×10^{-2}	−0.7117
2×10^{-3}	−0.5287	1×10^{-1}	−0.7375
5×10^{-3}	−0.5663		

3-11. Determine the anodic polarization curve from the cathodic data of Problem 3-10. Comment on the accuracy of the anodic curve obtained in this way.

3-12. Show the cathodic polarization curve which would result from the data of Problem 3-10 if the measured potential includes an additional ohmic contribution which results from all the polarizing current passing through an electrolyte resistance of 10Ω between reference and specimen or working electrodes.

3-13. Search the literature to find a method for mounting sheet specimens to expose a controlled surface area for electrochemical polarization studies without using paint or mounting resins. Inquire to commercial suppliers to find if such a device is available commercially.

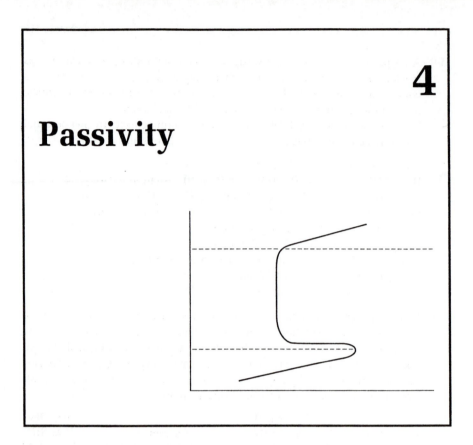

Passivity 4

4.1 Description

Passivity results when certain metals and alloys form very thin, oxidized, protective films on their surfaces in corrosive solutions. Most commercially available corrosion resistant alloys depend on passive films for their resistance. In this chapter we describe the desirable factors that contribute to formation of passivity and the resulting corrosion resistance.

4.1.1 Passive Films

Passivity is defined as a condition of corrosion resistance due to formation of thin surface films under oxidizing conditions with high anodic polarization. Some metals and alloys, having simple barrier films with reduced corrosion at active potentials and little anodic polarization, are not considered to be passive by this definition. Pourbaix diagrams (Section 2.2) for most metals show stability of one or more oxides at noble potentials in oxidizing solutions.[1] For example, the simplified diagram for iron in Figure 4.1 shows the oxides, Fe_2O_3 and Fe_3O_4, to be stable over wide ranges of potential and pH. A passive film forms by a direct electrochemical reaction, such as (Section 2.2.5)

$$Fe + 2H_2O \rightarrow Fe(OH)_2 + 2H^+ + 2e^-. \tag{1}$$

Insoluble compounds formed by dissolution and reprecipitation are not generally as tenacious and protective as oxides formed by oxidation in place on the surface. The corrosion rate is reduced substantially as a passive film begins to form, despite the fact that there is no visible evidence on the surface.

In the presence of concentrated "fuming" nitric acid, iron is virtually inert, despite the highly oxidizing conditions of the solution (Section 11.1.6). When the acid is diluted with water, the iron remains inert initially but corrodes vigorously, evolving brown, nitrous oxide gas when the surface is lightly scratched (Figure 4.2[2]). Faraday, who originated the experiment in the 1840s, suggested that an invisible surface oxide film, preformed in the concentrated acid, is unstable in the diluted solution and is destroyed when mechanically disturbed by scratching.

The structure of the film has eluded definition because it is so thin and fragile. Optical measurements indicate a compact, transparent film of 1 to 10 nm in thickness. Hydrogen has been detected in the film, indicating the presence of hydroxide or water of hydration. Mercury does not wet the passive surface, and an oxide film

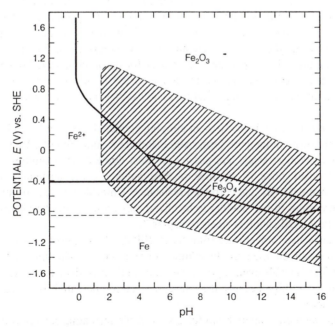

FIGURE 4.1 Pourbaix diagram for iron superimposed on the diagram for chromium designated by dashed lines. Shaded area indicates stability of Cr_2O_3. Approximately unit activity assumed for dissolved species. *(Adapted from Pourbaix.[1])*

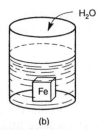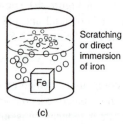

(a)   (b)   (c)

FIGURE 4.2   Faraday's demonstration of iron passivity in concentrated (fuming) nitric acid. *(From M. G. Fontana,* Corrosion Engineering, *3rd ed., McGraw-Hill, New York, p. 469, 1986. Reprinted by permission, McGraw-Hill Book Company.)*

has been isolated and removed from the substrate iron.[3] Although the most stable phases are usually the oxides, the Pourbaix diagram cannot rule out the formation of precursory condensed surface phases containing water and dissolved anions. Conflicting theories have suggested allotropic modifications, bulk oxide, adsorbed oxygen, adsorbed OH⁻, and adsorbed anions as the source of passivity. Results from modern electron spectroscopic measurements are compromised by removal of water in the instrument vacuum systems, necessarily altering the film structure and composition. Thus, conflicts in theory have been unresolved over the years.[4]

Usual corrosion conditions are not sufficiently oxidizing to form the passive state on pure iron, in contrast to concentrated fuming nitric acid (Figure 4.2), which is one of the most highly oxidizing solutions available. Chromium is noted for formation of very stable, thin, resistant, surface films in less oxidizing conditions, when alloyed with other metals, especially iron and nickel. The Pourbaix diagram for iron in Figure 4.1 is superimposed on that for chromium, which has an oxide forming at much lower potentials, as indicated by the crosshatched area. Thus, chromium additions provide the basis for stainless steels and other corrosion-resistant alloys.

Chromium is used exclusively as an alloying element because it is too brittle and difficult to fabricate in the pure form. Alloys having a minimum of 12% Cr in iron, known as the stainless steels, are passive in most dilute aerated solutions. Further additions of nickel above 8% stabilize the face-centered-cubic (austenite) phase and further enhance passivity and corrosion resistance. Chloride restricts the stability of oxidized chromium films,[1] leading to localized corrosion by film breakdown in the Fe–Cr and Fe–Cr–Ni stainless steels (Chapter 7).

Other metals also form surface films that fulfill the definition of passivity, including aluminum, silicon, titanium, tantalum, and niobium. For aluminum and silicon, the passive film is stable even at potentials characteristic of deaerated water. Anodic polarization is still high because the half-cell electrode potential for metal dissolution is very active.

Titanium, tantalum, and niobium form very stable insulating surface films which are resistant at very highly oxidizing potentials, leading to their use in anodes for impressed current cathodic protection systems. In a similar way, cast irons containing silicon, which is resistant at high potentials and low pH, are also used for impressed current anodes (Section 13.2.7).

4.1.2 Active–Passive Corrosion Behavior

Metals and alloys fulfilling the foregoing definition of passivity display distinctive behavior as potential and anodic polarization increase, as shown in Figure 4.3. At low potentials characteristic of deaerated acid solutions, corrosion rates measured by anodic current density are high and increase further with potential in the active state. Above the primary passive potential, E_{pp}, the passive film becomes stable, and corrosion rate falls to very low values in the passive state. The reduced corrosion rate in the passive state may be as much as 10^6 times lower than the maximum in the active state at i_c. Corrosion rate is plotted on a log scale in Figure 4.3 to accommodate these broad changes with oxidizing potential. At still higher potential, the passive film breaks down and the anodic rate increases in the transpassive state. For stainless steels and chromium-bearing nickel alloys, the transpassive breakdown occurs near the oxygen evolution potential, where the chromium-rich passive film is unstable, according to Figure 4.1. Only very strong oxidizers, rarely seen in practice, produce potentials in the transpassive region.

More severe conditions of higher acidity and temperature generally decrease the passive potential range and increase current densities and corrosion rates at all potentials, as indicated in Figure 4.4.

4.1.3 Oxidizer Concentration Effects

Increasing oxidizer concentration increases the potential of the redox half-cell reaction according to the Nernst equation (Section 2.1.3). The predictions of mixed potential theory (Section 3.3.3) are shown in Figure 4.5 for an active–passive alloy. As concentration increases from 1 to 2, corrosion rate and corrosion potential increase from A to B as for a conventional metal or alloy. At concentration 3, the alloy may exist in either the active state at C or the passive state at D. The "negative

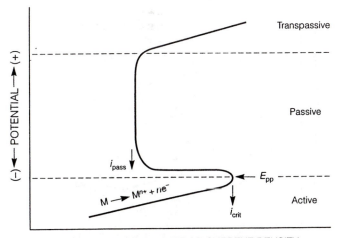

FIGURE 4.3 Schematic active–passive polarization behavior.

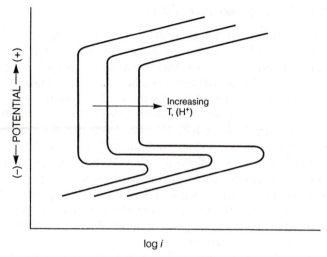

FIGURE 4.4   Effect of increasing acid concentration and temperature on passivity.

resistance" portion of the anodic curve at the active-to-passive transition, where current decreases with increasing potential, is an artifact of the instrumental measurement and is sometimes indicated by a dashed line as shown. Thus, the point x is not a stable potential state and is never observed when passivity is established by dissolved oxidizers, in practice. As concentration increases to 4 and 5, only the passive state is stable, and the corrosion rate drops to the low passive values near D. Still further increases to 7 and 8 cause a transition to the transpassive state and corrosion rate increases to E and F.

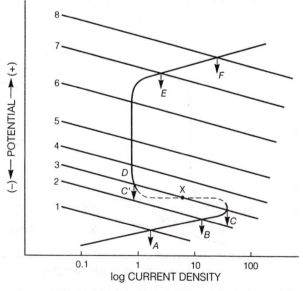

FIGURE 4.5   Effect of oxidizer concentration on corrosion of an active–passive alloy.

The resultant plot of corrosion rate versus oxidizer concentration is shown in Figure 4.6. As oxidizer concentration increases, corrosion rate increases in the active state initially. In the region BCD, either the passive or active states may be present, but the passive film may not form until concentration D is reached when only the passive state is stable and corrosion rate drops dramatically. Corrosion rate remains low in the passive state and increases in the transpassive state with further increases in concentration. When the process is reversed and concentration is decreased by dilution, the corrosion rate retraces its path from the transpassive back into the passive state. However, once the passive film has been formed, it is retained at concentrations lower than were needed for formation. Thus, in the region $DC'B$, "borderline passivity" is present in which any surface disturbance (e.g., scratching as in Figure 4.2) will destabilize the passive film and the corrosion can increase to the active state at, for example, C.

4.1.4 Solution Velocity Effects

Figure 4.7 shows the polarization diagram for corrosion of stainless steel in a dilute salt solution such as seawater. In deaerated solutions, the only available cathodic reduction reaction is

$$2H_2O + 2e^- \rightarrow H_2 + 2OH^-, \tag{2}$$

and the alloy corrodes in the active state. The rate of hydrogen evolution by (2) is quite low for most metals, and corrosion rate is quite low in deaerated neutral solutions, even in the active state. Upon partial aeration, reduction of dissolved oxygen,

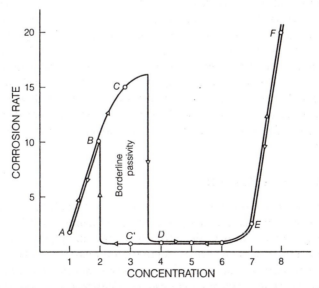

FIGURE 4.6   Effect of oxidizer concentration on corrosion rate in an active–passive metal or alloy.

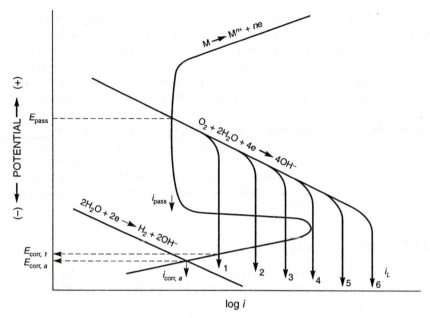

FIGURE 4.7   Effect of deaeration, aeration, and stirring on corrosion of active–passive stainless steel in neutral saltwater.

$$O_2 + 2H_2O + 4e^- \rightarrow 4OH^-, \tag{3}$$

predominates, and a borderline passivity condition ensues, with the alloy either in the passive state at E_{pass} or in the active state at $E_{corr,1}$. It is justifiable to assume conservatively that the active state will prevail since any surface damage will result in destruction of the passive film. Subsequent increase in solution velocity increases the limiting diffusion current from 1 progressively up to 6 (Section 3.3.4). Corrosion rate and corrosion potential in the active state also increase progressively up to 4, where the corrosion potential increases dramatically to E_{pass} and the corrosion rate falls from i_4 to i_{pass} in the passive state, as shown in Figure 4.8a. Figure 4.8b shows for comparison corrosion response with velocity or agitation for the usual nonpassive metal, in which corrosion rate increases to a plateau in changing from concentration to activation control, as shown in Figure 3.12.

4.1.5 Criterion for Passivation

Both oxidizer concentration and solution velocity have similar effects on the corrosion rate of an active–passive alloy, according to Figures 4.6 and 4.7. When the rate of cathodic reduction is made sufficiently high, the passive state is stable, and corrosion rate falls to low values. A criterion for passivation becomes apparent. *The passive state is stable when the rate (current density) for cathodic reduction is greater than the critical anodic current density, i_{crit}.* From the standpoint of alloy selection, alloys having lower i_{crit} and more active E_{pp} are more easily passivated.

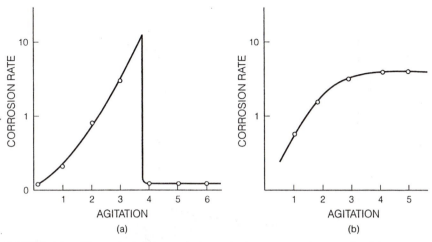

FIGURE 4.8   Effect of stirring or solution velocity on corrosion rate for (a) active–passive stainless steel, derived from Figure 4.6, and (b) normal active metal, reproduced from Figure 3.13b.

4.2 Experimental Apparatus and Procedures

Experimental polarization cells described in Section 3.5.3 are suitable for electrochemical studies of active–passive alloys. However, controlled current instrumental methods are not adequate for determining active–passive behavior, and controlled potential methods are required to show the entire anodic polarization curve.

4.2.1 Galvanostatic Anodic Polarization of Active–Passive Alloys

Polarization of an active–passive alloy using galvanostatic, controlled-current circuitry (Figure 3.17) is shown in Figure 4.9. As in Section 3.4.2, the applied anodic current density is $i_{app} = i_a - i_c$. Starting from E_{corr} in the active region, anodic overvoltage increases and i_{app} approaches i_a as i_a becomes $\gg i_c$. Above i_{crit}, however, i_{app} no longer follows the anodic curve in the passive region. The next higher current above i_{crit} increases potential into the transpassive region, and the passive loop of the active–passive curve is unrevealed. If the polarizing current is reduced from the transpassive region, i_{app} follows the transpassive curve down to i_{pass}, and potential drops into the active region at the next-lower current value.

   It is apparent that galvanostatic procedures are inadequate to define the active–passive curve properly because potential is not a single-valued function of current. However, current is a single-valued function of potential, and controlled potential procedures, described in the next section, are effective in studying the electrochemical behavior of active–passive alloys.

4.2.2 Potentiostatic and Potentiodynamic Procedures

A potentiostat automatically adjusts the applied polarizing current to control potential between working electrode (WE) and reference electrode (REF) at any

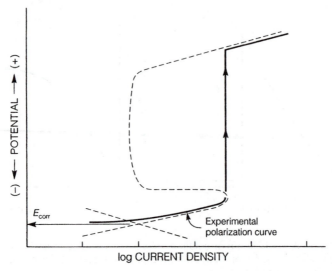

FIGURE 4.9 Polarization of an active–passive alloy with controlled (galvanostatic) current.

prescribed value. Figure 4.10 shows the potentiostatic circuit, which has potential and current measuring elements, as well as the polarization cell and electrodes, in common with the galvanostatic circuitry of Figure 3.17. Modern potentiostats often include the current and potential measuring instrumentation. The current I

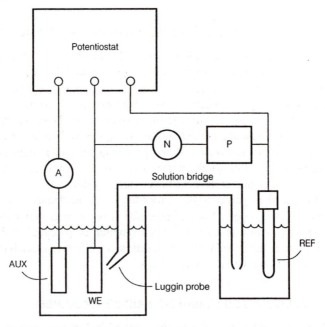

FIGURE 4.10 Controlled potential circuitry utilizing a potentiostat.

polarizes the WE to the prescribed potential with respect to REF, which remains at a constant potential with little or no current passing through the potential measuring circuit.

Polarization of an active–passive alloy by a potentiostat is shown in Figure 4.11. In potentiostatic procedures, potential is increased from E_{corr} in the active state in steps with current recorded at the end of an appropriate time at each step. Applied current increases with potential until it follows the anodic curve, just as in Figure 4.9 for galvanostatic control. At potentials above E_{pp}, contrary to galvanostatic procedures, the potentiostatic anodic polarization curve exactly follows the passive loop of the anodic curve. Current, the dependent variable, approaches a steady-state value at each controlled potential step. Thus, potentiostatic procedures allow detailed study of the important parameters affecting formation and growth of passive films and passivity.

Potentiostats may be programmed to increase potential continuously from E_{corr}. When potential is fed to an x-y recorder, along with current through a logarithmic converter, the anodic polarization curve of potential versus log applied current can be recorded automatically. The resulting *potentiodynamic* anodic polarization curve exactly matches the *potentiostatic* anodic polarization curve, obtained at an equivalent rate of polarization.[5] With widely available modern instrumentation, most anodic polarization curves are now obtained potentiodynamically. However, much of the illustrative data cited in this chapter and elsewhere have been obtained potentiostatically. Data obtained by the two methods are assumed to be equivalent.

A frequently used laboratory electrolyte for potentiostatic anodic polarization is dilute H_2SO_4, a common industrial chemical. One normal concentration with nominal unit activity H^+ is often selected for convenience. In a hydrogen-sparged solution, the hydrogen electrode is established on a platinum surface, as described in

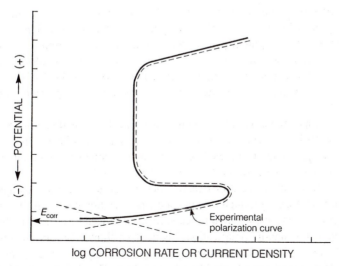

log CORROSION RATE OR CURRENT DENSITY

FIGURE 4.11   Polarization of an active–passive alloy with controlled (potentiostatic) potential.

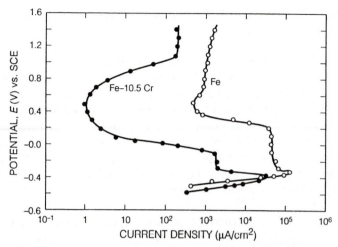

FIGURE 4.12 Potentiostatic anodic polarization of pure iron and iron-10.5% chromium alloy. *(Adapted from Steigerwald and Greene[6].)*

Section 2.3.1. Potentiostatic anodic polarization curves for pure iron and a binary iron chromium alloy in 1 N H$_2$SO$_4$ are compared in Figure 4.12[6] for illustration. Notwithstanding the passivity demonstrated in Figure 4.2 for highly oxidizing nitric acid, iron is only weakly passive in more usual reducing electrolytes such as 1 N H$_2$SO$_4$. Only when chromium is alloyed with iron is well-defined active–passive behavior developed in conventional dilute acids.

4.2.3 Variables Affecting Anodic Polarization Curves

Anodic polarization curves are very sensitive to rate of polarization and alloy composition.[7] Figure 4.13 shows the effect of polarization rate; slower polarization with longer time at each potentiostatic step results in lower currents at all potentials of the polarization curve, especially in the passive region. More recently, Mansfeld[8] has shown potentiodynamic polarization data for AISI 430 stainless steel (without Ni) which are dependent on scan rate only in the passive range and not in the active.

Effects of compositional variations within the same alloy specification are shown in Figure 4.14.[7] It is apparent that such variation can have significant effects on the details of anodic polarization curves. Even differing specimens from the same bar of the same heat show about twofold differences in Figure 4.14.

Thus, it is necessary to control experimental variables and alloy composition stringently to maximize reproducibility in anodic polarization curves. ASTM round-robin testing programs have resulted in recommended procedures[5] for potentiostatic and potentiodynamic anodic polarization. Figure 4.15 shows the reproducibility to be expected from different laboratories applying nominally the same potentiodynamic procedure to specimens from the same alloy heat. The similar scatter band for potentiostatic polarization is somewhat tighter than shown in Figure 4.15 for poten-

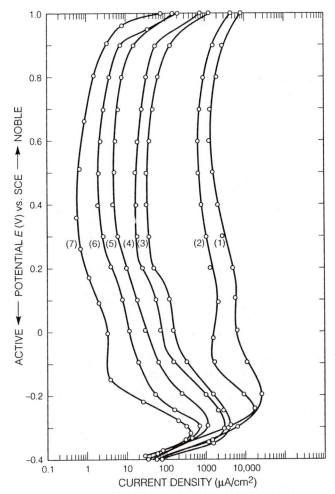

FIGURE 4.13   Effect of decreasing polarization rate (1) to (7) on potentiostatic anodic polarization curves for AISI 304 stainless steel in deaerated 1N H$_2$SO$_4$ at 25°C. *(From N. D. Greene and R. B. Leonard,* Electrochim. Acta, *Vol. 9, p. 45, 1964. Reprinted by permission, Pergamon Press.)*

tiodynamic polarization and about matches the difference between specimens of the same heat shown in Figure 4.14. These data are especially useful to validate equipment and procedures in new laboratories. The ASTM description of the recommended method[5] also describes sources of error causing deviation from the standard curves of Figure 4.15.

Low-conductivity electrolytes may lead to ohmic interferences in potentiostatic and potentiodynamic measurements.[8] Positive-feedback loops in the potentiostat have been used to compensate ohmic electrolyte resistance. However, feedback balance is very sensitive, and changes in R_Ω during polarization measurements can cause the potentiostat to lose potential control and ruin an experiment.[8]

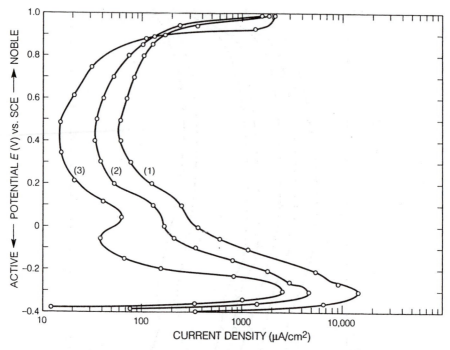

FIGURE 4.14 Effect of differing heats, (1) and (3), and differing specimens from the same heat, (1) and (2), on potentiostatic anodic polarization curves for AISI 304 stainless steel in deaerated 1*N* H_2SO_4 at 25°C; traverse rate 12 V/h. *(From N. D. Greene and R. B. Leonard,* Electrochim, Acta, *Vol. 9, p. 45, 1964. Reprinted by permission, Pergamon Press.)*

4.3 Applications of Potentiostatic Anodic Polarization

Potentiostatic anodic polarization curves may be used to judge the corrosion resistance of alloys and the corrosivity of solutions. A polarization curve can be obtained in a few hours, whereas weeks, months, or more are necessary to obtain the equivalent (e.g., Figure 4.6) by conventional testing. Time and labor are saved when developing new alloys or testing available alloys for new applications. An encyclopedic tabulation of anodic polarization data for various alloys in typical corrosive media would be useful but is presently not available.

Some examples of anodic polarization curves drawn from the literature are given below for illustration. Examples involving testing for galvanic corrosion, intergranular attack, and other forms of localized corrosion are discussed in the appropriate chapters that follow. Limitations of potentiostatic anodic polarization are discussed later in Section 4.3.3.

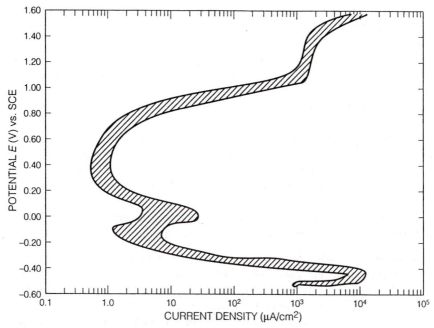

FIGURE 4.15   Reproducibility of potentiodynamic anodic polarization curves from different laboratories for AISI 430 stainless steel in deaerated $1N\ H_2SO_4$. *(From ASTM Standard Reference Test Method for Making Potentiostatic and Potentiodynamic Anodic Polarization Methods, Standard Method G5–87, Annual Book of ASTM Standards, Vol. 3.02, p. 97, ASTM, Philadelphia, 1988. Reprinted by permission, ASTM.)*

4.3.1 Alloy Evaluation

Potentiostatic anodic polarization is frequently used to compare the corrosion resistance of candidate alloys in a specific environment. A schematic comparison of hypothetical alloys A, B, C, and D is shown in Figure 4.16. For reducing conditions, as in number 1, either the non-passivating alloy A or the partially passivating alloy B is superior to the other two because A and B have lower corrosion rates in the active condition without oxidizers. That is, the corrosion rates or current densities, $i_{corr,1A}$ and $i_{corr,1B}$, labeled for simplicity 1A and 1B in Figure 4.16, are lower for alloys A and B, respectively, in condition 1 than either of the other two alloys. The alloying elements, especially chromium, needed to produce strong passivity make C or D far more expensive and thus unjustifiable for service in condition 1.

For modestly oxidizing conditions, number 2, the recommended alloy would be C because the reduction curve exceeds the critical current density for passivation and it is the only alloy in the stable passive condition. Although the reduction curve also exceeds the critical passivation current density for alloy B, the passive current at 2B is not low enough and the passive region is not broad enough to ensure good resistance in the passive state. Alloy D is in a state of borderline passivity with both active, 2D, and passive states possible. One always conservatively chooses the active state as the one most likely in cases of borderline passivity. The corrosion rate of the nonpassive alloy, A, is predictably high at 2A in oxidizing conditions of any degree.

The recommended alloy in highly oxidizing conditions is D, since now the reduction curve exceeds the critical current density for passivation, and the corrosion rate is low at 3D. Passivity breaks down for alloy C at E_c, and corrosion rate is increased to 3C, which is above the passive current density at 2C. Alloys A and B are not resistant to highly oxidizing conditions.

The breakdown of alloy C postulated at E_c may occur due to initiation of localized corrosion (Section 7.2.1) in pits or crevices. With no breakdown there may be little to choose from between alloys C and D because corrosion rates 2C and 3D in the passive state are both very low and either may be adequate for most applications.

In summary, the corrosion resistance of alloys may be judged by the following parameters derived from potentiostatic anodic polarization.

1. In the active state, corrosion rate is proportional to the anodic current density whether or not the alloy is of the active–passive type.
2. The current density (rate) of reduction must exceed the critical current density for passivation to ensure low corrosion rate in the passive state.
3. Borderline passivity should be avoided in which either the active or the passive state may be stable.

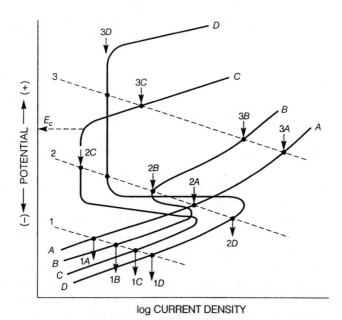

FIGURE 4.16 Schematic anodic polarization curves for hypothetical alloys A, B, C, and D, illustrating evaluation in various chemical conditions: 1, reducing; 2, moderately oxidizing; 3, highly oxidizing.

4. Breakdown of the passive film in oxidizing conditions due to transpassivity or initiation of localized corrosion should be avoided.
5. The passive state in oxidizing conditions is essential for corrosion resistance, but reasonably small variations in the passive current density may not be significant.

Potentiostatic anodic polarization curves for three nickel alloys of varying corrosion resistance (Figure 4.17) illustrate alloy evaluation with experimental data.[9] Alloy B (Ni-25Mo) shows only a hint of passivity. Alloy C (Ni-15Cr-15Mo-5Fe) shows low i_c and active E_{pp}, but the passive current increases steadily in the passive potential range. Alloy C-276 shows the benefits of the low-Si, low-C modification in which the passive current density remains low past the region of Cr_2O_3 stability (Section 4.1.1) near the oxygen evolution potential. The low Si and C also prevent precipitates that reduce the corrosion resistance of welds (Section 9.4.2). Anodic polarization is sensitive to these impurities even though the alloys of Figure 4.17 have been solution annealed to prevent precipitation. For reducing conditions there is little to choose from between the three alloys on the basis of corrosion resistance. Alloy B would normally be chosen because it has no alloyed chromium and therefore is the least expensive. At moderately oxidizing conditions, alloy C becomes useful because of its low i_c and active E_{pp}. However, in more highly oxidizing conditions, Alloy C-276 becomes superior because of its lower passive current density at higher potentials.

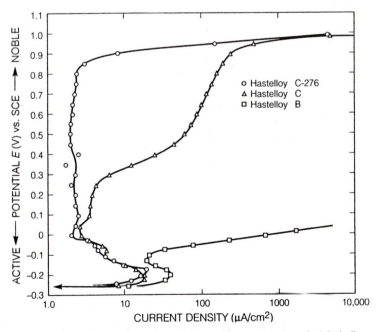

FIGURE 4.17 Comparison of potentiostatic anodic polarization of nickel alloys in 1 N H_2SO_4, ambient temperature. *(By courtesy of F. G. Hodge, Haynes International.)*

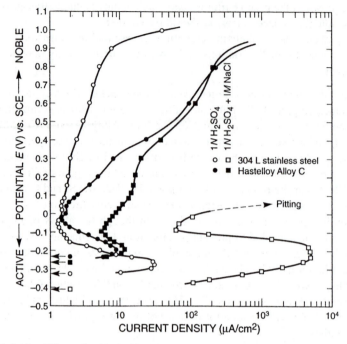

FIGURE 4.18 Effect of added chloride on potentiostatic anodic polarization of stainless steel and nickel Alloy C in sulfuric acid. *(Adapted from Green and Judd.[10])*

4.3.2 Solution Corrosivity

The corrosivity of a particular solution species can be judged by effects on anodic polarization curves. Figure 4.18 shows the effect of chloride ion on anodic polarization of AISI 304 stainless steel and Alloy C.[10] The stainless steel shows large current density increases with breakdown of the passive film and consequent pitting attack (Section 7.2.1). Chloride has much less effect on Alloy C, reflecting greater resistance to such solutions. On the other hand, AISI 304 stainless steel is more suitable (and economical) in chloride-free solutions in which the current densities are generally lower than for Alloy C. In the same study,[10] chloride had no effect on the polarization curve for titanium.

4.3.3 Limitations

The rapid evaluation of alloy and corrosive properties by potentiostatic anodic polarization is not obtained without a price. Laboratory chemicals are quite pure relative to industrial chemicals in plant process streams. The growth and dissolution of the passive film is dynamic; the passive current density or corrosion rate changes with time (Figure 4.13) and cannot be used as a quantitative guideline for long-term corrosion resistance. Surfaces in process equipment develop conditions over long periods of time that cannot be duplicated easily in short-term laboratory experiments. In short, plant conditions are difficult to characterize and duplicate in the laboratory. Therefore, caution must be used before predicting plant performance from short-term potentiostatic anodic polarization. At the very least, however, potentiostatic anodic polarization can provide useful preliminary guidelines for more extensive in-plant exposure tests.

4.4 Anodic Protection

If an active–passive alloy is maintained in the passive region with a potentiostat, its corrosion rate will be low at i_{pass}, as shown in Figure 4.19. When the electrical components are scaled up to control the corrosion of plant-size equipment, the method is called *anodic protection*.

To be suitable for anodic protection, an active–passive alloy should exhibit a broad range of passivity and a low passive corrosion rate in the process solution. These favorable circumstances are somewhat rare. Furthermore, the necessary electrical equipment is complex and expensive to install and maintain, and if control is temporarily lost, corrosion can be disastrously high should the control potential stray into the active region. Thus, anodic protection is not as widely applicable as are other corrosion mitigation methods, such as coatings and cathodic protection. Nevertheless, anodic protection has the unique ability to control corrosion economically in some aggressive media where no other prevention methods are suitable. As a result, it has assumed a significant role in corrosion control for some chemical process equipment discussed in Section 4.4.1.[11,12] More detailed comparisons are drawn between anodic and cathodic protection in Section 13.1.7.

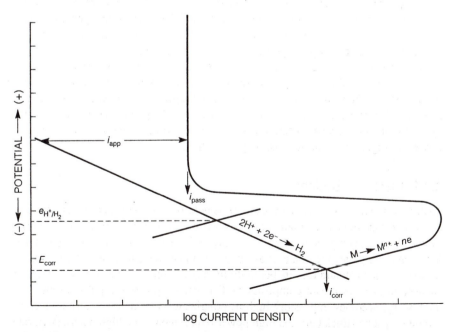

FIGURE 4.19   Potential control in the passive region for anodic protection.

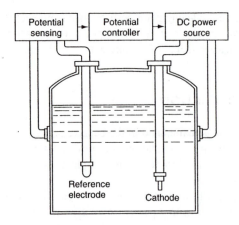

FIGURE 4.20   Schematic equipment for anodic protection of a storage tank. *(C. E. Locke,* Metals Handbook, *Vol. 13:* Corrosion, *9th ed., ASM International, p. 463, 1987. Reprinted by permission, ASM International.)*

4.4.1 System Description

The anodic protection system for an acid storage tank is shown schematically in Figure 4.20.[11] An amplifier controls a conventional dc rectifier to maintain the interior wetted tank surface at the necessary potential in the passive range, with respect to the reference electrode. Anodic protection has its greatest application for acid storage tanks because of the simple geometry and easy current access to all surfaces. Ordinary carbon steel can be used with anodic protection instead of more expensive stainless steels. In recent years, anodic protection has been employed to protect stainless steel heat exchangers in sulfuric acid manufacturing plants.[11] The heat exchangers are sold as a package with the anodic protection system.

Table 4.1 lists metals and alloys as well as corrosive solutions which have been at least considered for anodic protection. Use has not extended appreciably beyond sulfuric acid in the United States of America, although other applications are reported in the Soviet Union.[13] Anodic protection is usually precluded in chloride and hydrochloric acid because of the danger of passive film breakdown and localized corrosion (Figure 4.18) during anodic polarization.

4.4.2 Electrical Equipment

To initiate anodic protection, a metal must be polarized from the active state through E_{pp}, where current is a maximum at i_{crit}. Consequently, power requirements are high for the electrical components of an anodic protection system. Current supply is much lower once the passive region is attained, but extra current capabilities are necessary in case the metal should lose its passivity for any reason (e.g., brief power losses or changes in process stream composition). One cannot polarize rapidly through E_{pp} because i_{crit} becomes impossibly high as polarization rate increases. Thus, a compromise must be struck between high power requirements at a high polarization rate and excessive anodic dissolution at a low polarization rate (Figure 4.13). The cur-

TABLE 4.1   Alloys and Solutions for Anodic Protection

Solutions	Metals
Sulfuric acid	Steels
Phosphoric acid	Stainless steels
Nitric acid	Nickel
Nitrate solutions	Nickel alloys
Ammonia solutions	Titanium
Organic acids	
Caustic solutions	

rent to maintain passivity is thus several orders of magnitude lower than necessary to initiate it.

The controlled potential power supply (potentiostat) for anodic protection must provide the required voltage across the protected structure (anode) and the cathode. This voltage is made up of anodic overvoltage at the structure, cathodic overvoltage at the cathode(s), and ohmic $I_{app}R_{\Omega}$ losses through the solution between them (Section 3.5.4). In contrast to cathodic protection (Section 13.2.1), the $I_{app}R_{\Omega}$ drops during anodic protection are relatively low because of the high-conductivity solutions that are applicable. Overvoltages at anode and cathode are a few hundred millivolts during operation when passivity need only be maintained.

4.4.3 Cathodes and Potential Distribution

Because of the large surface area of the anode structure and high solution conductivity, the circuit resistance is controlled generally by the area of cathodes. Since there is a practical limitation on cathode area, cathode materials with low hydrogen overvoltage may be preferred. Platinum-clad brass was used in early installations,[11] but other, less expensive materials have been substituted in recent years, using more cathodes with higher surface area (Table 4.2[11]).

The "throwing power" is good for anodic protection. That is, potential is distributed widely on the structure from individual cathodes because of the high-conductivity solutions and low currents involved. Protection has been reported down long lengths of tubing.[9,14] However, for more complex structures such as heat exchangers, the cathodes should be distributed around the structure surfaces to ensure uniform protection and minimize circuit resistance.

A special problem of throwing power is anodic protection in the presence of crevices. Because of ohmic losses down the length of a constricted crevice, the metal at the bottom of the crevice may remain in the unprotected active state while the surface is fully protected. That anodic current which does flow at the bottom accelerates active metal dissolution, resulting in severe localized corrosion in the crevice. Crevices are unavoidably present between the structure and electrical insulation at the entries for cathode(s) and reference electrode. This and other such crevices should not be immersed in the corrosive solution. Crevices in the protected surface should be avoided, as much as possible, with flush welded joints and seams.[9] Alloys

TABLE 4.2   Cathode Materials for Anodic Protection

Cathode Material	Environment
Platinum on brass	Various
Steel	Kraft digester liquid
Illium G	Sulfuric acid (78–105%)
Silicon cast iron	Sulfuric acid (89–105%)
Copper	Hydroxylamine sulfate
Stainless steel	Liquid fertilizers (nitrate solutions)
Nickel-plated steel	Chemical nickel plating solutions
Hastelloy C	Liquid fertilizers (nitrate solutions)
	Sulfuric acid
	Kraft digester liquid

Source: C. E. Locke, Metals Handbook, 9th ed., Vol. 13: Corrosion, ASM International, p. 463, 1987. Reprinted by permission, ASM International.

with low i_{crit} (Figure 4.3) are easier to protect in crevices,[15] but anodic protection of complicated structures is difficult, in any case.

4.4.4 Reference Electrodes

Many of the conventional laboratory reference electrodes discussed in Section 2.3.1 have been used in anodic protection systems. In addition, certain corroding and inert metal electrodes have been used with some success. Corroding electrodes may contaminate the process solution. The reference potential of inert electrodes may be shifted by foreign redox systems in solution. Table 4.3 lists electrodes and media in which they have been used.[11] Construction and design of stable, long-life reference electrodes are major problems in the operation of anodic protection systems.[12]

TABLE 4.3   Reference Electrodes for Use in Anodic Protection

Electrode	Solution
Calomel	Sulfuric acid
Silver–silver chloride	Sulfuric acid, Kraft solutions Fertilizer solutions
Mo–MoO$_3$	Sodium carbonate solutions
Bismuth	Ammonium hydroxide solutions
Type 316 stainless steel	Fertilizer solutions, oleum
Hg–HgSO$_4$	Sulfuric acid Hydroxylamine sulfate
Pt–PtO	Sulfuric acid

Source: C. E. Locke, Metals Handbook, 9th ed., Vol. 13: Corrosion, ASM International, p. 463, 1987. Reprinted by permission, ASM International.

4.5 Properties of Passive Films

A lack of fundamental understanding of passive film properties has delayed the control and prevention of localized forms of corrosion that result from breakdown of the passive film. The most conveniently determined film properties involve chemical and electrochemical measurements which have been difficult to interpret objectively. Unfortunately, *ex situ* examinations for direct determination of structure and composition (e.g., in vacuum by electron spectroscopy) are likely to change the film structure by dehydration and precipitation of new phases on the surface.

Passivity has been the subject of curiosity and investigation for over 140 years.[4] It is beyond the scope of this book to review the historic body of work that has been devoted to the subject. Some of the more interesting properties of passive films can be discussed here only briefly. Some mechanisms from the literature are selected only for illustration of the principles involved, not because they have been proven correct or widely accepted. The emphasis is on iron and chromium, which are especially important from the standpoint of alloy selection for corrosion resistance. The interested reader is referred to a recent international symposium[4] for more in-depth information.

4.5.1 Structure and Composition

The structure favored by most investigators for the thin passive film on iron is an inner layer of Fe_3O_4 under an outer layer of $\gamma\text{-}Fe_2O_3$.[16] The inner Fe_3O_4 is thought to have a highly defective structure and is therefore conductive. The passive film usually has properties of an n-type semiconductor with excess negative charge carriers. Passive films on alloys may be enriched in some alloying elements, especially chromium, which forms a very stable oxide in preference to iron. Other pure metals and alloys have different properties, and all often vary as a function of applied potential and alloy composition.[16]

The foundation for the oxide film theory of passivity is the early work of Evans,[17] who isolated a bulk oxide film from the surface of air- or chromate-exposed passive iron. The two have the same apparent structure and composition, but the isolated chromate film may have been produced by decomposition of the actual passive film during the isolation process.[4]

Uhlig and co-workers[18] have stressed the importance of chemisorbed oxygen in establishing passivity. Adsorbed films are suggested to act not as a barrier to anodic dissolution but as a kinetic limitation reducing the exchange current density, i_o, for the dissolution reaction. Chemisoption of oxygen is favored by the presence of uncoupled d-electrons in the transition metals. In Fe–Cr alloys, chromium acts as an acceptor for uncoupled d-electrons from iron. When alloyed Cr is less than 12%, uncoupled d-electron vacancies in Cr are filled from the excess Fe and the alloys act like unalloyed iron, which is nonpassive in deaerated dilute acid solutions. Above 12% Cr the alloys are passive in such solutions, presumably because uncoupled d-electrons are available to foster adsorption.[19] During film thickening, metal cations are assumed to migrate into the film from the underlying metal, as well as protons (H^+ ions) from the solution.

Either H^+ as H_3O^+ or OH^- may fill a more direct role in passive film formation, however. Electrochemical measurements on steel in anhydrous ammonia, analogous to an aqueous electrolyte but free of water, showed no evidence of active–passive behavior, although an oxidation–reduction half-cell electrode potential could be measured readily for oxygen.[20] Active–passive behavior is absent in other nonaqueous electrolytes, as well.[21]

The overall anodic reaction (Section 2.2.5),

$$Fe + 2H_2O \rightarrow Fe(OH)_2 + 2H^+ + 2e^-, \tag{1}$$

is approximate and assumes a certain stable phase, $Fe(OH)_2$, which may or may not be attained eventually in the film structure. Lorenz et al.[22] assumed a series of anodic reactions involving adsorbed complexes:

$$Fe + H_2O \rightarrow (FeOH)_{ad} + H^+ + e^-$$

$$(FeOH)_{ad} \rightarrow (FeO)_{ad} + H^+ + e^- \tag{4}$$

$$(FeO)_{ad} + H_2O \rightarrow (FeOO)_{ad} + H^+ + e^-.$$

The "ad" subscript denotes an adsorbed complex. Adsorption seems likely in the early stages of film formation and growth. Whether the inhibition of the corrosion rate in the passive potential region is due to an oxide barrier film or a kinetic limitation by adsorption is moot at present. An oxide film could also reduce i_o, and an adsorbed film could conceivably form a barrier, as well. Experimental techniques to resolve the issue are not readily at hand.

4.5.2 Growth Kinetics

Figure 4.13 shows that the anodic polarization curve is highly time dependent not only in the passive, but in the active region, as well. Thus, at least in some instances, passive film formation is a dynamic process competing with active metal dissolution below the primary passive potential, E_{pp}. Once into the stable passive potential region above E_{pp}, the passive film begins to grow in thickness, as evidenced by the decreasing potentiostatic passive current density, i_{pass}, for lower polarization rate in Figure 4.13. Thickening of the passive film is evidenced by decrease in i_{pass} with time, t, in Figure 4.21. It appears that a steady state is being approached asymptotically. However, the passive current density still falls with time, but at a continually decreasing rate,[23] as shown in Figure 4.22, and no true steady state is ever attained.

Linearity in Figure 4.22 of log i_{pass} versus log time with a slope of -1 indicates that the passive corrosion rate is inversely proportional to time; that is, $i_{pass} = k/t$ where k is a proportionality constant. The passive current density, i_{pass}, must be proportional to the rate of passive film formation, and the rate of film thickening, dx/dt:[24]

$$i_{pass} = k' \frac{dx}{dt}$$

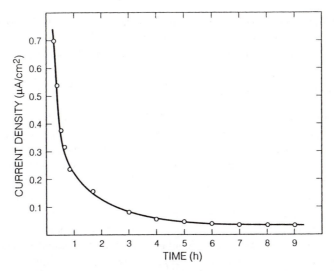

FIGURE 4.21   Decay of passive corrosion rate measured by potentiostatic current. *(Unpublished data of Jones and Greene.)*

where k' is again a proportionality constant. Integration produces the logarithmic film growth law of the form

$$x = A + B \log t, \tag{5}$$

where x is film thickness and A and B are constants.[24] The derivation of a logarithmic rate law has traditionally assumed a constant or increasing potential field across

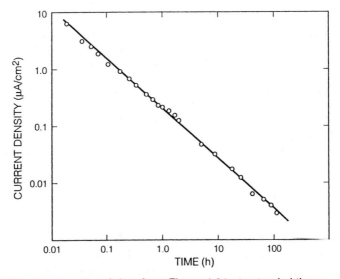

FIGURE 4.22   Log-log plot of data from Figure 4.21 at extended times. *(Unpublished data of Jones and Greene.)*

the film, which progressively decreases the film growth rate.[25] The logarithmic law has also been derived, assuming continuous adsorption,[26] with each adsorbed entity postulated to decrease the probability or area available for subsequent adsorption.

4.5.3 Electron Transfer Reactions

Although limited ionic mobility in passive films results in low anodic corrosion rate or current density, i_{pass}, electron conductivity has been found to be relatively high.[27] Easy electron transfer allows half-cell reactions to control surface potentials at noble values where the passive films are stable. The conduction behavior is explained by assuming semiconducting properties for the passive film. Half-cell reaction kinetics have been used to determine electron transfer across passive films. Asymmetry, in which electron transfer is easier in the cathodic than in the anodic direction, is common for many passive electrodes and is characteristic of n-type semiconductors. Tunneling mechanisms are often required to explain details. Schultze[27] discusses the various semiconducting mechanisms that may operate during electron transfer in passive films.

References

1. M. Pourbaix, *Atlas of Electrochemical Equilibria in Aqueous Solutions*, NACE, Houston, p. 263, 312, 1974.
2. M. G. Fontana, Corrosion Engineering, 3rd ed., McGraw-Hill, New York, p. 469, 1986.
3. T. P. Hoar, *Corrosion*, Vol. 1, 2nd edition, L. L. Shreir, ed., Newnes-Butterworths, Sevenoaks, Kent, England, p. 1:114, 1976.
4. H. H. Uhlig, *Passivity of Metals*, R. P. Frankenthal and J. Kruger, eds., Electrochemical Society, Princeton, NJ, p. 1, 1978.
5. ASTM Standard Reference Test Method for Making Potentiostatic and Potentiodynamic Anodic Polarization Curves, Standard Method G5-94, *Annual Book of ASTM Standards*, Vol. 3.02, p. 63, ASTM, Philadelphia, 1994.
6. R. F. Steigerwald and N. D. Greene, *J. Electrochem. Soc.*, Vol. 109, p. 1026, 1962.
7. N. D. Greene and R. B. Leonard, *Electrochim. Acta*, Vol. 9, p. 45, 1964.
8. F. Mansfeld, *Corrosion*, Vol. 44, p. 856, 1988.
9. D. A. Jones, *Ind. Eng. Prod. Res. Dev.*, Vol. 11, p. 12, 1972.
10. N. D. Greene and G. Judd, *Corrosion*, Vol. 21, p.15, 1965.
11. C. E. Locke, *Metals Handbook*, Vol. 13: *Corrosion*, 9th ed., ASM-International, p. 463, 1987.
12. O. L. Riggs and C. E. Locke, *Anodic Protection: Theory and Practice in the Prevention of Corrosion*, Plenum, New York, 1981.
13. V. Kozub and V. Novitskiy, *Proc. 9th Int. Cong. Metallic Corrosion*, Vol. 1, National Research Council of Canada, Toronto, p. 307, 1984.
14. C. Edeleanu and J. G. Gibson, *Chem. Ind.*, p. 301, Mar. 11, 1961.
15. W. D. France and N. D. Greene, *Corrosion*, Vol. 24, p. 247, 1968.
16. N. Sato, *Passivity of Metals*, R. P. Frankenthal and J. Kruger, eds., Electrochemical Society, Princeton, NJ, p. 29, 1978.
17. U. R. Evans, *J. Chem. Soc. (London)*, p. 1020, 1927.

18. H. H. Uhlig, *Passivity and Its Breakdown on Iron and Iron Base Alloys*, USA-Japan Seminar, NACE, Houston, p. 21, 1976.
19. H. H. Uhlig, *Z. Elektrochem.*, Vol. 62, p. 700, 1958.
20. D. A. Jones, C. D. Kim, and B. E. Wilde, *Corrosion*, Vol. 33, p. 50, 1977.
21. K. Schwabe, S. Herman, and W. Oelssner, *Passivity of Metals*, R. P. Frankenthal and J. Kruger, eds., Electrochemical Society, Princeton, NJ, p. 413, 1978.
22. A. A. Elmiligy, D. Geana, and W. J. Lorenz, *Electrochim. Acta*, Vol. 20, p. 273, 1975.
23. D. A. Jones and N. D. Greene, *Corrosion*, Vol. 22, p. 198, (1966).
24. M. Stern, *J. Electrochem. Soc.*, Vol. 106, p. 376, 1959.
25. M. G. Fontana, *Corrosion Engineering*, 3rd ed., McGraw-Hill, New York, p. 474, 1986.
26. D. A. Jones and R. E. Westerman, *Corrosion*, Vol. 21, p. 295, 1965.
27. J. W. Schultze, *Passivity of Metals*, R. P. Frankenthal and J. Kruger, eds., Electrochemical Society., Princeton, NJ, p. 82, 1978.

Exercises

4-1. (a) Plot schematically the polarization curve for anodic dissolution for the metal M that has the following electrochemical parameters: $E_{corr} = -0.500$ Vs vs. SCE, $i_{corr} = 10^{-4}$ A/cm^2, $E_{pp} = -0.400$ V vs. SCE, $\beta_a = +0.05$ V, $i_{pass} = 10^{-5}$ A/cm^2, $E_{tr} = +1.000$ V. (b) From the plot in (a) determine the critical current density for passivation, i_{crit}.

4-2. (a) Assuming that the measured exchange current density for dissolution of M in Problem 4-1 is $i_{o,M/M^+} = 10^{-9}$ A/cm^2, determine the reversible potential for dissolution, e_{M/M^+}, in this solution.
(b) Calculate the activity of M^+ in the corrosive solution, assuming that the anodic dissolution reaction is $M \rightarrow M^+ + e^-$, with $e^\circ_{M/M^+} = -0.250$ V vs. SHE, and that no other dissolved oxidizing species are present.

4-3. Assuming unit activity of H^+ ions in the corrosive electrolyte, determine the exchange current density for the reduction of H^+ on the surface of metal M in Problem 4-1. Estimate a value of β_c from Table 3.2 for use in your determination.

4-4. Plot schematically the effect of solution velocity on corrosion potential from Figure 4.7.

4-5. (a) Approximate the minimum concentration of Fe^{3+} oxidizer needed to passivate any of the Type 430 stainless steels of Figure 4.15 in $1N$ sulfuric acid, assuming that the acid is contaminated to a level of 10^{-3} moles/liter with corrosion product Fe^{2+}. Assume that the exchange current density on these steels is $i_{o,Fe^{3+}/Fe^{2+}} = 0.1$ μA/cm^2 independent of concentrations of Fe^{3+} and Fe^{2+} and that the reduction reaction is under activation control with $\beta_c = -0.100$ V.
(b) One might conclude from (a) that extremely low concentrations of Fe^{3+} and Fe^{2+} will passivate stainless steel. Will this always be true? Explain why or why not.

4-6. (a) Using the polarization curve for anodic dissolution in Problem 4-1, schematically plot the experimental potentiostatic anodic polarization curve, beginning at E_{corr}. **(b)** Do the same for the galvanostatic anodic polarization curve.

4-7. Using the potentiostatic curve (3) of Figure 4.14, predict the galvanostatic polarization curve ascending from E_{corr} followed by the similar curve descending from the maximum (most noble) potential shown for this sample.

4-8. For case 3 in Figure 4.5 (borderline passivity), show the appearance of the potentiostatic polarization curve ascending to higher (more noble) potentials from the corrosion potential. Carefully note changes in direction of current from anodic to cathodic.

4-9. Given active-passive alloys A and B having the following electro-chemical parameters:

	E_{corr}, V	i_{corr}, A	β_a, V	E_{pp}, V	i_{pass}, A	E_{tr}, A
Alloy A	−0.400	1×10^{-6}	+0.1	0	1×10^{-5}	+0.7
Alloy B	−0.200	7×10^{-7}	+0.1	+0.3	1×10^{-6}	+1.2

a. Which will be the more corrosion resistant in reducing conditions (active state)? Why?

b. Which will be the more corrosion resistant in the passive state? Why?

c. Which is more easily passivated by dissolved oxidizers? Why?

d. Which is more corrosion resistant in strongly oxidizing solutions? Why?

e. Which would be the more easily protected by anodic protection? Why?

4-10. Anodic protection is proposed for corrosion control in an existing acid storage tank. Corrosion on surfaces exposed to condensate in the vapor above the liquid is a particular problem. Will anodic protection be effective to mitigate this type of corrosion?

4-11. You have a titanium tank in your plant which is used to store 10% muriatic acid (commercial low cost hydrochloric acid), which contains as supplied 0.5% ferric chloride as an impurity. Careful measurements have shown that the tank is resistant to this media because it spontaneously passivates. Explain the following using appropriate polarization diagrams if necessary. **(a)** Your supplier offers you some 10% muriatic acid containing 2% ferric chloride at a big discount because other consumers reject it on the basis of its deep orange color. If ferric salts are tolerable in your process, will this acid corrode your tank more rapidly than the regular muriatic acid? **(b)** An excess of pure hydrochloric acid with no detectable ferric salts becomes available from another part of the plant. Can you take this acid into your storage tank without any risk of increasing corrosion rate?

5

Polarization Methods to Measure Corrosion Rate

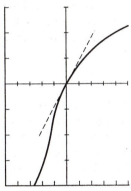

Two methods are available for measurement of corrosion by electrochemical polarization: Tafel extrapolation and polarization resistance. The former was introduced in Section 3.4.1 and is discussed in this chapter in somewhat more detail. The polarization resistance method deserves primary attention, since it has proven more useful in both research and engineering applications. Later in the chapter, more recent variations in techniques to obtain polarization resistance, including electrochemical impedance spectroscopy, are described.

Polarization methods to measure corrosion rates have inherent advantages. Usually only a few minutes are required to determine corrosion rate by polarization resistance, whereas conventional weight loss measurements (Section 1.6) require several days or more. Thus, a rapid, semicontinuous measurement of corrosion may be obtained which is very useful for kinetic studies or for corrosion monitoring in process plants. The methods are highly sensitive, and accelerating factors, such as elevated temperature, to increase rates in the laboratory are usually unnecessary. Polarization measurements are nondestructive and may be repeated numerous times to measure consecutive corrosion rates on the same electrode. Any of the polarization cells described in Section 3.5 are suitable for laboratory corrosion rate measurements. Corrosion rates can be monitored in process streams and industrial equipment with commercially available two- and three-electrode probes, described later in Section 5.4.

5.1 Tafel Extrapolation

In Section 3.4.1 the Tafel extrapolation method was introduced to illustrate the application of mixed potential theory to aqueous corrosion. Valid cathodic Tafel extrapolation requires a single reduction process that is activation controlled. This condition is most often found in deaerated strong acid solutions in which the reduction reaction is

$$2H^+ + 2e^- \rightarrow H_2. \tag{1}$$

Figure 5.1a[1] shows typical cathodic polarization data for iron in an acid solution, with extrapolation of the cathodic Tafel slope back to the corrosion potential, E_{corr}. The intersection gives the corrosion rate or corrosion current density, i_{corr}, as was demonstrated from mixed potential theory in Section 3.3.1.

Good Tafel behavior has also been observed in deaerated neutral electrolytes. Figure 5.1b[2] shows cathodic Tafel behavior for uncoated carbon steel and zinc and Al–Zn coatings on steel in deaerated sodium sulfate solution. In the absence of all other reduction reactions, cathodic polarization will be controlled by

$$2H_2O + 2e^- \rightarrow H_2 + 2OH^-. \tag{2}$$

This reaction is equivalent to reaction (1); adding $2OH^-$ to both sides of (1) produces (2). However, the kinetics of the two reactions are vastly different. The rate of (1) is relatively high, due to the ready availability of H^+, controlled by pH. The rate of (2) is limited by dissociation of the water molecule, which is quite low.

At least one decade of linearity on the semilog plots of Figure 5.1 is desirable for maximum accuracy to determine i_{corr} by Tafel extrapolation. This may be difficult to achieve in dilute solutions, where concentration polarization and ohmic resistance effects (Sections 3.2.2 and 3.5.4, respectively) are likely at higher current densities. However, adequate approximations are sometimes possible with limited Tafel behavior, and i_{corr} is always equal to the limiting current i_L for corroding systems with diffusion control.

Polarization potentials are generally time dependent, and considerable care is necessary to achieve a steady-state polarization curve truly representative of the pertinent corrosion reactions. In this respect, cathodic polarization is generally more rapid and reversible than anodic polarization, which shows considerable time dependence and hysteresis.[3] The anodic polarization curve of Figure 5.1a does not show well-defined Tafel behavior and cannot be used for determination of corrosion rate. Anodic Tafel constants may be obtained from cathodic data as described in Section 3.4.2. Galvanostatic (constant-current) methods are usually recommended for obtaining the cathodic polarization curves in Figure 5.1 (see Section 3.5.1).

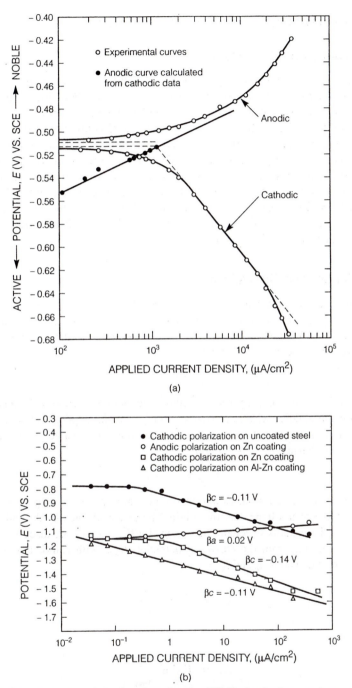

FIGURE 5.1 Cathodic polarization data showing Tafel behavior in room temperature deaerated (a) $1N\,H_2SO_4$ for carbon steel and (b) sodium sulfate solutions for carbon steel and carbon steel coated with zinc and zinc-aluminum. *(From R. Bandy and D. A. Jones, Corrosion, Vol. 32, p. 126, 1976; and D. A. Jones and N. R. Nair, Corrosion, Vol. 41, p. 357, 1985. Reprinted by permission, National Association of Corrosion Engineers.)*

5.2 Polarization Resistance Method

5.2.1 History and Description

Early researchers[4,5] observed experimentally that the degree of polarization at a given applied current was greater for a lower corrosion rate. Furthermore, an apparent linearity was observed at the origin of the polarization curve for overvoltages up to a few millivolts. Thus, the slope of this linear curve is inversely proportional to the corrosion rate. As a result, the method was initially, and sometimes still is, referred to as the linear polarization method. However, it was observed subsequently[6] that the linearity may be very restricted in range. Polarization resistance, defined as the slope of the polarization curve at the origin, is independent of the degree of linearity. Thus, the *polarization resistance method* has become the widely accepted name in recent years. Stern[6,7] popularized the method by correlating earlier theoretical[8] and experimental[4,5,9] observations.

5.2.2 Theory and Derivations

The polarization resistance method can be derived from the mixed potential theory described in Chapter 3. Considering the polarization curves for the activation controlled anodic and cathodic partial processes in Figure 5.2a, the expected experimental polarization curves can be derived from

$$i_{app,c} = i_c - i_a, \tag{3}$$

where i_c is the current density for the cathodic reduction reaction, i_a is the current density for the anodic oxidation reaction, and $i_{app,c}$ is the applied cathodic current density, all at the same potential, E. In a similar way (see Chapter 3), the applied anodic current density, $i_{app,a}$, is given by

$$i_{app,a} = i_a - i_c. \tag{4}$$

The graphically derived polarization curves in Figure 5.2 show apparently linear behavior at low overvoltage near the origin. The extent of linearity depends on the values of Tafel constants selected. The reasons for the linearity are discussed below.

When polarizing from the corrosion potential, E_{corr}, with cathodic and anodic applied current densities, i_c and i_a,

$$\varepsilon_c = \beta_c \log \frac{i_c}{i_{corr}}, \tag{5}$$

and

$$\varepsilon_a = \beta_a \log \frac{i_a}{i_{corr}}, \tag{6}$$

where ε_c and ε_a are the cathodic and anodic overvoltages, β_c and β_a are cathodic and anodic Tafel constants (activation polarization), respectively, and i_{corr} is the corrosion rate as current density. ε_c and ε_a are defined as overvoltages because they represent

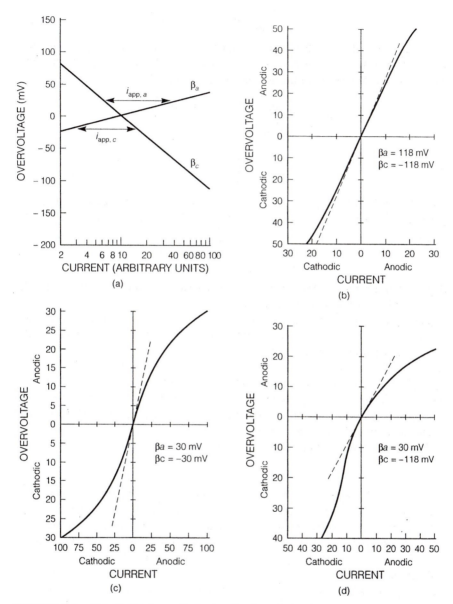

FIGURE 5.2   Hypothetical anodic and cathodic polarization curves (a), used to graphically derive simulated experimental polarization curves plotted on linear coordinates using indicated Tafel constants, β_a and β_c, in (b), (c), and (d).

potential changes from the steady-state corrosion potential, E_{corr}; $\varepsilon_{c/a} = E_{c/a} - E_{corr}$. They are to be distinguished from overvoltages previously designated by η (Section 3.2.1), which are similar potential displacements from half-cell potentials, e.

Converting (5) and (6) to exponential form and substituting into Equation (3), we obtain

$$i_{app,c} = i_{corr}(10^{-\varepsilon/\beta_c} - 10^{\varepsilon/\beta_a}).$$ (7)

Thus $i_{app,c}$ is the difference between two exponential functions which approaches linearity with overvoltage ε, as $\varepsilon \rightarrow 0$. Rearrangement of Equation (7) and differentiation gives an expression for the polarization resistance, R_p.[6,10,11]

$$R_p = \left[\frac{d\varepsilon}{di_{app}}\right]_{\varepsilon \rightarrow 0} = \left[\frac{\Delta\varepsilon}{\Delta i_{app}}\right]_{\varepsilon \rightarrow 0}$$

$$= \frac{\beta_a\beta_c}{2.3 i_{corr}(\beta_a + \beta_c)} = \frac{B}{i_{corr}}$$ (8)

Thus, the slope, $(d\varepsilon/di_{app})_{\varepsilon \rightarrow 0}$, at the origin of the polarization curve, defined as polarization resistance, R_p, is inversely proportional to the corrosion rate, where

$$B = \frac{\beta_a\beta_c}{2.3(\beta_a + \beta_c)}$$ (9)

is the proportionality constant. Note that the derivation results in β_a and β_c having positive values in Equations (8) and (9).

The linearity near the origin is especially pronounced when relatively high values are selected for the Tafel constants, $\beta_a = \beta_c = 0.118$ V in Figure 5.2b. However, departures from linearity appear at much lower overvoltages when absolute values of the Tafel constants are lower, as assumed in Figure 5.2c, where $\beta_a = \beta_c = 30$ mV. Unequal Tafel constants further compress the range of apparent linearity and also result in asymmetry about the origin, as seen in Figure 5.2d for $\beta_a = 30$ mV and $\beta_c = 118$ mV. The consequences of apparent nonlinearity at the origin are discussed further in Section 5.5.3.

Taking logarithms of Equation (8), one obtains

$$\log R_p = \log B - \log i_{corr}$$ (10)

Thus, a plot of $\log R_p$ versus $\log i_{corr}$ is linear with a slope of -1 and an intercept governed by the proportionality constant, B. Stern and Weisert[12] confirmed this relationship by plotting experimental corrosion rates and the corresponding polarization resistances over six orders of magnitude, as shown in Figure 5.3. The approximate linearity in this plot over such a wide range of data contributed substantially to general acceptance of the polarization resistance method. The inverse proportionality between polarization resistance and corrosion rate has been verified many times subsequently.

The data of Figure 5.3 form a scatter band enclosed by the straight lines of Equation (10) defined by experimentally extreme values of the Tafel constants. When approximate values of $\beta_a = \beta_c = 0.1$ volt are used, the corresponding straight proportionality line in Figure 5.3 falls near the middle of the scatter band. Thus, even when the actual values of the Tafel constants are unknown, inspection of Figure 5.3 shows that approximate values of 0.1 volt give a constant error in calculated corrosion rate of only a factor of two maximum.[12] Such an error is often

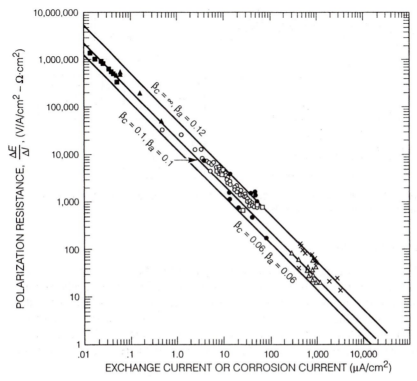

FIGURE 5.3   Experimental correlation between corrosion rate, i_{corr}, and polarization resistance. *(From H. Stern and E. D. Weisert, Proc. ASTM, Vol. 32, p. 1280, 1959. Reprinted by permission American Society for Testing and Materials.)*

within experimental scatter in plant corrosion measurements. In laboratory kinetic studies, relative corrosion rates are usually more significant than the absolute values, and any constant error in all measured corrosion rate values is of lesser importance.

A detailed review[11] of the theory and application of polarization resistance is recommended for present or potential users of the method.

5.3 Instrumental Methods for Polarization Resistance

Earliest instrumental methods to measure polarization resistance were galvanostatic or constant current (Section 3.5.1). Galvanostatic instrumentation is experimentally simpler and still favored for the commercial polarization resistance monitoring probes described in Section 5.4. Potentiostatic or potentiodynamic methods have been favored for laboratory use. Galvanostatic and potentiodynamic techniques are compared in the following sections.

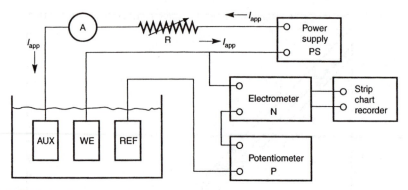

FIGURE 5.4 Galvanostatic, constant-current circuitry for determination of corrosion rates by polarization methods.

5.3.1 Galvanostatic

Simple constant-current, galvanostatic circuitry is repeated from Chapter 3 in Figure 5.4.[13] In the galvanostatic method, current steps are applied consecutively, and the resulting changes are monitored on the strip-chart recorder of Figure 5.4. Recorder results for zinc in dilute NaCl are shown in Figure 5.5.[13] The polarization resistance curve in Figure 5.6 is a plot of overvoltage, taken from Figure 5.5, plotted as a function of applied current at each current step. The polarization resistance is the slope

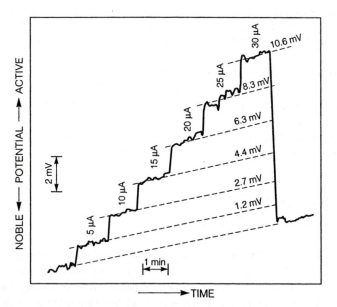

FIGURE 5.5 Potential-time strip-chart trace during polarization resistance measurement on zinc coating on steel in 0.1 N NaCl at room temperature. *(From D. A. Jones, Corrosion, Vol. 39, p. 444, 1983. Reprinted by permission, National Association of Corrosion Engineers.)*

at the origin of the curve. The definite nonlinearity of the curve for zinc is not unusual for systems in which the anodic and cathodic Tafel constants, β_a and β_c, are largely different in absolute value from one another, as shown in Figure 5.2d and discussed further in Section 5.5.2.

The apparent advantage of the galvanostatic method is that a small constant current applied to a freely corroding electrode superimposes a potential change (overvoltage) on the prevailing time trend of the unpolarized corrosion potential. After the applied current has been removed, the potential perturbation (overvoltage) decays rapidly back to a value consistent with the background time drift of the unpolarized corrosion potential. Potential drift, whether in an active (Figure 5.5) or noble direction, is apparently unaffected by the polarization resistance procedure. Galvanostatic polarization resistance measurements have been conducted successfully for crevice and pitting corrosion, erosion corrosion, and uniform corrosion over a wide range of corrosion rates.[13]

For an electrode corroding at an unknown rate, no prior knowledge is available as to the magnitude of the current steps which must be applied during a galvanostatic corrosion rate determination.[11] Thus, a preliminary trial-and-error procedure[13] is required, in which i_{app} is increased from zero in order-of-magnitude increments and finally adjusted until an initial current I_1 is determined which produces an overvoltage of 1 to 2 mV. Subsequent current steps, $I_n = nI_1$, are then applied consecutively and corresponding overvoltages recorded (Figure 5.6) until a total of 5 to 10 mV overvoltage has been achieved.

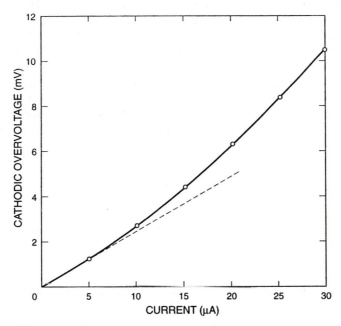

FIGURE 5.6   Polarization resistance curve derived from Figure 5.5. *(From D. A. Jones,* Corrosion, *Vol. 39, p. 444, 1983. Reprinted by permission, National Association of Corrosion Engineers.)*

Valid polarization measurements require that polarized potentials be measured at steady state. The described strip-chart recording technique allows continuous assessment of time needed to reach steady state at each galvanostatic current step. For example, considerable time may be required to reach steady state in the presence of a surface film at a relatively low corrosion rate. Figure 5.7[13,14] shows the time drift of potential during a polarization resistance measurement taken for stainless steel in oxygenated 1 N sulfuric acid with a passive film on the surface. About 4 minutes were required at each current step to obtain steady state and nearly 20 minutes for the entire polarization resistance procedure. Thicker passive film results in even longer measurement times. Methods to shorten measurement time in this situation are described in Section 5.6.2.

5.3.2 Potentiodynamic

Potentiostatic and potentiodynamic methods have some fairly obvious advantages. If a controlled overvoltage is impressed on a corroding electrode, the necessary i_{app} is then directly proportional to i_{corr}, according to

$$R_p = \frac{\Delta\varepsilon}{\Delta i_{app}} = \frac{B}{i_{corr}}. \tag{8}$$

Thus, the corrosion rate can be obtained directly from readings of the applied polarization current. If a potentiodynamic scan is applied, the necessary current, i_{app}, follows the controlled overvoltage and the polarization resistance curve can be plotted automatically. Figure 5.8[15] shows the polarization resistance curve obtained potentiodynamically for Type 430 stainless steel in deaerated 1 N sulfuric acid. The development of precise, stable potentiostats that can maintain potential within < 0.1 mV have made such techniques possible in recent years.

Potentiodynamic techniques require that the corrosion potential be stable and unchanging during the measurement. Otherwise, the applied overvoltage and current vary by an unknown amount as the background corrosion potential changes during potentiodynamic (or potentiostatic) measurement. Furthermore, the potentiodynam-

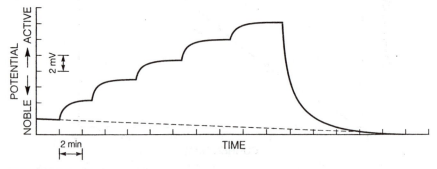

FIGURE 5.7 Potential-time strip-chart trace during polarization resistance measurement on stainless steel in oxygenated 1N sulfuric acid at room temperature. *(From D. A. Jones, Corrosion, Vol. 39, p. 444, 1983. Reprinted by permission, National Association of Corrosion Engineers.)*

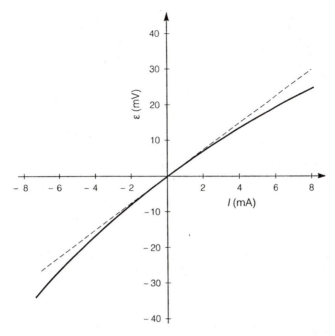

FIGURE 5.8   Polarization resistance curve generated potentiodynamically for Type 430 Stainless Steel in deaerated 1*N* sulfuric acid. *(Adapted from Mansfeld.[15])*

ic scan must be run slow enough to ensure steady-state behavior. Thus, the potentiodynamic techniques, which have been standardized for laboratory use,[16] have serious deficiencies for kinetic studies and in service conditions where corrosion potential drift and fluctuation are more typical.[13]

In round-robin tests conducted on behalf of ASTM,[15] it was found that potentiodynamic polarization resistance results were more reproducible than galvanostatic when comparing data from different laboratories on Type 430 stainless steel in deaerated 1 *N* H$_2$SO$_4$. This is a well-behaved system that comes to steady state rapidly with a relatively high corrosion rate. The scatter in potentiodynamic data was considerably higher for carbon steel in deaerated 1 *N* Na$_2$SO$_4$, where the corrosion rate was much lower. Still, the potentiodynamic method has the advantage that the procedure is instrumentally controlled and therefore more uniform from laboratory to laboratory, whereas galvanostatic procedures are more subject to variation in technique between different operators.

5.4 Commercial Corrosion Monitoring Probes

The remote, rapid, semicontinuous, nondestructive nature of polarization methods led to development of corrosion monitoring probes, used widely in chemical-process and cooling-water streams. Both two- and three-electrode instrumental configurations (Figure 5.9[1]) have been used successfully. Both use galvanostatic circuitry similar to that shown in Figure 5.4.

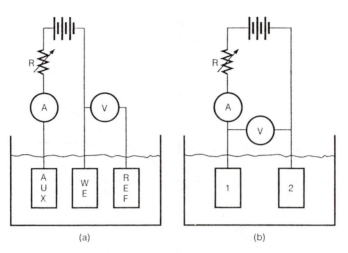

FIGURE 5.9 Schematic instrumental setups for (a) three-electrode and (b) two-electrode corrosion monitoring probes. *(From R. Bandy and D. A. Jones,* Corrosion, *Vol. 32, p. 126, 1976. Reprinted by permission, National Association of Corrosion Engineers.)*

A commercially available three-electrode probe is illustrated in Figure 5.10. The two-electrode version is of similar design with one less electrode. The probes are fitted with pipe threads or O-ring seals for convenient insertion into a process stream to

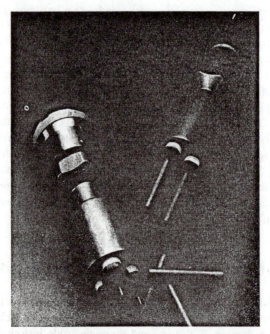

FIGURE 5.10 Commercial three-electrode polarization probe for *in situ* monitoring of corrosion rates in operating equipment. *(Photograph by courtesy of Petrolite Corp., Houston, TX)*

monitor corrosion under operating conditions. Two-electrode probes were widely adopted in industry after early aggressive marketing. However, three-electrode instruments have seen increased use in recent years. The three-electrode probes have found greatest use in chemical-process streams of high conductivity. On the other hand, two-electrode probes have been used in cooling-water applications of lower conductivity with ohmic-resistance compensation techniques described in Section 5.4.2.

5.4.1 Three-Electrode Procedures

Three-electrode probes (manufactured by Petrolite Corp., Houston, Texas) employ three identical corroding electrodes, the unpolarized corrosion potential of one serving as the reference electrode potential for polarization of the other two. The experimental arrangement is very similar to that shown in Figure 5.4, with the process stream as the electrolyte and the process equipment as the cell vessel. The corrosion rate of only the test electrode (WE) is measured, however, and the other two could in principle be a different alloy or composition. The polarization curve is assumed linear up to 10-mV overvoltage, and the current is adjusted automatically to obtain a 10-mV change in the potential of WE in Figure 5.9a. The current required to obtain the 10-mV change is proportional to the corrosion rate in

$$R_p = \frac{\Delta\varepsilon}{\Delta i_{app}} = \frac{B}{i_{corr}}, \tag{8}$$

and the current meter may be calibrated to read corrosion rate directly. Instrument packages are available to conduct the measurements and record the results automatically at preset time intervals during routine plant operations.

5.4.2 Two-Electrode Procedures

The two-electrode probe (manufactured by Rohrback Cosasco Systems, Santa Fe Springs, California) uses two identical corroding electrodes and eliminates the third reference electrode.[17] The two-electrode method has certain advantages. The potential between the two electrodes is measured directly, eliminating the need for a third reference electrode. The polarization procedure described below causes a cancellation of errors due to nonlinearity, making fine control of potential and current less critical. Instrumental techniques are available to compensate resistance interferences, as discussed in Section 5.5.1.

The instrumentation is galvanostatic or constant current, as shown schematically in Figure 5.9b. The following operational cycle is conducted for each two-electrode determination. The significance of each step will be explained subsequently.

1. Measure and record the short-circuit current between the two electrodes using a galvanic zero-resistance ammeter circuit.
2. Measure and record the difference between corrosion potentials of the two electrodes, $E_{diff} = E_{corr,1} - E_{corr,2}$.
3. Apply and record the current density, i_f, needed to superimpose a +20-mV potential change on E_{diff}.

4. Apply and record a reverse current density, i_r, to superimpose a -20-mV potential change on E_{diff}.
5. Apply a high-frequency 5-mV signal to the two electrodes to measure the solution resistance, R_Ω, between them.

In routine plant operation, this cycle is usually conducted under automatic computer control at preset intervals, although intermittent manual operation is also possible. The cycle time must be adjusted to give the minimum measured values of i_f and i_r and, thus, the minimum corrosion rate; 5 minutes total time with 1 minute for each of the five steps is typical. To calculate corrosion rate, the average i_{avg} of i_f and i_r is substituted for Δi_{app} in

$$ R_p \;=\; \frac{\Delta \varepsilon}{\Delta i_{app}} \;=\; \frac{B}{i_{corr}}, \tag{8} $$

using an average of 10 mV for $\Delta \varepsilon$ and a measured or estimated value for B. Unwanted contributions from $\varepsilon_\Omega = i_{avg} R_\Omega$ between the electrodes must be subtracted from $\Delta \varepsilon$ (see Section 5.5.1), using the value of R_Ω measured in Step 5 above. Again, the computations are usually conducted automatically by a microcomputer incorporated into the instrumentation.

A galvanic current measured in Step 1 with a zero-resistance ammeter circuit (Section 6.3.2) is related to preferential corrosion on one of the two electrodes, possibly due to pitting or incomplete inhibitor coverage. It may also suggest nonuniform metallurgical structure between the two electrodes or a malfunction of the probe (e.g., a loosened electrode).

When a polarizing current is applied between the two electrodes, one is polarized cathodically and the other anodically. In the usual case when $|\beta_a| \neq |\beta_c|$ (Figure 5.2d), one of the two electrodes polarizes more readily and takes a larger share of the 20 mV change, E_{diff}. Interestingly enough, the resulting asymmetry of anodic and cathodic polarization improves the accuracy somewhat. The deviation from linearity is negative for anodic polarization while it is positive for cathodic polarization, as shown in Figure 5.2d. These deviations partially cancel one another in the value of 10 mV taken for ΔE in Equation (8), and there is less error due to deviation from linearity.[18] Further discussion appears in Section 5.5.3.

5.5 Errors in Polarization Resistance Measurements

Errors in polarization resistance measurements are important in the accuracy of corrosion monitoring probes and laboratory studies. A significant experimental error is the inclusion of ohmic resistance interferences in the potential measurements. Other errors come about by assuming that the polarization curve is linear up to a specified overvoltage, usually 10 mV, as described for corrosion monitoring probes in Section 5.4. A few of the more important factors affecting accuracy of polarization resistance measurements are discussed below. Errors attributable to transient, non-steady-state, polarized potentials and unpolarized corrosion potentials, E_{corr}, are discussed in Section 5.3. The interested reader is referred to Callow et al.,[19] who have reviewed the considerable theoretical and experimental work devoted to the subject.

5.5.1 Ohmic Electrolyte Resistance

As described in Section 3.2.3, total overvoltage is the sum of activation, ε_a, concentration, ε_c, and ohmic resistance, ε_Ω, overvoltages. In conventional polarization measurements, ε_Ω does not become significant until relatively high currents are reached. Thus, ohmic resistance effects can often be neglected in polarization resistance measurements because the ohmic resistance of the solution electrolyte is low compared to the polarization resistance. That is, the total resistance, R, measured by polarization resistance procedures is the sum of the polarization resistance, R_p, and the solution resistance, R_Ω,

$$R = R_p + R_\Omega. \tag{11}$$

Normally, R is identical to R_p because $R_p \gg R_\Omega$. It is obvious from Equation (11), however, that a significant error will be present when R_Ω becomes a significant fraction of R_p. The error is non-conservative; a high R_Ω results in an apparent corrosion rate which is lower than the real one. Therefore, accurate polarization measurements in low-conductivity solutions require that R_Ω be either reduced by supporting electrolyte additions in the laboratory or compensated by any of the instrumental or procedural techniques described in Section 3.5.4. Fortunately, low-conductivity aqueous electrolytes of significant R_Ω are usually associated with low corrosion rates of high R_p, which are less affected by R_Ω.

In some polarization resistance monitoring probes, the third electrode used as reference is positioned in close proximity to the working electrode to reduce R_Ω. A high-frequency, low-amplitude signal measures R_Ω, which is used to reduce the R measured by polarization in Equation (11) to the desired value of R_p.

5.5.2 Uncertain Tafel Constants

To obtain i_{corr} accurately from

$$R_p = \frac{\Delta \varepsilon}{\Delta i_{app}} = \frac{\beta_a \beta_c}{2.3\, i_{corr}(\beta_a + \beta_c)}, \tag{8}$$

one must have reasonably accurate values for β_a and β_c. However, numerator and denominator of Equation (8) contain both β_a and β_c, and as a consequence, i_{corr} is not very sensitive to the values selected for β_a and β_c. Stern and Weisert[12] suggested that experimental β_a values range from 0.06 V to about 0.12 V, and β_c values range from 0.06 V to infinity, the latter corresponding to diffusion control by a dissolved oxidizer. Extreme values correspond to $\beta_a = 0.06$ V, $\beta_c = 0.06$ V; and $\beta_a = 0.12$ V, $\beta_c = $ infinity. These values substituted into Equation (9) define two lines in Figure 5.3, which differ only in their intercepts on the R_p axis, in agreement with Equation (10). Virtually all of the experimental data fall within the error band defined by these two lines on the graph. Furthermore, the line corresponding to $|\beta_a| = |\beta_c| = 0.1$ V lies between the two extremes, and the error resulting from using $|\beta_a| = |\beta_c| = 0.1$ V would not exceed a factor of 2 in any case. Hence, without any detailed knowledge of the system whatever, one can obtain a reasonably good estimate of the corrosion rate from polarization resistance measurements. Any more accurate estimates or measurements of β_a and β_c will reduce considerably the uncertainty of corrosion rates derived from R_p.

One should not confuse an error in corrosion rate with an error in R_p, which can be measured accurately and precisely. Thus, if relative values of corrosion rate are of greater interest than absolute values (which is often the case), the polarization resistance method becomes even more attractive. In the laboratory, the change in i_{corr} with time or the kinetics of the corrosion process may be of more interest than the absolute value of i_{corr}. The same is true industrially. Often, a large change of corrosion rate in a process stream is far more important than the exact value of corrosion rate, especially if the corrosion rate is known within a factor of 2.

5.5.3 Nonlinearity of Polarization Curves

The polarization curve in Figure 5.2b for $|\beta_a| = |\beta_c| = 118$ mV is symmetric about the origin and is apparently linear up to overvoltages above 10 mV. The deviation from linearity at any overvoltage increases as $|\beta_a| = |\beta_c|$ decreases, as illustrated in Figure 5.2c, where the linear region is much below 10mV for $|\beta_a| = |\beta_c| = 30$ mV. Thus an assumption of linearity at 10 mV may lead to somewhat inaccurate results in the use of the commercial probes described in Section 5.4. It is less ambiguous to define R_p as the slope or gradient, $(d\varepsilon/di_{app})_{\varepsilon \to 0}$, at the origin of the plot, as in

$$
R_p = \left[\frac{d\varepsilon}{di_{app}} \right]_{\varepsilon \to 0} = \frac{\Delta\varepsilon}{\Delta i_{app}}.
\tag{8}
$$

The polarization resistance is, thus, independent of the extent of linearity to be found in the polarization curve. However, the still-common practice of using the current at a single value of 10 mV overvoltage in corrosion monitoring probes leads to possible error, which has received considerable attention in the literature.

When $|\beta_a| \neq |\beta_c|$ the curve is not symmetric around the origin, as shown in Figure 5.2d for $\beta_a = 30$ mV and $\beta_c = 118$ mV. These values of the Tafel constants, taken from Figure 5.1, are more typical for many anodic and cathodic Tafel constants in acid and saline aqueous solutions.

The errors that accrue from assumed linearity of 10 mV are summarized in Figure 5.11[20] as a function of β_a and β_c. There is no error when $|\beta_a| = |\beta_c| = 120$ mV. The error is maximized for the greatest difference between β values, that is when one $\beta =$ infinity (usually β_c) while the other is 30 mV. It is notable that the errors are opposite in sign for assumed linearities, +10 mV anodic and −10 mV cathodic. Thus, when the polarized difference is measured in the two-electrode probe, the errors tend to cancel one another, and extended linearity and consequent reduced error may be observed in two-electrode measurements.[18]

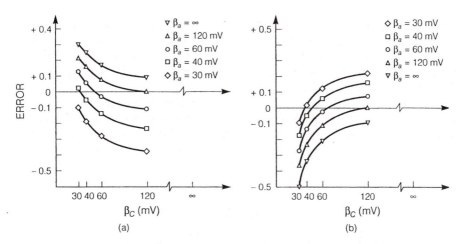

FIGURE 5.11 Errors in corrosion rate measurement due to assumed linearity of +10mV (anodic) and −10mV (cathodic) overvoltage for various values of β_a and β_c. *(From F. Mansfeld,* Corrosion, *Vol. 30, p. 92, 1974. Reprinted by permission, National Association of Corrosion Engineers.)*

5.5.4 Competing Redox Reactions

The polarization resistance method measures the total oxidation occurring at the corroding electrode. For most corroding metals, total oxidation is represented entirely by the anodic dissolution reaction of the corroding electrode, for example,

$$M \rightarrow M^{n+} + ne^-. \tag{12}$$

However, this is not always true, as illustrated by the schematic polarization diagrams for passive alloy, M, in a solution containing a hypothetical redox system, Z^{2+}/Z^+, in Figure 5.12. In the usual case, (a), the corrosion potential $E_{corr,1}$ is established several hundred millivolts active to the redox potential E_{Z^{2+}/Z^+}, and the corrosion rate $i_{pass,1}$ is an order of magnitude or more greater than the exchange current density, $i_{o,Z^{2+}/Z^+}$, for the redox reaction. As the passive surface film thickens, Case (b), the passive current density decreases (Section 4.5.2) to $i_{pass,2}$ near the exchange current density, $i_{o,Z^{2+}/Z^+}$, which is assumed constant for the purposes of this example. Now the rate of oxidation, $i_{Z^+ \rightarrow Z^{2+}}$, becomes significant compared to the passive corrosion rate, $i_{pass,2}$. The total rate of oxidation is given by the sum,

$$i_{ox} = i_{pass,2} + i_{Z^+ \rightarrow Z^{2+}}. \tag{13}$$

The corrosion potential is established at $E_{corr,2}$ where total oxidation, i_{ox}, equals total reduction, i_{red}, which in this case equals $i_{Z^{2+} \rightarrow Z^+}$. That is,

$$i_{ox} = i_{red} = i_{Z^{2+} \rightarrow Z^+}. \tag{14}$$

The polarization resistance method measures i_{ox}, and $i_{Z^+ \rightarrow Z^{2+}}$ is a significant fraction of i_{pass}. Considerable error may result if one assumes that $i_{ox} = i_{pass}$.

When the passive corrosion rate, i_{pass}, becomes quite low, Case (c), the total oxidation rate, i_{ox}, is essentially equal to the rate of Z^+ oxidation at the exchange current density, i_o. That is, $i_{ox} = i_{o,Z^{2+} \to Z^+}$. The corrosion potential draws very near to the redox potential, E_{Z^{2+}/Z^+}, until $E_{corr,3} = E_{Z^{2+}/Z^+}$. Polarization resistance essentially measures the exchange current density for the Z^{2+}/Z^+ redox system. In both cases, (b) and (c) of Figure 5.12, the corrosion rate is seriously overestimated by polarization resistance measurements.

Thus, parallel redox reactions can cause substantial errors in corrosion rates measured by polarization resistance. The possibility of this error can be detected with a platinum electrode in the same corrosive electrolyte. The redox potential is independent of the surface on which it occurs (Section 3.1.2), and the highly catalytic

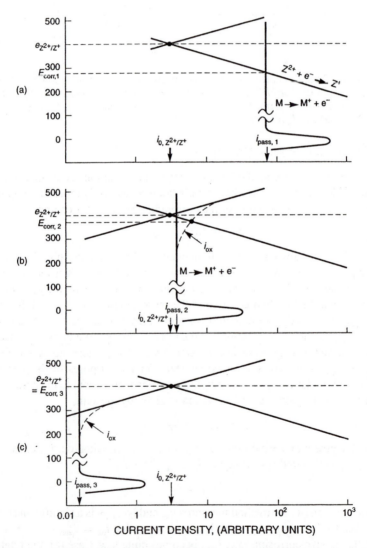

FIGURE 5.12 Schematic polarization diagrams showing alloy M passivated by a redox system Z^{2+}/Z^+: (a) $i_{pass} \gg i_{o,\ Z^{2+}/Z^+}$, (b) $i_{pass} \approx i_{o,\ Z^{2+}/Z^+}$, (c) $i_{pass} \ll i_{o,\ Z^{2+}/Z^+}$.

platinum surface readily assumes the potential of the major redox system present, with no competing oxidation currents due to corrosion. Thus, if the difference between the platinum redox potential and the specimen corrosion potential is quite low, a competing redox reaction is a strong possibility. Indig and Groot[21] were the first to observe errors due to parallel redox reactions in an elevated-temperature autoclave system in which the stainless-steel corrosion potential was only 18 mV removed from the H[+]/H$_2$ redox potential on platinum. They recommended that the potential difference should exceed 50 mV before the contribution from a parallel redox reaction can be neglected. Mansfeld and Oldham[22] showed analytically that a difference as low as 20 mV can be tolerated.

5.6 Other Methods to Determine Polarization Resistance

5.6.1 Analysis of Polarization Curve Shape

Various methods have been proposed for determining polarization resistance and Tafel constants directly from the low-overvoltage (non-Tafel) polarization curves by analysis of (Section 3.2.1)

$$i_{app,c} = i_{corr}(10^{\varepsilon/\beta_c} - 10^{\varepsilon/\beta_a}). \tag{7}$$

Such procedures are possible due to the sensitivity of the shape of these curves to the values of the Tafel constants or B in Equation (8). Barnartt[23] measured anodic overvoltage values of ε, 2ε, and a cathodic overvoltage of -2ε and substituted into a set of simple equations derived from fundamental principles to obtain the polarization resistance and Tafel slopes. Similarly, Oldham and Mansfeld[24] laid out the tangents to the polarization curves at two values, $+\varepsilon$ and $-\varepsilon$, and used the tangent slopes and intercepts to calculate the polarization resistance and Tafel slopes. Such manual calculations have been largely replaced more recently by computerized curve-fitting procedures.[25,26]

Curve shape analysis methods require that polarization data conform precisely to the expected functions derived from Equation (7), and very accurate determination of the polarization curve is necessary. Unfortunately, polarization data are often rather variable and time dependent. As a result, these analytical methods have been difficult to apply and not widely adopted.

5.6.2 Transient Methods

A step current, I_{app}, causes a change in activation overvoltage, ε_{act}, which often varies with time, t, according to a function

$$\varepsilon_{act} = I_{app}R_p(1 - e^{-t/R_pC}), \tag{15}$$

where R_p is the polarization resistance and C is the capacitance at the corroding surface. When $t = $ infinity, Equation (15) reduces to $\varepsilon_{act} = I_{app}R_p$, identical to Equation (8), when the steady-state value of ε_{act} is only a few millivolts. Knowing ε_{act} and I_{app},

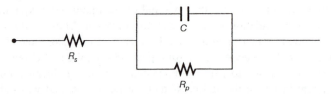

FIGURE 5.13   Equivalent electrical circuit to simulate time response of a corroding electrode with activation polarization.

R_p is readily obtained followed by corrosion rate from equation (8). Equation (15) can be derived assuming a simple equivalent circuit in which R_p and C are connected in parallel (Figure 5.13). This circuit is identical to the simple parallel circuit used to model impedance spectroscopy in Section 3.5.6.

Practically, ε_{act} will reach 98% of the steady-state value when $t = 4R_pC$. At usual corrosion rates, when R_p is a few thousand ohms and C is around 100 μF/cm^2, the steady-state value is reached in a matter of a few seconds or less, and time variations can be ignored. However, at very low corrosion rates (e.g., nuclear coolant loops, pharmaceutical preparations, and surgical implants), where contamination by corrosion products is important, R_p may reach several million ohms. In such instances, overvoltages can drift toward steady state over many minutes, which may be too long for experimental convenience. Also, the system variables, including the corrosion rate, may vary during the interval of the measurement. Jones and Greene[14] developed a trial-and-error graphic solution of Equation (15) in the early stages of the potential transient before attainment of steady state, thus shortening the measurement time substantially.

Under nonideal conditions, the total overvoltage measurement, ε, may include concentration overvoltage, ε_{conc}, and solution resistance, ε_{Ω}, components, as well as ε_{act}. That is,

$$\varepsilon = \varepsilon_{act} + \varepsilon_{conc} + \varepsilon_{\Omega}. \tag{16}$$

The concentration and solution resistance components of overvoltage must be removed before a valid calculation of corrosion rate can be obtained from ε_{act} using Equation (15). The solution resistance component reaches its maximum at very short times, and the concentration component continues to increase at very long times compared to ε_{act}. Thus, Giuliani[27] has been able to analyze overvoltage transients with a dedicated microcomputer to make a convenient separation for determination of corrosion rates from complex overvoltage data.

5.6.3 Electrochemical Impedance Spectroscopy

Electrochemical Impedance Spectroscopy (EIS) was introduced in Section 3.5.6. The polarization resistance, R_p, appears in the analog electrical circuits needed to model EIS behavior. Thus, R_p can be derived from impedance measurements, and the method has become quite popular for study of electrochemical mechanism and measurement of polarization resistance[28-30].

Examples of impedance spectra for a corroding system were shown in Section 3.5.6 (Figure 3.29). Figure 5.14[28] illustrates some difficulties as well as success in measuring R_p by impedance spectroscopy. The high-frequency end of the Nyquist plot (imaginary Z'' versus real Z' components of the impedance) gives the solution resistance, R_Ω (Figure 3.26), which can then be subtracted from the low-frequency limit of Z'' to obtain the polarization resistance, R_p, which has been shown[29] to have the usual inverse proportionality to corrosion rate determined by weight loss. A computerized curve-fitting procedure permitted extrapolation of the relatively high-frequency semicircular data to R_p at the zero-frequency limit. The polarization resistance decreased (corrosion rate increased) during the first 6 hours of exposure and then subsequently increased, as expected. Deviations at the low-frequency end of the semicircle were due to changes in impedance during the long times required to conduct measurements at low frequency. In a neutral water system such as this, low-frequency data are also often masked by the Warburg impedance line, which increases at a 45° angle, as shown in Figure 3.29.

The centers of the semicircles in Figure 5.14 are depressed below the horizontal Z' axis. This behavior is fairly typical and has been modeled and explained[30] assuming metal dissolution under a corrosion product film with oxygen reduction within the film pores.

Figure 5.15[29] shows impedance spectra of carbon steel in water inhibited with 25 ppm by weight of chromate added as potassium chromate. The low corrosion rate caused a very high R_p and a large imaginary contribution even at very low frequency. Thus, computerized extrapolation of the Nyquist plot must have limited accuracy in determination of R_p under such conditions. Two capacitive processes are apparent in the Bode plots. Computerized curve fitting was not able to give a close fit in all regions of the Bode plots, and it is not easy to ascertain the sources of the two phase-angle maxima.

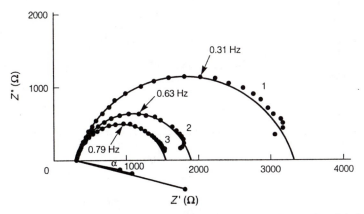

FIGURE 5.14 Nyquist plots for iron in tapwater for various times. Semicircles depressed by an angle α below the real impedance Z' axis. Electrode area, 1.2 cm^2; 1, 0.5 h; 2, 4.5 h; 3, 7.5 h. *(From F. Mansfeld, M. W. Kendig, and S. Tsai, Corrosion, Vol. 38, p. 570, 1982. Reprinted by permission, National Association of Corrosion Engineers.)*

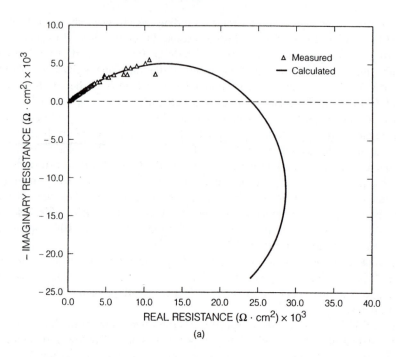

(a)

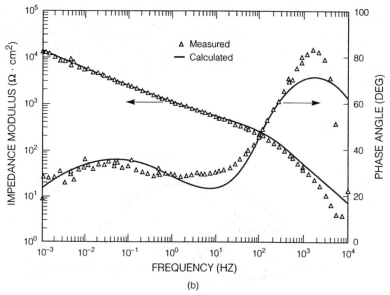

(b)

FIGURES 5.15 Nyquist (a) and Bode (b) plots for carbon steel in a water at 32°C containing 1000 ppm Cl^-, 10 ppm Zn^{2+}, and inhibited with 25 ppm chromate. *(From D. C. Silverman and J. E. Carrico,* Corrosion, *Vol. 44, p. 280, 1988. Reprinted by permission, National Association of Corrosion Engineers.)*

EIS measures invalid R_p when large exchange current densities for a redox reaction are present (Section 5.5.4)

In summary, EIS has the capability to give valid measurements of polarization resistance and corrosion rate corrected for ohmic interferences from solution resistance. Valuable information about mechanism may be revealed by frequency response of the corroding electrode. However, as shown in Figures 5.14 and 5.15, impedance measurements are not always totally unambiguous. At the low-frequency end of the spectra where R_p is measured, corrosion potential and corrosion rate may be changing during the measurement period. It may be possible in many cases to measure solution resistance more simply by other techniques (Section 3.5.4). Low-frequency measurements approach simple direct-current measurements in the limit of infinitely low frequency, and the required complex and expensive instrumentation of EIS may not always be justified.

References

1. R. Bandy and D. A. Jones, *Corrosion*, Vol. 32, p. 126, 1976.
2. D. A. Jones and N. R. Nair, *Corrosion*, Vol. 41, p. 357, 1985.
3. N. D. Greene and G. A. Saltzman, *Corrosion*, Vol. 20, p. 293t, 1964.
4. E. J. Simmons, *Corrosion*, Vol. 11, p. 255t, 1955.
5. R. V. Skold and T. E. Larsen, *Corrosion*, Vol. 13, p. 139t, 1957.
6. M. Stern and A. L. Geary, *J. Electrochem. Soc.*, Vol. 104, p. 56, 1957.
7. M. Stern, *Corrosion*, Vol. 14, p. 440t, 1958.
8. C. W. Wagner and W. Traud, *Z. Elektrochem.*, Vol. 44, p. 391, 1938.
9. K. F. Bonhoeffer and W. Jena, *Z. Elektrochem.*, Vol. 55, p. 151, 1951.
10. H. H. Uhlig and R. W. Revie, *Corrosion and Corrosion Control*, 3rd ed., Wiley, New York, p. 405, 1985.
11. F. Mansfeld, *Advances in Corrosion Science and Technology*, Vol. 6, M. G. Fontana and R. W. Staehle, eds. Plenum Press, New York, p. 163, 1976.
12. M. Stern and E. D. Weisert, *Proc. ASTM*, Vol. 32, p. 1280, 1959.
13. D. A. Jones, *Corrosion*, Vol. 39, p. 444, 1983.
14. D. A. Jones and N. D. Greene, *Corrosion*, Vol. 22, p. 198, 1966.
15. F. Mansfeld, *Electrochemical Techniques for Corrosion*, R. Baboian, ed., NACE, Houston, p. 18, 1977.
16. Standard Practice for Conducting Potentiodynamic Polarization Resistance Measurements, Standard Practice G5, *Annual Book of ASTM Standards*, ASTM, Philadelphia.
17. G. A. Marsh, *2nd Int. Cong. Metallic Corrosion, New York*, NACE, Houston, p. 936, 1963.
18. R. Bandy and D. A. Jones, *Br. Corros. J.*, Vol. 14, p. 202, 1980.
19. L. M. Callow, J. A. Richardson, and J. L. Dawson, *Br. Corros. J.*, Vol. 11, p. 132, 1976.
20. F. Mansfeld, *Corrosion*, Vol. 30, p. 92, 1974.
21. M. E. Indig and C. Groot, *Corrosion*, Vol. 25, p. 455, 1969.
22. F. Mansfeld and K. B. Oldham, *Corros. Sci.*, Vol. 11, p. 787, 1971.
23. S. Barnartt, *Electrochim. Acta*, Vol. 15, p. 1313, 1970.
24. K. B. Oldham and F. Mansfeld, *Corros. Sci.*, Vol. 13, p. 813, 1973.

25. F. Mansfeld, *Corrosion*, Vol. 29, p. 397, 1973.
26. N. D. Greene and R. H. Ghandi, *Mater. Perf.*, Vol. 21, No. 7, p. 34, 1982.
27. L. Giuliani, *Electrochemical Techniques for Corrosion Engineering*, R. Baboian, ed., NACE, Houston, p. 93, 1986.
28. F. Mansfeld, M. W. Kendig and S. Tsai, *Corrosion*, Vol. 38, p. 570, 1982.
29. D. C. Silverman and J. E. Carrico, *Corrosion*, Vol. 44, p. 280, 1988.
30. K. Juettner, W. J. Lorenz, M. W. Kendig, and F. Mansfeld, *J. Electrochem. Soc.*, Vol. 135, p. 332, 1988.

Exercises

5-1. Plot the following cathodic polarization data for carbon steel in 0.5 N H_2SO_4 on linear coordinates and determine the polarization resistance. From the shape of the plot, would you estimate that the absolute value of β_a is greater than or less than β_c?

Current density $(\mu A/cm^2)$	40	100	160	240	300
Cathodic Overvoltage (mV)	1.0	2.5	4.1	6.3	9.0

5-2. With the following anodic and cathodic polarization data—the same conditions as Problem 5.1, but larger currents—plot the polarization curves on semilogarithmic coordinates and determine β_a, β_c, E_{corr}, and i_{corr}.

Current Density, μA Anodic or Cathodic	Potential Anode, V(SCE)	Potential Cathode, V(SCE)
1.01×10^{-4}	−0.510	−0.520
2	−0.508	−0.522
3	−0.503	−0.530
5	−0.499	−0.540
7	−0.494	−0.549
1×10^{-3}	−0.490	−0.562
2	−0.477	−0.585
3	−0.470	−0.602
5	−0.458	−0.627
7	−0.448	−0.644
1×10^{-2}	−0.437	−0.660
2	−0.420	−0.688

5-3. From the results of Problem 5.2, calculate the value of the proportionality constant, B, between i_{corr} and R_p in Equation (8). Then calculate the i_{corr} corresponding to the R_p found in Problem 5.1.

5-4. A method of electrochemically determining the corrosion rate of passive alloys has been suggested. It consists of polarizing the alloy 400 mV in the noble direction. The resulting current is claimed to be the passive corrosion rate. Is this test valid? If so, explain the principles on which it is based. If it has limited use, state the exact conditions for its successful use. If the proposed method is invalid, explain why. Assume the Tafel constant for the reduction process is uniform at −0.1 volt on all alloys.

5-5. Calculate values of the proportionality constant, B, between polarization resistance and corrosion rate in equation (8) assuming the following values of the Tafel constants. Check your results against Figure 5.3.
   a. $\beta_a = 0.1$ volt, $\beta_c = 0.1$ volt.
   b. $\beta_a = 0.12$ volt, $\beta_c = \infty$
   c. $\beta_a = \infty$, $\beta_c = 0.12$ volt.
   d. $\beta_a = 0.03$ volt, $\beta_c = 0.12$ volt.

5-6. From Figure 5.3, estimate the proportionality constant, B, between polarization resistance and corrosion rate in Equation (8) for the following corrosion systems. Calculate B in each case and compare with your estimates.
   a. austenitic stainless steel in aerated sulfuric acid (passive)
   b. mild carbon steel in deaerated seawater
   c. mild carbon steel in aerated seawater

5-7. For the same corrosion rate, will R_p be the same for all conditions? Explain.

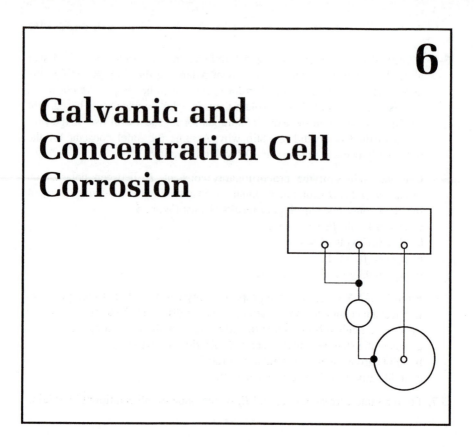

Galvanic and Concentration Cell Corrosion

6

A galvanic cell is formed when two dissimilar metals are connected electrically while both are immersed in a solution electrolyte. One of the two is corroded preferentially by galvanic corrosion. Increased corrosion also occurs in concentration cells when differing solution concentrations are present on the same metal surface. Corrosion in galvanic and concentration cells operate by similar principles, and the two are discussed together in this chapter. The emphasis is on galvanic corrosion, but differential aeration occurs frequently and is used here to introduce and illustrate concentration cell corrosion. Other concentration cells operate in many localized forms of corrosion discussed in later chapters.

6.1 Engineering Aspects

6.1.1 Recognition

Galvanic corrosion is the major suspect when attack at the junction between two dissimilar alloys is limited to only one of the two. A classic example was shown in

Figure 1.6 for steel coupled to stainless steel. Corrosion is greatest on the steel near the junction between the two alloys, because resistance in the electrolyte and a gasket limits the rate of anode attack to an area quite near the junction between the dissimilar alloys. *Contact corrosion* is consequently an alternative name for *galvanic corrosion*. The preferentially corroded alloy is the anode in the galvanic couple and the unattacked alloy is the cathode.

Not only is galvanic coupling harmful for the anode, but the cathode may be damaged by hydrogen (Section 10.1.2) or by cathodic polarization out of the passive and into the active region. In some rare cases, however, galvanic coupling may beneficially passivate the anode as described in Section 6.4.2 for titanium coupled to noble metals.

6.1.2 Galvanic Series

Any metal or alloy has a unique corrosion potential, E_{corr}, when immersed in a corrosive electrolyte (Section 3.3.1). When any two different alloys are coupled together, the one with the more negative or active E_{corr} has an excess activity of electrons, which are lost to the more positive alloy. In a couple between two metals, M and N, the anodic dissolution or corrosion reaction,

$$M \rightarrow M^{n+} + ne^-, \tag{1}$$

of the active metal, M, has its rate increased by loss of electrons. M thus becomes the anode in the galvanic cell. The more positive or noble alloy, N, has the rate of its anodic reaction,

$$N \rightarrow N^{m+} + me^-, \tag{2}$$

decreased due to the excess of electrons drawn from M. N is the cathode of the galvanic cell, and the corrosion-rate decrease is the basis of cathodic protection by a sacrificial anode alloy such as M (Section 13.1.3). m and n are integers denoting the number of electrons exchanged in the respective reactions. Rates of galvanic attack in the hypothetical M-N couple are discussed further in Section 6.2.4.

The Galvanic Series is a list of corrosion potentials for various useful alloys and pure metals. An abbreviated form was given in Table 1.1. Figure 6.1[1] shows more detail, including the measured potential range for each alloy. As discussed in Section 6.2.1, the electrolytic current in the couple is proportional to the rate of galvanic corrosion. The most active (negative) alloy in a couple is always attacked preferentially by galvanic corrosion. Selection of alloys with a minimum potential difference will minimize corrosion in galvanic couples.

The Galvanic Series should not be confused with the emf series discussed in Section 2.1.2. The emf series is a list of half-cell potentials proportional to the free-

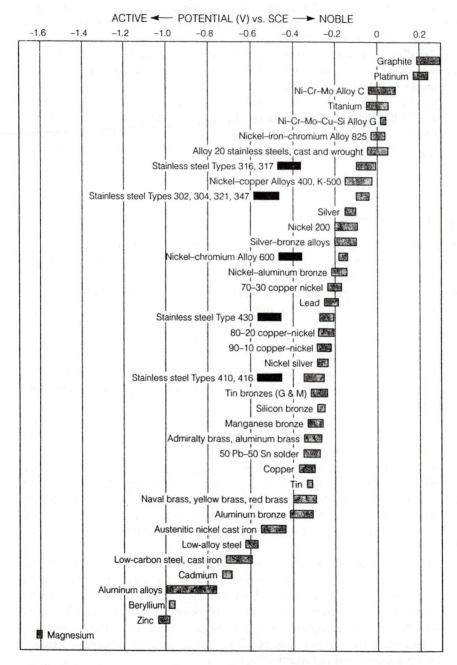

FIGURE 6.1   Galvanic Series for seawater. Dark boxes indicate active behavior for active–passive alloys. *(From H. P. Hack,* Metals Handbook, *Vol. 13,* Corrosion, *9th ed., ASM, Metals Park, OH, p. 234, 1987. Reprinted by permission, ASM International.)*

energy changes of the corresponding reversible half-cell reactions for standard-state (unit activity) conditions. The Galvanic Series is a list of corrosion potentials, each of which is formed by polarization of two or more half-cell reactions to a common mixed potential, E_{corr} (Section 3.3.1), on the corroding surface. Potentials in the emf series are listed with respect to the standard hydrogen electrode (SHE) described in Section 2.1.2. Potentials in the Galvanic Series are measured with respect to a secondary reference electrode (e.g., the saturated calomel electrode in Figure 6.1). In some tabulations, the measured potentials may not be included, as in Table 1.1, since only the qualitative differences are significant for prevention.

Corrosion potentials in the Galvanic Series are measured in real or simulated service conditions, whereas the emf series is valid only for unit activity of reactants and products. The listed order of half-cell potentials for the metal dissolution reactions in the emf series approximates the order measured for the same metals in the Galvanic Series. However, the order is not always the same, as shown in Section 6.2.2.

The Galvanic Series gives only tendencies for galvanic corrosion; it gives no information about the rate of attack. Rates of Galvanic corrosion derived from polarization measurements are discussed in Section 6.4.1. The series for aerated seawater in Figure 6.1 is sometimes used as an estimation for other environments, a practice that should be used only with caution. Changes in electrolyte composition and temperature can produce significant changes in potential positioning in the Galvanic Series. Each major change in environmental conditions would require that a new series be established. Thus, the Galvanic Series gives qualitative indications of the likelihood of galvanic corrosion, but quantitative predictions are impossible. As a result, extensive tabulations of the Galvanic Series for various environmental conditions have not been produced.

6.1.3 Area Effects

A large ratio of cathode-to-anode surface area is to be avoided because the galvanic attack is concentrated in small areas, and penetration of the anode thickness is hastened. Larger anode surface spreads the attack over a greater area and reduces the rate of penetration.

A classic example[2] illustrates the importance of area ratio in design and prevention. Several hundred large storage tanks were installed in a major plant expansion program. Older tanks were fabricated of carbon steel, and the inner surfaces were coated entirely with a baked phenolic paint. The old tanks had served well for many years, but coating damage on the bottoms caused product contamination. The new tank bottoms were upgraded with stainless steel–clad carbon steel for better service and reduced maintenance. The same phenolic coating was then applied over the tank wall surfaces, as with the old tanks, and over the welded junction with the tank bottom, as shown in Figure 6.2. The inherently resistant stainless steel bottoms were left uncoated. The plant operators were chagrinned and puzzled when leaks developed in the tank walls just above the welded junction after only a few months of service, despite their well-intentioned tank upgrades.

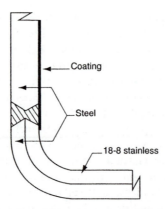

FIGURE 6.2 Painted steel storage tank which developed penetration through a coating due to galvanic coupling to a new stainless steel bottom. *(From M. G. Fontana, Corrosion Engineering, 3rd ed., McGraw-Hill, New York, pp. 46–49, 1986. Reprinted by permission, McGraw-Hill Book Company.)*

Subsequent investigation revealed that localized pinhole corrosion had developed through the coating. All coatings have defects in the form of pinholes, holidays, and mechanical damage (Chapter 14). No account had been taken of the galvanic couple formed between the stainless steel tank bottoms (cathode) and the carbon steel tank walls (anode). Anodic dissolution and galvanic corrosion of the carbon steel were concentrated at coating defects by a large cathode (stainless steel)/anode (steel) surface area ratio.

6.1.4 Prevention

The best prevention for galvanic corrosion is to eliminate the galvanic couple by design, if possible. Selection of materials near one another in the Galvanic Series eliminates the driving force for galvanic attack. Insulated flanges and connectors (Figure 6.3[2]) eliminate the galvanic couple, also. However, soluble corrosion products plating out on downstream active metals (e.g., copper on steel) can bypass insulated connectors and form *in situ* galvanic couples. If galvanic couples are unavoidable, a large anode area and/or a thickness allowance is advisable. The most economical measure may simply be to design the anode part for occasional easy replacement. Coatings must be applied with caution to avoid very small anode areas at coating defects. Strangely enough, one measure to alleviate the galvanic corrosion problem created in the tank example of the preceding section was to coat the corrosion-resistant stainless steel cathode to limit the area available for cathodic reduction, decrease the cathode/anode area ratio, and thereby reduce the galvanic current density. If a galvanic couple is to be coated, it should be applied only to the cathode, if at all. Coatings on the anode only serve to concentrate further the galvanic attack at coating defects.

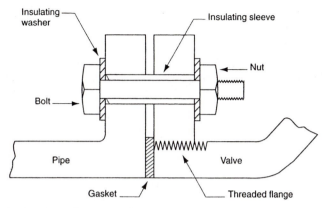

FIGURE 6.3 Insulated flange to eliminate a galvanic couple. *(From M. G. Fontana,* Corrosion Engineering, *3rd ed., McGraw-Hill, New York, pp. 46–49, 1986. Reprinted by permission, McGraw-Hill Book Company.)*

6.2 Fundamental Aspects

Whereas the potential difference in the Galvanic Series between a pair of alloys gives the driving force for galvanic corrosion, the current flowing in a galvanic couple measures the anode dissolution and mass rate of galvanic attack. Dividing galvanic current by the anode area results in current density, which is proportional to average corrosion or penetration rate (Section 3.1.1). Some knowledge of current distribution on the anode is vital to predict the rate of galvanic corrosion at any given point on the anode (Section 6.4.3).

6.2.1 Corroding Metal–Inert Metal Couple

Several experimental effects result from coupling zinc, a corroding metal, to platinum, an inert metal, in a dilute acid solution.

1. Corrosion potential of zinc is shifted to a more noble value.
2. Corrosion rate of zinc is increased.
3. Rate of hydrogen evolution on zinc is reduced.

These effects can be explained by anode and cathode polarization within the couple, as shown in Figure 6.4. In the uncoupled condition, the zinc arrives at a mixed potential, E_{corr}, and the platinum at the half-cell potential for the hydrogen reaction (Section 3.3.1). When the two are coupled or electrically connected together, electrons flow from the zinc with the more negative E_{corr} to the platinum with a more positive surface H^+/H_2 half-cell potential. The current from galvanic coupling causes polarization at the zinc and platinum surfaces, just as if supplied from an external source (Section 3.4). Electrons flowing from the zinc (anodic current) causes anodic polarization of the zinc dissolution reaction,

$$Zn \rightarrow Zn^{2+} + 2e^-. \tag{3}$$

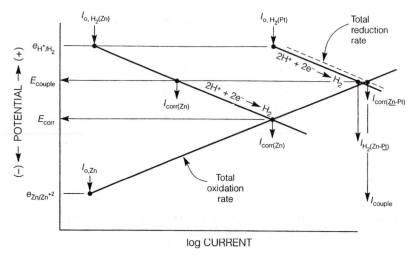

FIGURE 6.4 Schematic polarization in a galvanic couple between zinc and platinum in dilute acid solution.

The same electrons enter the platinum (cathodic current) and cathodically polarize the hydrogen reduction reaction

$$2H^+ + 2e^- \rightarrow H_2. \tag{4}$$

The two surfaces polarize until they reach the same potential, E_{couple}, where the total reduction current is equal to the total oxidation current. The galvanic current, I_{couple}, flowing at steady state, is exactly analogous to the i_{corr} between half-cell reactions at the mixed potential on the surface of a single corroding metal (Section 3.3.1). I_{couple} is equal to the corrosion current of zinc in the galvanic couple, $I_{corr(Zn)}$, and to the sum of reduction currents for hydrogen (4) on Zn ($I_{H_2(Zn-Pt)}$) and on Pt ($I_{H_2(Zn-Pt)}$). Because $I_{H_2(Zn-Pt)} \gg I_{H_2(Zn-Pt)}$, $I_{H_2(Zn-Pt)}$ is also essentially equal to I_{couple}. Reduction and oxidation currents equalize to avoid charge accumulation, as required by mixed potential theory.

Current, rather than current density, must be used in polarization diagrams of galvanic couples in this and following sections. The anodic current at the alloy anode must be divided by anode area to obtain current density, which is proportional to penetration rate.

6.2.2 Effect of Exchange Current Density

The emf series shows gold with a more noble half-cell potential, 1.5 V, than platinum, 1.2 V (Table 2.1). However, the order is reversed in the Galvanic Series of Table 1.1; the corrosion potential for gold is active to that of platinum. Figure 6.5 explains the reversal with the polarization diagrams for zinc coupled to gold and zinc coupled to platinum, again in dilute acid solution. It is apparent from Figure 6.5 that the half-cell potentials for the dissolution reactions for gold and platinum do not determine the respective couple potentials. E_{couple} for Zn-Au is active to

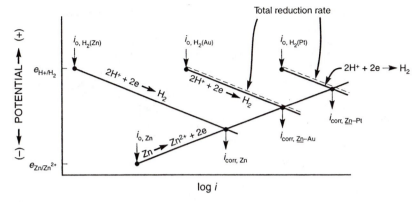

FIGURE 6.5   Effects of platinum and gold galvanically coupled to zinc in dilute acid solution.

E_{couple} for Zn-Pt due to major differences in exchange current densities for the cathodic hydrogen reduction reaction (4) on platinum and gold. The half-cell potential of (4) is the same on both (and all) surfaces. The cathodic polarization curve for gold intersects the anodic dissolution curve for zinc at a more active value because the exchange current density for reduction of hydrogen is considerably lower on gold than on platinum.

6.2.3 Effect of Surface Area

Relative surface area influences the rate of galvanic corrosion, as well. Larger cathode area provides more surface for the reduction reaction, and the anodic dissolution current (rate) must increase to compensate. Effect of increased platinum surface in the Zn-Pt couple is shown in Figure 6.6. To keep oxidation current equal to reduction current, the couple potential becomes more noble and the couple current must increase as a result. Inasmuch as the couple current is equal to the anodic dissolution current, the galvanic corrosion or penetration rate also increases if the anode surface area remains constant. Again, the penetration rate is obtained by dividing anode area into the couple or galvanic current.

6.2.4 Corroding Metal–Corroding Metal Couple

The hypothetical anode metal, M, and cathode metal, N, provide an illustration, as in Section 6.1.2. The polarization diagram (Figure 6.7) shows the two with separate uncoupled corrosion potentials, $E_{corr(M)}$ and $E_{corr(N)}$, which for actual alloys are listed in the Galvanic Series of Figure 6.1. The coupled potential, E_{couple}, is again determined by the point at which total oxidation is equal to total reduction, according to mixed potential theory. With corroding metals coupled, the total oxidation rate must be considered, as well as the total reduction rate or current. In examples using Zn-Pt couples (Figures 6.4 to 6.6), only the oxidation or dissolution of zinc need be considered because platinum is inert. At E_{couple} the anodic dissolution rate for M, the

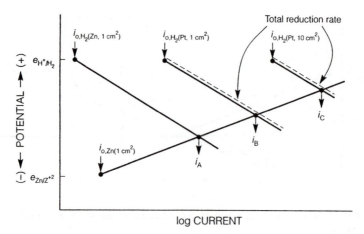

FIGURE 6.6 Effect of increased cathode surface area on the galvanic interaction between zinc and platinum in dilute acid solution.

anode in the couple, has increased from $i_{corr(M)}$ to $i_{corr(\underline{M}-N)}$, and that for N, the cathode, has decreased from $i_{corr(N)}$ to $i_{corr(M-\underline{N})}$.

In galvanic couples involving two corroding metals, the potential of the couple always falls between the uncoupled corrosion potentials of the two metals. The corrosion rate of the metal with the more active corrosion potential, the anode, is always increased, while the corrosion rate of the one with the more noble corrosion potential, the cathode, is always decreased. Decreased corrosion of the cathode at the expense of increased anode corrosion is the basis for cathodic protection by sacrificial anodes, discussed in Section 13.1.3.

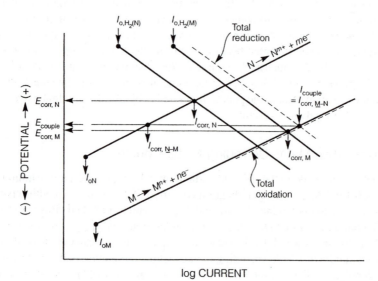

FIGURE 6.7 Schematic polarization in a galvanic couple between corroding metals M (anode) and N (cathode).

It is sometimes generalized that the corrosion rate of the less resistant alloy in a galvanic couple is always increased, while that of the more resistant alloy is decreased. Although this statement is true in many instances, corrosion potential, not the uncoupled corrosion rate, determines the possibility of galvanic corrosion. For example, in atmospheric exposures, zinc (galvanized) coatings corrode at lower rates than the underlying steel in the surface water film formed during atmospheric exposures. However, the zinc, with a more active corrosion potential, is galvanically corroded at breaks in the coating to protect the exposed steel cathodically (Section 14.3.1).

6.3 Experimental Measurements

6.3.1 Polarization in Galvanic Couples

When two dissimilar metals or alloys are coupled in an aqueous environment, the resultant galvanic current polarizes their respective surfaces as anode and cathode just as if the polarizing current were supplied by an external power supply. Schematic *experimental* polarization curves in a galvanic couple are shown by the solid lines in Figure 6.8. The dashed lines represent the polarization curves for the half-cell reactions defining the uncoupled corrosion potentials, $E_{corr,A}$ and $E_{corr,C}$, for anode and cathode, respectively. The couple potential, E_{couple}, is established where anode and cathode are polarized to equal potentials by the same current, I_{couple}, as discussed in Section 6.2. Reduction reactions on the anode metal are assumed to be

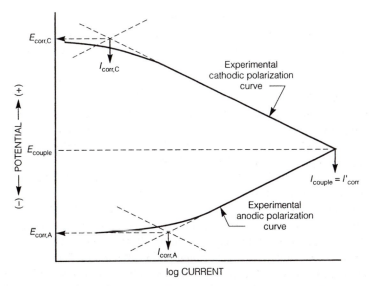

FIGURE 6.8   Schematic experimental polarization of anode and cathode in a galvanic couple.

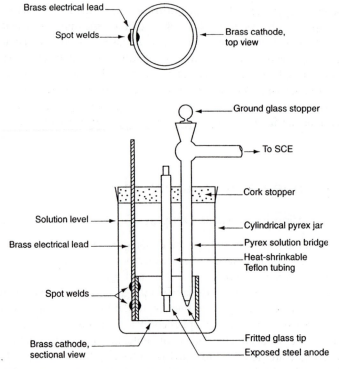

FIGURE 6.9 Cell for electrochemical study of steel anode, A, and brass cathode, C. *(From D. A. Jones,* Corrosion, *Vol. 40, p. 181, 1984. Reprinted by permission, National Association of Corrosion Engineers.)*

very small, compared to such reactions on the cathode in Figure 6.8—that is, $I_{couple} \gg I_{corr,A}$. This is a good assumption for most practical galvanic couples.

With minor modifications, the apparatus and methods described in Sections 3.5.1 to 3.5.3 can be used to measure polarization in galvanic couples directly. An illustration from the author's research[3] is shown in Figure 6.9. A central steel rod anode was surrounded by a brass tube cathode, and potentials of both were measured with respect to a saturated calomel electrode (SCE) through the indicated solution bridge. This configuration was chosen to give uniform current distribution between steel and brass. The high surface-area ratio, cathode to anode, of approximately 12:1, is common for couples accenting galvanic attack.

The brass-tube cathode replaced the usual auxiliary electrode in a conventional galvanostatic polarization set up, as shown schematically in Figure 6.10. DC current from the power supply, PS, simultaneously polarized the steel as an anode and the brass as cathode with stepwise increasing currents. Potential was recorded at steady state for each current step for both the brass and the steel. The resulting polarization curves of current versus potential in aerated 20% NaCl for each metal are plotted in Figure 6.11. It is notable that the cathode, showing evidence of concentration polarization of dissolved oxygen, is more strongly polarized than the steel anode. That is, the galvanic current is controlled by diffusion of dissolved

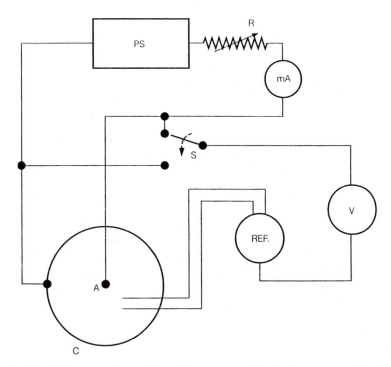

FIGURE 6.10 Galvanostatic polarization circuit for electrochemical measurements on the brass (cathode), steel (anode) galvanic couple shown in Figure 6.9. *(From D. A. Jones,* Corrosion, *Vol. 40, p. 181, 1984. Reprinted by permission, National Association of Corrosion Engineers.)*

oxygen to the cathode, as is typical for galvanic couples in aerated dilute neutral solutions.

Area ratios other than the experimental one of 12:1 were simulated simply by moving the experimental cathodic curve to higher or lower values with respect to the anodic curve in Figure 6.11. Intersection of the two curves in each case shows the couple potential at the indicated surface-area ratio. These graphically determined couple potentials are plotted as the solid line in Figure 6.12. The data points show experimentally determined couple potentials using the same indicated surface-area ratios. The agreement is good, considering the varying sources of brass and steel that were needed to make up the experimental couple ratios.[3] Individual experimental polarization curves of anode and cathode thus provide a reasonable prediction of polarization in the corresponding galvanic couple.

6.3.2 Zero Resistance Ammeters

As indicated in Figures 6.8 and 6.11, the couple current is found at the crossover of the anodic and cathodic polarization curves, where potentials of anode and cathode are equal. Anode and cathode potentials are also equal in a shorted galvanic couple with no intermediate instrumentation. Therefore, I_{couple} in Figure 6.11 is the current

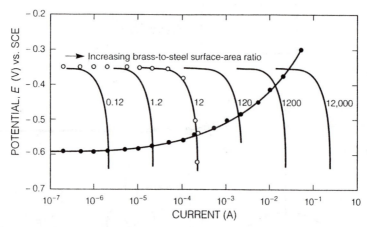

FIGURE 6.11 Experimental cathodic (O) and anodic (●) polarization curves in aerated 20% NaCl at room temperature for brass and steel, respectively, having surface area ratio of 12:1. Other surface area ratios simulated by moving cathodic curve with respect to the anodic curve. *(From D. A. Jones, Corrosion, Vol. 40, p. 181, 1984, Reprinted by permission, National Association of Corrosion Engineers.)*

in a shorted galvanic couple. A simple ammeter placed between anode and cathode does not measure I_{couple} at short circuit because there is an IR_m drop across the ammeter, where R_m is the meter resistance. Thus, anode and cathode potentials are separated by IR_m, and I is less than the desired I_{couple}, as indicated in Figure 6.13.

The circuit of Figure 6.10 may be defined as a zero resistance ammeter (ZRA), because the instrumentation can be adjusted so that there is no resistance, in effect, between anode and cathode. The power supply, PS, compensates for the usual IR_m

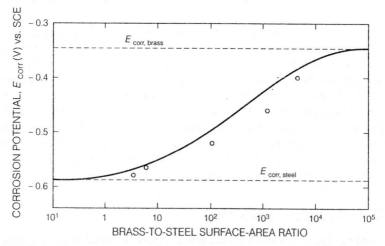

FIGURE 6.12 Graphically simulated couple potential (———) compared with experimental measurements (O) versus cathode/anode area ratio. *(From D. A. Jones, Corrosion, Vol. 40, p. 181, 1984. Reprinted by permission, National Association of Corrosion Engineers.)*

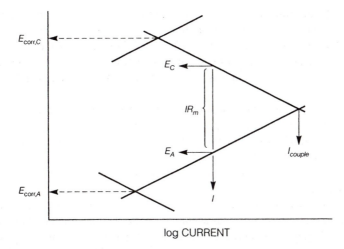

FIGURE 6.13 Separation of anode and cathode potentials when measuring galvanic current I with ammeter of resistance R_m.

ohmic loss, and I_{couple} is measured where the potential difference between anode, A, and cathode, C, is zero. A high-impedance voltmeter, V, is adequate to monitor the potential difference between anode and cathode in a simple ZRA. The reference electrode, REF, in Figure 6.10 is useful to obtain the polarization curves of Figure 6.11.

The galvanostatic ZRA is useful for periodic measurements of I_{couple} in a galvanic couple. However, it cannot be used to record couple current continuously. The polarization behavior of both anode and cathode usually vary with time along with the couple current. Thus, continuous adjustments to change the measured current are necessary to maintain anode and cathode potential difference at zero. On the other hand, a potentiostat can be utilized readily in a circuit to measure I_{couple} continuously and automatically.[4] The potentiostat senses a difference between the REF and WE terminals of the instrument and controls the difference at a preset value by automatically varying the current between the WE and AUX terminals. When WE and REF terminals are shorted, as shown in Figure 6.14, the potentiostat will control potential between anode A and cathode C at any specified value. If that value is set at zero, the circuit will continuously and automatically read the I_{couple} at short circuit on the milliammeter, mA. Figure 6.15 shows continuous recording of short-circuit current with time, as affected by dichromate inhibitor additions.[3]

6.3.3 Galvanic Current by Polarization Resistance

In the case of a uniformly corroding metal, M, it is generally assumed that anodic and cathodic reaction sites are distributed uniformly over the surface when determining corrosion rate (current density) by polarization resistance. Anodic and cathodic sites are not necessarily stationary but may migrate randomly over the surface, although metallurgical inhomogeneities tend to fix positions. If a second, more noble metal, N, is coupled to M, the anodic dissolution rate of M will be increased, as shown in Figure 6.7. Anodic dissolution is largely restricted to M, and cathodic reduction of some oxidizer occurs on both M and N. Conservation of charge still dictates that each electron liberated by anodic dissolution of an M atom must be balanced by simultaneous consumption of another electron by reduction at the surface of N.

Both anodic and cathodic reactions are governed by charge-transfer or diffusion-controlled processes during galvanic corrosion, just as they are in the usual uniform corrosion of a single metal. The derivation of the polarization resistance method (Section 5.2.2) does not restrict the distribution of anode and cathode sites.[4,5] Thus, the galvanic current or galvanic corrosion can be measured by polarization resistance.

Polarization resistance has not been widely exploited to measure galvanic currents. However, the feasibility and accuracy of the method for this purpose have been studied. Figure 6.16[4] shows a comparison of galvanic current, I_{couple}, between steel and aluminum in wet portland cement, as measured by polarization resistance and more conventional methods (Section 5.5). The agreement is good, demonstrating the decrease of I_{couple} with time.

The advantage of measuring I_{couple} by polarization resistance is that no instrumentation need be inserted between the anode and cathode as in the other methods described in Chapter 5. Galvanic currents can be measured *in situ* without interrupting the physical and geometric conditions that exist in the couple. For example, the galvanic currents flowing between active metallic (e.g., Zn and Al) coatings and cathodically protected substrate steel at breaks in the coating have been measured by polarization resistance.[6] Such currents can be measured *in situ* only by polarization resistance.

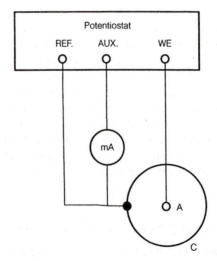

FIGURE 6.14 Potentiostatic zero resistance ammeter for continuous and automatic measurement of couple current at short circuit. *(From D. A. Jones, Corrosion, Vol. 40, p. 181, 1984. Reprinted by permission, National Association of Corrosion Engineers.)*

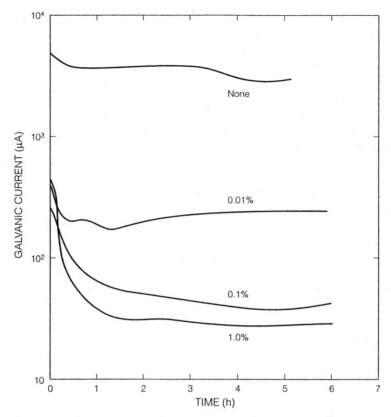

FIGURE 6.15   Galvanic couple currents between brass and steel recorded automatically from a potentiostatic zero resistance ammeter. Effect of added $K_2Cr_2O_7$ in aerated 20% NaCl. *(From D. A. Jones, Corrosion, Vol. 40, p. 181 1984. Reprinted by permission, National Association of Corrosion Engineers.)*

6.4 Determining Rates of Galvanic Corrosion

The rate of galvanic attack in a dissimilar metal couple is measured by the galvanic current density at the anode. The anodic current density at any given point on the anode in turn depends on polarization of both anode and cathode (Section 6.3.1), conductivity of the solution, and physical geometry of the couple, including the cathode/anode surface-area ratio. These factors are discussed in this section from the viewpoint of predicting galvanic corrosion rates.

6.4.1 Galvanic Current Diagrams

The usefulness of anodic and cathodic polarization curves to visualize polarization within a galvanic couple was demonstrated in Figure 6.11. The polarization curves in Figure 6.11 were obtained simultaneously in a single experiment. However, electrochemical polarization is not dependent on the source of polarizing current as long as current distribution is uniform. Thus, the experimental polarization curves for anode and cathode may be measured independently by the conventional galvanostatic and potentiostatic methods described in Sections 3.5.1 and 4.2.2, respectively, and used to predict the galvanic current in a galvanic couple with uniform current distribution.

Bennett and Greene[7] were the first to suggest a composite polarization diagram to predict current in galvanic couples. The diagram consists of anodic and cathodic polarization curves for numerous alloys in a particular solution electrolyte. Their collection of polarization curves for this purpose, measured in aerated 3% NaCl and deaerated 1 N H_2SO_4, is shown in Figures 6.17 and 6.18,[7] respectively. The corrosion potentials on the vertical axis constitute the galvanic series for the listed alloys in the specified environment. The polarization curves add a new dimension of current or reaction rate to the galvanic series.

Polarization curves from Figure 6.17 for any pair of alloys can be used to predict the galvanic current passing when equal 1-cm^2 areas of the two are galvanically coupled in aerated 3% NaCl. For example, the galvanic current in an aluminum and copper couple is approximately 70 $\mu A/cm^2$ using the polarization curves in Figure 6.19, taken from Figure 6.17. The vertical dashed lines represent the spread of values for the limiting current for reduction of dissolved oxygen, which is essentially the same for all alloys listed in Figure 6.17. The anodic current density is directly proportional to the penetration rate, assuming that the current distribution is uniform. The effects of a 10:1 surface-area ratio can be predicted by graphically increasing the cathode currents in Figure 6.19, as shown in Figure 6.11. The new galvanic current

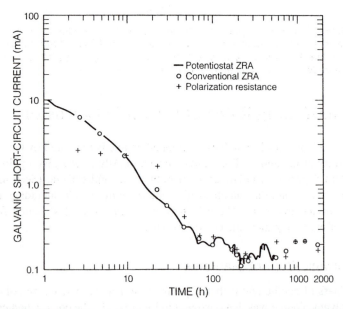

FIGURE 6.16 Comparison of galvanic currents measured by polarization resistance and conventional zero resistance ammeters. *(From D. A. Jones,* Electrochem. Technol., *Vol. 6, p. 241, 1968. Reprinted by permission, The Electrochemical Society.)*

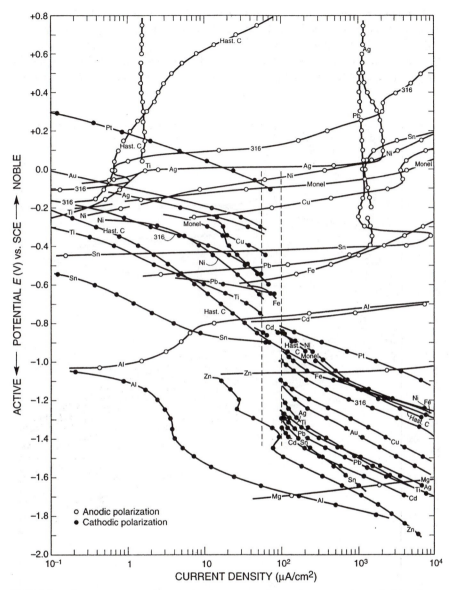

FIGURE 6.17 Potentiostatic polarization curves for various metals and alloys for prediction of galvanic corrosion in aerated 3% NaCl. *(From Bennett and Greene[7].)*

is approximately 700 μA, a corresponding increase in galvanic current by a factor of 10. Since the anode surface area has remained at 1 cm^2, the anodic current density has increased to 700 μA/cm^2. Cathodic current density at the copper remains at 70 μA/cm^2 because the cathode surface area has been increased. This emphasizes the need to use current rather than current density in the electrochemical analysis of galvanic corrosion.

The galvanic current diagrams of Figures 6.17 and 6.18 share the same limitations as anodic (and cathodic) polarization data for evaluation of general corrosion and passivity described in Section 4.3.3. Polarization data taken in laboratory purity electrolytes in short periods of time, compared to service exposure times, often do not give a reasonable measure of service performance. Over long periods of time and

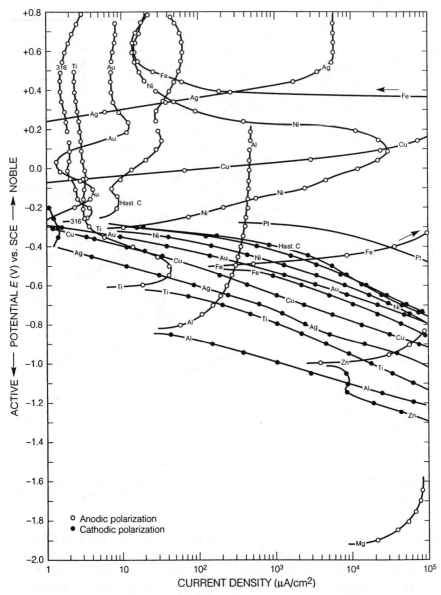

FIGURE 6.18 Potentiostatic polarization curves for various metals and alloys for prediction of galvanic corrosion in deaerated $1N\,H_2SO_4$. *(From Bennett and Greene[7])*

in the presence of undetermined impurities, surface conditions can be expected to form which cannot be predicted from short term laboratory electrochemical testing. Nevertheless, the examples cited in this and the following sections show that the galvanic current diagrams give useful information and confirm known galvanic effects. Hack and Scully[8] have shown that polarization curves on preexposed electrodes at very low scan rates give improved correlation with galvanic behavior in long-term seawater exposures.

6.4.2 Galvanic Passivation of Titanium

In dilute acid solutions, titanium corrodes at a high rate in the active state. When coupled to platinum, the corrosion rate of the titanium anode surprisingly decreases rather than increases as for zinc in Figure 6.4 or aluminum in Figure 6.19. This novel behavior is due to the passivation effects shown by the polarization curves in Figure 6.20, taken from Figure 6.18. Platinum increases the cathodic reduction of hydrogen to a level that stabilizes the passive film according to the criterion described in Section 4.1.5. The reduction current is higher than the critical anodic current density for passivation, i_c. In fact, the exchange current density for the hydrogen reaction on platinum, $i_o,H_2(Pt)$, is so high that the potential is controlled essentially at the half-cell potential for the hydrogen redox reaction. The corrosion rate for Ti is

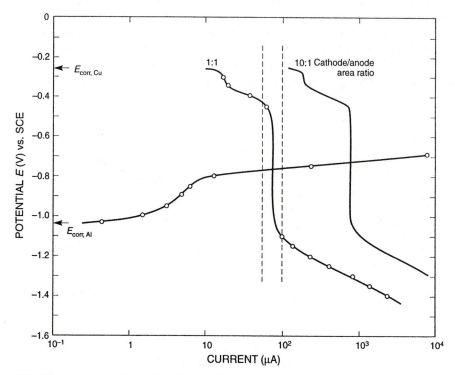

FIGURE 6.19   Predicting galvanic corrosion rates of aluminum-copper couples: Effect of Cu:Al surface area ratio. Polarization curves for Cu and Al taken from Figure 6.17.

reduced to $i_{couple(Ti-Pt)}$. Equal areas of Ti and Pt are assumed for simplicity so that current densities, i, may be used for these estimates.

The half-cell potential of the oxidizer (e.g., E_{H_2/H^+} in Figure 6.20) must be more noble than the primary passivation potential, E_{pp}, of the active-passive alloy. The exchange current density for the oxidizer reaction on the cathode must be very high also in order for the total reduction current to exceed i_c at E_{pp}. These conditions are met only for titanium and chromium in air-free acid solutions when coupled to noble metal cathodes such as platinum, palladium, rhodium, and iridium.[9] However, Figure 6.18 shows that when coupled to titanium, nickel may also have the requisite properties for passivation.

Small amounts of noble metals alloyed with titanium successfully increase the corrosion resistance in hot acid solutions, in which dissolved oxidizers such as oxygen have low solubility. Although not sufficient in bulk concentration, the alloying element enriches on the surface because it is essentially inert in the acid solutions. Soon there is enough surface area of the noble metal to facilitate passivation. If the enriched surface is abraded or otherwise damaged, the enrichment process repeats itself until the damaged area is again passivated. Commercial purity Ti with additions of 0.12 to 0.25% Pd are commercially available and have maximum corrosion resistance to hot acid solutions and crevice corrosion. Less expensive Ti-Ni alloys have met with considerable commercial success.[10] Titanium, Code-12, contains optimally 0.8% Ni and 0.3% Mo. Molybdenum reduces the critical current density needed for passivation[11] and thereby the necessary amount of alloyed nickel.

Chromium is also passivated by noble metal additions, but all chromium alloys are too brittle for practical use. Stainless steels with noble metal alloying additions are more readily passivated in aerated acids but have increased corrosion in deaerated acids because E_{pp} is active to e_{O_2/OH^-} but not e_{H_2/H^+}. The loss of resistance in

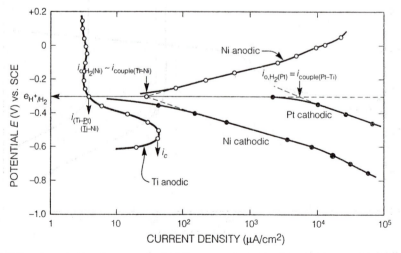

FIGURE 6.20   Polarization curves taken from Figure 6.17 illustrating galvanic passivation of Ti when coupled to Pt or Ni.

deaerated acid far offsets the improvement in aerated acid, and noble metal alloying of stainless steels is not technically feasible.

6.4.3 Current and Potential Distributions

Many useful predictions can be made assuming uniform current distribution, as indicated in the two preceding sections. However, many galvanic couples, in practice, do not conform to such an ideal. Quantitative predictions of galvanic corrosion often require further considerations of current and potential distribution on the surface of the anode.

The current and potential distributions between two galvanically coupled dissimilar metals depend on the conductivity of the electrolyte and the physical geometry of the couple, as well as polarization of anode and cathode. The effect of electrolyte or solution conductivity is similar to that of meter resistance, R_m, in Figure 6.13. The effective solution resistance, R_Ω, between any pair of points on anode and cathode separates the potential between those points by an amount, IR_Ω, with a galvanic current, $I < I_{couple}$, flowing in the couple.

The early data of Copson[12] in Figure 6.21 illustrate the potential distribution in a coplanar couple between steel (iron) and nickel exposed for several weeks in aerated tap water. Potential was measured as a function of position in the electrolyte between a stationary reference electrode near the steel surface remote from the junction and an identical movable reference electrode. Lines of constant potential are shown on a plane perpendicular to the linear junction between the two metals. The tap-water electrolyte is of relatively low conductivity, creating substantial ohmic losses between remote points on the two metals. The nickel cathode is highly polar-

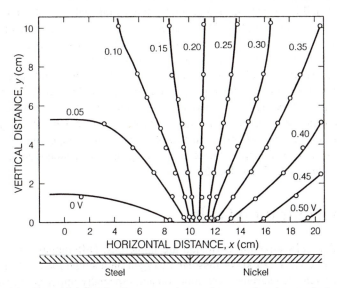

FIGURE 6.21   Potential distribution around a coplanar galvanic couple between iron and nickel. *(From H. R. Copson, Trans. Electrochem. Soc., Vol. 84, p. 71, 1943. Reprinted by permission, The Electrochemical Society.)*

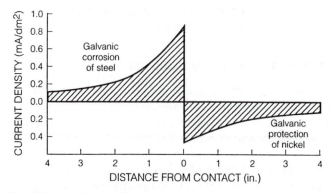

FIGURE 6.22   Effect of distance from the junction on attack at the anode in a galvanic couple. *(From H. R. Copson,* Trans. Electrochem. Soc., *Vol. 84, p. 71, 1943. Reprinted by permission, The Electrochemical Society.)*

ized, whereas the steel anode has very low polarization, as in the brass-steel couple of Figure 6.11. As a result, the nickel cathode potential is affected by lower currents at a much greater distance from the junction, as compared to the steel anode.

Strong concentration polarization is typical of the oxygen reduction reaction and largely independent of the metal surface because the limiting diffusion currents are the same for all. Similar concentration polarization is shown in Figure 6.11 for cathodic polarization in aerated salt solution on a brass cathode. The low anodic polarization on the steel anode concentrates anodic current near the junction, and the attack decreases to the background or self-corrosion rate, characteristic of the environment at points remote from the junction, as shown in Figure 6.22. Lower solution conductivity and lower polarization generally concentrate galvanic attack nearer the junction in a galvanic couple.[13] Thus, galvanic attack is increased further by concentration of galvanic current near the contact line between dissimilar metals, due to the usual lower polarization of anode, as compared to cathode.

Mathematical analysis of current and potential distributions in galvanic cells has been conducted for simplified geometries and polarization functions.[13,14] However, more complex systems have remained intractable until recently, when modern finite element[15] and boundary element[16] methods have been applied. Current and potential distributions on cathode structures are also of great practical importance on a larger physical scale in cathodic protection systems using sacrificial anodes and impressed currents (Sections 13.2.3 and 13.3.5).

6.5 Concentration Cells

Concentration cells affecting corrosion are numerous. Differential aeration cells, with differing concentrations of dissolved oxygen on a single metal surface or on electrically connected surfaces, are common and especially important to an understanding of corrosion behavior for many metals and environments. Such cells are of special interest for iron and carbon steel and are discussed for illustration in this

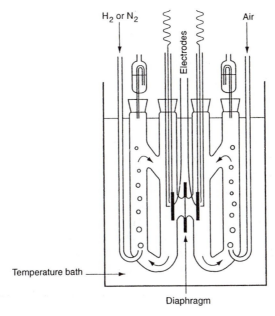

H$_2$ or N$_2$ Air

Electrodes

Temperature bath

Diaphragm

FIGURE 6.23   Differential aeration cell. *(From H. Grubitsch, cited by U. R. Evans, The Corrosion and Oxidation of Metals, Arnold, London, p. 129, 1960.)*

section. Acid-chloride concentration cells are important for the initiation and growth of pitting and crevice corrosion in stainless alloys and are discussed in Sections 7.2 and 7.3.

6.5.1 Differential Aeration Cell

The experimental differential aeration cell of Grubitsch[17] is shown in Figure 6.23. Identical iron electrodes are immersed in 0.1N NaCl solutions on both sides of the porous diaphragm, which allows charge transfer but not sufficient mass transfer to affect the dissolved oxygen content in the solutions on either side. On one side, nitrogen (or hydrogen) is purged through the solution and on the other, compressed air, forming a differential aeration cell between the two electrodes. The rising bubbles in both compartments continuously recirculate the solutions to maintain uniform solution composition at the electrode surfaces.

The mixed potential analysis of the differential aeration cell is shown in Figure 6.24 for electrodes of equal area. With the two electrodes uncoupled, a mixed or corrosion potential, $E_{corr,A}$, is established on the nitrogen side, where the anodic reaction is

$$Fe \rightarrow Fe^{2+} + 2e^-, \tag{5}$$

and the cathodic reduction reaction is

$$2H_2O + 2e^- \rightarrow 2OH^- + H_2. \tag{6}$$

Very low dissolved oxygen content reduces the limiting current for oxygen reduction, $I_{L,A}$, to very low levels, precluding reduction of dissolved oxygen as the cathodic reaction. On the aerated side, the anodic reaction is still (5), and with equal electrode areas, the anodic polarization curve is congruent with that of the deaerated side. But reduction of dissolved oxygen by

$$O_2 + 2H_2O + 4e^- \rightarrow 4OH^- \tag{7}$$

is the cathodic reduction reaction. Corrosion rate of the uncoupled cathode is controlled by concentration polarization and diffusion of dissolved oxygen at $I_{L,C}$. The uncoupled corrosion potential on the aerated electrode is then $E_{corr,C}$.

When the two electrodes are coupled together, the area and current of the anode reaction (5) doubles, but the cathode current is still controlled by the reduction current in the aerated electrolyte because the corresponding current in the deaerated electrolyte is comparatively negligible. The overall couple potential comes to a value, E_{couple}, intermediate between $E_{corr,A}$ and $E_{corr,C}$, where total oxidation equals total reduction. The corrosion current increases from $I_{corr,A}$ to $I'_{corr,A}$ at the anode and decreases from $I_{corr,C}$ to $I'_{corr,C} = I'_{corr,A}$ at the cathode. Anode current densities are the

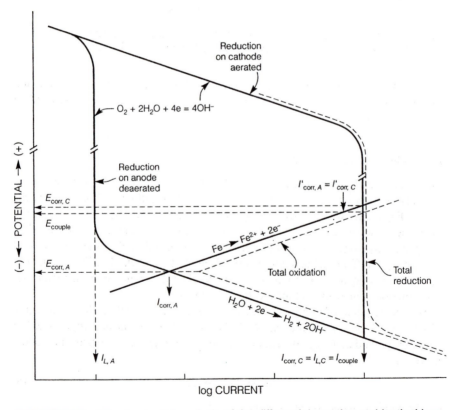

FIGURE 6.24 Mixed potential analysis of the differential aeration on identical iron electrodes of equal area, assuming uniform constant solution concentration and current distribution.

same obviously for identical electrodes at the same potential, and there should be no difference in corrosion rates between anode and cathode,[18] assuming uniform current distribution in Figure 6.23. The corrosion of the anode is increased but only to match the cathode corrosion rate, which is decreased by cathodic polarization in the differential aeration cell. Anode and cathode corrosion rates are identical, and there is no preferential anode corrosion, assuming uniform current distribution.

However, polarization differences force nonuniform current distributions in common differential aeration cells, in practice. For example, it has already been demonstrated that anodic current will concentrate near boundaries between coplanar anodic and cathodic areas because of the comparatively low polarization of the anode as compared to the cathode (Figure 6.21). This will also be true even when anode and cathode areas are caused by differential aeration on the same metal. Thus, the penetration rate will be greater in the deaerated anode near the boundary between aerated and deaerated areas, because galvanic current will be concentrated by the low overvoltage behavior of the anode.

Corrosion rate differences between anode and cathode in shorted differential aeration cells are still further accented by concentration changes developing with time. As reduction of dissolved oxygen by reaction (7) continues in the cathode compartment, the pH is observed to increase, due to the liberation of OH^-. At the same time, as anodic dissolution of iron by reaction (5) continues in the anode compartment, the pH decreases. Fe^{2+} cations hydrolyze to form a soluble weak base, $Fe(OH)_2$, leaving behind excess H^+.

$$Fe^{2+} + 2H_2O \rightarrow Fe(OH)_2 + 2H^+. \tag{8}$$

Increased alkalinity passivates the cathode, and increased acidity increases the anodic activity of the anode. Polarization in this acid-alkaline cell is shown schematically in Figure 6.25. Corrosion current at the anode is still controlled by limiting diffusion of dissolved oxygen at $I'_{corr,A}$, but the cathode is passivated at a very low $I'_{corr,C}$, and there is a clear acceleration of anode over cathode corrosion.

This simplified analysis shows the source of accelerated corrosion by differential aeration. It is complicated by solution resistance effects and variations in current and potential distribution, which are unique for each cell geometry.

6.5.2 Partial Immersion Cells—Waterline Corrosion

Concentration changes in differential aeration cells explain accelerated corrosion near the waterline on metals, especially steel and zinc. Corrosion depletes the solution of dissolved oxygen, which is most readily replaced near the surface in an unstirred solution. Depleted oxygen creates an anode at depth with the cathode formed at the waterline by reduction of the excess dissolved oxygen, as shown in Figure 6.26.[19] The surface at the waterline is passivated by formation of local alkalinity from reduction of dissolved oxygen by reaction (7). However, the metal is strongly attacked nearby, where dissolved oxygen is less accessible, but solution resistance is minimized between adjacent surfaces.

The well-known Evans water drop experiment[20] is an even more striking result of differential aeration, forming anode and cathode areas on a steel surface. An air-

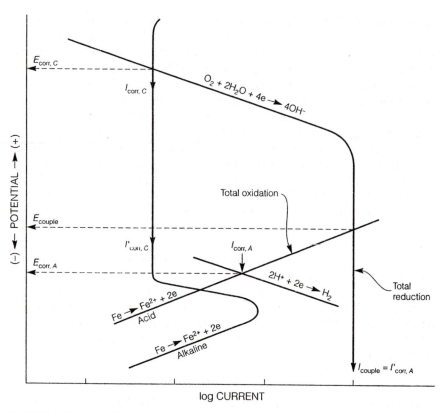

FIGURE 6.25 Mixed potential analysis of a differential aeration cell on iron with passivated cathode and acidified anode.

saturated drop of dilute NaCl solution containing small amounts of phenolphthalein and ferricyanide indicators was placed on an abraded horizontal steel surface. The phenolphthalein indicates formation of OH^- at cathodes by a pink color and the ferricyanide shows liberation of Fe^{2+} at anodes by a dark blue. Anode (blue) and cathode (pink) sites initially developed uniformly beneath the drop (Figure 6.27a), with the anode areas generally congregating on the abrasion lines. As oxygen was depleted in the central area, the blue anode expanded, while the pink cathode segregated in the surrounding areas near the edges of the drop, where dissolved oxygen was more accessible. In an intermediate annular ring between the blue and pink zones, Fe^{2+} from the central anode migrated to the outer alkaline cathode area, reacted there with dissolved oxygen, and precipitated as $Fe(OH)_3$, rust. The final distribution is sketched in Figure 6.27b.

These and similar experiments years ago confirmed the electrochemical mechanism of corrosion in aqueous solutions.

6.5.3 Full-Immersion Cells

Differential aeration cells can form on the same metal surface even in full immersion, in either quiescent or agitated solutions, but longer times are usually required.

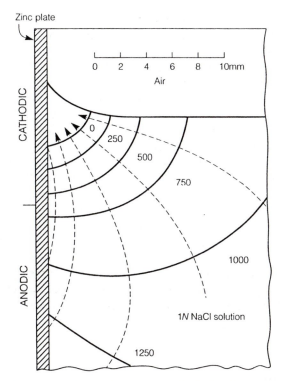

FIGURE 6.26 Potential (solid) and current (dashed) distributions on a vertically immersed zinc plate at the water line. *(From J. N. Agar and U. R. Evans, cited by Kaesche[18].)*

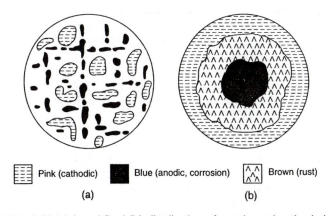

FIGURE 6.27 Initial (a) and final (b) distribution of anode and cathode in a water drop on a horizontal iron surface. *(From U. R. Evans,* An Introduction to Metallic Corrosion, *Arnold, London, p. 36, 1981. Reprinted by permission Edward Arnold Publishers Ltd.*

Cleary[21] used microelectrode techniques to demonstrate that anode areas develop on horizontal surfaces in aerated saline water over periods of days and weeks as a result of local accumulation of ferrous corrosion products. Values of pH as low as 6 were measured under the anode deposits. Nearby cathodic areas remained relatively bright, free of corrosion products, and were characterized by surface pH values up to 10.

Even differences in oxygen mass transport can develop anode and cathode areas as a result of differential aeration. Spinning disks developed bright unattacked cathodes at their outer edges, where dissolved oxygen access was greatest, and anode rust deposits in the center where access was least.[22] Apparently, the anode (5) and cathode (7) reactions can form separate areas, even within the surface adjacent boundary layer, which is unaffected by nearby solution flow. Further evidence is given by the experiments of LaQue and May[23] in flowing aerated seawater. Steel specimens were placed in one wall of a rectangular flow channel. Bright (cathode) and rusted (anode) areas formed again, and cathodic polarization increased the bright cathodic areas at the expense of the anodic rusted ones. The last surviving anodic areas were in the corners of the flow channel, where again dissolved oxygen had least access, even under turbulent flow conditions. It is consistent with our discussions in this and the preceding section that cathodic polarization increases the formation of OH⁻ by reduction of dissolved oxygen (7) to expand the cathodic areas and passivate the nearby anodic ones.[24]

References

1. H. P. Hack, *Metals Handbook*, Vol. 13, *Corrosion*, 9th ed., ASM, Metals Park, OH, p. 234, 1987.
2. M. G. Fontana, *Corrosion Engineering*, 3rd ed., McGraw-Hill, New York, pp. 46–49, 1986.
3. D. A. Jones, *Corrosion*, Vol. 40, p. 181, 1984.
4. D. A. Jones, *Electrochem. Technol.*, Vol. 6, p. 241, 1968.
5. M. Stern and A. L. Geary, *J. Electrochem. Soc.*, Vol. 104, p. 56, 1957.
6. D. A. Jones and N. R. Nair, *Corrosion*, Vol. 41, p. 357, 1985.
7. D. C. Bennett, M.S. thesis, University of Connecticut, 1972.
8. H. P. Hack and J. R. Scully, *Corrosion*, Vol. 42, p. 79, 1986.
9. M. Stern and H. Wissenberg, *J. Electrochem. Soc.*, p. 759, 1959.
10. R. W. Schutz and D. E. Thomas, *Metals Handbook*, Vol. 13, *Corrosion*, 9th ed., ASM International, Metals Park, OH, p. 669, 1987.
11. J. C. Griess, *Corrosion*, Vol. 24, p. 96, 1968.
12. H. R. Copson, *Trans. Electrochem. Soc.*, Vol. 84, p. 71, 1943.
13. H. Kaesche, *Metallic Corrosion*, R. A. Rapp, transl., NACE, Houston, p. 301, 1985.
14. J. T. Waber et al, *J. Electrochem. Soc.*, Vol. 101, p. 271, 1954; Vol. 102, p. 344, 1955; Vol. 102, p. 420, 1955; Vol. 103, p. 64, 1956, Vol. 103, p. 138, 1956; Vol. 103, p. 567, 1956.
15. J. W. Fu, *Corrosion*, Vol. 38, p. 296, 1982.
16. S. Aoki, K. Kishimoto, and M. Miyasaka, *Corrosion*, Vol. 44, p. 926, 1988.
17. H. Grubitsch, cited by U. R. Evans, *The Corrosion and Oxidation of Metals*, Arnold, London, p. 129, 1960.

18. H. Kaesche, *Metallic Corrosion*, R. A. Rapp, transl., NACE, Houston, pp. 313–319, 1985.
19. J. N. Agar and U. R. Evans, cited by Kaesche.[18]
20. U. R. Evans, *An Introduction to Metallic Corrosion*, Arnold, London, p. 36, 1981.
21. H. J. Cleary, *J. Met.*, Vol. 22, No. 3, p. 39, 1970.
22. F. L. LaQue, *Corrosion*, Vol. 13, p. 303t, 1957.
23. F. L. LaQue and T. P. May, *Mater. Prot.*, Vol. 21, No. 5, p. 18, 1982.
24. D. A. Jones, *Corrosion*, Vol. 42, p. 430, 1986.

Exercises

6-1. Compare the potentials of the following listed elements measured in the emf series (Table 2.1) and the Galvanic Series (Figure 6.1), and explain any differences: **(a)** iron, **(b)** platinum, **(c)** nickel, **(d)** aluminum, **(e)** chromium. Take note in your answer of the difference in reference electrodes usually used to measure the two. Comment on any similarities between the emf series and the Galvanic Series.

6-2. The corrosion potential of low-carbon steel in aerated seawater is measured as –0.695 volts vs. SCE. The same measurement for commercially pure nickel gives a corrosion potential of –0.141 volts. Compare these values with similar ones in the Galvanic Series (Figure 6.1) and the emf series (Table 2.1), using the same reference electrode for all. Use polarization diagrams to show why compared values are near or far apart.

6-3. Predict the possibility of galvanic corrosion in seawater for the following coupled pairs of alloys, and indicate which alloy in each pair would be corroded: **(a)** magnesium and low-carbon steel, **(b)** aluminum and zinc, **(c)** aluminum alloy 6061 and yellow brass, **(d)** cast iron and low-alloy steel, **(e)** lead-tin solder and 90-10 copper-nickel, **(f)** nickel Alloy G and Type 430 stainless steel, **(g)** low-alloy steel and Type 304 stainless steel, **(h)** Type 321 stainless steel and Type 430 stainless steel, **(i)** graphite and aluminum alloy 7075.

6-4. It is important to be able to recognize galvanic corrosion. In the following list of cases, explain whether galvanic corrosion is a probable cause of the corrosion.
a. Steel rivets in aluminum drain gutters leak after 2 years of service.
b. Aluminum drain plug in a steel automotive oil pan leaks one month after installation.
c. Testing of unconnected aluminum, copper, and steel specimens exposed in a laboratory reactor shows accelerated corrosion of aluminum.
d. Graphite fiber reinforced aluminum composite specimens show delamination of the fiber after exposure to a salt spray test.
e. Carbon steel pipe leaks near the weld to stainless steel pipe of nearly the same diameter.

6-5. Equal 1-cm^2 areas of metals P and Q are immersed together in sulfuric acid of unit H$^+$ activity having no other dissolved oxidizers. Using the electrochemical parameters listed below: **(a)** Determine corrosion potential and corrosion rate for each, when they are uncoupled. **(b)** When P and Q are electrically coupled, determine the potential and current passing in the galvanic couple so formed. **(c)** Determine the corrosion rates for P and Q when the two are galvanically coupled. **(d)** Which metal is cathodically protected upon galvanic coupling? **(e)** On which metal is the hydrogen evolution rate increased upon galvanic coupling?
Exchange current densities for hydrogen evolution on P, 1×10^{-6}, and on Q, 1×10^{-5} A/cm^2. Exchange current densities for metal dissolution on P, 3×10^{-7}, and on Q, 2×10^{-6} A/cm^2. Reversible potentials for anodic dissolution for P, -0.500, and for Q, -0.600 volt vs. SHE. Assume all Tafel constants are ± 0.1 volt.

6-6. **(a)** Figure 6.7 shows an example in which the uncoupled corrosion rate, $I_{corr,N}$, of cathode is less than the uncoupled corrosion rate, $I_{corr,M}$, of the anode in a galvanic couple. Show in a similar diagram the effect of having $I_{corr,N} > I_{corr,M}$. **(b)** By appropriate changes in electrochemical parameters, use polarization diagrams to show conditions in which the couple potential between two metals will not fall between the uncoupled corrosion potentials of the two metals. For such conditions, which of the two metals is cathodically protected? Which of the two is galvanically corroded?

6-7. Plot the polarization curves and determine the galvanic corrosion rate **(a)** when 1-cm^2 of metal C is coupled to 1-cm^2 of metal A in a corrosive solution; **(b)** when 10-cm^2 of C is coupled to 1-cm^2 of A in the same solution; **(c)** when a solution resistance of 3000 ohms is present between A and C in (b).
For A: $\beta_c = 0.1$ V, $\beta_a = 0.05$ V, $i_{corr} = 1 \times 10^{-7}$ A/cm^2.
For C: $\beta_c = 0.1$ V, $\beta_a = 0.04$ V, $i_{corr} = 1 \times 10^{-6}$ A/cm^2.
$E_{corr,C} - E_{corr,A} = 0.500$ V with C noble to A.

6-8. How might the apparatus of Figure 6.9 be modified to measure effects of a low conductivity corrosive solution in the cell?

6-9. Galvanic corrosion of nickel and nickel alloys is seldom a problem in most environments. Show why this is so from Figures 6.17 and 6.18.

6-10. Determine the galvanic corrosion rate in aerated 3% NaCl of the anode in a couple between equal areas of **(a)** carbon steel (iron) and 316 stainless steel and **(b)** between equal areas of carbon steel and titanium. **(c)** Show the effect of a 15{ra}1 ratio of cathode to anode in (a). **(d)** Explain the difference, if any, between the galvanic corrosion rates in (a) and (b).

6-11. Determine the galvanic corrosion rate in deaerated $1N$ H$_2$SO$_4$ of the anode in a couple between equal areas of **(a)** iron and Hastelloy C and **(b)** iron and copper. For (b), what conditions would cause a greater rate of galvanic corrosion between iron and copper?

Pitting and Crevice Corrosion

7

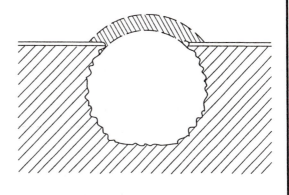

Pitting and crevice corrosion are localized forms of attack that result in relatively rapid penetration at small discrete areas. Pits are often quite small at the surface and easily hidden by apparently inoffensive corrosion products (Figure 1.9). Similarly, localized attack is usually shielded from view within crevices created on metal parts under deposits or between metal and other metal or nonmetal parts. Thus, both pitting and crevice corrosion often remain undetected until leaks result from penetration of the wall thickness. The two are easily confused because the crevice attack under deposits removed by cleaning can be in the form of pits. Both are insidious and unpredictable and often share similar growth processes.

Pitting and crevice corrosion of the stainless alloys containing various proportions of iron, chromium, nickel, and sometimes molybdenum are of the greatest practical interest and are emphasized here. Most failures in stainless alloys occur in neutral-to-acid solutions with chloride or ions containing chlorine. Such conditions are of prime importance in the marine and chemical process industries.

Iron and aluminum pit in alkaline chloride solutions by mechanisms similar to the stainless alloys but in less aggressive conditions. Both corrode at high rates by general corrosion in acid chloride solutions where the stainless alloys are subject to pitting. Considerable available information on the pitting of pure iron in chloride-containing solutions buffered near pH 9[1] is utilized in the sections that follow.

Still another type of crevice corrosion may result from retention of corrosive water or solutions within crevices formed by gaskets, insulation or joints. Examples are discussed in Section 7.4. Thus, crevice corrosion may be caused simply by retaining water at the alloy surface or by creation of concentration cells of aggressive acid solutions in the crevice. Many practical cases of crevice corrosion may be a combination of the two.

Localized deposits from potable waters can foster pitting in copper, and similar deposits from the cooling waters of nuclear steam generators can pit nickel Alloy 600 heat exchanger tubing. These are discussed further in Section 7.5.

7.1 Examination and Evaluation

7.1.1 Conditions Causing Pitting and Crevice Corrosion

Pitting and crevice corrosion result from a failure of the passive film. Thus both service and test solutions must be sufficiently oxidizing to favor passivity, and chloride is usually present as an essential ingredient to break down the passive film and initiate localized corrosion. For example, a 6% $FeCl_3$ (10% $FeCl_3 \cdot 6H_2O$) solution is a common testing media for both pitting and crevice corrosion.[2] The ferric ion acts as the passivating oxidizer by reduction to ferrous ions, and the chloride, of course, is the pitting agent. Hydrolysis of the 6% salt solution produces an acid pH of 1.2. The combination of a strong oxidizer to maintain passivity, an acid solution, and considerable chloride results in an aggressive environment for testing the resistance of the stainless alloys toward pitting and crevice corrosion.

Susceptibility increases with temperature in chloride solutions with a strong oxidizer. A minimum temperature to cause pitting is sometimes used to characterize resistance to pitting[3] and crevice[4] corrosion. Dissolved oxygen is sufficient to passivate stainless steels and induce pitting in the presence of chlorides. In this case, however, higher temperature may reduce pitting due to decreasing solubility of dissolved oxygen.

7.1.2 Evaluation of Pitting Corrosion

Pitting is unpredictable, especially in conditions forming deep pits. The rate is variable, depending on uncertain migration of corrodents into and out of the pit. Pits may be initiated by a number of surface discontinuities, including sulfide inclusions, insufficient inhibitor coverage, holidays or scratches in coatings, and deposits of slag, scale, dust, mud, or sand.

Depending on the metallurgy of the alloy and chemistry of the environment, pits may be shallow, elliptical, deep, undercut, or subsurface and may follow metallurgical features. Typical pit morphologies are sketched in Figure 7.1.[5]

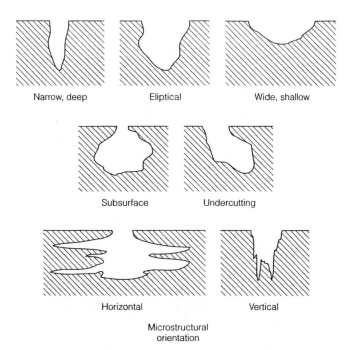

Narrow, deep   Eliptical   Wide, shallow

Subsurface   Undercutting

Horizontal   Vertical

Microstructural
orientation

FIGURE 7.1 Variations in cross sectional shape of pits. *(From* Standard Practice G 46–76, Annual Book of ASTM Standards, *Vol. 3.02, ASTM, Philadelphia, p. 197, 1988. Reprinted by permission, American Society for Testing and Materials.)*

Figure 7.2 shows examples of various pitting modes:[3] (a) shallow pits, (b) deep pits, and (c) deep closely spaced pits bordering on an irregular type of uniform corrosion in some areas. For the stainless steels, a crevice may in some instances be thought of as the preferred initiation site for pitting. Pitted surfaces are often covered by deposits from the process stream and corrosion product precipitates. Thus, the pits in Figure 7.2c may well be the result of deposit corrosion described above. Conversely, the pit mouth may be covered by insoluble deposits formed by corrosion products leaking from the pits. Photographs such as those in Figure 7.2 are often taken after the surface has been cleaned of all deposits, whether precipitated from within the pits or settled out of the process stream.

Pit density (spacing), surface size, and depth may be compared using standard charts. For example, Figure 7.3[5] is useful for comparisons and quantifications for data storage. A pitted specimen characterized as A-4, B-3, C-2 would have pits with average values of surface density, 1×10^5 pits/m^2; surface opening, 8 mm^2; and depth, 0.8 mm. However, full evaluation by this procedure soon becomes tedious and time consuming for any significant number of specimens. Furthermore, maximum values, especially pit depth, are usually more significant than averages.

Clearly, weight-loss methods are inadequate for pitting evaluations because even a very small weight loss can be concentrated in a few pits, with those of maximum depth penetrating the wall thickness to produce failure by leakage. Thus,

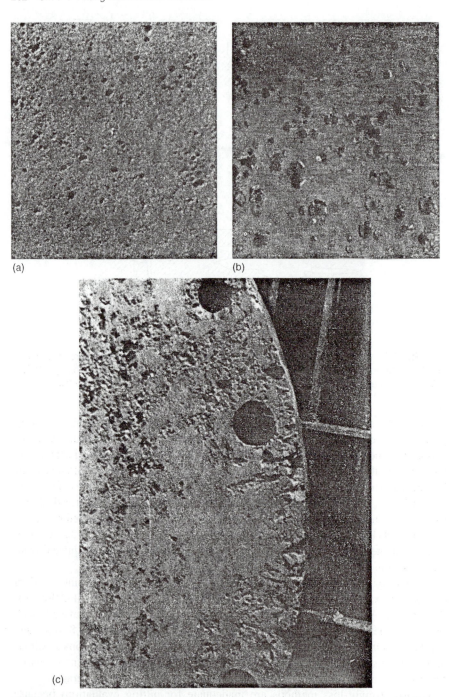

(a)

(b)

(c)

FIGURE 7.2   Examples of pitting in stainless steel: (a) shallow; (b) deep; (c) deep and closely spaced in some areas. *(From A. I. Asphahani and W. L. Silence,* Metals Handbook, *Vol. 13,* Corrosion, *9th ed., ASM, Metals Park, OH, p. 113, 1987. Reprinted by permission, ASM International.)*

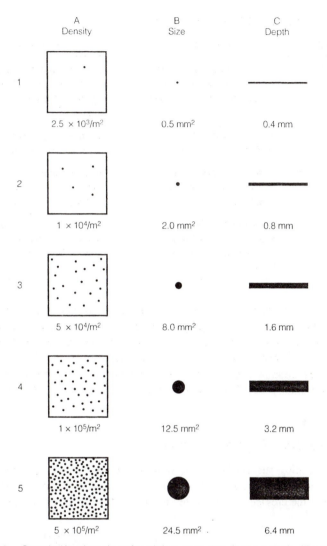

	A Density	B Size	C Depth
1	$2.5 \times 10^3/m^2$	$0.5\ mm^2$	$0.4\ mm$
2	$1 \times 10^4/m^2$	$2.0\ mm^2$	$0.8\ mm$
3	$5 \times 10^4/m^2$	$8.0\ mm^2$	$1.6\ mm$
4	$1 \times 10^5/m^2$	$12.5\ mm^2$	$3.2\ mm$
5	$5 \times 10^5/m^2$	$24.5\ mm^2$	$6.4\ mm$

FIGURE 7.3   Standard rating chart for pitting corrosion (actual size). *(From Standard Practice G 46–76, Annual Book of ASTM Standards, Vol. 3.02, ASTM, Philadelphia, p. 197, 1988. Reprinted by permission, American Society for Testing and Materials.)*

maximum pit depth measurements are usually preferred. Pit depth measurements may be conducted by a number of methods, which are summarized in Table 7.1. Further details on procedures are given by the ASTM.[5] To somewhat quantify the extent of pitting as compared to general attack, a pitting factor, p/d, is sometimes measured in the laboratory, where p is the maximum penetration by microscopy and d is the average penetration by specimen weight loss (Figure 7.4). A pitting factor of unity indicates uniform corrosion. However, the pitting factor goes to infinity, and is thus inappropriate, for cases where general penetration is very low or near zero.

TABLE 7.1 Methods of Measuring Pit Depth

Method	Description	Remarks
Metallographic	Section and polish through selected pit followed by microscopic measurements.	Time consuming. Large uncertainty in selecting deepest pits and sectioning at maximum depth.
Machining	Measure depth where no evidence of pits remains.	Requires sample of regular shape. Sample destroyed.
Micrometer Depth Gauge	Compare readings between surface and pit bottoms with needle probe.	Pits must have large opening. Cannot be used for undercut or directionally oriented pits.
Microscopic	Use calibrated fine focus to determine depth difference between surface and pit bottoms.	Light must reach pit bottom. May not be useful for undercut or subsurface pits.

Maximum pit depth increases not only with time, as would be expected, but also with surface area. The probability of finding a pit of any given depth increases to 100% as the exposed area increases, as shown schematically in Figure 7.5.[6] Thus, it would be very unwise to predict plant life from the results of small laboratory size test coupons. However, the pitting resistance of various alloys can be compared reasonably well from maximum pit depth measurements in the laboratory.

7.1.3 Evaluation of Crevice Corrosion

Depending on alloy susceptibility and solution aggressiveness, crevice corrosion may have various forms. General penetration or broad shallow depressions may be present within the crevice. Pits are often present only near the mouth of very tight

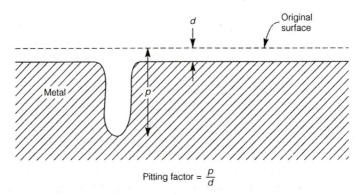

Pitting factor = $\frac{p}{d}$

FIGURE 7.4 Schematic diagram to define the pitting factor, p/d.

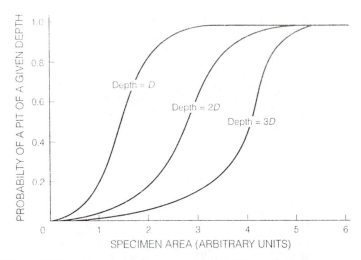

FIGURE 7.5   Effect of exposed area on pit depth. *(From M. G. Fontana, Corrosion Engineering, 3rd ed., McGraw-Hill, New York, p. 72, 1986. Reprinted by permission, McGraw-Hill Book Company.)*

crevices. Figures 1.8 and 7.6[7] show "classical" morphology; attack is greatest at the mouth of the crevice and declines into the interior, while the outer surfaces remain passive and resistant.

The degree of crevice attack is dependent on crevice geometry and dimensions, which are difficult to control quantitatively. Tight crevices usually produce more attack because less crevice volume is required for enrichment mechanisms (Section 7.3). Metal-polymer crevices can be tightened to give less volume and more attack than metal-metal crevices. However, in the latter, galvanic factors may accelerate the attack between dissimilar metals, and more metal area is available to provide solute enrichment within the crevice. The factors affecting test conditions are not well understood, and as a result, crevice testing procedures are largely empirical.

Considerable testing in industrial environments has been conducted with the spool racks, shown in Figure 1.20a. A natural crevice is formed at the polytetrafluoroethylene (PTFE) spacers between specimen coupons. Although the racks were not designed with crevice testing in mind, they do provide a good comparison of materials in the test environment for screening purposes. Assembly procedures and examination schedules are generally not precise enough to yield any information about initiation times or propagation rates of crevice corrosion. Crevice conditions may be more severe in some test assemblies than in service. Alloys that have served satisfactorily for up to 20 years in pulp and paper bleach plant washers showed up to 0.46 mm (18-mil) crevice penetration after only 3 months' exposure on spool-rack specimens in the same environment.[8]

Laboratory testing by ASTM G48[2] in FeCl$_3$ solutions includes a crevice assembly consisting of PTFE blocks held in place on both sides of a sheet specimen with

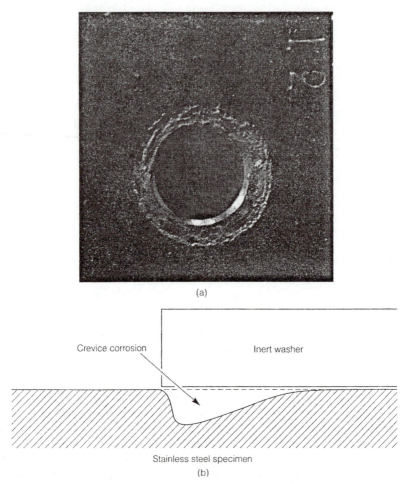

FIGURE 7.6   (a) Crevice corrosion of Type 316 stainless steel in acid condensate under a PTFE spacer; (b) typical schematic morphology with attack greatest at the mouth of the crevice. *(From R. M. Kain, Metals Handbook, Corrosion, Vol. 13, 9th ed., ASM, Metals Park, OH, p. 109, 1987. Photograph by courtesy of R. M. Kain, LaQue Center for Corrosion Technology. Reprinted by permission, ASM International.)*

rubber bands (Figure 7.7). Attack is evaluated by measuring the depth of attack at the crevices under the PTFE blocks and under the rubber band contacts.

A recent advance in testing technology has been the development of multiple crevice test assemblies.[9] Acetal resin or PTFE serrated washers,[10] shown in Figure 7.8, are bolted to either side of a sheet specimen at a specified torque (e.g., 0.28 N-m or 2.5 in.-lb.). The washers make a number of contact sites on either side of the specimen. Figure 7.9[11] shows typical attack patterns after exposure. Additional attack due to corrosion products seeping from crevices or pits (Figure 7.9c) is not unusual for less resistant alloys. The number or percentage of sites showing attack in a given time gives the resistance to initiation, and the average or maximum depth

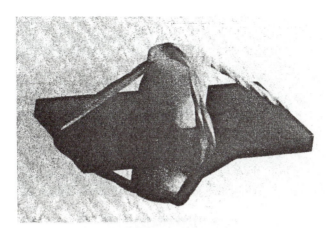

FIGURE 7.7 Crevice corrosion test assembly. *(From ASTM G 48. Photograph by courtesy of T. A. DeBold, Carpenter Technology Corp.)*

of attack indicates the rate of propagation. The large number of sites in duplicate or triplicate specimens is amenable to analysis by extreme-value statistics.[9]

Iron and nickel corrosion-resistant alloys have been compared by minimum temperature and chloride levels to cause crevice attack in multiple-crevice assemblies exposed to $FeCl_3$ solutions.[12] These tests are considered valid for alloy development studies, but alloy performance and ranking in service are difficult to predict from laboratory tests. Kain[10] reports on extensive round-robin testing of multiple-crevice

FIGURE 7.8 Serrated washers used in multiple crevice testing. *(From R. M. Kain, Metals Handbook, Vol. 13, Corrosion, 9th ed., ASM, Metals Park, OH, p. 303, 1987. Photograph by courtesy of R.M. Kain, LaQue Center for Corrosion Technology. Reprinted by permission, ASM International.)*

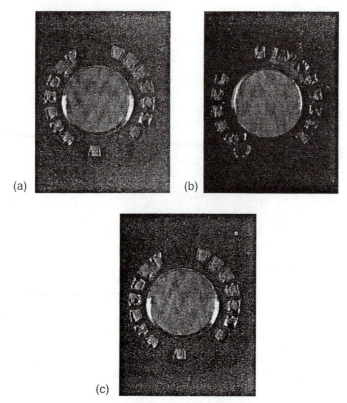

(a)   (b)   (c)

FIGURE 7.9   Crevice attack under serrated washers. (a) Alloy 825, (b) Alloy 20Cb-3, (c) Type 304 stainless steel. *(From R. M. Kain, Corrosion/80, Paper 74, NACE, Houston, 1980. Photograph by courtesy of R. M. Kain, LaQue Center for Corrosion Technology. Reprinted by permission, ASM International.)*

procedures in seawater environments. Variability from site to site was considerable but less at any individual site. This variability is considered to be typical for crevice testing,[10] and multiple-crevice testing has become very popular, replacing the less controllable rubber band attachment methods of Figure 7.7.

7.1.4 Prevention

Methods of prevention for pitting and crevice corrosion are similar and involve either decreasing the aggressiveness of the environment or increasing the resistance of materials. Solution aggressiveness is reduced by decreasing the chloride content, acidity, and temperature, separately or together. Stagnate process streams allow deposits to form, and the accumulated corrosion products hydrolyze and reduce pH. Thus, regular uniform flow, elimination of deadlegs and other stagnant areas, and occasional surface cleaning, whenever feasible, will reduce the incidence of pitting and crevice corrosion. Suspended solids should be removed by filtering or settling to minimize formation of deposits. Equipment should be designed for complete drainage, avoiding areas that retain standing solutions (Figure 15.13).

Inhibitors may be helpful but must be used with caution, because insufficient inhibitor content in any part of the equipment may aggravate the damage by making fewer but deeper pits. Cathodic protection can also stop both pitting and crevice corrosion in marine applications (Section 13.3.2) but is not usually a feasible option in more aggressive chemical-process streams.

Alloys resistant to pitting are also resistant to crevice corrosion. Increased chromium, nickel, molybdenum, and nitrogen increase the resistance to both pitting and crevice corrosion of stainless steels. Nickel alloys of equivalent chromium and molybdenum content are more resistant (and expensive) than the stainless steels. Carbon and sulfur impurities are usually detrimental to the resistance of both iron- and nickel-based stainless alloys. Titanium alloys are resistant to pitting in even the most aggressive plant environments but are subject to crevice corrosion in chloride and other halide solutions above 70°C.[13] See also Section 15.1.6.

Crevice corrosion can be controlled by design to remove crevices as far as possible. Butt-welded joints are preferable to riveted or bolted joints in the design of new structures. Welding or caulking may be used to close crevices in new or existing equipment. Impervious gasketing materials are preferable to porous ones.

Various equipment design measures to control pitting and crevice corrosion are described in more detail in Section 15.2.3.

7.2 Mechanism of Pitting Corrosion

The differential aeration cell, described in Section 6.5.2 by the Evans water-drop experiment, may be considered a macromodel for the initiation of pitting and crevice corrosion.[1] Corrosion at the center of the water drop leads to deaeration, acidification, and formation of a localized anode. Areas at the perimeter become alkaline by cathodic reduction of dissolved oxygen, which has greater access to the outer surfaces of the drop. These processes form localized anodes within pits and crevices, which are supported by cathodes on the surrounding surfaces.

Once initiated, the pit provides a sheltered area that prevents easy mass transport between the pit interior and the exterior bulk solution. Hydrolysis of the corrosion product, chloride, results in acidic solutions that destroy passivity locally and create an active corroding anode within the pit. The pit anode is supported by reduction of dissolved oxidizers on the surrounding cathode surfaces.

This mechanism, described above in general terms, as well as others thought to occur in copper and nickel alloys under localized deposits, is discussed in more detail in the following sections.

7.2.1 Initiation: Pitting Potential

Pitting initiates at a critical pitting potential, E_{pit}, which is used as a measure of resistance to pitting corrosion. The presence of chloride in an acid solution generally increases potentiostatic or potentiodynamic anodic currents at all potentials, but the most singular feature is the dramatic increase in current at the critical potential, E_{pit}, as shown schematically in Figure 7.10. This increase in current density above E_{pit} measures low-overvoltage anodic dissolution within pits, which initiate and become visible at the critical potential. The more noble is E_{pit}, the more resistant is the alloy to pitting. Good correlation has been demonstrated between E_{pit} and the resistance of a series of Fe-Cr alloys to pitting in long-term exposures to seawater in the absence of crevices.[14] As the chromium content increases, E_{pit} becomes more noble, and the alloys become more resistant to pitting.

Pits also initiate above E_{pit} when the potential (i.e., corrosion potential) is established chemically by a dissolved oxidizer. Figure 7.11 shows the results of potential measurements in $FeCl_3$ solutions for a series of alloys with varying E_{pit} and resistance to pitting.[15] The potential increased steadily with time after immersion, and pitting was initiated in any alloy when E_{pit} was exceeded. However, the time and potential above E_{pit} before initiation were not always reproducible. Table 7.2[15] compares E_{pit} measurements with maximum and minimum corrosion potentials for the alloys of Figure 7.11. In every case, pitting was observed when the maximum measured potential exceeded E_{pit}. Pit initiation arrested the increasing corrosion potential, which decreased thereafter in the active direction to values active to E_{pit} during pit propagation. However, Hastelloy C did not pit because the maximum corrosion potential, the solution redox potential for Fe^{3+}/Fe^{2+}, was below E_{pit} for this alloy.

The actual mechanism of pit initiation at E_{pit} is not well understood,[16] but the following sequence of events[17] is supported by considerable experimental data. As potential increases and approaches E_{pit}, the concentration of Cl^- increases at a passive stainless steel surface, as measured with Cl^- sensitive microelectrodes, whether

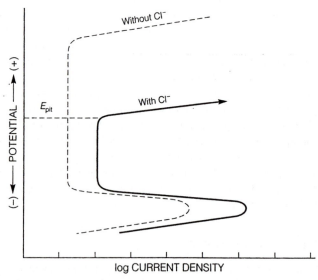

FIGURE 7.10   Schematic determination of critical pitting potential, E_{pit}, from anodic polarization.

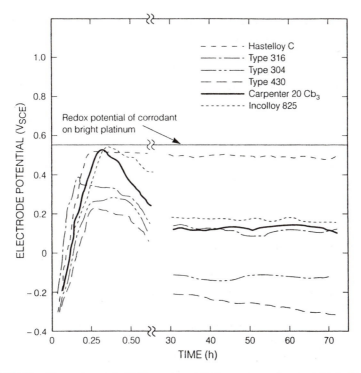

FIGURE 7.11   Corrosion potential measurements on corrosion resistant alloys during pitting in acidified FeCl₃ solutions (1.185 *N* in Cl ⁻). *(From B. E. Wilde and E. Williams, J. Electrochem. Soc., Vol. 117, p. 775, 1970. Reprinted by permission, The Electrochemical Society.)*

the potential is induced chemically by a redox reaction or electronically with a potentiostat.[18] This results from the electrostatic attraction between the positively charged surface and the negatively charged Cl⁻ anion. Electron microprobe measurements[19] have revealed the accumulation of relatively thick chloride salt

TABLE 7.2   Corrosion Potential Data Compared with Pitting Potentials (Volts vs. SCE)

Alloy Designation	E_{pit}	Corrosion Potential	
		Maximum	**Minimum**
Type 430 stainless steel	−0.130	0.230	−0.310
Type 304 stainless steel	−0.020	0.280	−0.140
Type 316 stainless steel	0.100	0.385	0.090
Carpenter 20 Ch	0.0500	0.520	0.120
Incoloy 825	0.525	0.530	0.180
Hastelloy C	>0.900	0.530	0.530

From B. E. Wilde and E. Williams, J. Electrochem. Soc., Vol. 117, p. 775, 1970. Reprinted by permission, The Electrochemical Society.

"islands" on the surface of iron at potentials even below E_{pit}. The Cl⁻ must be fairly strongly bound because simple surface rotation has no effect on E_{pit}.[20] However, the accumulation of Cl⁻ is apparently dispersed by ultrasonic vibration, which makes E_{pit} more active.[21] A high-chloride, low-pH microenvironment may be developed beneath an island by the hydrolysis reaction, with cation corrosion products from the alloy exemplified by

$$Fe^{2+} + 2H_2O + 2Cl^- \rightarrow Fe(OH)_2 + 2HCl. \tag{1}$$

$Fe(OH)_2$ is a weak base and HCl a strong acid, leading to the suggested low pH. Other anions of strong acids, such as sulfate and nitrate, will also hydrate to acid pH, but chloride is far more mobile in solution and more aggressive in both pitting and crevice corrosion. The acid hydrolysis reaction is still more apparent during the propagation of pits (Section 7.2.2).

It is not known exactly how chloride interacts with the passive film, but the rotating ring-disk experiments of Heusler and Fischer[22] are instructive. In Figure 7.12, an increase of Fe^{3+} emanating from a potentiostatically controlled rotating iron disk was detected after introduction of Cl⁻. Enhanced reduction current for Fe^{3+} was measured at the adjacent annular ring electrode (Section 3.5.5). No corresponding increase in the anodic current at the electrode was apparent when the chloride was introduced. With the disk held at a potential above E_{pit} in the presence of the added chloride, initiation of pitting produced the expected increase in the anodic dissolution current, coincident with the appearance of an oxidation current for Fe^{2+} to Fe^{3+} at the annular ring. An explanation of this experimental behavior follows.

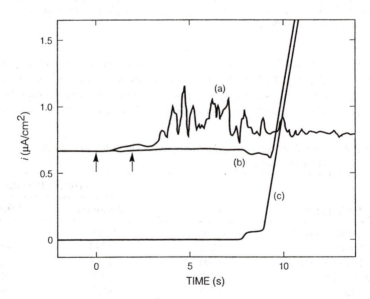

FIGURE 7.12 Generation of (a) Fe^{3+} after introduction of Cl⁻ (arrows) with constant (b) potentiostatic anodic current which increases along with generation of (c) Fe^{2+} upon initiation of pitting. *(From K. E. Heusler and L. Fisher, Werkst. Korros., Vol. 27, pp. 551, 697, 788, 1976. Reprinted by permission Werkstoffe und Korrosion.)*

In the absence of chloride, the passive film dissolves slowly, as ferric ions (Figure 7.13a),

$$FeOOH + H_2O \rightarrow Fe^{3+} + 3OH^-, \qquad (2)$$

where FeOOH represents the hydrated passive film, with iron in the ferric oxidation state. Chloride has been proposed[17] to catalyze the liberation of Fe^{3+} by displacement of the outer layers of the passive film (Figure 7.13b);

$$FeOOH + Cl^- \rightarrow FeOCl + OH^-$$
$$FeOCl + H_2O \rightarrow Fe^{3+} + Cl^- + 2OH^-, \qquad (3)$$

where FeOCl approximates the composition of the salt islands on the passive film and presumably dissociates to yield the enhanced Fe^{3+} measured by Heusler and Fischer.[22] Reactions (3), or others that are similar, thin or remove the passive film (Figure 7.13c) at a preferred site until direct anodic dissolution to Fe^{2+} initiates a pit. Auger electron spectroscopy (AES) has indicated that chloride adsorbs on the outer surfaces of the passive film with little or no penetration.[23,24]

Preferred sites for pit initiation are often related to sulfide inclusions. Mixed $(MnFe)S_x$ sulfides, especially those associated with oxides of aluminum or chromium, have been found to be the most potent nucleants.[25,26,27] The mechanism of nucleation at inclusions is uncertain. A microcrevice may be created by dissolution of the inclusion; the inclusion may be electrochemically active and corrode preferentially.[17] However, recent studies[28] have revealed that mineral sulfides are generally noble to passive stainless steel. Therefore, in some cases at least, sulfides may cre-

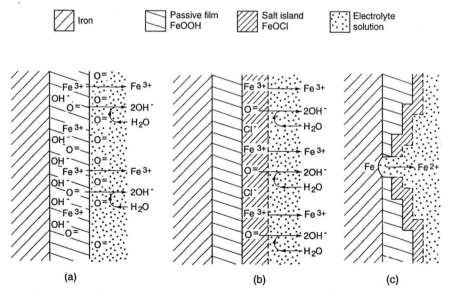

FIGURE 7.13 (a) Slow dissolution of passive film forming Fe^{3+}; (b) accelerated dissolution at a soluble salt island forming Fe^{3+}; (c) direct anodic dissolution at a pit initiation site forming Fe^{2+}.

ate a microgalvanic couple which locally accelerates anodic dissolution of the nearby metal. Precipitation, segregation, cold work, and heat treatments can affect size and distribution of pits.

At E_{pit}, sufficient chloride has concentrated in salt islands at the surface to start a new anodic reaction at the initiation sites. The increase in anodic current above E_{pit} represents the polarization curve for the new anodic process characteristic of the pit environment, as discussed in the next section. The experimental value of E_{pit} depends very much on procedure. Rapid potential scanning produces pitting sooner and elevates E_{pit}. Slow scanning allows more time for chloride accumulation and initiation to lower E_{pit} but requires longer testing time. *In situ* surface-scratching methods have been developed to determine the most active value of E_{pit} in the shortest time.[29]

These principles are similar to the localized acidification theory developed by Galvele,[30,31] based primarily on data for non-ferrous alloys. Space is not available in this textbook to mention all of the considerable effort which has been devoted to the theory of pitting corrosion over the years. A more complete review with historical perspectives by Smialowska[16] is recommended.

7.2.2 Propagation: Protection Potential

Figure 7.14 for pitting of iron in a slightly alkaline chloride solution serves as a simplified model for stainless steel. Copious anodic production of positively charged Fe^{2+} attracts negative anions, e.g., Cl^-, to the initiation site. Hydrolysis by

$$Fe^{2+} + 2H_2O + 2Cl^- \rightarrow Fe(OH)_2 + 2HCl \tag{1}$$

produces local pH reductions at the initiation site. The result is a self-propagating or *autocatalytic* mechanism of pit growth. The acid chloride solution further accelerates anodic dissolution, which in turn further concentrates chloride in the pit. An

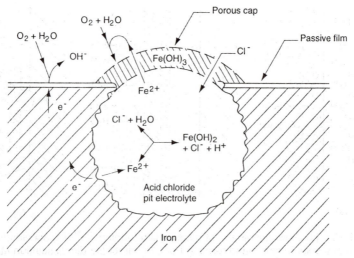

FIGURE 7.14 Schematic of processes occurring at an actively growing pit in iron.

insoluble cap of $Fe(OH)_3$ corrosion products collect at the pit mouth when Fe^{2+} diffuses out of the acid pit interior to the exterior, where it is oxidized to Fe^{3+} and precipitates in the neutral bulk solution. The cap impedes easy escape of Fe^{2+} but is sufficiently porous to permit migration of Cl^- into the pit, thereby sustaining a high acid chloride concentration in the pit. Anodic polarization of the pit interior occurs by coupling to the exterior passive cathode surfaces. Cathodic reduction of a dissolved oxidizer such as oxygen consumes the electrons liberated by the anodic pit reaction.

For stainless steels, the additional anodic reactions for nickel and chromium are similar to those for iron in the pit. Both chromium and nickel cations hydrolyze to acid chlorides by reactions similar to (1). Although chromium strongly passivates the outer surfaces, it aggravates pitting and crevice corrosion by hydrolyzing to even lower pH within the pit than iron, as discussed further, later in this section. Typical undercut pit morphology for stainless steel pitting is shown in Figure 7.15. The morphology is similar to that in Figure 1.9, but on a smaller scale.

The importance of the cathodic reaction to sustain pitting should be understood. Pit growth cannot continue without a cathodic reduction reaction to consume the electrons liberated by (i.e., to polarize anodically) the pit anode reaction. Thus, pits are widely spaced in aerated salt solutions, because oxygen has limited solubility, and a large surrounding area is needed to provide enough reduction capability to support the central pit anode. Any pit initiating within the cathodic area of a larger pit is suppressed by cathodic protection. The greater solubility of Fe^{3+} yields greater reduction rates on a smaller surface area, and pits can survive much closer to one

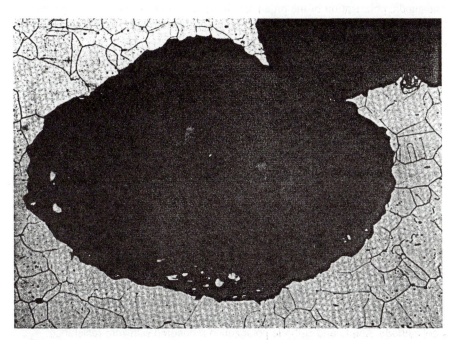

FIGURE 7.15   Pitting in Type 304 stainless steel. *(Photograph by courtesy of C. D. Kim, USX Corp.)*

another in FeCl$_3$ test solutions. Pits sustained above E_{pit} by potentiostatic anodic polarization are also very closely spaced because the instrumentation provides the necessary anodic polarization, removing the cathodic reaction to the auxiliary or counter electrode.

The electrochemical mechanism of pit growth has been clarified by separating pit anode and cathode, as shown in Figure 7.16.[32] A single pit could be initiated at the mouth of the tight crevice in Figure 7.16 by applying a slight anodic current for a few seconds. Nearby passive surfaces of the pit anode were masked with a resistant lacquer coating. A separate, pit-free, passive cathode of the identical Type 304 stainless steel was coupled to the pit anode electrode. The crevice area of the cathode was coated to ensure that no localized pit anodes could easily initiate. The pit anode and adjoining cathode on surrounding surfaces were thus separated to enable individual study of the polarization processes on each. The crevice pit, once initiated, was able to grow by simple shorting to the separated cathode with the switch, S, closed (Figure 7.16).

The galvanostatic (controlled-current) circuit of Figure 6.10 was operated initially as a zero resistance ammeter by applying sufficient current to maintain potential of anode and cathode at the same potential, E_{couple}, with S open in Figure 7.16. Separate anode and cathode polarization curves were produced simultaneously by reducing the current in the circuit, while measuring potentials at anode and cathode for each current step. Figure 7.17[32] shows the polarization of the pit anode and the passive surface cathode for stainless steel in oxygenated 3% NaCl at 90°C. The passive surface was strongly polarized as a cathode by reduction of dissolved oxygen. The anodic polarization of the pit anode shows low overvoltage at noble potentials typical of polarization in concentrated chloride solutions (see Figure 7.18).

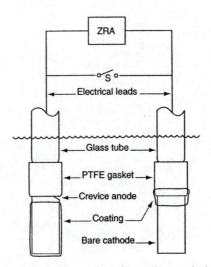

FIGURE 7.16 Cell consisting of a crevice pit anode coupled to a passive stainless steel cathode. Anode and cathode polarization measured with zero resistance ammeter (ZRA) circuit of Figure 6.10. *(From D. A. Jones and B. E. Wilde,* Corros. Sci., *Vol. 18, p. 631, 1978. Reprinted by permission, Pergamon Press.)*

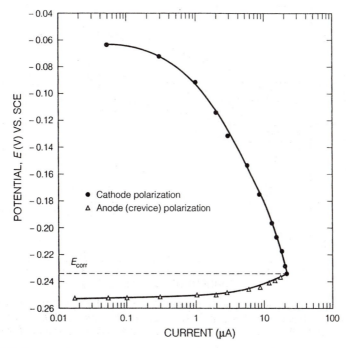

FIGURE 7.17   Anodic and cathodic polarization curves produced from the crevice pit cell of Figure 7.16.   3% NaCl, 90°C, O_2 purge. *(From D. A. Jones and B. E. Wilde, Corros. Sci., Vol. 18, p. 631, 1978. Reprinted by permission, Pergamon Press.)*

Susceptibility to pitting and crevice corrosion in stainless steel is maximized near 90°C in dilute chloride solutions.[33]

Concentrated chloride solutions of nickel, iron, and chromium produce very acidic pH, as indicated in Table 7.3.[32] Polarization in such strong acid chloride solutions accounts for noble potentials in the active state, with low overvoltage for the pit-cell anode of Figure 7.17. Although chromium confers passivity on the stainless alloys, it also enhances susceptibility to localized pitting and crevice attack, because it hydrolyzes so strongly to very low (even negative) pH. Nickel does not hydrolyze as strongly as the other two, possibly accounting for the generally improved pitting resistance of the higher-nickel alloys.

TABLE 7.3   Room Temperature pH of Concentrated Salt Solutions

Salt	1 N	3 N	Saturated
$NiCl_2$	3.0	2.7	2.7
$FeCl_2$	2.1	0.8	0.2
$CrCl_3$	1.1	−0.3	−1.4

From D. A. Jones and B. E. Wilde, Corrosion Sci., Vol. 18, p. 631, 1978. Reprinted by permission, Pergamon Press.

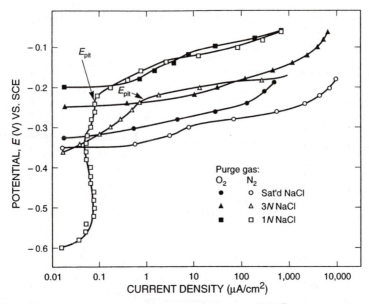

FIGURE 7.18 Potentiostatic anodic polarization curves for Type 304 stainless steel in sodium chloride solutions at 90°C. *(From D. A. Jones and B. E. Wilde, Corros. Sci., Vol. 18, p. 631, 1978. Reprinted by permission, Pergamon Press.)*

In advanced stages, pits may become deep enough that reduction of H^+ on the pit walls near the outer surface is possible, while the pit bottom still sustains anodic dissolution. Thus, hydrogen bubbles emanating from pits have been observed.

Potentiostatic anodic polarization curves for Type 304 stainless steel in concentrated solutions of NaCl are shown in Figure 7.18. Pitting was observed in the 1 *N* and 3 *N* solutions, but only general corrosion in the saturated solution. Once passivity has been lost, even at only discrete points on the surface, anodic polarization is controlled by the new reactions in the pits growing at those points. Similar behavior was observed in acidic solutions formed by concentrated nickel, ferrous, and chromium chlorides,[32] and it was concluded that chloride concentration is more important than acidity in the initiation and growth of pits. Concentrated chloride solutions place active anodic polarization behavior at very noble values, compared to similar curves in more dilute solutions. The separated pit anode polarization curve in Figure 7.17 falls in the range of noble potential and low overvoltage defined by the polarization curves in concentrated NaCl solutions. It is apparently not necessary to postulate a large ohmic IR_Ω potential drop through the electrolyte between pit anode and surface cathode[34] to maintain the former in the active state.

A *protection potential*, E_{prot}, has been defined by a cyclic potentiostatic or potentiodynamic procedure, as shown schematically in Figure 7.19. After some degree of anodic polarization above E_{pit}, the direction of polarization is reversed in a cyclic polarization test, and hysteresis is observed in which the return polarization curve follows an active path, compared to the initial anodic one. The crossover at the passive current density defines E_{prot} below which established pits presumably cannot continue to grow. By contrast, new pits initiate only above E_{pit}. Between E_{prot} and

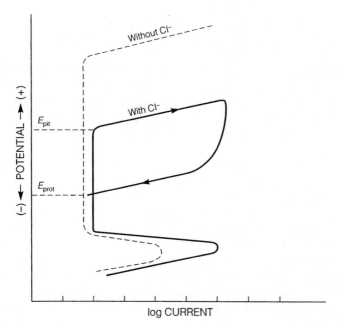

FIGURE 7.19   Cyclic polarization to give pitting, E_{pit}, and protection, E_{prot}, potentials (schematic).

E_{pit}, new pits cannot initiate, but old ones can still grow. An alloy that is resistant to pitting shows no hysteresis, whereas susceptible alloys show increasing hysteresis, as shown in Figure 7.20.[35]

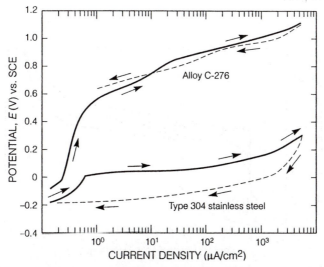

FIGURE 7.20   Cyclic polarization for alloys resistant (Hastelloy C) and susceptible (Type 304 stainless steel) to pitting. *(From* Standard Method G61–86, Annual Book of ASTM Standards, *Vol. 3.02, ASTM, Philadelphia, p. 254, 1988. Reprinted by permission, American Society for Testing and Materials.)*

E_{prot} is very much a function of experimental procedure, as is E_{pit}. Longer time above E_{pit} or slower potential scan allows greater chloride concentration within the pits. The reverse polarization curve, characteristic of the occluded pit electrolyte, is depressed to more active values of potential as chloride becomes more concentrated in the pit, as shown in Figure 7.21.[36] The depression of anodic polarization to more active values by higher chloride concentration was demonstrated in Figure 7.18. Crevices create geometric features on the surface to enhance chloride concentration and further depress E_{prot}, as discussed in the next section.

Thus, the reverse polarization curve in the cyclic procedure and the anodic polarization curve in the pit cell of Figure 7.17 are representations of the anodic process in the pit anode. The unpolarized $E_{corr,A}$ for the pit anode is identical to E_{prot}. Any potential active to E_{prot} would cathodically polarize the pit anode and would be expected to suppress or stop corrosion within the pits. This agrees with the observation[37] that pit growth occurs only at potentials more noble than the open-circuit anode potential of the pit interior. E_{prot} becomes more active as chloride concentration increases, but concentration is limited at the condition where diffusion out of the pit balances migration into the pit. Nevertheless, the most conservative value of E_{prot} would be E_{corr} in a saturated solution of the chloride salts of the metals constituting the alloy.[17]

As in any electrochemical cell, there are potential and current distributions around a pit dictated by geometry and solution electrolyte conductivity. The potential profile across a pit from precision microprobe measurements is shown in Figure 7.22.[38] Again, strong polarization of the cathode is apparent. Polarization does not extend to a great distance outward from the pit center, because of the high cathode polarization of the Fe^{3+}/Fe^{2+} redox system, despite the high conductivity of the electrolyte (Section 6.4.3).

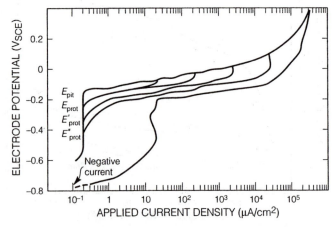

FIGURE 7.21   Effect of longer polarization above E_{pit} on E_{prot} determined by cyclic polarization. *(From B. E. Wilde, Corrosion, Vol. 28, p. 283, 1972. Reprinted by permission, National Association of Corrosion Engineers.)*

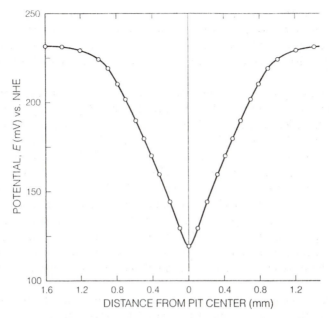

FIGURE 7.22 Potential determinations across a pit by precision microprobe measurements. *(Adapted from Rozenfeld and Danilov.[38])*

7.3 Mechanism of Crevice Corrosion

A crevice shields part of the surface and thereby enhances the formation of differential aeration and chloride concentration cells. Both play a large role in the initiation and propagation of crevice corrosion, as they do in pitting.

7.3.1 Initiation by Differential Aeration

Corrosion consumes the dissolved oxygen in the small volume of a crevice, impairing passivity and increasing the concentration of metal cations, which attract negatively charged cations such as Cl^- from the bulk solution. Thus, the crevice serves as a ready-made initiation site for localized corrosion. The initiation process is again rather similar to that in the Evans water-drop experiment (Section 6.5.2).

The potential for initiation of crevice corrosion is more active than E_{pit} for the same alloy, because of the favorable geometric conditions for deaeration and chloride concentration. Thus, a specimen with a crevice will initiate attack in the crevice at potentials active to E_{pit}. However, the initiation or crevice breakdown potential is very dependent on crevice geometry and tightness, and no reproducible values have been established for crevice corrosion.

7.3.2 Propagation by Chloride Concentration

Once initiated, crevice corrosion grows much as it does within a pit. Chloride concentration and acid hydrolysis result in a concentrated acid chloride solution within the crevice by reaction (1), creating a localized anode coupled to a large surface area cathode on the surrounding surfaces. Crevice corrosion grows autocatalytically as more chloride is attracted to the crevice, promoting further hydrolysis and consequent acidity. The similarities to growth processes within a pit (Figure 7.15) are apparent.

As shown in Figure 7.23, the resistance to crevice corrosion in seawater exposures has been correlated with the difference between E_{pit} and E_{prot} in the cyclic polarization test (Section 7.2.1).[36] The rationale for the correlation is that a pit once initiated contains the same conditions as within actively corroding crevices. A test that shows resistance to propagation of pitting should do the same for crevice corrosion. Some alloys of intermediate resistance between Hastelloy C and the austenitic stainless steels may not show a well-defined E_{pit}. Such alloys (e.g., Incoloy 825 and Carpenter 20 Cb3) have shown an initiation or crevice breakdown potential, E_{cb}, with a crevice present. The difference, $E_{cb} - E_{prot}$, in the cyclic polarization test again correlates reasonably well with resistance to crevice corrosion in limited seawater exposure testing (Table 7.4[36]). However, alloys may be ranked in a different order if the crevice geometry and tightness do not exactly duplicate those in service.[39]

Again, due to geometric constraints, E_{prot} with a crevice should be more active than the equivalent value without the crevice. The crevice allows more extensive chloride concentration and acid hydrolysis, which pushes the unpolarized corrosion potential within the crevice to more active levels. Any potential active to the unpolarized corrosion potential cathodically protects the anode and stops corrosion in the crevice. The limit for E_{prot} is still the unpolarized anode corrosion potential in a solution saturated in the chloride salts of cations from the alloy (Section 7.2.2). Crevice corrosion or pitting can never initiate or propagate at any potential active to that value because the occluded anode concentration cannot exceed saturation.

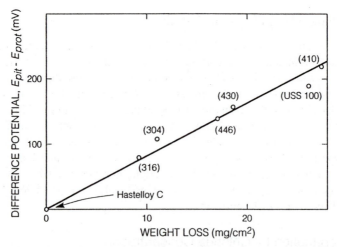

FIGURE 7.23 Correlation between difference potential, $E_{pit} - E_{prot}$, and weight loss due to crevice attack after 4.25 years in seawater. *(From B. E. Wilde, Corrosion, Vol. 28, p. 283, 1972. Reprinted by permission, National Association of Corrosion Engineers.)*

TABLE 7.4   Comparison of Electrochemical and Two-Year Seawater Exposure
Data for Crevice Corrosion

Alloy Designation	$E_{cb} - E_{prot}$ (V)	Weight Loss (mg/cm²)
Hastelloy C	0.00	0.16
Incoloy 825	0.02	4.10
Carpenter 20 Cb3	1.00	26.10

From B. E. Wilde, Corrosion, Vol. 28, p. 283, 1972. Reprinted by permission, National Association of Corrosion Engineers.

It should be noted that the unpolarized corrosion potential of the anode in Figure 7.17 is not stable. When uncoupled from the surrounding surface cathode, the pit interior is no longer subject to anodic current, and the interior acid chlorides eventually escape from the pit, which eventually passivates at the noble potential of the unpolarized cathode.

7.4 Other Forms of Crevice Corrosion

Crevice corrosion may result simply from retention of water or corrosive solutions within a crevice, while the outer surfaces are dried out or exposed only to a relatively noncorrosive environment. The severe attack resulting from the chloride concentration and acid hydrolysis mechanism (Section 7.3) may be absent, but the effects are often significant, nevertheless. Uniform or somewhat nonuniform general attack is present in the crevice, inasmuch as the crevice environment is relatively uniform.

For crevices formed by easily deformable insulation, packing, or gaskets, corrosion products accumulated within the crevice are usually harmless. However, when the crevice materials are more rigid, accumulated corrosion products can cause distortion in the crevice. Damage from distortion may be merely cosmetic, as in filiform corrosion, or may have more serious structural or operational consequences, as discussed in later examples in this section.

7.4.1 Deposit and Gasket Corrosion

A corrosive environment may be formed by deposits or wet packing materials. A recently critical example of the latter has been newly discovered corrosion under thermal insulation coverings used to maintain temperature in piping and containment vessels (Figure 7.24). Thermal cycling, which leads to condensation, evaporation, and concentration of solutes, leads to general corrosion of carbon steel and even stress corrosion cracking of stainless steel under the insulation.[40] No measures have been found to keep moisture from penetrating the insulation. Coatings on the steel surfaces and inhibitor-impregnated insulation have proven ineffective. Since vessel and piping walls are relatively thick and corrosion rates moderate, common practice is simply to monitor attack ultrasonically through "windows" cut periodically in the insulation and to replace, when necessary, those parts suffering excessive damage.

FIGURE 7.24 Corrosion beneath insulation coverings. *(From W. G. Ashbaugh, Process Industries Corrosion, NACE, Houston, p. 759, 1986. Photograph by courtesy of W. G. Ashbaugh, Cortest Engineering Services, Inc. Reprinted by permission, National Association of Corrosion Engineers.)*

An example of attack under a gasket is shown in Figure 7.25.[41] Attack takes the form of broad areas of general corrosion, in this case without chloride. The cause is uncertain. It may be deaeration and loss of passivity within the gasket crevice, as observed in other acetic acid service.[42] However, it is always difficult to justify a deaeration mechanism of crevice corrosion electrochemically without impurity concentration effects in the crevice (see Exercise 7-6).

7.4.2 Filiform Corrosion

Another manifestation of differential aeration and crevice corrosion is filiform corrosion, which has been observed under thin organic coatings on steel, aluminum and magnesium food and beverage cans, packaging, and aircraft structures exposed to humid atmospheres. An example under the paint coating on aluminum aircraft skin is shown in Figure 7.26.[43] Filiform corrosion usually initiates at scratches or other defects in the coating and propagates laterally as narrow, 0.05- to 3-mm-wide filaments under the coating. Penetration into the substrate metal is usually only superficial. The filaments consist of an actively corroding head followed by an inactive tail filled with corrosion products. On steel, the head of the filament is usually blue, blue-green, or gray, and the tail is a rusty red, indicating that the head is deaerated while the tail is aerated.

Oxygen is consumed by active corrosion at the head and is accompanied by hydrolysis and acidification to a pH of 1 to 4.[44] Filiform corrosion is enhanced by atmospheric constituents such as soluble chlorides, sulfates, sulfides, or carbon diox-

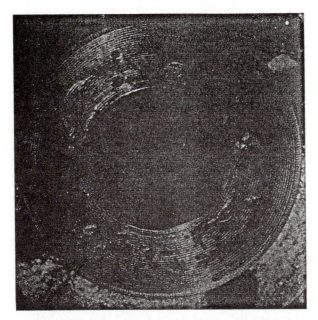

FIGURE 7.25   Gasket corrosion on a stainless steel pipe flange carrying acetic and formic acid solutions at 49°C. *(From E. V. Kunkel,* Corrosion, *Vol. 10, p. 260, 1954. Reprinted by permission, National Association of Corrosion Engineers.)*

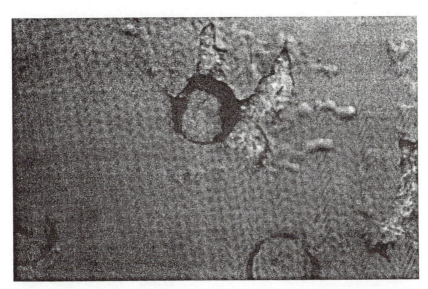

FIGURE 7.26   Filiform corrosion under paint coating on aluminum aircraft skin. *(From M. L. Bauccio,* Metals Handbook, *Vol. 13,* Corrosion, *9th ed., ASM, Metals Park, OH, p. 1019, 1987. Reprinted by permission, ASM International.)*

ide, which assist acidification during differential aeration. Precipitation of rusty red $Fe(OH)_3$ occurs when Fe^{2+} from the head contacts aerated conditions in the tail of the filament. Essential water and oxygen migrate to the corroding filament head through porosity or microcracks in the coating over the filament tail. $Fe(OH)_3$ decomposes to $Fe_2O_3 \cdot 3H_2O$ as the filament advances. The processes are shown schematically in Figure 7.27.

Sealing the tail against water and oxygen transport deactivates a filament.[45] A propagating filament cannot cross the path of another, presumably because the tail of the earlier one provides an oxygen source that disrupts the propagating filament cell. Coating permeability and composition have little or no effect on the initiation and growth characteristics of filiform corrosion. Growing filaments are unaffected by type or thickness of the coating. Thus, a typical differential aeration cell, as in the Evans water-drop experiment (Section 6.5.2), is set up between a deaerated acidified anode at the filament head and a nearby alkaline cathode fed by water and oxygen through the filament tail.[45,46] Mechanisms for aluminum and magnesium are parallel to that of steel.

The only sure way to prevent filiform corrosion on steel, aluminum, or magnesium is to reduce the relative humidity below nominally 60%, thereby dehydrating the filament cell. Improved coating quality, multiple coatings, and inhibited coatings and primers will retard, but not totally prevent filiform corrosion in operating equipment, where the atmosphere cannot be controlled. Corrosion resistant substrates of stainless steel, titanium, or copper do not exhibit filiform corrosion.

7.4.3 Atmospheric Crevice Corrosion

Structural crevices often retain water and other solutions, while the outer surfaces are washed clean and become relatively dry after periodic wetting. The inner crevice surfaces are thus exposed to corrosion for longer periods and are therefore subject to greater attack. An example of a bolt that had been severely corroded in the crevice area by retained water is shown in Figure 1.7.

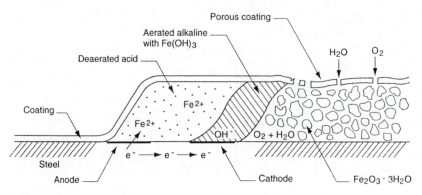

FIGURE 7.27   Chemical processes in a filament cell during filiform corrosion of steel.

Certain compositions of high-strength low-alloy weathering steels have been developed for atmospheric corrosion resistance.[47] Small amounts of Cu, Cr, Ni, and Si (at levels of about 0.2 to 1.2%) result in a tenacious oxide film (Section 12.1.2) that confers improved atmospheric corrosion resistance if the surfaces are allowed to dry periodically in normal service. However, bolted lap joints that retain water in a humid environment for longer times may develop distortion due to accumulation of corrosion products in the joint. A severely distorted joint is shown in Figure 7.28. Crevice attack may be aggravated by alternating currents leaking to the ground through the legs of transmission towers fabricated from weathering steel.[48]

Designers have derived minimum bolt-spacing guidelines[49] to ensure adequate mechanical resistance to joint distortion. Caulking or other sealants may be appropriate for some larger overlapping joints.[50]

7.4.4 Denting in Nuclear Steam Generators

Nickel Alloy 600 heat-exchanger tubing in the steam generators of pressurized water nuclear reactors have been subject to distortion (denting) by accumulation of corrosion products in the crevice between the tubing and carbon-steel support plates.[51,52] Primary cooling water recirculates from the reactor core through Alloy 600 tubing in the steam generators. The secondary water is heated in the steam generator to 290°C by the hot surfaces of the Alloy 600 tubing to form steam at 1000 psi, which drives the power-generating turbines. The spent steam from the turbines is condensed, purified in resin beds (in some designs), and recycled to the steam genera-

FIGURE 7.28 Distortion in a bolted joint of high strength low alloy steel exposed for several years in a humid atmosphere.

tors. The secondary water picks up impurities from the condensers, resin beds, makeup water system, and additives for control of dissolved oxygen, pH, and general water chemistry. Denting is generally limited to nuclear plants using chloride bearing waters (seawater and brackish waters) to cool the condensers.

Several consequences may result from uncontrolled denting. These include tube cracking and leakage of primary radioactive water into the secondary cooling system, distortion of the support plates, loss of dimensional stability in the tube assembly, and reduced heat transfer and steam-generating efficiency by gross deformation of the tubes.

Denting is apparently due to an occluded cell in the crevice between the carbon-steel support sheet and the Alloy 600 heat-exchanger tubing. Chloride impurities concentrate in the crevice by migration to neutralize the cation corrosion products, as described earlier for crevice corrosion in stainless alloys. Boiling and evaporation due to limited heat transfer in the crevice further concentrate impurities, and galvanic corrosion between carbon steel and Alloy 600 accelerates cation concentration in the crevice. The carbon-steel galvanic anode produces bulky insoluble ferrous oxide corrosion products in hot chloride-bearing waters, according to the studies of Potter and Mann.[53] Accumulation of the corrosion products in the crevice produces sufficient pressure to cause denting of the Alloy 600 tubing.

Denting has been mitigated in steam generators by improved control of secondary water chemistry, which includes reducing condenser inleakage and adding boric acid to the secondary water. The boric acid is thought to neutralize the acid chloride crevice solution. New and replacement steam generators have introduced broached support plates with a less constricted contact area at the tubes and/or replaced carbon steel with more corrosion-resistant stainless steel (Types 405 and 409) in the support plates.[51]

7.5 Pitting Under Localized Deposits

Pitting may also occur under deposits or films caused by precipitates from neutral water. An occluded anode is still present within the pit, but the cathode has been identified at the exterior surface of the deposit. Such pitting has been observed in copper in potable waters at ambient and slightly elevated temperature and nickel Alloy 600 in nuclear steam generators in pressurized water at about 300°C. It also may operate in pitting under tubercules generated by microbiologically induced corrosion (Section 11.3.3).

The key to the mechanism of underdeposit pitting is the conductivity of the oxide-hydroxide and sulfide deposits on many metals. For example, copper, iron, and nickel oxides and sulfides are known to be electrically conductive. Thus, their outer surfaces can serve as the cathode for reduction of dissolved oxygen and other dissolved oxidizers from the bulk solution, as shown schematically for the generalized metal, M, in Figure 7.29. The occluded cell is formed by differential aeration, and the cell electrolyte may be acidified by hydrolysis of chloride or sulfide, if available. The deposit over the pit is the insoluble hydroxide, oxide, or sulfide of M^+, which is electrically conductive and supports reduction of dissolved oxygen as the

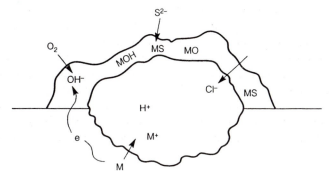

FIGURE 7.29   Schematic mechanism of the occluded cell in underdeposit pitting of metal M.

deposit thickens. Underdeposit pitting might be thought of as a case of crevice corrosion under an electrically conductive porous deposit. Corrosion is localized under the deposits, but pits need not be narrow and deep as they are for stainless steels in salt solutions described in Section 7.2. Details of the mechanism for different alloys in different conditions are uncertain, but a conductive deposit in aerated neutral water seems common.

7.5.1 Copper in Fresh Waters

Wall penetration within 5 years by pitting has been observed commonly in copper plumbing supply tubes at ambient temperature. The attack is apparently restricted to certain well waters of high hardness (high dissolved solids) at pH 7 to 8.2.[54] Below this pH range, corrosion is more rapid and generalized. Above, protective scales are formed. The attack is sporadic and unpredictable; in adjoining buildings, one may experience extensive pitting and the other none.

Surfaces showing pitting are characterized by an outer hydrated cupric carbonate [$Cu_2CO_3(OH)_2$, malachite] layer over an inner cuprous oxide (Cu_2O, cuprite) layer on the metal substrate. Pits apparently initiate under the cuprite inner layer and generate surface nodules at the pit site by excretion of corrosion products through a pin hole in the cuprite. The nodules consist of basic cupric salts, mostly malachite and calcium carbonate. A bright metallic surface in the pit is revealed when the oxide layer is broken or removed. Multiple pits are often aligned longitudinally along the bottom of horizontal tubes.

The mechanism of cold-water pitting is uncertain. $CaCO_3$ in the pit nodule indicates that cathodic reduction of dissolved oxygen is localized immediately over the pit site. Reduction of dissolved oxygen locally increases pH, which stimulates precipitation of $CaCO_3$ (Section 13.1.5). The cuprite layer, a p-type semiconductor, must then carry sufficient charge to support reduction on the oxide surface. Hydrolysis of copper chlorides has been postulated to form a typical acidified occluded anode in the pit,[55] but no chloride has been observed in pits in the United States of America.[54] Dissolved carbon dioxide correlates with the incidence of pitting[56] but may be present merely as an inverse function of pH.

Carbon contamination on the inner surfaces of copper tubing has been associated in Europe with initiation of pitting.[57] Although U.S. manufacturers claim that their methods preclude carbon contamination, pitting persists, nevertheless.

Prevention involves both surface conditioning and water treatment. Baking, pickling, and abrasive cleaning have been used to remove carbon contamination.[54] Water conditioning to raise pH (or concurrently reduce dissolved CO_2) has been used successfully.[56] Widespread scaling by precipitation of $CaCO_3$ and related compounds may result from waters of high hardness when pH is increased.[54] The sporadic nature of pitting and uncertain control in most systems make evaluation of prevention measures difficult.

7.5.2 Alloy 600 Nuclear Steam Generator Tubes

Sludge deposits collect on top of the lower tubesheet in vertical steam generators of pressurized water nuclear power plants. The deposits are acidic and have significant chloride content from condenser inleakage. The same generators that experience denting (Section 7.4.4) are also susceptible to pitting of the tubes within the sludge pile. Pits are randomly scattered on the surface and may be found very close to one another. Figure 7.30 shows typical morphology within a pit.[58] The pits are filled with a laminated deposit consisting of chromium-rich oxide layers uniformly dispersed with copper. Microseams of metallic copper are often present in the fissures between the laminate layers.

Despite large uncertainties, some reasonable suggestions can be made as to the mechanism. Local boiling on the tube surfaces concentrates solutes from the sludge pile, especially chloride, at the surface. Cations dissolved from the alloy hydrolyze to produce a strongly acidic solution at the surface, which further promotes corro-

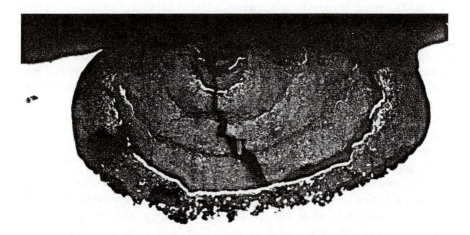

FIGURE 7.30 Pit in Alloy 600 steam generator tube. *(From A. K. Agrawal, W. N. Stiegelmeyer, J. F. Sykes, and W. E. Berry,* Report NP-3029, *Electric Power Research Institute, Palo Alto, CA, May 1983. Photograph by courtesy of A. K. Agrawal, Battelle-Columbus Laboratories.)*

sion in an autocatalytic manner (Section 7.2.2). Corrosion products precipitate as oxides as they migrate into the bulk environment of higher pH. Chromium is presumably enriched in the resulting deposits because of lower solubility at the operating temperature of about 300°C. The layered structure presumably results from differing growth rates and thermal cracking of the oxide during operational maintenance interruptions. The presence of interspersed and layered copper suggests that cathodic reduction reactions, including $Cu^{2+} + 2e^- \rightarrow Cu$, are present within or at the outer surface of the oxide deposit as the pit grows. Cathodic reduction proceeds at oxide surfaces which are exposed to solutions bearing Cu^{2+}, CuO, or dissolved oxygen, either in fissures or at the outer surfaces adjacent to the sludge pile.

Pitting in Alloy 600 would be prevented by elimination of the sludge pile. Although total elimination is difficult, if not impossible, some measures have been used to control sludge buildup. Evacuation procedures (sludge lancing) and chemical cleaning are in use to remove the sludge.[51] Better secondary water control reduces chloride and dissolved oxygen and limits the entry of suspended solids which settle out in the sludge pile. Chlorides and dissolved oxygen cause increased corrosion throughout the system, as well as increased pitting. Improved maintenance of brass condensers and replacement with titanium have reduced dissolved copper in steam generators and inleakage of contaminants, especially chloride, through the condensers.

References

1. H. Kaesche, *Metallic Corrosion*, R. A. Rapp, transl., NACE, Houston, pp. 321–54, 1985.
2. Standard Method G48-76, *Annual Book of ASTM Standards*, Vol. 3.02, ASTM, Philadelphia, p. 209, 1988.
3. A. I. Asphahani and W. L. Silence, *Metals Handbook*, Vol. 13, *Corrosion*, 9th ed., ASM International, Metals Park, OH, p. 113, 1987.
4. E. L. Hibner, *Mater. Perform.*, Vol. 26, No. 3, p. 37, 1987.
5. Standard Practice G 46-76, *Annual Book of ASTM Standards*, Vol. 3.02, ASTM, Philadelphia, p. 197, 1988.
6. M. G. Fontana, *Corrosion Engineering*, 3rd ed., McGraw-Hill, New York, p. 72, 1986.
7. R. M. Kain, *Metals Handbook*, Vol. 13, *Corrosion*, 9th ed., ASM International, Metals Park, OH, p. 109, 1987.
8. A. H. Tuthill, *Mater. Perform.*, Vol. 24, No. 9, p. 43, 1985.
9. D. B. Andersen, *ASTM STP 576*, ASTM, Philadelphia, p. 261, 1976.
10. R. M. Kain, *Metals Handbook*, Vol. 13, *Corrosion*, 9th ed., ASM International, Metals Park, OH, p. 303, 1987.
11. R. M. Kain, *CORROSION/80*, Paper 74, NACE, Houston, 1980.
12. R. S. Treseder and E. A. Kachik, *ASTM STP 866*, ASTM, Philadelphia, p. 373, 1985.
13. R. W. Schutz and D. E. Thomas, *Metals Handbook*, Vol. 13, *Corrosion*, 9th ed., ASM International, Metals Park, OH, p. 303, 1987.
14. B. E. Wilde and E. Williams, *J. Electrochem. Soc.*, Vol. 118, p. 1057, 1971.
15. B. E. Wilde and E. Williams, *J. Electrochem. Soc.*, Vol. 117, p. 775, 1970.
16. Z. Szklarska-Smialowska, *Pitting Corrosion of Metals*, NACE, Houston, p. 377, 1986.
17. D. A. Jones, *Corrosion Processes*, R. N. Parkins, ed., Applied Science, Englewood, NJ, p. 180, 1982.

18. B. E. Wilde, *Passivity and Its Breakdown in Iron Base Alloys*, NACE, Houston, p. 129, 1976.

19. M. Janik-Czakor, A. Szummer, and Z. Szklarska-Smialowska, *Corros. Sci.*, Vol. 15, p. 775, 1975.

20. F. Mansfeld and J. V. Kenkel, *Corrosion*, Vol. 35, p. 43, 1979.

21. T. Nakama and K. Sasa, *Corrosion*, Vol. 32., p. 283, 1976.

22. K. E. Heusler and L. Fischer, *Werkst. Korros.*, Vol. 27, pp. 551, 697, 788, 1976.

23. Z. Szklarska-Smialowska, H. Viefhaus, and M. Janik-Czakor, *Corros. Sci.*, Vol. 16, p. 649, 1976.

24. J. B. Lumsden and R. W. Staehle, *Scr. Metall.*, Vol. 6, p. 1205, 1972.

25. Z. Szlarska-Smialowska, *Corrosion*, Vol. 28, p. 388, 1972.

26. M. Janik-Czakor, A. Szummer, and Z. Szlarska-Smialowska, *Br. Corros. J.*, Vol. 7, p. 90, 1972.

27. Z. Szlarska-Smialowska, A. Szummer, and M. Janik-Czakor, *Br. Corros. J.*, Vol. 5, p. 159, 1970.

28. D. A. Jones and A. J. P. Paul, *Hydrometallurgical Reactor Design and Kinetics*, R. G. Bautista, R. J. Wesley, and G. W. Warren, eds., TMS-AIME, Warrendale, PA, p. 293, 1987.

29. N. Pessal and C. Liu, *Electrochim. Acta*, Vol. 16, p. 1987, 1971.

30. J. R. Galvele, *J. Electrochem. Soc.*, Vol. 123, p. 464, 1976.

31. J. R. Galvele, J. B. Lumsden, and R. W. Staehle, *J. Electrochem. Soc.*, Vol. 1204, 1978.

32. D. A. Jones and B. E. Wilde, *Corros. Sci.*, Vol. 18, p. 631, 1978.

33. H. H. Uhlig and R. W. Revie, *Corrosion and Corrosion Control*, 3rd ed., Wiley, New York, p. 313, 1985.

34. H. W. Pickering and R. P. Frankenthal, *J. Electrochem. Soc.*, Vol. 119, p. 1297, 1972.

35. Standard Method G61-86, *Annual Book of ASTM Standards*, Vol. 3.02, ASTM, Philadelphia, p. 254, 1988.

36. B. E. Wilde, *Corrosion*, Vol. 28, p. 283, 1972.

37. Y. Kitamura and H. Suzuki, *Proc. 4th Int. Cong. Metallic Corrosion*, NACE, Houston, p. 716, 1972.

38. I. L. Rozenfeld and I. S. Danilov, *Corros. Sci.*, Vol. 7, p. 129, 1967.

39. R. M. Kain, *Corrosion/80*, Paper 74, NACE, Houston, 1980.

40. W. G. Ashbaugh, *Process Industries Corrosion*, NACE, Houston, p. 759, 1986.

41. E. V. Kunkel, *Corrosion*, Vol. 10, p. 260, 1954.

42. D. A. Jones, *The Forms of Corrosion, Recognition and Prevention*, C. P. Dillon, ed., NACE, Houston, p. 19, 1982.

43. M. L. Bauccio, *Metals Handbook*, Vol. 13, *Corrosion*, 9th ed., ASM International, Metals Park, OH, p. 1019, 1987.

44. C. Hahin, *Metals Handbook*, Vol. 13, *Corrosion*, 9th ed., ASM International, Metals Park, OH, p. 104, 1987.

45. R. T. Ruggeri and T. R. Beck, *Corrosion*, Vol. 39, p. 452, 1983.

46. H. Kaesche, *Werkst. Korros.*, Vol. 11, p. 668, 1959.

47. Standard Specification A588, *Annual Book of ASTM Standards*, ASTM, Philadelphia.

48. D. A. Jones, *Corrosion*, Vol. 43, p. 66, 1987.

49. R. L. Brockenbraugh and R. J. Schmitt, *Preprint C75 041-9*, Winter Meeting IEEE, New York, Jan. 1975. J. B. Vrable, R. T. Jones, and E. H. Phelps, *Mater. Perform.*, Vol. 18, p. 39, 1979.

50. S. K. Coburn and Y. Kim, *Metals Handbook*, Vol. 13, *Corrosion*, 9th ed., ASM International, Metals Park, OH, p. 515, 1987.

51. S. J. Green, *Metals Handbook*, Vol. 13, *Corrosion*, 9th ed., ASM International, Metals Park, OH, p. 937, 1987.
52. A. R. Vaia, G. Economy, and M. J. Wooten, *Mater. Perform.*, Vol. 19, p. 9, Feb. 1980.
53. E. C. Potter and G. M. W. Mann, *Br. Corros. J.*, Vol. 1, p. 26, 1965.
54. H. Cruce, O. von Franque, and R. D. Pomeroy, *Internal Corrosion of Water Distribution Systems*, American Water Works Association, Denver, p. 317, 1985.
55. V. F. Lucey, *Br. Corros. J.*, Vol. 2, p. 175, 1967.
56. A. Cohen and J. R. Myers, *Corrosion/84*, Paper 153, NACE, Houston, 1984.
57. H. S. Campbell, *J. Inst. Met.*, Vol. 77, p. 345, 1950.
58. A. K. Agrawal, W. N. Stiegelmeyer, J. F. Sykes, and W. E. Berry, *Report NP-3029*, Electric Power Research Institute, Palo Alto, CA, May 1983.

Exercises

7-1. For a pitted specimen designated as A-3, B-3, and C-5 (Figure 7.3), sketch the pit density, indicate pit diameter, and give the pit depth.

7-2. Assuming that Figure 7.2a has the proper magnification, estimate the pit size and pit density of the pits in Figure 7.2a, using Figure 7.3 for comparison.

7-3. Sketch the cyclic polarization curve for Type 304 stainless steel, showing exact values of the pitting and protection potentials. Compare this sketch with the trace of potential vs. time in Figure 7.11 for Type 304 stainless steel and indicate the significance of pitting and protection potentials on the potential-time plot.

7-4. Judging from the discussion of the pit initiation mechanism in Section 7.2.1, what is the effect of scan rate on pitting potential in a potentiodynamic anodic polarization experiment? Would you classify pitting potential as a unique material property? Explain.

7-5. **(a)** From Figure 7.11, what is the corrosion potential of Type 430 stainless steel after 40 hours in this acidified $FeCl_3$ solution? What would happen to pitting if this specimen were cathodically polarized at this time? Could we call this the protection potential? Is it constant? **(b)** Referring to Figure 7.21, how does the corrosion potential in (a) compare to the protection potentials obtained for Type 430 stainless steel by cyclic polarization? Can we conclude that protection potential is a unique material property?

7-6. It is suggested in Section 7.4.1 that crevice corrosion in Figure 7.25 is caused by a loss of passivity due to consumption of dissolved oxygen in the crevice. **(a)** Justify the suggested mechanism with a diagram showing the polarization curves for the electrochemical reactions occurring in a crevice corrosion cell with loss of passivity in the crevice due to differential aeration. **(b)** Is the chloride concentration cell mechanism possible, as shown for pitting in Figure 7.14? Explain.

7-7. Is it possible to have crevice corrosion resulting from differential acidity with no other dissolved oxidizers? In such a case, the crevice anode reaction would be anodic dissolution of the active metal in a depleted acid solution in the crevice due to consumption of H^+ by corrosion, and the outer surfaces would act as cathodes for reduction of bulk solution H^+. As usual, the sheltered crevice anode surface is much smaller than the large surrounding cathode surface. Use polarization diagrams to justify your answer assuming **(a)** a metal that is always in the active state at all concentrations of H^+, and **(b)** a metal that passivates at high concentrations of H^+.

7-8. Make a sketch similar to Figure 7.29 for the underdeposit pitting corrosion of copper in hard waters, using the discussion in Section 7.5.1 for your guide. Include in your sketch the composition and morphology of all deposits and the flow of all charge carriers.

7-9. Furuya and Soga (*Corros. Engr.*, Vol. 39, p. 79, 1990) claim that the pitting potential of aluminum alloys in chloride solutions can be measured by the corrosion potential when dissolved Cu^{2+} is present. Is this test valid? Explain. Would there be an effect of exposure time on E_{corr}?

8

Environmentally Induced Cracking

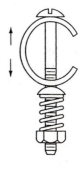

Environmentally induced cracking (EIC) is a general term for brittle mechanical failures that result from a synergism between tensile stress and a corrosive environment. Corrosion rates are usually quite low, and design stresses to cause EIC are often below the yield stress. However, when tensile stress and a corrosive environment are combined, EIC can result. EIC includes stress corrosion cracking (SCC), corrosion fatigue cracking (CFC), and hydrogen embrittlement or hydrogen induced cracking (HIC). In any given situation, more than one of the three may be operative, further complicating analysis of failures and the determination of appropriate prevention methods. HIC is more correctly classified as a part of hydrogen damage effects, which are discussed in Chapter 10. However, HIC is included here because of its similarity to other forms of EIC and the consequent comparisons in the literature and in practice.

Liquid metal embrittlement (LME) and solid metal embrittlement (SME) of alloys may also occur in the presence of tensile stress while in contact with liquid and solid metals, respectively. These are considered in this book primarily as metallurgical rather than corrosion effects. However, future study of LME and SME may lead to better understanding of the conventional forms of EIC discussed in this chapter, and further mention is made in Section 8.8.1.

HIC has a number of synonyms in addition to hydrogen embrittlement used above. These include hydrogen-assisted cracking, hydrogen-assisted stress corrosion cracking, hydrogen stress cracking, and sulfide cracking. The latter two are used particularly in the presence of sulfides and sulfur-bearing compounds (Section 11.2.1). Unfortunately, usages of these terms vary.

This chapter begins with a definition of SCC, a description of the critical effects of electrochemical potential, and a brief description of the more widely known combinations of alloy and environment in which SCC has been observed under constant load or deformation. CFC under cyclic stress is defined and discussed in a similar manner. HIC is then described and compared to the other two forms of EIC. Methods of prevention are also described for all forms of EIC. The chapter continues with a discussion of testing methods, including the principles and techniques of modern fracture mechanics. The concluding discussion is devoted to current theories on the mechanism of SCC and relationships with CFC and HIC.

New designs and technology continually demand higher-performance alloys which must be exposed to still more severe conditions of stress, temperature, and corrosion. Modern alloys of high strength and corrosion resistance are often more susceptible to EIC. As a result, the incidence of EIC has risen rapidly in recent years. SCC, CFC, and HIC are perhaps the most pernicious forms of corrosion, because of the elusiveness of their mechanisms and the resultant unpredictability of their occurrence.

It is difficult to make unqualified statements regarding the characteristics of any of the forms of EIC because exceptions are not unusual. Nevertheless, some common characteristics of SCC, CFC, and HIC are summarized in Table 8.1 for purposes of comparison in the following sections. *The characteristics listed must be regarded only as general, but useful guidelines.* For example, SCC and HIC, as well as CFC, may have corrosion products in the cracks if the failed parts are exposed for long periods after the cracks have formed. Also, fatigue failures in vacuum or inert atmospheres are increased in some cases even by the very low corrosion rates resulting from exposure to humid air. On rare occasions (see, for example, Section 8.4.2), cathodic polarization may move an alloy into a potential region of SCC susceptibility. Failure analysis is further complicated by the fact that two or more may operate simultaneously. Again, it should be emphasized that these characteristics apply to *most* cases of practical importance. An exhaustive discussion of minor exceptions is beyond the scope of this chapter.

8.1 Characteristics of Stress Corrosion Cracking

8.1.1 Definition and Description

Stress corrosion cracking (SCC) is the brittle failure at relatively low constant tensile stress of an alloy exposed to a corrosive environment. SCC was apparently first reported as the so-called season cracking of brass in ammonia-bearing environments in the early twentieth century and was a serious problem in the failure of cartridge

TABLE 8.1   Characteristics of Environmentally Induced Cracking[a]

Characteristic	Stress Corrosion Cracking	Corrosion Fatigue Cracking	Hydrogen Induced Cracking
stress	static tensile	cyclic with some tensile	static tensile
aqueous corrosive environment	specific to the alloy	any	any
temperature increase	accelerates	accelerates	increases to RT, then decreases
pure metal	more resistant	susceptible	susceptible
crack morphology	TG or IG branched sharp tip	TG unbranched blunt tip	TG or IG unbranched sharp tip
corrosion products in the crack	absent (usually)	present	absent (usually)
crack surface appearance	cleavage like	beach marks and/or striations	cleavage like
cathodic polarization	suppresses (usually)	suppresses	accelerates
near maximum strength level	susceptible, but HIC often predominates	accelerates	accelerates

[a]RT: room temperature, TG: transgranular, IG: intergranular

cases for firearms during both world wars. Caustic cracking with resulting explosions, of carbon-steel steam-engine boilers became a serious and dangerous problem in the 1920s.

Historically, it has been thought that three conditions must be present simultaneously to produce SCC: a critical environment, a susceptible alloy, and some component of tensile stress (Figure 8.1). Environmental species are often specific to the alloy system and may not have an effect on other alloys of different type. For example, hot aqueous chloride solutions readily crack stainless steels but do not have the same effect on carbon steels, aluminum, or other nonferrous alloys. Not all environments cause cracking of any particular alloy, but new alloy-environment combina-

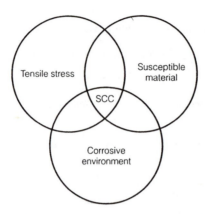

FIGURE 8.1 Simultaneous tensile stress, susceptible metallurgical condition, and critical corrosive solution required for stress corrosion cracking.

tions resulting in SCC are being discovered on a regular basis. Thus, many researchers are of the opinion that a specific environment is not required for SCC. Nevertheless, the engineer should be familiar with those alloy-environment combinations which are known to produce SCC, in order to avoid them in design. Various well-known alloy-environment combinations susceptible to SCC are discussed in Section 8.4.

Although the three factors of Figure 8.1 are not usually present together, time and service conditions may conspire to produce the necessary combinations that result in surprising and expensive failures. Boiling and evaporation can concentrate the critical solutes in very dilute and otherwise nonaggressive solutions. Tensile stresses even below yield are sufficient to cause SCC and may result from bolting and fastening parts that fit together imperfectly. Uneven thermal expansion and contraction can leave residual tensile stresses after welding and other heat treatments (Section 9.3.1).

SCC is normally associated with static tensile stresses. However, only slight, long-term variations in loading (e.g., even once a week) are known to accelerate the onset of SCC. It is uncertain whether such effects should be attributed to SCC or corrosion fatigue. Further discussion of low-cycle loading effects appears in Section 8.2.1.

8.1.2 Metallurgical Effects

Pure metals are more resistant to SCC than alloys of the same base metal but are not immune. For example, pure copper has been induced to crack[1] in slow strain rate tests (Section 8.6.3), but these conditions are quite severe in comparison to service conditions. Virtually all alloys are susceptible to some degree in the appropriate environments, and susceptibility increases with strength in any given alloy class. However, even alloys of low yield strength are susceptible (e.g. brasses and the stainless steels). A complete understanding of SCC requires an explanation of the high resistance of pure metals, as compared to the alloys of the same metals.

SCC may be either transgranular or intergranular, but the crack follows a general macroscopic path that is always normal to the tensile component of stress. In transgranular failures, the cracks propagate across the grains usually in specific crystal planes, usually having low indices such as {100}, {110}, and {210}.[2] The cracks follow grain boundaries in the intergranular mode. Transgranular failures are less common than the intergranular ones, but both may exist in the same system or even in the same failed part, depending on conditions. Example metallographic sections of the two crack morphologies are shown in Figure 8.2.[3] Scanning electron micrographs of fracture surfaces appear in Figure 8.3. Facing halves of recently failed specimens often exactly match one another, whether intergranular or transgranular, indicating that cracking is primarily by mechanical fracture, with little electrochemical dissolution or corrosion during the fracture process. However, anodic dissolution does play a significant role, as discussed in the following Section 8.1.3.

The intergranular failure mode suggests some inhomogeneity at the grain boundaries. For example, segregation of sulfur and phosphorous at grain boundaries is the probable cause of intergranular SCC of low alloy steels. And in fact, intergranular stress corrosion cracking may be the result of stress-assisted intergranular corrosion since most alloys showing such failures also show at least weak evidence of intergranular corrosion (Sections 9.1 and 9.2) without stress.[4]

8.1.3 Electrochemical Effects

Electrochemical potential has a critical effect on stress corrosion cracking (SCC). Figure 8.4[5,6] shows the schematic potentiodynamic anodic polarization curve for a typical active-passive corrosion-resistant alloy, with crosshatched zones where SCC occurs in susceptible alloy-solution combinations. The passive film is an apparent prerequisite for SCC, but the two zones of susceptibility appear at the potential boundaries where the passive film is less stable. In zone 1, SCC and pitting are associated in adjacent or overlapping potential ranges. The common example of zone-1 SCC is austenitic stainless steel in hot $MgCl_2$ solutions.[6] SCC occurs in a narrow potential range, with pitting present at slightly noble potentials and passivity, with no cracking at slightly active potentials. Although stress corrosion cracks may initiate at pits due to stress intensification, they are not necessarily a prerequisite for SCC, even in zone 1. However, in some instances, potent solutions or oxides that are unstable on the exposed surface can accumulate within pits and initiate cracks. For example, cracking of carbon steels exposed to hot water or nitrates initiates in pits where magnetite can accumulate.[4]

In zone 2, far from the pitting potential range, SCC occurs where the passive film is relatively weak at active potentials barely adequate to form the film. Zone-2 SCC is typified by carbon steel in hot carbonate/bicarbonate solutions.[7] SCC has been observed even in the active region, for example, for carbon steel in strong caustic solutions.[7] However, because anodic currents decrease with time, film formation and

(a)

(b)

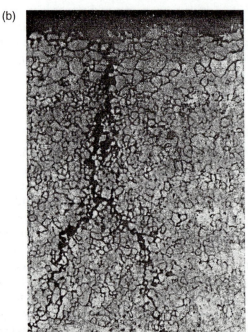

FIGURE 8.2 Metallographic sections of stress corrosion cracking: (a) transgranular in brass. *(Photograph by courtesy of A. Cohen, Copper Development Association.)* (b) intergranular in ASTM A245 carbon steel. *(From B. E. Wilde, Metals Handbook, Failure Analysis, Vol. 11, 9th ed., ASM, Metals Park, OH, p. 203, 1986. Reprinted by permission, ASM International.)*

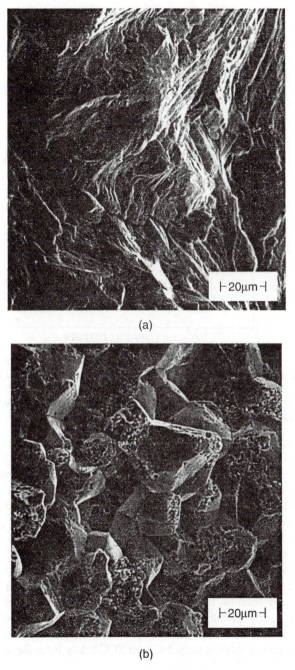

(a)

(b)

FIGURE 8.3   Fracture surfaces by scanning electron microscopy of (a) transgranular stress corrosion cracking of austenitic stainless steel in hot chloride solution and (b) intergranular stress corrosion cracking of carbon steel in hot nitrate solution, surface cleaned with inhibited HCl. Grain surfaces in (b) were corroded during exposure after crack growth.

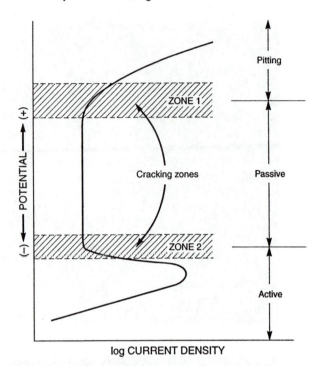

FIGURE 8.4 Schematic anodic polarization curve showing zones of susceptibility to stress corrosion cracking. *(From R. W. Staehle,* Stress Corrosion Cracking and Hydrogen Embrittlement of Iron Base Alloys, *R. W. Staehle et al, eds., NACE-5, NACE, Houston, p. 193, 1977. Reprinted by permission, National Association of Corrosion Engineers.)*

growth is probably present even in active potential ranges. Anodic polarization curves at slow and fast scan rates are shown in Figure 8.5,[7] along with the active potential range where SCC occurred. Figure 4.13 shows similar results for stainless steel in acid solution. Also, what may appear to be SCC in the active region may actually be occurring in Zone 2 because the potential of the active-peak current tends to become more active with time. That is, a potential in the active potential range during short-term potentiodynamic polarization may actually be in Zone 2 during longer-term SCC exposures.

It is notable that nonsusceptible alloy-environment combinations will not crack even if held in one of the potential zones described by Figure 8.4. Thus, temperature and solution composition (including pH and dissolved oxidizers, aggressive ions, and inhibitors or passivators) can modify the anodic polarization behavior to permit SCC, as well as control corrosion potential in a critical region. Hence, one cannot predict susceptibility to SCC solely from the anodic polarization curve.

Crack growth rates are proportional to anodic dissolution currents at straining electrode surfaces, as shown in Figure 8.6.[8] Agreement between crack growth rates and anodic current densities is generally good. However, some systems, particularly those with fast transgranular cracking (e.g., austenitic stainless steels and alpha

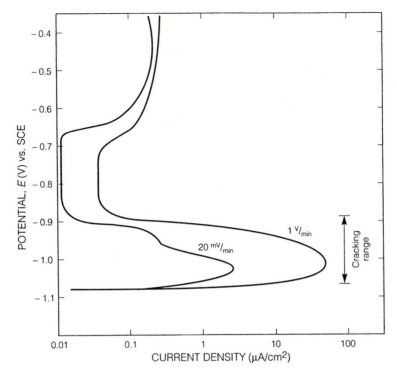

FIGURE 8.5   Zone 2 cracking (Figure 8.4) of carbon steel at or near the active state shown by fast and slow scan rate potentiodynamic anodic polarization in boiling 35% NaOH. *(From R. N. Parkins, Proc. 5th Symposium on Line Pipe Research, Am. Gas Assoc., Arlington, VA, p. U-1, 1974. Reprinted by permission, American Gas Association.)*

brasses in Figure 8.6), have crack growth rates higher than can be accounted for by simple electrochemical dissolution. The photomicrographs of Figure 8.3 indicate also that topographic features of the brittle crack surfaces have been preserved and thus were not subject to extreme anodic dissolution (Section 8.1.1). Thus, electrochemical anodic dissolution probably initiates mechanical fracture processes,[5] accounting for both anodic current proportionality and brittle crack topography (Section 8.8.4).

There is evidence that the solution within the microvolume of the crack becomes acidified, probably by hydrolysis reactions similar to those which occur in pits (Section 7.2.2). During free corrosion with no imposed polarizing potentials, there must be cathodic reduction reactions to compensate for the anodic dissolution reactions within the crack. For austenitic stainless steels exposed to hot $MgCl_2$ solutions, where hydrolysis and acidification are known to be strong, evolution of hydrogen from cracks indicates cathodic reduction of H^+ on the crack walls. The presence of hydrogen in the crack and the brittle cleavage characteristics of transgranular cracks have prompted suggestions that hydrogen causes SCC, as described later, in Section 8.8.1.

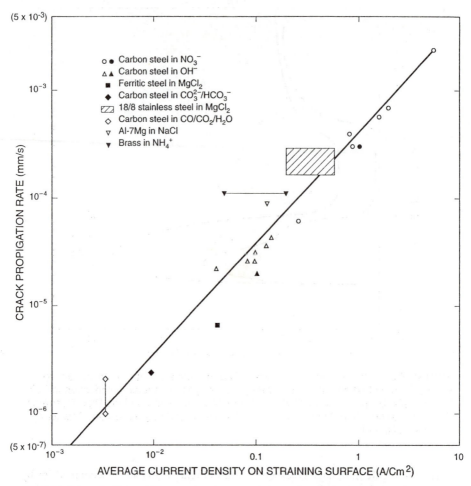

FIGURE 8.6   Proportionality between crack growth rate and anodic dissolution current at straining electrode surfaces. *(From R. N. Parkins, Br. Corrosion J., Vol. 14, p. 5, 1979. Reprinted by permission, Pergamon Press.)*

8.2 Characteristics of Corrosion Fatigue Cracking

8.2.1 Definition and Description

Corrosion fatigue cracking (CFC) is brittle failure of an alloy caused by fluctuating stress in a corrosive environment. An example of CFC in a carbon-steel boiler tube is shown in Figure 8.7,[9] which shows corrosion products typically present in cracks that grow slowly in service. Fracture surfaces from CFC, as well as from air fatigue, sometimes show macroscopic beach marks (Figure 1.11) where corrosion products accumulate at discontinuous crack-advance fronts. On the microscopic scale, striations are often evident where each cycle produces a discontinuous advance of the crack front, as illustrated in Figure 8.8.[10]

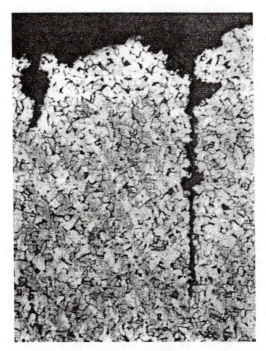

FIGURE 8.7   Corrosion fatigue cracking in carbon steel boiler tube. Nital etch. 250x. Corrosion products present along length of crack. *(From P. S. Pao and R. P. Wei, Metals Handbook, Failure Analysis, Vol. 11, 9th ed., ASM, Metals Park, OH, p. 253, 1986. Reprinted by permission, ASM International.)*

The frequency of cyclic stress is important also, as indicated in Figure 8.9[11,12] for AISI 4340 steel. Lower frequency leads to greater crack propagation per cycle, *da/dN*. Only slight, long-term variations in loading (e.g., even once a week) are known to accelerate the onset of SCC. Very high frequencies eliminate the effects of the corrosive environment. Plots of crack growth rate per stress cycle, as in Figure 8.9, require further description of fracture mechanics technology, which is given in Section 8.7. Presently, it is sufficient to comment that stress intensity, K, is proportional to stress for a given crack length.

Increasing the ratio, R (Section 8.6.4), of the minimum to the maximum stress in the cycle generally decreases the resistance to corrosion fatigue. The effect of R is much lower or has no effect on fatigue in noncorrosive environments at ambient temperature. R can also have an effect at higher temperature when creep is possible.

Stress raisers such as notches or surface roughness increase susceptibility to corrosion fatigue. Severe notches have a greater effect on the fatigue life than does corrosion alone. Figure 8.10[11,13] shows that corrosion has a lesser effect on the reduced properties of notched than on smooth fatigue specimens. Surface roughening by preexposure to a corrosive environment is well known to degrade subsequent fatigue life. Cracks have often been observed to initiate from corrosion pits, which again serve as surface stress concentrators. However, surface topographic modifications are not necessary precursors to the effects of CFC.

FIGURE 8.8 Fatigue striations on the surfaces of nickel Alloy 800 exposed to untreated boiler water containing sulfates and chlorides. *(From L. Engel and H. Klingele, Atlas of Metal Damage, Prentice-Hall, p. 120, 1981. Reprinted by permission, Springer Verlag, Munich.)*

The usual "endurance limit," or minimum stress to cause fatigue failure for low carbon ferritic steels is eliminated in a corrosive environment. In any alloy system, corrosion reduces the stress amplitude and shortens the time or number of stress cycles to failure (Figure 8.10). An unnoticed low-amplitude and/or low-frequency cyclic stress superimposed on a high-tensile stress can be a critical agent in causing failures originally thought to be purely SCC under constant stress. Only slight, long-term variations in loading (e.g., even once a week) are known to accelerate the onset of SCC. Thus, the environmentally induced cracking of buried carbon-steel linepipe carrying natural gas has been traced[7] to alternating stresses due to vibrations near compressor stations.

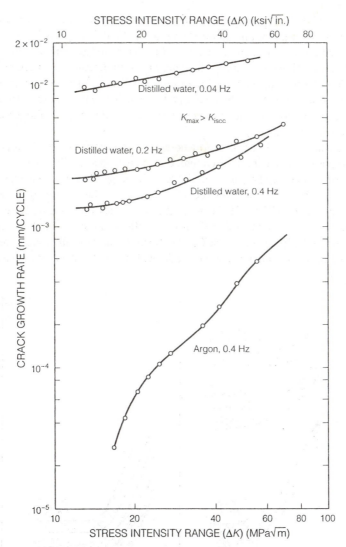

FIGURE 8.9 Corrosion fatigue of AISI 4340 steel in aqueous sodium chloride at 23°C. *(From C. S. Kortovich, "Corrosion Fatigue of 4340 and D6AC Steels below* K_{Iscc}*,"* Proc. 1974 Tri-Service Corrosion Conference on Corrosion of Military Equipment, *AFML-TR-75-42, Air Force Materials Laboratory, Wright Patterson Air Force Base, 1975.)*

8.2.2 Comparison with Stress Corrosion Cracking

Corrosion fatigue cracking (CFC) is similar to stress corrosion cracking (SCC) inasmuch as a corrosive solution induces brittle fracture in an alloy that is normally ductile in a noncorrosive environment. The stress is cyclic rather than constant, but must have at least some tensile component, as in SCC. As mentioned earlier, CFC cracks propagate perpendicular to the principal tensile stress, as in SCC. Beyond these similarities, however, the differences are striking.

In contrast to SCC, CFC requires neither a specific corrodent nor a very low corrosion rate. A minimum required corrosion rate of about 1 mpy has been observed for steel,[14] and very low corrosion rates often have no accelerating effect over fatigue life in air. However, simply changing the environment from vacuum to dry or moist air will also reduce fatigue life in some systems.[11,15] The fatigue life of passive alloys is often decreased when exposed to corrosion in the passive state at very low rates. Exposure to any type of corrosive solution will accelerate fatigue failures of both pure metals and alloys.

Corrosion fatigue cracks usually form more slowly and corrosion products are more likely to be present in the crack. Therefore, corrosion products will usually be absent in stress corrosion and hydrogen-induced cracks, if the part has not been exposed for some time after the cracks were formed. CFC cracks are often relatively blunt at the tip, in comparison to SCC and HIC cracks. The blunt crack tip in Figure 8.7 is apparent. Another example of a blunt-tipped corrosion fatigue crack in a weldment appears later in Figure 9.22. In contrast to SCC, CFC fractures are typically transgranular, as are those for air fatigue failures. Fatigue cracks with and without a corrosive environ-

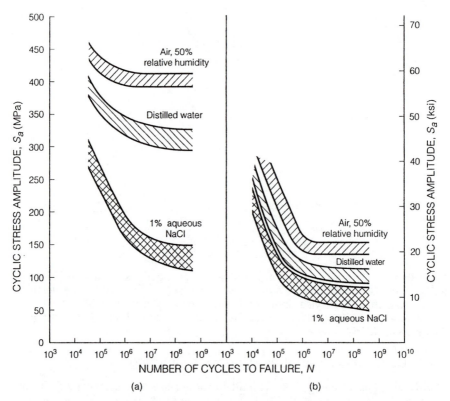

FIGURE 8.10 Fatigue curves showing reduced service life (number of cycles to failure) in the presence of (a) corrosive environment and (b) added notches. *(From M. O. Speidel, Proc. Int. Conf. on Stress Corrosion and Hydrogen Embrittlement of Iron Base Alloys, NACE, Houston, p. 1071, (1977). Reprinted by permission, National Association of Corrosion Engineers.)*

ment initiate and grow at areas of localized slip within grains exposed to the surface. Thus, it is perhaps not surprising that CFC and air fatigue cracks are confined to crystallographic features of the grains and do not usually follow grain boundaries. Further discussion of the CFC mechanism is given in Section 8.8.7.

8.3 Characteristics of Hydrogen Induced Cracking

8.3.1 Definition and Description

Hydrogen induced cracking (HIC) is brittle mechanical fracture caused by penetration and diffusion of atomic hydrogen into the crystal structure of an alloy. Hydrogen may be present from reduction of water or acid by

$$H_2O + e^- \rightarrow H + OH^- \tag{1}$$

and

$$H^+ + e^- \rightarrow H, \tag{2}$$

in neutral and acidic solutions, respectively. These reactions are similar to those described in Chapter 3, which showed the hydrogen molecule as the cathodic reaction product. However, before the H_2 molecule can be formed, there is a significant residence time of the nascent H atom on the surface, especially if cathodic poisons such as S^{2-} and As^{3+} are present to delay the recombination rate. Because of its small size, atomic hydrogen can then enter the lattice to produce HIC. Reactions (1) and (2) are further accelerated and the danger of HIC increased during cathodic protection and electroplating. The necessary atomic hydrogen can also be provided by dissociation of hydrogen gas on the surface during exposure to elevated temperature gases. Various proposed mechanisms of HIC are discussed in Section 10.1.2. If hydrogen entry in the metal crystal lattice is allowed to continue, additional irreversible hydrogen-damage mechanisms will ensue, as described further in Sections 10.1.4 through 10.1.7.

HIC effects often are reversible. Figure 8.11[16] shows failure times of cathodically charged SAE 4340 (0.4% C) steel that had been baked for various times at 300°F (150°C). Low-temperature baking treatments subsequent to cathodic charging allowed dissolved hydrogen to escape, restoring the original properties. This figure also shows that there is an incubation time before cracking which decreases with increasing applied stress, and that there is a minimum stress below which HIC will not occur. The incubation time and minimum stress increased as the baking time increased and the resulting content of dissolved hydrogen decreased. Increasing hardness or tensile strength of steel also decreases the incubation time and minimum stress for HIC. Hydrogen damage effects from further hydrogen charging cannot be restored by removal of dissolved hydrogen, however.

Aqueous hydrogen sulfide, H_2S, dramatically accelerates hydrogen entry and hydrogen damage in most alloys. The S^{2-} anion slows (poisons) the recombination reaction,

$$H + H \rightarrow H_2,$$

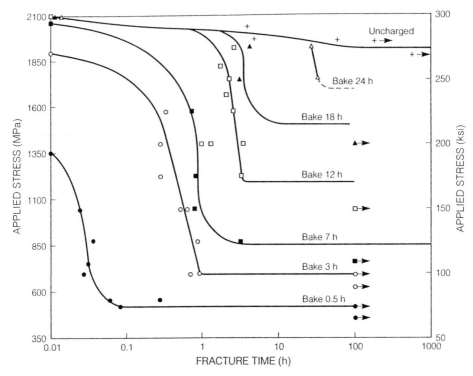

FIGURE 8.11 Fracture of AISI 4340 steel baked at 150°C after initial cathodic hydrogen charging. *(From A. R. Troiano, Trans. ASM, Vol. 52, p. 54, 1960. Reprinted by permission, ASM International.)*

and provides a greater activity of nascent atomic H on the surface. The result is sometimes called sulfide stress corrosion cracking, although it is really a form of HIC.

HIC is especially prevalent in iron alloys because of the restricted slip capabilities in the predominantly body-centered cubic (BCC) structure. HIC is generally limited to steels having a hardness of 22 or greater on the Rockwell C scale.[17] The face-centered cubic (FCC) stainless steels and FCC alloys of copper, aluminum, and nickel are more resistant because of their inherent high ductility and lower diffusivity for hydrogen, but all can become susceptible if highly cold worked.[18] The BCC and FCC stainless steels are also resistant because of low strength, but again are made susceptible by cold working. Reactive alloys of titanium, zirconium, vanadium, niobium, and tantalum, which are embrittled by insoluble or soluble hydrides, are discussed in Chapter 10.

8.3.2 Comparison with Stress Corrosion Cracking

Hydrogen induced cracking (HIC) is similar to stress corrosion cracking (SCC) inasmuch as brittle fracture occurs in a corrosive environment under constant tensile stress. However, cathodic polarization initiates or enhances HIC but suppresses or

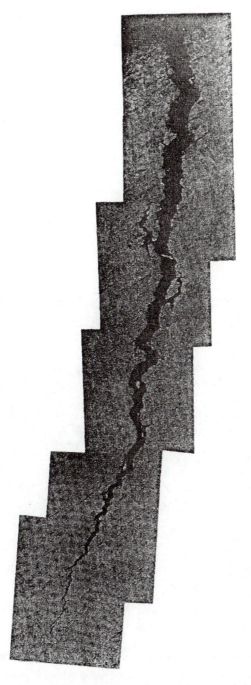

FIGURE 8.12   Failure of cold-worked Type 301 stainless steel due to hydrogen induced cracking. *(From G. Schick,* Metals Handbook, Corrosion, *Vol. 13, 9th ed., ASM, Metals Park, OH, p. 1132, 1987. Photograph by courtesy of G. Schick, Bell Communications Research. Reprinted by permission, ASM International.)*

stops SCC. Cathodic polarization occurs in service during cathodic protection, metal plating, and on galvanized steel at breaks in the coating. SCC cracks are usually branched, whereas CFC and HIC cracks are unbranched or only slightly branched. An example of an unbranched HIC crack in cold-worked Type 301 austenitic stainless steel is shown in Figure 8.12.[19]

Transgranular HIC cracks show very brittle cleavage-like features, as shown in Figure 8.13.[20] Characteristic micropores caused presumably by hydrogen bubbles and ductile hairlines due to microplastic mechanisms (Sections 8.6.7 and 10.1.2) are also evident. Similar ductile hairline markers are also present on intergranular HIC surfaces (Figure 8.14[20]). Surface crack morphologies are often very similar. Consequently, some researchers have suggested that hydrogen embrittlement processes are required for SCC. However, comparisons of characteristics in Table 8.1 suggest that separate mechanisms are more probable.

To further compare SCC with HIC, cracks from HIC are highly brittle, fast growing, and usually unbranched. SCC cracks are characteristically branched and grow at slower rates. HIC cracks are more often transgranular than intergranular, in

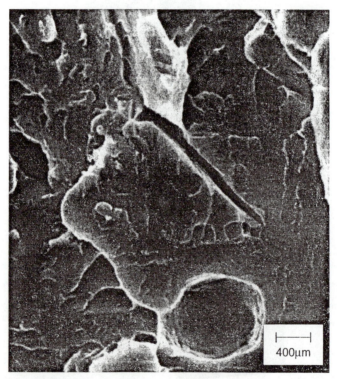

FIGURE 8.13 Transgranular hydrogen induced cracking in precipitation-hardened steel showing cleavage-like fracture planes, micropores, and ductile hairlines. *(From L. Engel and H. Klingele, Atlas of Metal Damage, Prentice-Hall, pp. 125–128, 1981. Reprinted by permission, Springer Verlag, Munich.)*

FIGURE 8.14 Intergranular hydrogen induced cracking of gas carburized steel screw. *(From L. Engel and H. Klingele,* Atlas of Metal Damage, *Prentice-Hall, pp. 125–128, 1981. Reprinted by permission, Springer Verlag, Munich.)*

contrast to SCC, in which intergranular cracking predominates. Notable exceptions are many cold-worked nickel alloys, which crack intergranularly by HIC. Any corrosive solution may produce HIC in a susceptible alloy, if hydrogen is liberated on the surface. On the other hand, SCC requires a specific and usually different dissolved species for each alloy. HIC has been observed in pure metals more often than SCC, but alloys are usually more susceptible. HIC is suppressed by anodic polarization, in contrast to SCC, which is often enhanced. However, SCC may be suppressed also, if the potential is moved out of the critical potential zone for cracking (Figure 8.4).

Failures by HIC are usually maximized at or near room temperature, as shown for AISI type 4340 steel in gaseous hydrogen in Figure 8.15a.[18] In contrast, failure times for SCC are usually shortened by higher temperature. Figure 8.15b shows crack-growth rates for a 3% Ni steel in water as a function of temperature. The crack-growth rates show a peak at room temperature, characteristic of HIC, but at higher temperatures the crack-growth rates increase again, suggesting that an SCC mechanism has assumed control.

Both SCC and HIC are strain-rate dependent. At high strain rates, embrittlement disappears, because both SCC and HIC are time-dependent processes. At low strain rates, hydrogen maintains embrittlement effects. In principle, SCC should be suppressed at very slow strain rates due to repassivation of ruptured films. However, some aggressive systems still show SCC, even at very low strain rates. Section 8.6.3 further describes slow strain-rate testing.

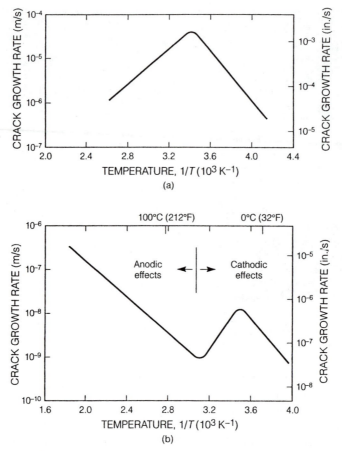

FIGURE 8.15 Effect of temperature on (a) hydrogen induced cracking of Type 4340 steel in hydrogen and (b) stress corrosion cracking of 3% Ni steel in water. *(From B. D. Craig,* Metals Handbook, Corrosion, *Vol. 13, 9th ed., ASM, Metals Park, OH, pp. 164–166, 1987. Reprinted by permission, ASM International.)*

8.4 Typical Cases of Stress Corrosion Cracking

Alloy-environment combinations causing stress corrosion cracking (SCC) are many and varied. Table 8.2 summarizes environments known to cause stress corrosion cracking (SCC) for some significant engineering alloy systems, along with usual temperatures for failure. This listing provides a reference for further discussion in this section and can be used as an initial guide for alloy selection. Liberal use has been made of the ASM *Metals Handbook*, 9th edition, Volume 13 *Corrosion*, which is highly recommended as a general reference. Sufficient space is not available in this book to include SCC environments for all possible alloy systems. Not all environments cause SCC for a given alloy, although more combinations of alloy and environment that result in SCC are being discovered. More detail and information on additional systems can also be found in the *Metals Handbook*.

TABLE 8.2   Environment-Alloy Combinations Known to Produce Stress Corrosion Cracking

Alloy	Environment	Temp.	Ref.
Austenitic Stainless Steels	Hot acid chloride solutions such as $MgCl_2$ and $BaCl_2$	60°C– 200°C	25
	$NaCl$–H_2O_2 solutions		23
	Neutral halides: Br^-, I^-, F^-		25
	Alkaline $CaCl_2$		25
	Seawater		23
	Concentrated caustic solutions	>120°C	25
	$NaOH$–H_2S solutions		23
	Condensing steam from chloride waters		23
	For sensitized alloys:		
	Polythionic acids ($H_2S_nO_6$)	RT	30
	Sulfurous acid	RT	30
	Pressurized hot water containing 2 ppm dissolved oxygen	300°C	29
Ferritic Stainless Steels	H_2S, NH_4Cl, NH_4NO_3, hypochlorite. (Resistant to most environments if free of nickel but may fail by other modes of corrosion in same media.)		30 25
Duplex Stainless Steels	Susceptible to same environments as Austenitic stainless steels but more resistant. (Immune to intergranular SCC in polythionic acid. Also greater resistance than ferritic stainless steels to other forms of corrosion.)		30
Martensitic Stainless Steel	Caustic NaOH solutions. (Resistant to SCC in hot chlorides. Susceptible to hydrogen embrittlement. See Chapter 10.)		30
Carbon Steels	Caustic NaOH solutions	>50°C	30
	$NaOH$–$NaSiO_2$ solutions	>255°C	23
	Calcium, ammonium, and sodium nitrate solutions	Boiling	30
	Mixed Acids (H_2SO_4–HNO_3)	RT	23
	HCN solutions, acidified	Warm	30
	Acidic H_2S solutions		23
	Scawater		23
	Anhydrous liquid ammonia	RT	36
	Carbonate/bicarbonate		30
	Amines	All	30
	CO/CO_2 solutions		28

TABLE 8.2   (Cont.)

Alloy	Environment	Temp.	Ref.
Nickel- Cr-Fe Alloys 600,800, 690	High-temperature chloride solutions aggravating factors: pH < 4, oxi- dizing species, such as dissolved oxygen, H_2S, free sulfur	>205°C	37
	Polythionic acids and thiosulfate solutions, sensitized alloys with excess carbon	RT	37 28
	Caustic alkaline solutions	315°C	37
Ni-Cu	Acidic fluoride solutions	RT	28
Monel	Hydrofluoric acid	RT	28

Alloy
400
	Hydrofluosilicic acid
	  Susceptible in cold-worked
	  state. Resistant in
	  stress relieved state
		RT	23

	H_2S		28
Nickel Alloy 200, 201	Caustic alkaline solutions	290°C	28
Copper- Zinc Alloys (Brass) >15% Zn	Ammonia vapors in water	RT	28
	Amines in water	RT	28
	Nitrites in water	RT	28
	Water, Water vapor alone (45–50% Zn, β or $\beta + \tau$ alloys)	RT	39
	Nitrate solutions		28
	Some sulfate solutions		40
Aluminum Alloys	Air with water vapor	RT	21
	Potable waters	RT	21
	Seawater	RT	21
	NaCl solutions	RT	21
	$NaCl$-H_2O_2 solutions	RT	21
Titanium Alloys	Red fuming nitric acid	RT	39
	Hot salts, molten salts	>260°C	39
	N_2O_4	30–75°C	39
	Methanol/halide	RT	39

Elevated temperature accelerates SCC in most of the systems shown in Table 8.2 and is essential in many of them. In some alloy-environment systems, SCC observed in slow strain rate testing has not been detected by other testing methods and has never been reported in service.[21,22] Such systems are not included in Table 8.2. Table 8.2 should be used only as a guide for further study, evaluation, and investigation before final alloy selection. Further comments on the most common examples of SCC listed in Table 8.2 are given in subsequent sections.

8.4.1 Stainless Steel

Stress corrosion cracking (SCC) of austenitic stainless steel in hot chlorides is perhaps the most widely known and intensely studied example of SCC.[23] Chlorides are ubiquitous in seawater, mammalian body fluids, and industrial process streams. Although the hazard is lessened in lower concentrations of chloride, boiling at the heat-transfer surfaces of heat exchangers and condensers or simple evaporation to the ambient atmosphere can concentrate local solutions at the surface sufficiently to initiate SCC. As a result, failures occur frequently in apparently benign environments containing only a few ppm chlorides or less. Susceptibility increases with temperature, but any minimum temperature is ill defined or even nonexistent. Less aggressive conditions (lower temperature) lead to intergranular cracking in service conditions. Truman[24] has defined the boundaries of temperature and chloride content causing SCC of austenitic stainless steel; some of his results appear in Figure 8.16. At lower temperature and chloride content, SCC is replaced by pitting and crevice corrosion. Other halides will also produce SCC in conditions similar to those for chloride.[25]

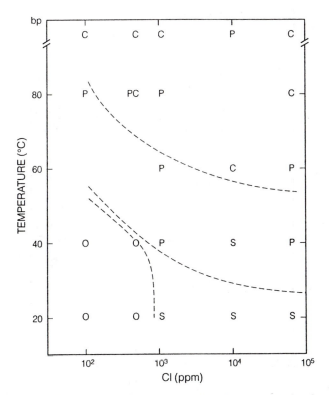

FIGURE 8.16 Conditions of temperature and chloride concentration for stress corrosion cracking of austenitic stainless steel. O, no effect; S, staining; P, pitting; C, cracking. *(From J. E. Truman,* Corrosion Science, *Vol. 17, p. 737, 1977. Reprinted by permission, Pergamon Press.)*

Although relatively rare, austenitic stainless steel can suffer SCC even at ambient temperature in the presence of concentrated chlorides and strong oxidizers.[26,27] Such conditions can arise in the thin condensed aqueous films which can form in humid seacoast and enclosed swimming-pool atmospheres. Ferric ions from surface rust and chlorine compounds from biocides can supply strong oxidizers in addition to usual easy access of oxygen in the condensed films.

SCC in austenitic stainless steels is often transgranular. The scanning electron fractograph of Figure 8.3a shows a generally cleavage-like character illustrating the brittle nature of SCC in this system, in which the crack path clearly crosses grain boundaries and passes through the grains. Elevated temperature and chloride concentration are required to induce transgranular SCC in this alloy.

Austenitic stainless steels are frequent candidates for use in boilers and heat exchangers of fossil- and nuclear-fueled power plants. Although austenitic stainless steels are generally resistant in such environments, careful control must be exercised over chloride and dissolved oxygen, as illustrated in Figure 8.17.[28] Dissolved oxygen is clearly required, and minimizing both dissolved oxygen and Cl⁻ is necessary

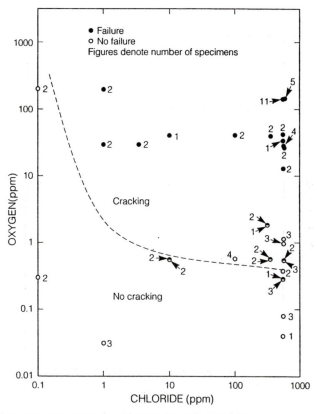

FIGURE 8.17   Conditions of dissolved oxygen and chloride concentration for stress corrosion cracking of austenitic stainless steel exposed to the steam phase and intermittent wetting in alkaline phosphate boiler water. *(From W. L. Williams,* Corrosion, *Vol. 13, p. 539t, 1959. Reprinted by permission, National Association of Corrosion Engineers.)*

for safe operation. Intergranular SCC of sensitized austenitic stainless steel has been a recurring and expensive problem in the cooling-water piping of boiling-water nuclear power plants.[29]

Polythionic acids ($H_2S_nO_6$; n = 2 to 5) and sulfurous acid, in which sulfur exists in a reduced oxidation state, cause intergranular SCC in sensitized (Section 9.2.1) austenitic stainless steels,[30] especially in petrochemical refinery equipment. Polythionic acids are formed by reaction of sulfide surface scales with moisture and oxygen.[31] This type of failure was also found extensively in the steam generators of the Three Mile Island pressurized-water nuclear reactor (Section 11.2.3) after layup during the downtime following the nuclear accident there in 1983.[32]

Ferritic stainless steels generally resist SCC in most common service environments that attack austenitic stainless steels. Small amounts of nickel, as low as 1.5 to 2%, are sufficient to induce susceptibility.[33] However, absence of nickel reduces the general corrosion resistance, and the ferritic stainless steels are susceptible to other forms of corrosion in many such environments.[30] The ferritic stainless steels are also reported to be susceptible in H_2S, NH_4Cl, NH_4NO_3, and $HgCl_2$ solutions.[25]

The duplex stainless steels generally contain more chromium to maintain corrosion resistance of austenitic stainless steels, and less nickel to increase the ferrite content and improve the resistance to SCC in hot chloride environments. Thus, the duplex stainless steels are more resistant to SCC than the austenitic stainless steels but are not totally immune.[30] See also Sections 9.2.1 and 9.4.3.

8.4.2 Carbon and Low-Alloy Steels

One of the first and best-known examples of stress corrosion cracking (SCC) was observed in boiler waters in the crevices created by rivets and bolt heads.[34] Iron and steel are resistant to general corrosion in slightly alkaline solutions, and boiler waters were controlled at pH 9.5 to 11.5 by additions of alkali (NaOH). Local boiling concentrates caustic in the crevices, causing the highly cold-worked rivets to fail by SCC. Cracking often appears at weldments because of residual stress concentrations (Section 9.3.1) and the susceptible heat-affected zone.[35] Caustic cracking has been alleviated by the addition of phosphate buffers to control free caustic in crevices. Also, ammonium hydroxide has been substituted for NaOH. The ammonium ion, NH_4^+, decomposes at higher temperature to volatile ammonia, which escapes from a boiling crevice, depriving the OH^- of the necessary cation in the crevice to maintain alkalinity.

As indicated in Table 8.2, carbon steels are susceptible to SCC in a number of other corrosive environments. Perhaps the best known are nitrates and cyanides, which are often present in chemical-process streams. The SCC is intergranular, as shown in Figure 8.3b. Similar failures occur in aqueous solutions and moist gases containing mixtures of carbon monoxide and carbon dioxide.[30] Carbon and nitrogen interstitials,[34] and more recently phosphorus,[5] have been found to segregate at the grain boundaries and cause cracking. Unfortunately, it is often impractical to reduce any of these to acceptable levels in commercial alloys.

Intergranular SCC has been observed on the external surfaces of buried carbon and low-alloy steel linepipe used to transport petroleum and natural gas. The critical environment is a mixture of carbonate and bicarbonate at 75°C. These anions are found in the presence of mill scale at holidays and beneath paint coatings in the presence of cathodic protection. Parkins[7] found that cathodic protection puts the steel into a critical potential region (i.e., zone 2 in Figure 8.4), where the material is more susceptible to SCC. Cracking is minimized if the mill scale is removed before painting.

Similar intergranular SCC has been found in carbon and alloy steel tanks holding liquid anhydrous (water-free) ammonia.[36] Liquid ammonia kept free of air contamination does not cause SCC. Stress-relief annealing to reduce residual welding stresses is helpful, and water additions of 0.2% are recommended as an inhibitor.[36]

8.4.3 Nickel Alloys

Increasing nickel content above about 8% has a favorable effect on resistance to stress corrosion cracking (SCC) of austenitic alloys in hot chloride solutions (Figure 8.18[37]). Consequently, high-nickel alloys (e.g., Inconel) are often the choice in

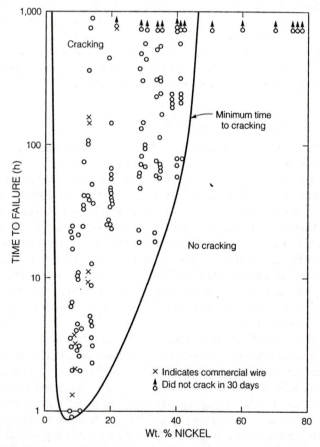

FIGURE 8.18   Effect of nickel content on stress corrosion cracking in Fe-Cr-Ni alloys exposed to boiling 42% magnesium chloride solutions at 154°C. *(From H. R. Copson, Physical Metallurgy of Stress Corrosion Fracture, T. N. Rhodin, ed., Interscience, p. 247, 1959. Reprinted by permission, The Metallurgical Society of AIME.)*

chemical-process plants and in fossil- and nuclear-fueled power plants. Molybdenum additions also improve SCC resistance. Generally, the same factors that cause SCC in austenitic stainless steels also cause failure in the nickel alloys, but at much higher concentrations and temperatures.[38] H_2S has an accelerating influence on SCC in deep, hot, sour gas wells at temperatures where hydrogen has a lower solubility, and the usual HIC effects associated with H_2S are less effective.

The Ni-Cr-Fe Alloy 600 is used extensively in heat-exchanger tubes of nuclear steam generators. Although more resistant than type 304 stainless steel, Alloy 600 has still experienced intergranular SCC in high-temperature pressurized water at temperatures above 300°C, especially in concentrated caustic solutions that develop in the tubesheet crevice. Considerable research has been conducted on heat treatments to achieve semicontinuous carbide precipitation and minimize chromium depletion at grain boundaries.[39] Alloys 690 and 800 have also experienced SCC in this service but are more resistant than Alloy 600.

Polythionic acid will also produce intergranular SCC at or near ambient temperature in the Ni-Cr-Fe alloys.[30] However, the alloys usually have excess carbon content and must be strongly sensitized to intergranular corrosion (Section 9.4.2) before such failures will occur.

Considerable study of the stainless alloys have been conducted at temperatures near 315°C in caustic alkaline solutions known to accumulate in the tubesheet crevices of nuclear steam generators. Austenitic stainless steel is much less resistant to SCC than are the nickel alloys, of which Alloy 690 is favored for this service.[39]

8.4.4 Copper Alloys

So-called "season cracking" of brass (Cu-Zn alloy) was first observed in cold-worked and stressed cartridge cases in tropical environments. The critical species causing stress corrosion cracking (SCC) has been identified as the ammonium ion NH_4^+.[40] Water, together with some oxidizer (usually dissolved O_2), is also necessary. A widely used test solution attributed to Mattson[41] contains $CuSO_4$ and $(NH_4)_2SO_4$. Cu^{2+} serves as an oxidizer, together with the necessary NH_4^+. If temperature is high enough and dissolved oxidizers are present, brass can crack in dilute sulfate solutions.[42] Sulfite, SO_3^{2-}, can decompose to SO_2 and H_2S to crack admiralty brass. Cracking is most often intergranular in the prevalent solutions, which cause a tarnish film. However, SCC may be transgranular if tarnish films are absent or if the alloy is heavily cold-worked. Brasses with a zinc content below 15% are resistant to SCC.

Only small amounts of NH_4^+ are necessary to cause SCC. A frequent source is the leaching and decomposition of organic and agricultural wastes, including nitrogen-rich fertilizers. NH_4^+ may also be derived from reaction of nitrogen oxides with the brass surface.[42] Frequent cases of SCC of brass in polluted moist air containing nitrogen oxides have been traced to this source. Sufficient NH_4^+ may also accumulate in a closed environment from polymers decomposing to amines over an extended period of time, especially within cabinets housing electronic apparatus. Amine inhibitors will also decompose to ammonia and cause SCC of copper alloys.

8.4.5 Aluminum

Stress corrosion cracking (SCC) of aluminum is especially sensitive to strength level, as affected by alloy content and heat treatment. Pure aluminum and alloys tempered to less than maximum strength are more resistant. Resistance of wrought alloys is at a minimum in the short transverse direction, which is perpendicular to the plane of rolled sheet or plate. SCC in aluminum alloys is almost exclusively intergranular. Grain-boundary precipitates or depletion are thought[43] to cause anodic regions at the grain boundaries. Cathodic protection will eliminate or retard SCC in many alloys, but some of the highest-strength, age-hardened alloys are thought to fail by hydrogen-induced cracking. Aluminum alloys are susceptible to humid air, seawater, and potable water in service. The most accepted laboratory environment uses sodium chloride solutions with oxidizers such as peroxide (H_2O_2) under alternate immersion conditions (Figure 1.23a) to further concentrate the chloride on the surface.

It is notable that corrosion fatigue failures of aluminum alloys are transgranular, notwithstanding the fact that SCC failures are exclusively intergranular.[43]

8.4.6 Titanium

Titanium and its alloys are highly corrosion resistant, lightweight, and resistant to stress corrosion cracking in nearly all aqueous environments. However, a number of nonaqueous agents, listed in Table 8.2, cause SCC of unalloyed titanium and other α-phase (hexagonal close packed) alloys.[44] Small amounts of water ($\{dsim\}2\%$), if stable in the environment, inhibit SCC. Laboratory testing of more complex aerospace alloys with precracked fracture toughness specimens (Section 8.7) has shown additional susceptibility in distilled water and aqueous halide solutions, but no service failures in these environments have been reported.

8.5 Methods of Prevention

8.5.1 Prevention of Stress Corrosion Cracking

Prevention of stress corrosion cracking (SCC) generally requires elimination of one of the three factors shown in Figure 8.1: tensile stress, critical environment, or susceptible alloy. Redesign may allow the elimination of service tensile stresses in critical parts. Shotpeening can put the surface into a state of compressive stress. Removal of residual tensile stresses may be accomplished by stress-relief anneal, which allows sufficient creep to relax loads due to fabrication by fastening or welding. Annealing may be impractical for some stainless steels, which sensitize and become susceptible to intergranular attack.

Control of the environment by lowering oxidizing agents (e.g., dissolved oxygen) or removing the critical species is probably the most popular method for controlling SCC. Inhibitors may be effective in some instances, as well. Coatings are often ineffective or impractical because they will not withstand the aggressive chemical and/or physical environments associated with SCC.

Changing the proportion of alloying elements in an alloy system to lower strength or alter the metallurgical structure can increase resistance to SCC. Choosing a different alloy resistant to the particular environment is, of course, a popular option.

Cathodic protection will usually stop SCC but will accelerate hydrogen induced cracking (HIC). Thus, the alloys listed in Table 8.2, all of which crack by an anodic mechanism (Section 8.8), allow control of SCC by cathodic protection. Caution must be exercised with higher-strength alloys, which may fail by HIC when cathodic protection is applied.

8.5.2 Prevention of Corrosion Fatigue Cracking

Corrosion fatigue cracking (CFC) can be mitigated by any means that reduces the general corrosion rate, including inhibitors, cathodic protection (when hydrogen-induced cracking is not a danger), reduction of oxidizers, or increase in pH. A change to a more corrosion-resistant alloy will of course be effective as well, but only if the corrosion rate is sufficiently low. Barrier coatings to exclude the corrosive solution from the alloy surface and sacrificial zinc (galvanized) coatings to cathodically protect steel at breaks in the coating can also prevent CFC.

Redesigning to reduce or remove cyclic stresses will, of course, prevent or improve resistance to CFC. Shotpeening may be helpful to reduce tensile stresses.

8.5.3 Prevention of Hydrogen Induced Cracking

It is possible to eliminate hydrogen induced cracking (HIC) by removing the hydrogen source, decreasing the tensile stress, reducing the strength level by alloying or heat treatment, or by selecting a more resistant alloy. There are numerous sources of hydrogen that may result from manufacturing and processing, which are beyond the scope of this book, but which have been reviewed elsewhere.[18] Inert coatings to exclude the hydrogen-bearing environment can be effective in service environments. Inhibitors or an increase in pH (for carbon or low-alloy steels) will reduce the corrosion rate and consequent degree of hydrogen production on the surface. Stopping or reducing cathodic protection, including removal or replacement of galvanized coatings, will also limit hydrogen at the surface. Annealing or baking can increase the mobility, and permit the escape, of dissolved hydrogen as well as relieve residual tensile stresses. Redesigning the equipment or shotpeening to remove or reduce tensile stress can be effective. Replacement of a ferritic steel with a more corrosion-resistant face-centered cubic (FCC) alloy that is also more resistant to HIC is a frequently used option. Nickel alloys are known as replacements for ferritic steels for this purpose. However, it must be recognized that most alloys, even if FCC, are susceptible to HIC to some degree.

8.6 Testing Methods

Stress corrosion cracking and hydrogen induced cracking involve constant stress, and testing methods described in Sections 8.6.1 through 8.6.3 are generally common to both. Corrosion fatigue testing with alternating stresses is discussed in Section 8.6.4. Recent interest in fracture mechanics testing merits a separate discussion in Section 8.7.

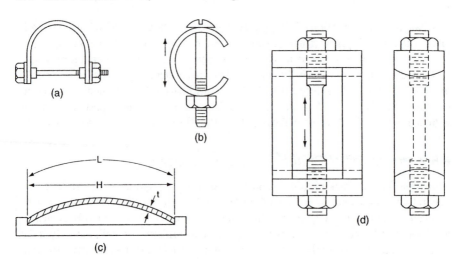

FIGURE 8.19 Constant deformation specimens and loading methods for stress corrosion cracking testing: (a) U-bend; (b) C-ring; (c) bent beam; (d) tensile.

8.6.1 Constant-Deformation Tests

Historically, the parameter used to measure resistance to stress corrosion cracking (SCC) has been time to failure. Much useful information has been derived from time-to-failure of various constant-deformation (constant-strain) specimens pictured in Figure 8.19.[45] With the U-bend configuration (Figure 8.19a), a sheet or bar specimen is plastically deformed over a mandrel (usually 180° bend) and the ends are constrained by a bolt through the ends. A U-bend specimen failed by SCC appears in Figure 1.10. The U-bend specimen configuration is quite convenient experimentally and forms one of the most severe smooth specimen tests for SCC. However, the stress distribution within the specimen is not well known, and the stress conditions are difficult to duplicate from specimen to specimen. Still, the U-bend specimen has found frequent use to evaluate qualitatively the resistance of various alloys to various solution conditions. Further detail for fabrication of U-bend specimens is given in standard method ASTM G30.

The C-ring specimen (Figure 8.19b) can be machined from thick (e.g. >1-inch) plate and the bolt again applies a tensile stress to the outer surfaces. More conveniently, a C-ring can be fabricated from a section of appropriately sized tubing. The C-ring specimen has the advantage that variable stresses both above and below yield strength may be controlled by the amount of applied deformation.

Very thin rectangular sheet coupons can be readily formed into bent beam specimens. One type is shown in Figure 8.19c. When stressed below the yield strength, the bent-beam specimen permits precise calculation of the maximum stress in the outer fibers of the bent beam. Experimental standards for C-ring and bent beam specimens are given in ASTM G38 and ASTM G39, respectively.

The most sophisticated constant-deformation specimen uses a conventional tensile specimen in a small loading frame with nuts on the threaded ends of the specimen (Figure 8.19d). This configuration gives the greatest control of applied stress and permits easy stress measurement and analysis. Further details are given in ASTM G49.

Constant-deformation specimens are immersed in the solution of interest, removed periodically, and inspected for visual evidence of stress corrosion cracks. The time to failure, although arbitrary, is usually judged to be the first appearance of cracks on the tension surface. Constant-displacement specimens are portable, compact, easily fabricated, and can readily be placed in plant process streams for in-plant testing. All components of the loading fixtures must be insulated from the specimen to eliminate galvanic interferences.

These and other such constant-displacement tests suffer from certain disadvantages. As cracks initiate and grow, the load or driving force for SCC decreases. Thus, the specimens must be somewhat overloaded to generate visually detectable cracks before the load has decayed to the point that SCC cracks will no longer grow. Specimen preparation, placement, and subsequent inspections are labor intensive and consequently expensive. It is difficult to determine the maximum failure time because it is always possible that cracks may appear in the next exposure increment.

8.6.2 Sustained Load Tests

A reasonably constant load is applied to the specimen throughout the duration of the test. The simplest configuration is a dead-weight load hung from one end of the specimen. However, the apparatus for dead-weight loading is bulky, and more compact constant or sustained loading configurations have been devised. Figure 8.20[45] shows a C-ring specimen adapted to sustained load using a compressed spring. Upon initiation of a crack, the net cross-sectional area decreases and the stress increases under sustained load as the crack(s) grows, driving an ordinary smooth tensile specimen to total fracture, eliminating the growth portion of cracking from the measured time to failure. Most of the constant deformation specimens shown in Figure 8.19 can be adapted to constant load by adding a spring, as in Figure 8.20.

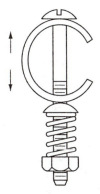

FIGURE 8.20   C-ring specimen adapted to sustained load testing with a compressed spring.

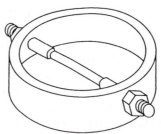

FIGURE 8.21   Proving ring assembly for application of sustained load to a tensile specimen.

Figure 8.21[46] shows an apparatus in which a proving ring is used to apply the necessary load. Because there is negligible plastic deformation before SCC fracture, the device essentially applies a constant test load, which can be accurately and conveniently measured by the elastic deformation of the proving ring. At fracture, displacement of the specimen holder may be designed to trip a microswitch to record time to failure accurately.[46]

8.6.3 Slow Strain-Rate Testing

A recent advance in stress corrosion cracking (SCC) testing technology is the slow strain rate or continuous extension rate test.[47] A conventional smooth bar (round or rectangular cross section) tensile specimen is slowly pulled at a constant strain rate to failure, while exposed to a corrosive environment. The equipment consists of a motor-driven crosshead in a load frame with a load cell to measure the load on the specimen continuously. A diagram of the benchtop apparatus is shown in Figure 8.22.[48] Time is proportional to crosshead travel, which is a measure of deformation or strain in the tensile specimen. The result is a load versus time or stress versus strain curve at a very low rate of loading.

Whereas a conventional stress-strain test requires only a few minutes to complete at a strain rate of about 10^{-2} sec^{-1}, a slow strain rate SCC test at 10^{-6} sec^{-1} may take nearly two days to complete. Figure 8.23[47] shows a pair of stress-strain curves for carbon-manganese steel, one conducted in sodium nitrate solution and the other in oil, both at 80°C and a strain rate of 10^{-6} sec^{-1}. The specimen exhibiting SCC in the nitrate showed a much lower time or strain to failure. While the testing time is long compared to a conventional tensile test, the previously described constant-deformation and constant-load SCC tests are indeterminate in length. They may require months or even years of exposure, with no certainty that failure would not be observed at a still longer time. The slow strain rate test usually yields a firm result in a matter of 2 days maximum.

Figure 8.24[47] shows the appearance of failed specimens susceptible and resistant to SCC. Cracks are numerous in the area of reduced cross section, and the final crack has multiple branches, as in SCC cracks in the usual service and laboratory tests. Examination of fractured specimen surfaces shows features identical to those of

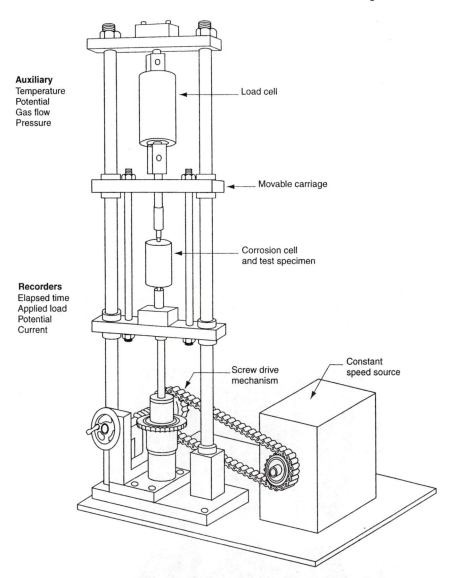

Auxiliary
Temperature
Potential
Gas flow
Pressure

Load cell

Movable carriage

Corrosion cell
and test specimen

Recorders
Elapsed time
Applied load
Potential
Current

Screw drive
mechanism

Constant
speed source

FIGURE 8.22   Slow strain rate testing apparatus. *(From J. H. Payer, W. E. Berry, and W. K. Boyd,* ASTM STP 610, *p. 82, 1976. Reprinted by permission, American Society for Testing and Materials.)*

SCC in Figures 8.2 and 8.3. The ductility of the specimens shown in Figure 8.24 can also be measured conventionally by percent reduction in area. Thus, any one of several parameters can be used to judge SCC failure in the slow strain-rate test, including time to failure, elongation or strain at failure, and percent reduction in area.

Some experimentation is necessary to determine the critical strain rate for SCC testing—usually near 10^{-6} sec^{-1} for iron, aluminum, and copper alloys. A schematic

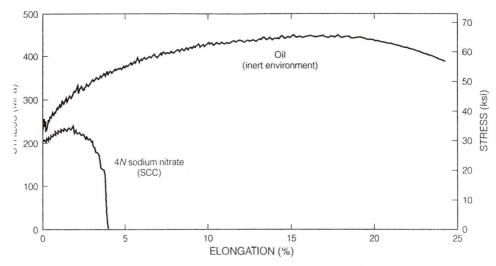

FIGURE 8.23 Stress vs. strain curves for carbon steel compared in an inert (hot oil) environment and a boiling 4*N* sodium nitrate solution causing SCC at the same temperature. *(From R. N. Parkins,* Stress Corrosion Cracking—The Slow Strain Rate Technique, *ASTM STP 665, p. 5, 1979. Reprinted by permission, American*

FIGURE 8.24 Fractured specimens from slow strain rate testing showing (a) immunity and (b) susceptibility to SCC. *(From D. O. Sprowls,* Metals Handbook, Vol. 13, Corrosion, *9th ed., ASM International, Metals Park, OH, p. 263, 1987. Photograph by courtesy of D. O. Sprowls. Reprinted by permission, ASM International.)*

plot of ductility versus controlled strain rate is shown in Figure 8.25.[49] At a certain critical strain rate, a minimum ductility is observed characteristic of SCC. Above the critical strain rate, film formation cannot keep pace with the mechanical plastic strain, and the test specimen fails by ordinary cup-and-cone ductile rupture (Figure 8.24a). Below the critical strain rate, the film formation kinetics are sufficiently rapid that film ruptures are healed before corrosive events can occur, and ductile failure is again observed. For hydrogen induced cracking (HIC), hydrogen can readily enter the lattice at all of the lower strain rates, and no ductility minimum is expected.

The slow strain-rate test has the particular advantage of speed, while retaining correlation with long-term constant-displacement tests. For example, Deegan and Wilde[50] were able to confirm in a matter of days the SCC results of Loginow and Phelps[36] for low-alloy structural steels in liquid ammonia. The latter studies, using a constant-deformation tuning fork specimen, required about 3 years to complete.

Constant-displacement and constant-load methods often require elevated temperature or solution concentration to accelerate laboratory testing. In the slow strain-rate test, continuous strain is the accelerating agent. Thus, susceptibility in simulated service conditions can be determined in the laboratory. However, the slow strain-rate test is generally somewhat more aggressive than either constant-displacement or constant-load tests. Thus, the slow strain-rate test is conservative; an alloy showing resistance in this test should be resistant in service with similar corrosive conditions. *But an alloy showing SCC in a slow strain-rate test would not necessarily fail in service, where a forced, continuous strain is absent.* However, in some instances, brittle cracking in the slow strain-rate test may indicate a very long failure time in service.

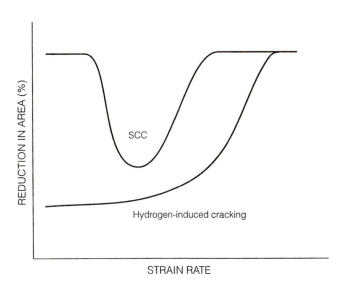

FIGURE 8.25 Schematic effect of strain rate on ductility in slow strain rate testing for stress corrosion cracking and hydrogen induced cracking. *(From C. D. Kim and B. E. Wilde, ASTM STP 665, p. 97, 1979. Reprinted by permission, American Society for Testing and Materials.)*

The slow strain-rate method has been most useful to evaluate the resistance of a given alloy in a series of environments. Specimens must begin with identical mechanical properties for the results to be easily comparable. Thus, it is difficult to compare different alloys, or the same alloy with different heat treatment and strength, using slow strain-rate testing. However, percent loss in ductility has been used to make such comparisons.

The various proposed mechanisms of SCC (see Section 8.8) generally consider dynamic strain processes at the crack tip. Therefore, it seems reasonable to find that controlled strain rate provides an effective means to accelerate SCC in laboratory testing. Conversely, any successful mechanism must adequately explain the accelerating effects of continuous strain on SCC.

8.6.4 Corrosion Fatigue Testing

Fatigue tests are conventionally conducted with smooth tensile or cantilever beam specimens. Axially loaded specimens are of the common type used for tensile testing and are tested in screw-driven or hydraulically driven test machines. Axial loading gives easier analysis of the stress state and better control of frequency and stresses in the fatigue cycle. However, cantilever beam testing is common for routine evaluations because the apparatus is compact and convenient, and more specimens can be tested at less expense.

A sinusoidal stress function (Figure 8.26) is common. The stress ratio, R, of minimum to maximum stress ($R = S_{min}/S_{max}$) is used to characterize the stress cycle. Tensile stress is taken as positive and compressive as negative. If the stress is fully reversed ($S_{max} = -S_{min}$), $R = -1$. If the stress is only partially reversed, R is a negative fraction. If stress is cycled from tension to no load, $R = 0$. R is a positive fraction if both maximum and minimum stresses are tensile. Primarily or fully compressive cycles are avoided because they have reduced or no effect, respectively.

A common rotating cantilever beam apparatus outfitted for corrosion fatigue testing is shown schematically in Figure 8.27. Each rotation of the bent-beam specimen constitutes a fatigue cycle and brings the outer fibers at a particular position alternately into tensile and compressive stress ($R = -1$). The results are com-

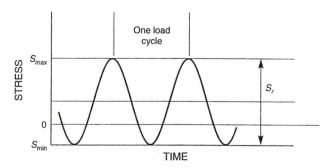

FIGURE 8.26   Sinusoidal stress function for fatigue and corrosion fatigue testing.

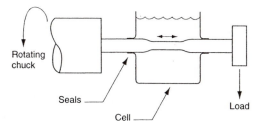

FIGURE 8.27   Schematic cantilever beam corrosion fatigue testing apparatus.

monly plotted as the *S-N* curve, the maximum tensile stress as a function of number of cycles to failure (Figure 8.10). For body-centered cubic (BCC) steels exposed only to dry air, a fatigue limit occurs below which no failure is observed after 10^8 cycles. For nonferrous alloys the fatigue limit is not as readily apparent. In any case, the addition of a corrosive environment considerably decreases the number of cycles to failure at any stress and eliminates the fatigue limit for the BCC steels.

8.7 Fracture Mechanics Testing

In recent years, the concepts and methods of linear elastic fracture mechanics have been applied to the evaluation of environmentally induced cracking (EIC). In this section, these concepts and methods are examined and their usefulness evaluated for EIC testing.

8.7.1 Introduction and Principles

Linear elastic fracture mechanics (LEFM) has been developed over the last 30 years as a means for design against unstable fast fracture. Conventional design relies on the onset of plastic deformation (yield) as the criterion for failure. However, unexpected brittle failures at stresses nominally below yield of large steel structures, especially seagoing cargo ships during World War II, focused attention on fracture as the failure criterion under conditions of large cross section, high-yield strength, low temperature, and rapid or impact loading. The designer usually uses load P and stress σ as design parameters for comparison with yield strength σ_y as the design criterion. In using fracture mechanics, load is again used, but the parameter calculated from load is the stress intensity K_I rather than stress σ. The design criterion is then the critical stress intensity K_{Ic} necessary to propagate an incipient crack or flaw. The material property K_{Ic} is called fracture toughness. Stress intensity and fracture toughness clearly require further explanation which follows.

The stress intensity K is a term relating the magnitude of stress σ at the tip of a crack or flaw to the geometry of the specimen or part in which the flaw exists:

$$K = Y\sigma\sqrt{a},$$

where Y is a function of specimen and crack geometry and a is crack length. A similar function for the stress at any point ahead of a linear crack front is given by

$$\sigma = g(\theta)\frac{K}{r},$$

where r is distance from the crack tip to the point and θ is the angle between the plane of the crack and a line normal to the crack tip passing through the point. Thus, K is a measure of the "intensity" of stress at any point ahead of the crack tip.

The subscript "I" in K_{Ic} denotes the mode I crack displacement condition, in which the crack propagates in a plane perpendicular to the direction of the applied stress, as shown in Figure 8.28. Modes II and III, also shown in Figure 8.28, denote loading configurations, in which shear stresses cause cracks to propagate by shear and tearing, respectively. Mode I conditions are promoted at the expense of modes II and III as specimen thickness increases because a triaxial stress state is generated at the crack tip by constraints from the outer elements of the specimen near the surfaces. Thicker specimens restrict deformation or strain to zero in the lateral or z-direction of Figure 8.28, causing a stress normal to other stresses in the x- and y-directions. Deformation or strain is thus confined to the x-y plane, defining so-called *plane strain conditions*. Mode I assumes overall importance because K_{Ic} is lower than K_c for either of the other two modes. For example, the critical stress intensity

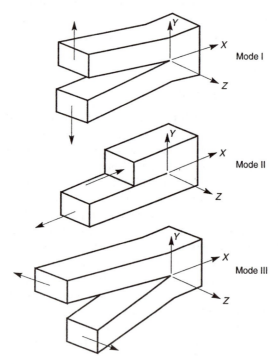

FIGURE 8.28   Crack surface displacement modes during crack growth.

K_c to cause crack propagation approaches a minimum with increasing thickness, as shown for aluminum alloy 7075, Figure 8.29.[51]

Thus to avoid rapid brittle fracture, the designer must maintain the Mode I stress intensity K_I at levels below the critical value, K_{Ic}, of the selected constructional alloy.

Calculation of K_I values in complex structures is difficult, just as determination of stresses in such structures is difficult. However, expressions for K_I are simplified in simple crack geometries incorporated into test specimens. Thus, testing procedures for LEFM have been developed extensively to determine K_{Ic}, but application of the results to design of real structures has proven to be more difficult.

There is an inverse relationship between yield strength and fracture toughness, and modern high-strength alloys consequently have low fracture toughness. LEFM was developed initially for application to high-yield-strength carbon and alloy steels and was later applied to other high-strength alloys of, for example, titanium and aluminum, as well. LEFM has been especially effective in quantifying the effects of temperature, rate of loading, and plate thickness on the fracture toughness K_{Ic} of high-strength alloys.

The minimum specimen thickness, B, necessary to produce the required lateral constraints and resulting plane strain conditions, has been experimentally determined[52] to be

$$B \geq 2.5\left[\frac{K_{Ic}}{\sigma_{ys}}\right]^2. \qquad (3)$$

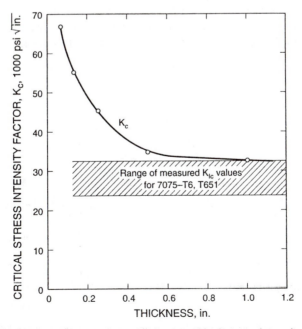

FIGURE 8.29   Effect of specimen thickness on critical stress intensity to propagate a brittle crack in aluminum alloy 7075. *(From J. G. Kaufman, ASTM STP 463, p. 7, 1970. Reprinted by permission, American Society for Testing and Materials.)*

Thus, materials of low fracture toughness K_{Ic} and high-yield strength σ_{ys} require relatively low thickness to satisfy the requirements for plane strain conditions. Fracture behavior of normally ductile alloys of relatively low-yield strength and high fracture toughness cannot be analyzed by LEFM because the thickness requirements become prohibitively high. However, stress corrosion, hydrogen embrittlement, fatigue, and corrosion fatigue reduce the stress intensity necessary to propagate brittle cracks to values much below the usual K_{Ic}. Thus, it has been possible in recent years to apply the methods of LEFM to the testing and evaluation of environmentally induced cracking (EIC) in ductile alloys of high fracture toughness.

Figure 8.30a shows a schematic representation of the effect of stress intensity, K, on the crack growth rate, da/dt, in the presence of either hydrogen induced cracking (HIC) or stress corrosion cracking (SCC). At high constant stress intensity, cracks grow at "critical" velocity above the fracture toughness value, K_{Ic}, in region III. A corrosive environment may cause "subcritical" crack growth at a reduced rate above a threshold stress intensity below K_{Ic}. Subcritical crack growth may be caused by either HIC or SCC, and threshold values are labeled here K_{Ihic} and K_{Iscc}. The effects of the two on LEFM tests are difficult to distinguish, and both K_{Iscc} and K_{Ihic} have sometimes been erroneously classified together as simply K_{Iscc}. HIC in high-strength alloys can readily be analyzed and measured by LEFM methods, but SCC in ductile alloys is more difficult because K_{Ic} is high and K_{Iscc} is often quite low. High K_{Ic} requires very high specimen thickness, as explained above. Very low K_{Iscc} substituted for K_{Ic} in equation (3) relieves the thickness requirement somewhat, but experimental application of K even slightly above K_{Iscc} allows secondary cracks to initiate and grow outside the region of maximum plane strain in the test specimens, invalidating the LEFM analysis of the results.

The combined effects of fatigue and corrosion on crack growth rate is shown schematically in Figure 8.30b. For fatigue, the increment of crack growth during

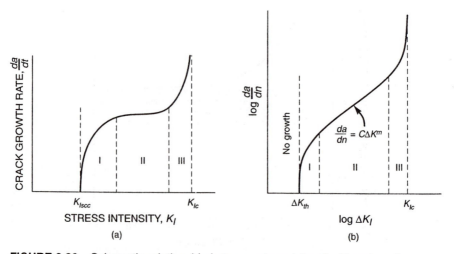

FIGURE 8.30   Schematic relationship between stress intensity K_I and crack growth rate (a) da/dt per unit time during stress corrosion cracking or hydrogen induced cracking and (b) da/dN per stress cycle during corrosion fatigue cracking.

each fatigue cycle, da/dN, is defined as the crack growth rate, which is plotted vs. the stress-intensity amplitude ΔK. Crack growth rate follows the well known power-law relationship

$$\frac{da}{dN} = C(\Delta K)^n,$$

(4)

for both fatigue (no corrosion) and corrosion fatigue in region II. Ideally, K_{Ic} is reduced to threshold values K_{Ith} and K_{Icfc} for fatigue and corrosion fatigue, respectively, below the slow crack growth region II. It is expected that K_{Icfc} with corrosion is reduced below K_{Ith}. However, a very slow crack growth rate in region I is difficult to measure within reasonable laboratory time, and the threshold values are not easily identified experimentally.

8.7.2 Test Methods and Typical Results

Specimen designs and loading to satisfy the plane strain requirements of linear elastic fracture mechanics (LEFM) are numerous.[53] Most specimen designs employ an air fatigue precrack at the root of a notch to form the sharpest possible defect and often side grooves to control the path of crack growth and enhance plane strain conditions. Most employ a loading configuration in which the stress and stress intensity either rise or decline continually as the crack grows during the test.

Historically, the rising-stress cantilever beam configuration, introduced by Brown and Beachem,[54] was first used for LEFM analysis of hydrogen induced cracking (HIC) in high-strength alloys (Figure 8.31). In this method, a dead-weight load is hung from the end of the notched and side-grooved cantilever beam. The notched area of the specimen is enclosed in a vessel (cell) holding the corrosive solution. The stress intensity K_I is given by[55]

$$K_I = \frac{6M}{(B \cdot B_N)^{1/2}(W-a)^{3/2}} F,$$

where
$\quad M$ = bending moment,
$\quad B$ = specimen width,
$\quad B_N$ = net specimen width subtracting side grooves,
$\quad a$ = crack length,
$\quad W$ = specimen depth.

$F(a/W)$ =	0.36	for a/W =	0.05
	0.49		0.10
	0.60		0.20
	0.66		0.30
	0.69		0.40
	0.72		0.50
	0.73		0.60 and larger.

As the crack grows (a increases), the net cross section decreases and the stress intensity increases during the test. When the initial stress intensity is above K_{Ihic}, the crack

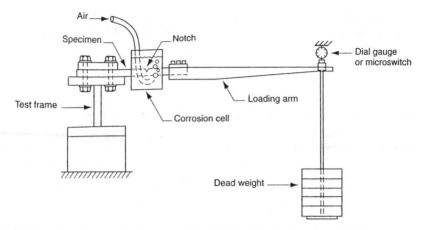

FIGURE 8.31   Schematic apparatus for cantilever beam testing for hydrogen induced cracking. *(From B. F. Brown and C. D. Beachem,* Corrosion Science, *Vol. 5, 1965. Reprinted by permission, Pergamon Press.)*

grows and the stress intensity continually rises until it reaches K_{Ic}, when critical fast cracking rapidly completes specimen fracture. The microswitch in Figure 8.31 records failure time. As the initial stress intensity, K_{Ii}, is reduced and approaches K_{Ihic}, cracking or failure time increases until it becomes very long at and below K_{Ihic}. The results are plotted as initial stress intensity versus failure time, which defines K_{Ihic}, as shown in Figure 8.32.[54]

The wedge-opening-loading (WOL) specimen shown in Figure 8.33 is loaded in a conventional screw or hydraulically activated test machine. The configuration of

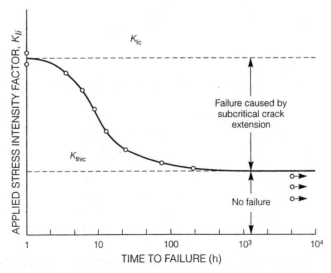

FIGURE 8.32   Determination of K_{Ihic} by cantilever beam testing.

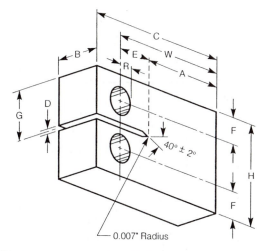

FIGURE 8.33   Wedge opening loading (WOL) specimen configuration for fracture mechanics testing.

Figure 8.33 is sometimes referred to as a compact tension (CT) specimen. The stress intensity K_I in the WOL specimen is given by[55]

$$K_I = \frac{C_3 P}{B\sqrt{a}}.$$

(5)

where C_3 is a function of a/W, P is the applied load, and B and a are specimen thickness and crack length as defined above. In WOL testing, the specimen displacement is increased in steps, and the load P (or K_I) decreases at each step as the crack grows (as a increases). This configuration is convenient to determine the value of the fracture toughness K_{Ic} at the load step where, once initiated, the crack grows very rapidly across the specimen. Unfortunately, the WOL configuration is less convenient for slower crack growth by HIC and less so for stress corrosion cracking (SCC), because the cracks grow slowly at each load step, and the test machine must be occupied for extended periods. Results for HIC of AISI type 4340 steel appear in Figure 8.34,[56] which shows effects of strength (hardness) and solution corrosivity. Because the fracture toughness K_{Ic} value shown in Figure 8.30 is of limited interest in testing for EIC, the K_{Ic} value is usually not determined and consequently is omitted from actual data plots such as Figure 8.34.

The slow bend version of the WOL specimen has one of the loading pins replaced by a threaded bolt, which permits self-loading, as shown in Figure 8.35. The specimen is first calibrated to determine load at any given displacement and crack length. The specimen is bolt loaded to constant displacement at $K_I > K_{Iscc/hic}$ and exposed to the corrosive solution. As the crack grows, the load relaxes and the stress intensity approaches $K_{Iscc/hic}$, at which time the crack arrests or stops growing. Because the modified WOL specimen is self-loaded and compact, $K_{Iscc/hic}$ can in principle be determined in any solution or service conditions from crack length at arrest in a single experiment. However, it shares the disadvantage of most constant-displacement specimens; very long exposure times are required (5000 hours usually recommended), and it is difficult

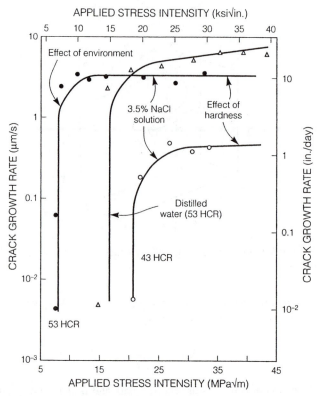

FIGURE 8.34 Effect of solution corrosivity and alloy hardness on crack growth rate in AISI 4340 steel determined by linear elastic fracture mechanics. *(From L. Raymond, Metals Handbook, Corrosion, Vol. 13, 9th ed., ASM, Metals Park, OH, p. 286, 1987. Reprinted by permission, ASM International.)*

to ascertain when crack arrest has been reached. Furthermore, in long exposures, corrosion products accumulated in the crack may produce extraneous loading and consequent erroneous crack-length and stress-intensity measurements.

These LEFM specimens have been useful to determine K_{Ihic} for high-strength steels and other high-strength alloys which are subject to HIC in dilute solutions at ambient temperature. LEFM testing has been of limited value to determine K_{Iscc} in ductile alloys because of extensive crack branching, which invalidates the LEFM analysis. Very low K_{Iscc} values, which are often present in such alloys, permit cracks to initiate at multiple locations at very low stresses in areas around the notch, where stress intensity is at the maximum in an LEFM specimen. Also, it is difficult to design and fabricate vessels or cells enclosing LEFM specimens of large size and complicated shape and to contain and control the hot corrosive solutions that are often required.

Fatigue and corrosion fatigue testing by LEFM most often employ the WOL specimen of Figure 8.33 cyclically loaded in a servohydraulic or equivalent testing machine. Figure 8.9 shows results for type 4340 steel. Crack-growth rates are increased by corrosion, lower frequency, and higher stress-intensity amplitude ΔK. Frequency generally has little or no effect without corrosion.

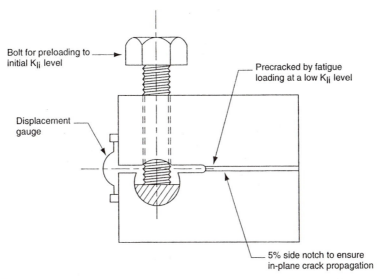

FIGURE 8.35 Self loaded configuration for wedge opening loading specimen. *(From L. Raymond,* Metals Handbook, Corrosion, *Vol. 13, 9th ed., ASM, Metals Park, OH, p. 286, 1987. Reprinted by permission, ASM International.)*

8.8 Proposed Mechanisms of Stress Corrosion Cracking

Extensive experimental investigation and theoretical analysis have been devoted to stress corrosion cracking (SCC). Consequently, numerous mechanisms have been proposed, and those receiving the greatest current attention are discussed in this section. Schematic diagrams for reference in later sections are given in Figure 8.36. Any satisfactory mechanism of SCC should explain many, if not most of the factors discussed in Section 8.1. Space is not available to discuss thoroughly the merits and demerits of each proposed mechanism. The interested reader is referred to recent reviews[4,5,6] for more exhaustive discussions and historical perspective on the mechanism of stress corrosion cracking.

Because of recurrent suggestions that hydrogen induced cracking may be a universal cause of SCC, it is discussed in this section. Atomic mechanisms of hydrogen effects are discussed in Chapter 10. The mechanism of corrosion fatigue cracking is discussed in Section 8.8.7.

8.8.1 Hydrogen Embrittlement

Numerous investigators have suggested that the brittle nature of stress corrosion cracking (SCC) can only be accounted for by a mechanism controlled by hydrogen induced cracking, which has brittle cleavage-like features very similar to those of SCC. It is generally assumed that hydrogen acts to weaken interatomic bonds in the plane strain region at the crack tip (Figure 8.36a). The result is a singular crack with

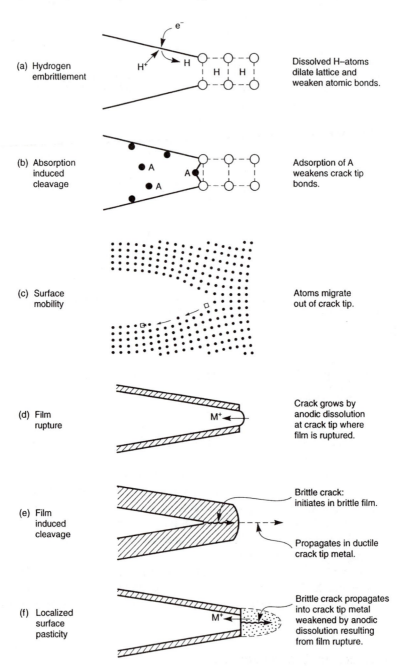

FIGURE 8.36 Schematic summary of some proposed mechanisms of stress corrosion crack growth.

little branching.[18] Unfortunately, such a simplified explanation has been difficult to support. In some instances, copious cathodic formation of hydrogen and subsequent measured entry into the metal lattice has stopped growing stress corrosion cracks.[57] Proponents have suggested that in some way hydrogen may enter the lattice at the crack tip, despite contrary surface potentials. Nevertheless, very deep, growing stress corrosion cracks have been stopped effectively in low-strength ductile alloys by cathodic polarization with accompanying hydrogen evolution. Hence, hydrogen is not generally accepted as the cause of stress corrosion cracking in most low-strength ductile alloys. However, hydrogen induced cracking (HIC) has generally been accepted as the controlling mechanism of failure for the less ductile high-strength iron, titanium, and some aluminum alloys.

A recent suggestion of possible hydrogen effects on dislocation morphology at and ahead of the stress corrosion crack tip in austenitic stainless steel is worthy of note. Jani et al.[58] observed predominantly coplanar deformation at the crack tip but homogeneous deformation at some distance (2 mm) ahead of it. Although low levels of plastic deformation in this alloy are usually coplanar, the larger deformations found at the crack tip usually form cellular dislocation arrays. The authors concluded that environmental factors, namely hydrogen, must have decreased the stacking fault energy to induce coplanar deformation at the crack tip. Embrittlement would then result from Lomer-Cottrell (L-C) supersessile dislocations on intersecting slip planes. L-C locks lying on the (100) planes act as barriers to slip, and fracture would occur above a critical stress.

They further suggest that cathodic polarization, which enhances hydrogen evolution and suppresses cracking in this system,[57] may cause passivation effects by removing potential from the critical zone for cracking (Figure 8.4). However, any such passivation effects do not hinder hydrogen entry into the lattice because Wilde and Kim[57] observed copious solid-state hydrogen transport during the cathodic polarization that stopped growth of stress corrosion cracks.

The mechanism of embrittlement by hydrogen may be similar to that which is proposed to occur in liquid metal embrittlement (LME), in which specific liquid metal atoms cause embrittlement of alloys.[59] The atoms are again thought to weaken interatomic bonds in the plane strain region at the crack tip. However, in LME, entry into the bulk crystal is unlikely because aggressive liquid metal atoms are usually insoluble in the embrittled alloy.

8.8.2 Adsorption Induced Cleavage

Various investigators have suggested that metal bonds at the crack tip are weakened by adsorption of critical anions from solution (Figure 8.36b). Uhlig originally proposed that specific aggressive dissolved species adsorb at "mobile defect sites,"[60] weakening the cohesive bonds between adjacent atoms at the crack tip. This mechanism, titled *stress sorption cracking*, has some appeal, inasmuch as the described bond weakening is similar to the effect postulated for hydrogen during hydrogen induced cracking (HIC) and for liquid metal atoms during liquid metal embrittlement. Also, the environmental cracking of polymers by specific organic solvents has been thought to occur by such an adsorption mechanism. Adsorption in this mechanism is assumed[60] to be potential dependent, to account for cathodic suppression of stress corrosion cracking below a "critical potential." Effects of inhibitor species are explained by assuming competitive adsorption between aggressive and inhibiting anions. The effect of a specific dissolved species on stress corrosion cracking (SCC) of a particular alloy is conveniently explained by assuming specific adsorption of such species to cause bond weakening at the crack tip.

An explanation of metallurgical effects on SCC requires the assumption of adsorption at mobile defect sites. The nature and character of such defects have not been specified. Pure metals are thought to be resistant because the defect sites ". . . move into and out of the surface areas at the root of a notch too rapidly for adsorption to succeed."[61] No explanation is offered as to why such undefined defects should be more mobile in pure metals than in alloys nor why some adsorbed species inhibit, while others promote, SCC. Furthermore, there is no independent evidence of preferential adsorption at such sites. It has been argued[62] that adsorption is not specific at any particular potential but continually increases above the zero point of charge where positive and negative charges are neutralized on the surface. An apparent critical potential for cracking may appear simply as the result of anodically controlled crack growth which decreases exponentially with potential, as does anodic current density in the Tafel equation (Section 3.2.1). Others[5] claim that the model fails to explain measurable, but limited plasticity ahead of an atomically sharp crack tip in a ductile metal and the discontinuous nature of crack propagation. Parkins[4] has summarized these and other arguments against the stress sorption mechanism.

8.8.3 Atomic Surface Mobility

Galvele[63,64] has suggested that many forms of environmentally induced cracking grow by the capture of surface vacancies at the crack tip and counter-current surface diffusion of atoms away from the crack tip, as shown schematically in Figure 8.36c. The mechanism predicts that stress corrosion cracking should be prevalent at temperatures below 0.5 times the melting point of the metal, T_m, and in the presence of low melting surface compounds, when surface mobility is maximized in preference to bulk diffusion in the metal crystal. The effect of critical environments to cause stress corrosion cracking (SCC) is explained by low melting compounds which enhance surface mobility.

Parkins[4] points out that although carbon steel cracks in nitrates which form low melting compounds with iron, it also cracks in the presence of high melting point Fe_3O_4. Oriani[65] shows that the flow of surface atoms should be toward the crack tip rather than away from it in the presence of a stress field. Galvele[66] argues that surface atoms exist in the absence of a stress field. However, it is then debatable as to whether a driving force is available for surface vacancy formation and diffusion.

8.8.4 Film Rupture

Film rupture or *slip-step dissolution*[67,68] was one of the first proposed stress corrosion cracking (SCC) mechanisms and still receives considerable support. Tensile stress is assumed to produce sufficient strain to rupture the surface film at an emerging slip band, and a crack then grows by anodic dissolution of the unfilmed surface at the rupture site (Figure 8.36d). Most investigators now agree that film rupture is essential to initiate cracking, but considerable controversy persists as to how a stress corrosion crack grows thereafter. For example, transgranular cracking would be expected to grow on the active slip plane, where deformation would cause continual film rupture, contrary to the experimental fact that cracks grow on planes of the type {100} or {110} (Section 8.1.2).

Parkins[8] has convincingly demonstrated a correlation between crack growth velocity and anodic current density on straining electrode surfaces (Figure 8.6). However, a crack growing by electrochemical dissolution at an unfilmed crack tip should leave a relatively smooth, featureless surface, contrary to the crystallographic cleavage (transgranular SCC) and well-defined grain boundary (intergranular SCC) features typically present on the fracture surfaces (Figure 8.3). Such topographic features on opposing brittle fracture surfaces often match exactly, indicating further that the surfaces were formed by brittle mechanical cleavage-like processes with very little surface dissolution.

The proportionality demonstrated in Figure 8.6 does not guarantee that the actual cracking process is electrochemical. Electrochemical (charge transfer) reactions may trigger a brittle mechanical cracking process that accounts for crack growth. Indeed, crack growth in some cases has been observed to proceed discontinuously, in steps,[69] and corresponding periodic crack arrest markings (striations) are frequently present on fracture surfaces,[70] contrary to the expected smooth increase in crack length expected for electrochemical growth.

8.8.5 Film-Induced Cleavage

The *film-induced cleavage* (FIC) mechanism[71] has been proposed to explain discontinuous transgranular crack growth and high transgranular crack growth rates. FIC postulates the presence of brittle surface films, to include dealloyed films on certain alloys and oxides on pure metals and other alloys. A crack growing in the film may propagate further into the underlying metal if it reaches sufficient velocity at the film/metal interface (Figure 8.36e). Anodic dissolution and corrosion are not necessary to propagate cracks but only to form the required brittle surface film. Thus, a relatively low anodic dissolution rate forms a brittle surface film, and a fast-growing crack in the film may penetrate for some distance into the ductile substrate metal before its growth is arrested. The surface film must reform at the crack tip surface before a new burst of brittle crack growth is possible. Thus, a low anodic current at a strained surface can foster rapid, discontinuous, brittle cracking, according to the FIC mechanism.

Unfortunately, there is no *a priori* reason to expect that surface films or layers arc sufficiently brittle to support a brittle crack or that such a crack will be sustained after crossing the interface into the underlying ductile matrix. Most passive films are very thin and hydrated. While they can be ruptured, it seems unlikely that they would be sufficiently brittle to support the suppositions of FIC. Dealloyed layers are unlikely to have sufficient adhesion to the substrate or inherent brittleness to sustain cracking. Fritz et al.[72] observed that a stress corrosion crack could not propagate beyond the dealloyed layer in brass without an externally imposed anodic current.

8.8.6 Localized Surface Plasticity

Metallurgical effects on stress corrosion cracking (SCC) are not easily explained by models that pertain only to reactions on the surface. It would seem that a valid mechanism must include effects of the corrosive environment on the properties of the underlying metal.

Mechanical creep precedes the initiation of SCC[73] and is accelerated by anodic currents.[74] Such anodic currents can originate, of course, from the usual anodic corrosion reactions (Section 3.3.1). Creep accelerated by anodic current implies softening by a surface defect structure which attenuates the normal strain hardening that accompanies primary creep. Corrosion induced relief of strain hardening may be one symptom of a critical corrosion effect on metallurgical properties during SCC.[75] That is, SCC may result from a defect structure ahead of the crack tip, as proposed in the mechanism, *localized surface plasticity* (LSP).[76]

In the LSP mechanism, film rupture supposedly initiates large anodic currents at the rupture site by galvanic coupling of the unfilmed active surface to surrounding noble passive surfaces (Figure 8.36f). The passive film must be in a weakened state to prevent rapid repassivation which would suppress crack growth. These weakened states usually occur at critical potentials near the boundaries of the passive potential region (Figure 8.4) and are probably influenced by the presence of critical anions such as Cl^-. These galvanic anodic currents produce a softened defect structure locally at the rupture site, which coincides with the crack tip for a growing crack. The softened structure, which would normally deform plastically, can do so to only a limited extent in the microscopic volume ahead of the crack, when constrained by the surrounding material of full strength and hardness. Microstrain within the softened, yet constrained, crack-tip volume produces a triaxial stress state (plane-strain condition), which suppresses plastic slip. Continued strain can only propagate a brittle crack. The condition at the crack tip has been likened to that of soft solder joining the ends of two steel bars.[76] Loading perpendicular to the thin solder joint causes brittle failure of the soft solder material due to the triaxial stress state, which prevents plastic deformation of the normally ductile solder.

Forty[69] originated the proposal that vacancies injected into the crack tip microvolume during dealloying cause a defect crack-tip structure responsible for SCC of brass. Efforts to demonstrate embrittlement by vacancy production in foils by quenching or deformation have been unsuccessful.[77] According to the LSP mechanism, this is not unexpected. Mechanically tested foils will nearly always be in a biaxial stress state, which permits easy plastic deformation. Only when the softened material is constrained by the plane-strain condition at the crack tip is it subject to brittle cracking.

A vacancy-saturated defect structure is one that may be responsible for SCC. Others may be possible as well, and the LSP mechanism is not restricted to any specific defect structure type. However, it is notable that vacancies have been credited with causing increased creep by anodic reactions[74] and by cyclic loading.[78]

The LSP mechanism seems to explain several metallurgically related effects in SCC. Resistance of pure metals is explained by the lower degree of strain hardening in pure metals, as compared to alloys. Thus, extensive deformation and ductility in pure metals are permissible before relief of strain hardening is sufficient to initiate cracks. Highly strain hardenable alloys, even of low yield strength (e.g., stainless steels and brasses), are often highly susceptible. Anodic currents are thought to trigger a brittle mechanical cleavage process, which occurs on prismatic planes of the {100} or {110} type. These planes have the least atomic density and greatest atomic spacing and therefore are most susceptible to cleavage when plasticity is suppressed at the plane-strain crack tip. Discontinuous growth of cracks is explained by a postulated brittle crack burst into the softened crack-tip volume and possibly beyond. The softened crack-tip defect structure must reform at an arrested crack tip before a new brittle burst is possible.

8.8.7 Correlation with Corrosion Fatigue and Hydrogen Induced Cracking

Microscopic plastic processes may be a common precursor to all forms of environmentally induced cracking. Fatigue in the absence of corrosion has also been observed to increase primary creep. Figure 8.37 shows recent data[79] for carbon manganese steel, in which primary creep was enhanced by low-cycle fatigue loading. Cyclic loading apparently acts also to attenuate strain hardening during corrosion fatigue cracking (CFC), as do anodic currents[74] during stress corrosion cracking (SCC). Despite the brittle appearance of hydrogen induced cracking (HIC), hydrogen is known to increase plasticity on a local microscopic scale, based on fractographic examinations[80] of fracture surfaces. Thus, it seems reasonable that cathodically generated hydrogen dissolved in the lattice may produce local surface plasticity in HIC, as do anodic currents in SCC.

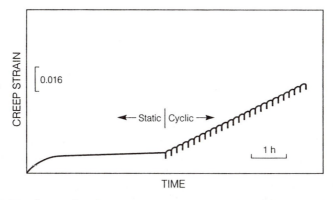

FIGURE 8.37 Creep of carbon-manganese steel accelerated by cyclic loading. *(From J. T. Evans and R. N. Parkins, Acta Metall., Vol. 24, p. 511, 1976. Reprinted by permission, Pergamon Press.)*

8.8.8 General Comments

Hydrogen and adsorbed anions have not been accepted as causitive agents for stress corrosion cracking (SCC) of low strength, ductile alloys. Film rupture has been generally accepted as the necessary prerequisite for SCC, although the mechanism of subsequent brittle crack growth needs clarification. Sections 8.8.5–6 describe two proposals for this mechanism, film induced cleavage and localized surface plasticity. Not enough experimental data is available to confirm or deny either, although it is probably obvious that the author prefers the latter. In fact, however, neither may be valid; the truth may still be awaiting discovery; or as some believe,[4] several mechanisms may be operating in the many and varied systems in which stress corrosion cracking has been observed.

References

1. S. P. Pednekar, A. K. Agrawal, H. E. Chaung, and R. W. Staehle, *J. Electrochem. Soc.*, Vol. 126, p. 701, 1979. K. Sieradzki, R. L. Sabatini, and R. C. Newman, *Metall. Trans. A*, Vol. 15A, p. 1941, 1984. E. I. Melitis and R. F. Hochman, *Corrosion Sci.*, Vol. 24, p. 843, 1984.
2. R. Liu, N. Narita, C. Alstetter, H. Birnbaum, and E. N. Pugh, *Metall. Trans. A*, Vol. 11A, p. 1563, 1980.
3. B. E. Wilde, *Metals Handbook*, Vol. 11, *Failure Analysis*, 9th ed., ASM International, Metals Park, OH, p. 203, 1986.
4. R. N. Parkins, *Environment-Induced Cracking of Metals*, R. P. Gangloff and M. B. Ives, eds., NACE, Houston, p. 1, 1990.
5. R. H. Jones and R. E. Ricker, *Metals Handbook*, Vol. 13, *Corrosion*, 9th ed., ASM International, Metals Park, OH, p. 145, 1987.
6. R. W. Staehle, *Stress Corrosion Cracking and Hydrogen Embrittlement of Iron Base Alloys*, R. W. Staehle et al, eds., NACE-5, NACE, Houston, p. 193, 1977.
7. R. N. Parkins, *Proc. 5th Symposium on Line Pipe Research*, Am. Gas Assoc., Arlington, VA, p. U-1, 1974.
8. R. N. Parkins, *Br. Corrosion J.*, Vol. 14, p. 5, 1979.
9. P. S. Pao and R. P. Wei, *Metals Handbook*, Vol. 11, *Failure Analysis*, 9th ed., ASM International, Metals Park, OH, p. 253, 1986.
10. L. Engel and H. Klingele, *Atlas of Metal Damage*, Prentice-Hall, p. 120, 1981.
11. D. O. Sprowls, *Metals Handbook*, Vol. 13, *Corrosion*, 9th ed., ASM International, Metals Park, OH, p. 296, 1987.
12. C. S. Kortovich, "Corrosion Fatigue of 4340 and D6AC Steels below K_{Iscc}," *Proc. 1974 Tri-Service Corrosion Conference on Corrosion of Military Equipment," AFML-TR-75-42*, Air Force Materials Laboratory, Wright Patterson Air Force Base, 1975.
13. M. O. Speidel, *Proc. Int. Conf. on Stress Corrosion and Hydrogen Embrittlement of Iron Base Alloys*, NACE, Houston, p. 1071, (1977).
14. D. J. Duquette and H. H. Uhlig, *Trans. Am. Soc. Metals*, Vol. 61, p. 449, 1968.
15. W. Glaeser and I. G. Wright, *Metals Handbook*, Vol. 13, *Corrosion*, 9th ed., ASM International, Metals Park, OH, p. 142, (1987).
16. A. R. Troiano, *Trans. ASM*, Vol. 52, p. 54, 1960.

17. G. Kobrin, *Metals Handbook*, Vol. 13, *Corrosion*, 9th ed., ASM International, Metals Park, OH, pp. 329–31, 1987.
18. B. D. Craig, *Metals Handbook*, Vol. 13, *Corrosion*, 9th ed., ASM International, Metals Park, OH, pp. 164–6, 1987.
19. G. Schick, *Metals Handbook*, Vol. 13, *Corrosion*, 9th ed., ASM International, Metals Park, OH, p. 1132, 1987.
20. L. Engel and H. Klingele, *Atlas of Metal Damage*, Prentice-Hall, pp. 125–8, 1981.
21. J. H. Payer, W. E. Berry, and W. K. Boyd, *ASTM STP 665*, p. 61, 1979.
22. C. D. Kim and B. E. Wilde, *ASTM STP 665*, p. 97, 1979.
23. M. G. Fontana, *Corrosion Engineering*, 3rd ed., McGraw-Hill, p. 119, 1986.
24. J. E. Truman, *Corrosion Science*, Vol. 17, p. 737, 1977.
25. D. L. Graver, ed., *Corrosion Data Survey*, Metals Section, 6th ed., NACE, 1985.
26. D. R. McIntyre and C. P. Dillon, MTI Publication No. 15, Materials Technology Institute of the Chemical Process Industries, St. Louis, p. 15, 1985.
27. C. P. Dillon, *Materials Performance*, Vol. 29, No. 12, p. 66, 1990.
28. W. L. Williams, *Corrosion*, Vol. 13, p. 539t, 1959.
29. B. M. Gordon, *Metals Handbook*, Vol. 13, *Corrosion*, 9th ed., ASM International, Metals Park, OH, p. 928, 1987.
30. G. Kobrin, *Metals Handbook*, Vol. 13, *Corrosion*, 9th ed., ASM International, Metals Park, OH, p. 326, 1987.
31. D. O. Sprowls, *Metals Handbook*, Vol. 13, *Corrosion*, 9th ed., ASM International, Metals Park, OH, p. 273, 1987.
32. R. L. Jones, R. L. Long, and J. S. Olszewski, *CORROSION/83*, Paper No. 141, NACE, Houston, 1983.
33. A. Bond and H. J. Dundas, *Corrosion*, Vol. 24, 344, 1968.
34. H. H. Uhlig and R. W. Revie, *Corrosion and Corrosion Control*, Wiley, 3rd ed., pp. 125–9, 1985.
35. K. F. Krysiak, *Metals Handbook*, Vol. 13, *Corrosion*, 9th ed., ASM International, Metals Park, OH, p. 353, 1987.
36. A. Loginow and E. H. Phelps, *Corrosion*, Vol. 18, p. 299t, 1962.
37. H. R. Copson, *Physical Metallurgy of Stress Corrosion Fracture*, T. N. Rhodin, ed., Interscience, p. 247, 1959.
38. J. Kolts, *Metals Handbook*, Vol. 13, *Corrosion*, 9th ed., ASM International, Metals Park, OH, p. 647, 1987.
39. N. Pessal, G. P. Airey, and B. P. Lingenfelter, *Corrosion*, Vol. 35, p. 100, 1979.
40. N. W. Polan, *Metals Handbook*, Vol. 13, *Corrosion*, 9th ed., ASM International, Metals Park, OH, p. 610, 1987.
41. E. Mattson, *Electrochim. Acta*, Vol. 3, p. 279, 1960–61.
42. H. H. Uhlig and R. W. Revie, *Corrosion and Corrosion Control*, 3rd ed., Wiley, p. 334, 1985.
43. E. H. Hollingsworth and H. Y. Hunsicker, *Metals Handbook*, Vol. 13, *Corrosion*, 9th ed., ASM International, Metals Park, OH, p. 590, 1987.
44. R. W. Schutz and D. E. Thomas, *Metals Handbook*, Vol. 13, Corrosion, 9th ed., ASM International, Metals Park, OH, p. 674, 1987.
45. D. O. Sprowls, *Metals Handbook*, Vol. 13, *Corrosion*, 9th ed., ASM International, Metals Park, OH, pp. 246–50, 1987.
46. A. W. Loginow, *Materials Performance*, Vol. 5, No. 5, p. 33, May, 1966.
47. R. N. Parkins, *Stress Corrosion Cracking—The Slow Strain Rate Technique*, ASTM STP 665, p. 5, 1979.
48. J. H. Payer, W. E. Berry, and W. K. Boyd, *ASTM STP 610*, p. 82, 1976.

49. C. D. Kim and B. E. Wilde, *ASTM STP 665*, p. 97, 1979.

50. D. C. Deegan and B. E. Wilde, *Stress Corrosion Cracking and Hydrogen Embrittlement of Iron Base Alloys*, R. W. Staehle et al., eds., NACE-5, NACE, Houston, p. 663, 1977.

51. J. G. Kaufman, *ASTM STP 463*, p. 7, 1970.

52. W. F. Brown, Jr., and J. E. Srawley, *ASTM STP 463*, 1967.

53. D. O. Sprowls, *Metals Handbook*, Vol. 13, *Corrosion*, 9th ed., ASM International, Metals Park, OH, pp. 253–63, 1987.

54. B. F. Brown and C. D. Beachem, *Corrosion Science*, Vol. 5, 1965.

55. S. T. Rolfe and J. M. Barsom, *Fracture and Fatigue Control in Structures*, Prentice-Hall, p. 294–9, 1977.

56. L. Raymond, *Metals Handbook*, Vol. 13, *Corrosion*, 9th ed., ASM International, Metals Park, OH, p. 286, 1987.

57. B. E. Wilde and C. D. Kim, *Corrosion*, Vol. 28, p. 350, 1972.

58. S. C. Jani, M. Marek, R. F. Hochman, and E. I. Meletis, *Environment-Induced Cracking of Metals*, R. P. Gangloff and M. B. Ives, eds., NACE, Houston, p. 541, 1990. *Metall. Trans. A*, Vol. 22A, p. 1453, 1991.

59. M. H. Kamdar, *Metals Handbook*, Vol. 11, *Failure Analysis*, 9th ed., ASM International, Metals Park, OH, p. 171, 1987.

60. H. H. Uhlig, *Physical Metallurgy of Stress Corrosion Fracture*, T. N. Rhodin, ed., Interscience, p. 1, 1959.

61. H. H. Uhlig and R. W. Revie, *Corrosion and Corrosion Control*, Wiley, 3rd ed., p. 134, 1985.

62. J. O'M. Bockris, *Stress Corrosion Cracking and Hydrogen Embrittlement of Iron Base Alloys*, R. W. Staehle et al, eds., NACE-5, NACE, Houston, p. 177, 1977.

63. J. R. Galvele, *Corrosion Science*, Vol. 27, p. 1, 1987.

64. G. S. Duffo and J. R. Galvele, *Environment-Induced Cracking of Metals*, R. P. Gangloff and M. B. Ives, eds., NACE, Houston, p. 261, 1990.

65. R. A. Oriani, *Environment-Induced Cracking of Metals*, R. P. Gangloff and M. B. Ives, eds., NACE, Houston, p. 263, 1990.

66. J. R. Galvele, *Environment-Induced Cracking of Metals*, R. P. Gangloff and M. B. Ives, eds., NACE, Houston, p. 264, 1990.

67. F. A. Champion, *Symposium on Internal Stresses in Metals and Alloys*, Inst. Metals, London, p. 468, 1948.

68. H. L. Logan, *J. Res. Natl. Bur. Stand.*, Vol. 48, p. 99, 1952.

69. A. J. Forty and P. Humble, *Philos. Mag.*, Vol. 8, p. 247, 1963.

70. E. N. Pugh, *Corrosion*, Vol. 41, p. 517, 1985.

71. K. Sieradzki and R. C. Newman, *Philos. Mag. A*, Vol. 51 (No. 1), p. 95, 1985.

72. J. D. Fritz, B. W. Parks, and H. W. Pickering, *Scripta Metall.*, Vol. 22, p. 1063, 1988.

73. M. C. Petit and D. Desjardins, *Stress Corrosion Cracking and Hydrogen Embrittlement of Iron Base Alloys*, R. W. Staehle et al, eds., NACE-5, NACE, Houston, p. 1205, 1977.

74. R. W. Revie and H. H. Uhlig, *Acta Metall.*, Vol. 22, p. 619, 1974.

75. D. A. Jones, *Metall. Trans. A*, Vol. 16A, p. 1133, 1985.

76. D. A. Jones, *Environment-Induced Cracking of Metals*, R. P. Gangloff and M. B. Ives, eds., NACE, Houston, p. 265, 1990.

77. E. N. Pugh, *Environment-Induced Cracking of Metals*, R. P. Gangloff and M. B. Ives, eds., NACE, Houston, p. 284, 1990.

78. C. E. Feltner and G. M. Sinclair, *Int. Conf. on Creep*, London, pp. 3–9, 1963.

79. J. T. Evans and R. N. Parkins, *Acta Metall.*, Vol. 24, p. 511, 1976.

80. C. D. Beachem, *Metall. Trans.*, Vol. 3, p. 437, 1972.

Exercises

8-1. With the evidence given in each case listed below, tell whether stress-corrosion cracking, corrosion-fatigue cracking, hydrogen-induced cracking or a combination of them may be the cause of failure in each case. Assume any information not given is unavailable. Comment on your reasoning in each case.

    a. alloy under constant stress, transgranular branched cracks, sharp crack tips;

    b. pure metal under tensile stress with superimposed cyclic stress, blunt crack tip, transgranular unbranched crack;

    c. high-strength alloy, transgranular branched cracks, sharp crack tips, room temperature failure

    d. alloy under constant stress, intergranular branched cracks, sharp crack tips, high temperature failure, corrosion products in the cracks

    e. highly-cold-worked alloy, intergranular unbranched crack, sharp crack tip, room temperature failure.

8-2. Determine candidate alloys which would be resistant to stress corrosion cracking in the following environments. Refer to Chapter 11 and cross-out the candidates that would not be resistant to other forms of corrosion. Underline the one candidate which you conclude to be the most suitable based on engineering properties and cost.

    a. hot acid chloride solutions at 150°C

    b. hot caustic solutions above 120°C requiring high strength

    c. polythionic acids at room temperature

    d. dilute hydrofluoric acid at room temperature

    e. aqueous amine solutions at room temperature

8-3. Give at least 3 methods to prevent stress-corrosion cracking in the following systems. Assume that the environmental conditions causing the cracking cannot be changed appreciably. When prescribing alloys substitutions or modifications, give specific alloys when possible.

    a. austenitic stainless steels in hot chloride solutions

    b. carbon steel in hot boiler waters

    c. austenitic stainless steel in polythionic acid

    d. brass (30% Zn) in aqueous ammonia solutions

    e. titanium in methanol/chloride solutions

8-4. Indicate the most appropriate stress corrosion cracking test(s) for each of the following purposes. Explain why other tests may not be appropriate.

    a. in-plant evaluation of a number of alloys for resistance in process streams

    b. determination of minimum critical stress for SCC in plant process streams

    c. laboratory alloy development of new alloys for resistance to SCC

    d. laboratory measurement of SCC crack growth rate

8-5. For the conditions and requirements listed in the following cases, briefly describe your best recommendation for a laboratory testing program.

    a. evaluation of a number of alloys for resistance to SCC in hot, acidic process streams containing about 1% chloride

    b. determination of the environmental causes of SCC of a carbon steel specified for use in boiler waters and the environmental modifications needed to prevent it

    c. determination of the contributions of pitting, hydrogen, and fatigue to the environmental cracking of a high strength steel used in the rotating shaft of a plant compressor

8-6. **(a)** Calculate the minimum specimen width necessary for valid fracture mechanics testing for a constructional steel of yield strength, 100 ksi and fracture toughness of 150 ksi$\sqrt{in}$. Would it be practical to measure the fracture toughness of this material?

    (b) Supposing a corrosive environment makes hydrogen embrittlement possible with K_{Ihic} of 20 ksi$\sqrt{in}$. what is the minimum specimen width?

8-7. Calculate K_{Ihic} for a high-strength steel exposed to 3% NaCl in the cantilever beam apparatus of Figure 8.31, if a minimum load of 50 lb. at a maximum crack length of 1.3-in. causes failure when applied to a momentum arm of 6-ft, assuming the following specimen dimensions: depth, 2-in.; width, 1-in.; reduced width across side grooves, 0.900-in.

8-8. Describe an original experimental plan which would prove or at least support any one of the SCC mechanisms in section 8.8. An original plan is one which has not already been described in this text or elsewhere.

Effects of Metallurgical Structure on Corrosion

9

Metallurgical structure and properties often have strong effects on corrosion. Some have already been described. For example, the role of sulfide inclusions in initiation of pitting in stainless steels was discussed in Section 7.2.1. Grain boundary segregation of carbon, nitrogen, and phosphorus can induce intergranular stress corrosion cracking of carbon steels (Section 8.4.2). Unfavorable metallurgical structures, especially martensite, reduce the resistance of steels to hydrogen induced cracking (Section 8.3).

Crystallographic orientation of grains have an affect on corrosion resistance, as evidenced by metallographic etching rates. Grain boundaries have an inherently imperfect, high-energy structure, which is expected to have lower corrosion resistance. Alloying elements may segregate to affect corrosion resistance at grain boundaries. Different phases in an alloy system have differing corrosion resistances. Cold work can affect corrosion rate by adding energy to the alloy or changing the crystal orientation and phase composition. Thus, metallurgical effects on corrosion are numerous and specific to the alloy system. Space is available in this chapter to discuss only the technologically more important of such effects.

This chapter begins with a description of intergranular corrosion (IGC), which is perhaps the best-known example of metallurgical effects on corrosion. Observations of IGC in austenitic stainless steel and a number of other alloys are described. Welding is discussed, in which altered metallurgical structures have serious effects on corrosion resistance. A common corrosive effect on metallurgical structure is dealloying. Dezincification of brass, described in this chapter, is the best-known example.

9.1 Intergranular Corrosion of Austenitic Stainless Steels

9.1.1 Grain Boundary Chromium Depletion

Intergranular corrosion (IGC) of stainless steels was introduced in Section 1.5.7. An example of IGC in an austenitic stainless steel is shown in Figure 1.13. In an intermediate temperature range of 425° to 815°C (800° to 1500°F) chromium carbides, $(Fe,Cr)_{23}C_6$, are insoluble and precipitate at grain boundaries. Above 815°C the chromium carbides are soluble; below 425°C the diffusion rate of carbon is too low to permit formation of the carbides. The chromium carbide precipitates are very high in chromium, but the matrix alloy is depleted of chromium in the grain boundaries

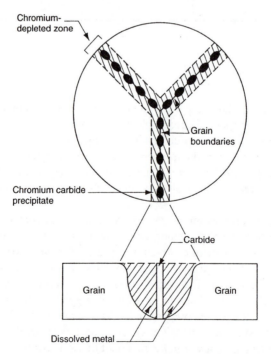

FIGURE 9.1 Schematic representation of carbide precipitation at a grain boundary during sensitization to intergranular corrosion in stainless steel.

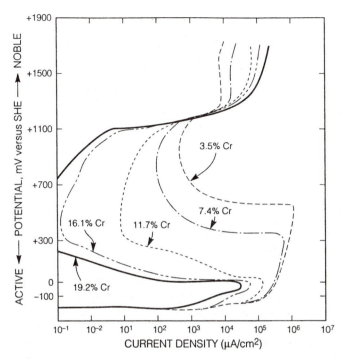

FIGURE 9.2 Effect of chromium content on anodic polarization of Fe-Ni alloys of 8.3 to 9.8% Ni in $2N\,H_2SO_4$ at 90°C. *(From K. Osozawa and H. J. Engell,* Corros. Sci., *Vol. 6, p. 389, 1966. Reprinted by permission, Pergamon Press.)*

(Figure 9.1). Although the effect of heat treatments on IGC strongly implies grain boundary chromium depletion, direct experimental evidence has been provided only recently by electron spectroscopy.[1,2]

The chromium-depleted alloy in the grain boundaries is much less corrosion resistant than the surrounding grains. Figure 9.2 shows anodic polarization curves for alloys of approximately constant Ni and variable Cr content[3] in hot sulfuric acid typical of chemical testing solutions (Section 9.1.4). Below about 12% Cr, the passive potential region is substantially constricted. Thus, in moderately oxidizing conditions, the chromium-depleted grain boundary alloy, in the active state, is coupled to the undepleted passive alloy at the grains. The low anode/cathode surface area ratio (Section 6.2.3) results in rapid microscopic galvanic attack and IGC at the grain boundaries. The chromium carbide is resistant, but the surrounding chromium-depleted alloy in the grain boundary is aggressively corroded.

Stainless steel, experiencing a thermal treatment that produces grain boundary carbides and chromium-depleted grain boundaries, is known to be "sensitized" to IGC. As the carbon content is reduced, longer times are required to sensitize austenitic stainless steel. Figure 9.3 shows the sensitization diagram for Type 304 stainless steel for various carbon compositions.[4] Sensitization was measured by an acid exposure test accelerated with copper, as described in Section 9.1.4.

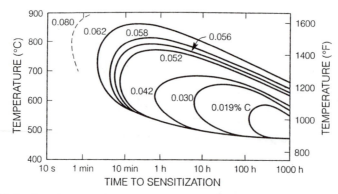

FIGURE 9.3 Sensitization diagram for 18Cr-8Ni stainless steel of varying carbon content. *(From R. M. Davison, T. DeBold, and M. J. Johnson,* Metals Handbook, *Vol. 13,* Corrosion, *9th ed., ASM, Metals Park, OH, p. 547, 1987. Reprinted by permission, ASM International.)*

It is apparent from the preceding discussion that chromium and carbon are the primary elements causing sensitization. Other elements can have secondary effects as well, however. Nickel increases the activity of carbon in solid solution, facilitates the precipitation of carbides, and thereby enhances sensitization. Molybdenum behaves rather like chromium; it precipitates as a carbide at grain boundaries and the depletion contributes to sensitization. Molybdenum has much less effect, however, because of its correspondingly lower concentration in the alloys.

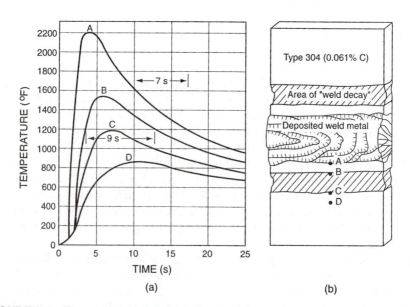

FIGURE 9.4 Thermal transients producing weld decay during welding of austenitic stainless steel: (a) temperature-time relationships; (b) location of thermocouples.

9.1.2 Weld Decay and Knifeline Attack

Sensitization of austenitic stainless steels during welding is known as weld decay. The classic form of weld decay results in intergranular corrosion (IGC) within the heat affected zone (HAZ) somewhat removed from the weld bead itself. The thermal transients producing weld decay are shown schematically in Figure 9.4.[5] At the base metal/weld bead interface, A, the alloy is in the critical temperature range for an insufficient time to produce sensitization. Remote from the weld at C and D, the alloy does not reach the critical temperature range and remains unsensitized. However, at an intermediate position, B, the alloy is in the critical temperature range for sufficient time to produce sensitization. The exact position of the weld decay region where IGC occurs within the HAZ and the critical temperature interval is sensitive to metallurgical history, plate thickness, and the various welding parameters of heat input, cooling rate, and so on. The IGC zone may not always be removed from the weld bead as is commonly portrayed in Figure 9.4. Nevertheless, the classic weld decay morphology is not uncommon, and an example is shown in Figure 9.5.[6]

IGC is related to the intergranular stress corrosion cracking (IGSCC) of sensitized austenitic stainless steels, particularly in nuclear reactor cooling systems, and in petrochemical systems containing polythionic acid (Section 8.4.1). Although significant IGC is not present in unstressed parts, the presence of tensile stress produces intergranular cracking. The same measures used to mitigate IGC are also used for IGSCC.

Knifeline attack (KLA) is a highly localized form of IGC that occurs for only a few grain diameters immediately adjacent to the weld bead in Types 321 and 347

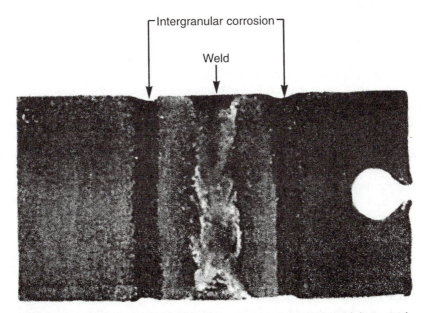

FIGURE 9.5 Weld decay in the heat affected zone of a weld in stainless steel. *(From H. H. Uhlig and R. W. Revie, Corrosion and Corrosion Control, 3rd ed., Wiley, New York, p. 307, 1985. Reprinted by permission, John Wiley & Sons.)*

TABLE 9.1   Temperature Ranges for Sensitization to Intergranular Corrosion in Austenitic Stainless Steels

Temperature	Metallurgical Reactions	Remarks	
		Unstabilized Nb/Ti Absent (Type 304)	Stabilized Nb/Ti Present (Type 321, 347)
Melting Point –			
A	All carbides dissolve.	Rapid cooling prevents IGC.	Rapid cooling and reheating to C causes KLA. Reheating to B prevents KLA.
1230°C– –			
(2250°F) B	Niobium carbide precipitates. Chromium carbide dissolves.	Rapid cooling prevents IGC.	IGC prevented by precipitating dissolved carbide uniformly.
815°C –			
(1500°F) C	Chromium carbide precipitates at grain boundaries.	Sensitization to IGC caused.	No sensitization. Nb/Ti Carbides precipitated at B.
425°C –			
(800°F) D	No reactions.	Temperature too low for adequate diffusion.	

austenitic stainless steels. These "stabilized" stainless steels contain titanium and niobium, respectively, which react with carbon to prevent sensitization (Section 9.1.3). In KLA, titanium or niobium carbides dissolve in solid solution at high temperature, >1230°C, next to the weld bead, followed by rapid cooling, which retains all carbides in solid solution. Such conditions are more likely in thin welded sheet, which allows very rapid cooling. Later weld passes or stress relief annealing in the chromium carbide sensitization range allows conventional sensitization because the titanium or columbium has not had any opportunity to react with the carbon. Thus, failure occurs by conventional IGC in the narrow sensitized region next to the weld. The relevant temperature ranges are summarized in Table 9.1. Niobium in Type 347 stainless steel is taken as the example; titanium behaves similarly when alloyed in Type 321 stainless steel.

9.1.3 Prevention

Generally, a strongly oxidizing condition is needed to generate intergranular corrosion (IGC). Many weakly corrosive conditions do not cause IGC in sensitized microstructures.[7] Lower acidity and less oxidizing conditions will generally reduce the susceptibility to IGC. In nuclear reactor coolant piping, a reduction in dissolved oxygen by injection of hydrogen has been effective to mitigate intergranular stress corrosion cracking (IGSCC) of sensitized austenitic stainless steel.[8] Highly aggressive conditions are inherent in the chemical process industry, however, and altering the corrosive conditions often is not a feasible option.

Metallurgical measures are more common for prevention of IGC. These fall into three categories: (a) solution annealing, (b) low carbon alloy modifications, and (c) stabilized alloys containing niobium or titanium. Solution annealing consists of heating the alloy above 815°C (Table 9.1), where all chromium carbides are dissolved, followed by rapid cooling (often by water immersion) to retain the carbides in solution. Since no carbides are allowed to precipitate, the alloy is in the unsensitized condition, and no IGC will occur.

Low-carbon alloy modifications allow welding and other heat treatments in the critical range without sensitization because there is insufficient carbon available for chromium carbide precipitation at the grain boundaries. Type 304L stainless steel containing <0.03% C was developed for the nuclear applications but is now used widely throughout industry.

Stabilized stainless steels, Types 347 and 321, contain niobium and titanium, respectively, which react with carbon above 815°C to precipitate the respective carbides. This removes the carbon from solution, and the carbides are precipitated randomly in the grains at the higher temperature. The alloy matrix becomes carbon free, and no carbon is available to precipitate with chromium at the grain boundaries in the critical range of 425° to 815°C.

Knifeline attack can be prevented simply by heating above 815°C, where the chromium carbides dissolve, and the carbon reprecipitates as the niobium or titanium carbide. Subsequent cooling rate is unimportant because the carbon is unavailable for reaction with chromium at lower temperatures. In welded equipment, where the reheating step may be impractical, the low-carbon modifications are recommended.

9.1.4 Chemical Testing

Sensitized austenitic stainless steels show intergranular corrosion (IGC) in a number of highly oxidizing acidic media[9] used in testing for sensitization. Testing is necessary to qualify alloy heats for acceptance in construction and repair of chemical process and other industrial plants. Table 9.2 summarizes the major testing procedures that have been used. Most were originally developed to simulate specific plant conditions.[10] However, questionable correlations with plant conditions have reduced their use in most cases to detection of the sensitized metallurgical structure.

The sulfuric acid-copper sulfate or Strauss test (Table 9.2) was historically the first for detecting sensitization in austenitic stainless steel. Dissolved Cu^{2+} acts as an oxidizer in sulfuric acid to passivate the grains and attack the chromium-depleted grain boundaries. However, the rate of attack is relatively low, and the method now is recommended only for highly sensitized alloys. The copper-sulfuric acid-copper sulfate or copper-accelerated Strauss test increases the rate in the same solution, by making the stainless steel an anode in a galvanic couple with copper metal. Both the Strauss tests are evaluated by bending the plate specimen over a mandrel and examining at about 20× for fissures caused by IGC.

TABLE 9.2   Standard Intergranular Corrosion Tests for Stainless Steels

ASTM Standard (Common Name)	Species Environment	Exposure	Evaluation	Attacked
A708-86 (Strauss)	16% H_2SO_4 +6% $CuSO_4$. Boiling.	One 72-hour period.	Macroscopic appearance after bending.	Chromium-depleted area
A262-86 Practice A (Oxalic Etch)	10% $H_2C_2O_4$.	1.5 min. Anodic at one A/cm^2. Ambient temp.	Microscopic type of attack.	Various carbides.
A262-86 Practice B (Streicher)	50% H_2SO_4 +2.5% $Fe_2(SO_4)_3$. Boiling.	One 120 hour period.	Weight loss per unit area.	Chromium-depleted area.
A262-86 Practice C (Huey)	65% HNO_3. Boiling.	Five 48 hour periods. Fresh solution each period.	Average weight loss per unit area.	Chromium-depleted area, σ phase and carbides.
A262-86 Practice D (Warren)	10% HNO_3 +3% HF. 70C	Two 2-hour periods.	Weight loss per unit area.	Chromium-depleted area in Mo bearing steels.
A262-86 Practice E (Copper Accelerated Strauss)	16% H_2SO_4 +6% $CuSO_4$. Boiling. Specimen in contact with copper metal.	One 24-hour period.	Macroscopic appearance after bending.	Chromium-depleted area.

The sulfuric acid-ferric sulfate or Streicher test (Table 9.2) is accelerated by increasing the acid concentration to 50% and substituting Fe^{3+}, a stronger oxidizer, for the Cu^{2+} in the Strauss test. Attack on sensitized material is evaluated by weight loss, which is more quantitative than visual inspection of bent specimens in the Strauss tests.

The copper-accelerated Strauss and Streicher tests are the most popular for evaluation of sensitization by chromium depletion. However, both take several days to complete and the oxalic etch test (Table 9.2) was developed as a quick screening method. A polished specimen is anodically etched at 1 A/cm^2 for 1 minute in 10% oxalic acid at room temperature. A step or dual structure by microscopic examination of the etched surface (Figures 9.6a and 9.6b[9]) indicates a nonsensitized condition, which is acceptable without further testing. The ditch structure in Figure 9.6c,

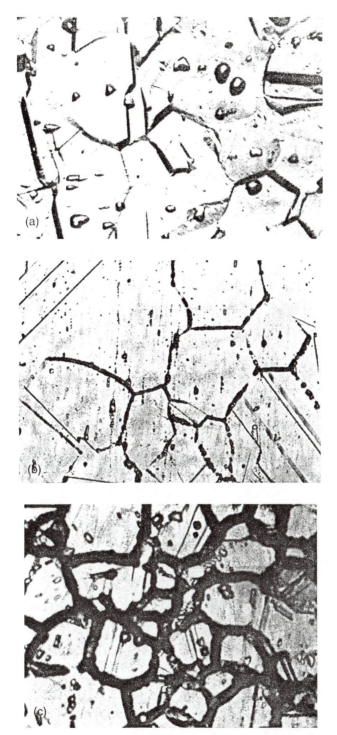

FIGURE 9.6 Oxalic etch structure on Type 304 stainless steel: (a) step; (b) dual; (c) ditch. *(From ASTM Standard Practice A262, Annual Book of ASTM Standards, Vol 3.02, p. 4, 1988. Reprinted by permission, American Society for Testing and Materials.)*

on the other hand, indicates some degree of sensitization, which requires further evaluation, usually by the Streicher or the copper-accelerated Strauss test.

The boiling 65% nitric acid or Huey test has long been the most popular test method for detecting sensitization. However, it has certain disadvantages, mostly related to excessive severity. First, it requires longer test time than the more recently developed copper-accelerated Strauss and Streicher tests. Second, attack is accelerated by buildup of hexavalent chromium during each of the four 48-hour test exposures. Third, it is also sensitive to presence of σ phase, which is significant only for nitric acid service.

The Strauss test, the copper-accelerated Strauss test, and the Huey test share an interesting experimental oddity. All are sensitive to the type of condenser used in the experimental apparatus.[10] The condenser is used to return condensed steam to the solution and thereby prevent loss of water and increase in the boiling test temperature and acid concentration during a test. Applicable condensers are of two types, shown in Figure 9.7: The cold finger type makes a loose fit in the wide mouth of an Erlenmeyer flask, and the Allihn type makes a tight fit in a ground glass joint of a similar flask. The cold finger type is recommended in the Huey test because escape of gaseous reducing agents prevents buildup of hexavalent chromium corrosion products, which cause irrelevantly high general attack at the end of each 48-hour test period. On the other hand, greater access of dissolved oxygen around the cold finger condenser in the Strauss tests tends to passivate the grain boundaries and reduce IGC in hot sulfuric acid. Thus, the Allihn condenser is generally recommended for Strauss testing.

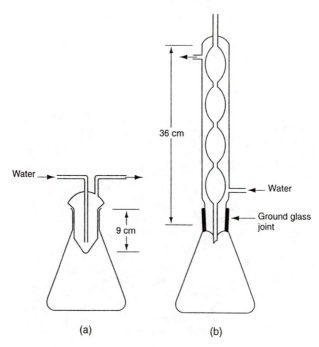

FIGURE 9.7 (a) Cold finger and (b) Allihn (45/50 joint) condensers used in boiling acid testing for intergranular corrosion (1-L Erlenmeyer flasks).

The nitric-hydrofluoric acid or Warren test, included in Table 9.2 for completeness, is used sparingly because hydrofluoric acid is hazardous and attacks conventional glass-testing vessels. Other tests run at the boiling temperature of the solution, but the Warren test requires inconvenient measures for temperature control at 70°C. The copper-accelerated Strauss test and the Streicher test are equally effective in detecting sensitization by chromium depletion, although both require longer testing times than the Warren test.

The ASTM specifications make no comment on acceptable levels of attack for any of the tests in Table 9.2. Acceptance criteria are somewhat arbitrary and depend on the risk and expense of a failure in any given application or industry. They are usually negotiated between buyer and seller. The ASM Handbook[11] and the review by Sedricks[12] offer some guidelines.

9.1.5 Electrochemical Testing

The value of the oxalic acid anodic etch test as a screening tool was discussed in Section 9.1.4. Figure 9.2 shows that there is a large difference in the electrochemical behavior between grains and the chromium-depleted grain boundaries as oxidizing potential increases. Potential can be instrumentally controlled more precisely and conveniently with a potentiostat than by addition of soluble oxidizing agents, e.g., ferric and cupric ions, in chemical tests. As a result, attempts have been made

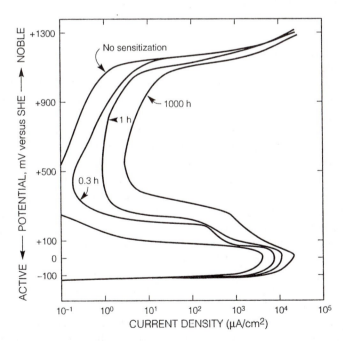

FIGURE 9.8 Effect of degree of sensitization on anodic polarization of Type 304 stainless steel in 2N sulfuric acid at 90°C. *(From K. Osozawa, K. Bohnenkamp, and H. J. Engell, Corros. Sci., Vol. 6, p. 421, 1966. Reprinted by permission, Pergamon Press.)*

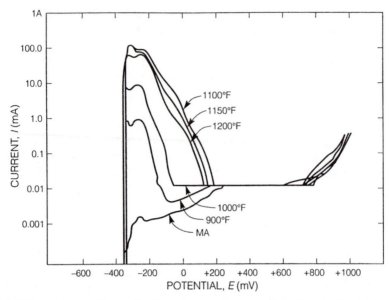

FIGURE 9.9 Reactivation cathodic polarization curves from the passive potential region in 0.5M H$_2$SO$_4$ + 0.01M KSCN for Type 304 stainless steel sensitized 4 hours at various temperatures. *(From W. L. Clarke, R. L. Cowan, and W. L. Walker, ASTM STP 656, ASTM, Philadelphia, p. 99, 1978. Reprinted by permission, American Society for Testing and Materials.)*

to develop electrochemical test procedures that simulate service conditions and determine degree of sensitization to intergranular corrosion (IGC).

Figure 9.8[13] shows the effect of sensitization on anodic polarization of Type 304 stainless steel. Despite the fact that anodic current is affected only at the grain boundaries, a significant anodic current increase is apparent over the entire passive range, especially just above the critical or maximum current density, i_c, for passivation (Section 4.1.2). There seems to be nothing in principle to prevent electrochemical evaluation of susceptibility to sensitization.[12] Anodic current could be used to evaluate the degree of attack at chromium-depleted grain boundaries more rapidly and conveniently than by weight loss or microscopic examination. The only practical limitation seems to be the expense and education required to properly operate and maintain the required instrumentation.

The electrochemical potentiokinetic repassivation (EPR) procedure has been developed[14] for laboratory and field determinations of sensitization, based on the original work of Cihal[15]. A reverse or "reactivation" potential scan from a passive potential back to the corrosion potential in hot concentrated sulfuric acid improves the discrimination between sensitized and solution-quenched alloys. The inconvenience of high-temperature solutions has been eliminated by the addition of potassium thiocyanate (KSCN), which apparently activates the chromium-depleted grain boundaries. Reactivation scans in deaerated 0.5M H$_2$SO$_4$ + 0.01M KSCN at 30°C are shown (Figure 9.9) for Type 304 stainless steel sensitized for 4 hours at various temperatures.[16] Sensitized material displays a current peak which increases with degree (temperature) of sensitization. In contrast, unsensitized material gives no reactivation peak. The area under the reactivation peak in coulombs measures the degree of sensitization in the EPR procedure.

The standardized EPR procedure is summarized in Figure 9.10. A polished, controlled surface area of stainless steel is passivated at +0.200 V vs. saturated calomel electrode (SCE) for 2 minutes. Potential is then scanned in the active direction at a rate of 6 V/hr, down to the corrosion potential, where no further current is indicated. The charge, Q, or the shaded area under the reactivation peak is measured automatically with a coulometer in the polarization circuit. Q (coulombs), a measure of sensitization, is normalized by the grain boundary area GBA (cm^2) to a sensitization number, P_a, where

$$P_a \ (C/cm^2) = Q/GBA.$$

GBA is calculated from

$$GBA = As[5.095 \times 10^{-3} \exp(0.347x)]$$

where As is the sample area in cm^2 and x is the ASTM grain size at 100× magnification.

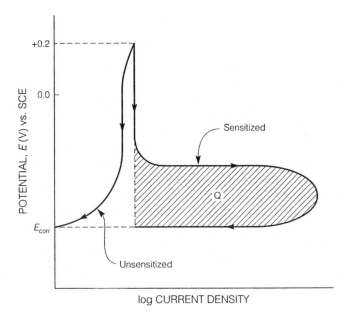

log CURRENT DENSITY

FIGURE 9.10   Summary of electrochemical potentiokinetic reactivation (EPR) procedure for determination of sensitization to intergranular corrosion. Electrolyte, 0.5 $M\,H_2SO_4$ + 0.01 $M\,KSCN$; deaeration, purified N_2; temperature, 30°C; specimen finish, 1 μm (diamond paste); passivation potential, +0.200 V (SCE); passivation time, 2 min; reactivation sweep rate, 6 V/h; sensitization P_a (C/cm^2) = Q (C)/GBA (cm^2); grain boundary area, GBA = As[5.095 × 10^3 exp(0.347x)]; As = sample area (cm^2); x = ASTM grain size at 100×. *(From W. L. Clarke, R. L. Cowan, and W. L. Walker, ASTM STP 656, ASTM, Philadelphia, p. 99, 1978. Reprinted by permission, American Society for Testing and Materials.)*

The EPR test correlates reasonably well with other methods of testing for sensitization and is much more sensitive to low levels of sensitization than the ASTM acid exposure tests of Table 9.2. The following general guidelines have been given in proposed ASTM standard G EPR-90, which was under consideration at the time of this writing.[16]

P_a Value	General Interpretation
<2	Unsensitized microstructure, no pitting
2–5	Slightly sensitized microstructure; Pitting and limited intergranular attack
5–15	Sensitized microstructure; Pitting and attack of entire grain boundaries
>15	Heavily sensitized; Strong grain boundary attack

However, because of the broad range of applications of the austenitic stainless steels, acceptance limits must be established by the user or by agreement between user and supplier.

Only slightly sensitized Type 304 stainless steel may be subject to intergranular stress corrosion cracking in nuclear cooling systems, while other applications can tolerate moderately sensitized material without failures.[17] Furthermore, the EPR test is nondestructive and operates at ambient temperature with more modest acid concentrations than the conventional ASTM A262 chemical test procedures in Table 9.2. A polarization cell and portable instrumentation have been developed to apply the test in the field on operating equipment.[18] EPR is currently under consideration for adoption as a standard test procedure by ASTM. An early version of the proposed test method appears in the appendix of Clarke et al.[16]

An early attempt was made[19] to simulate service conditions electrochemically using controlled potential (potentiostatic) corrosion testing. Figure 9.11 is a map of potential versus acid concentration showing regions where IGC was observed in sensitized Type 304 stainless steel after 24-hours exposure at constant potential. It is apparent that IGC is much more prevalent at 90°C than at 25°C. However, IGC may indeed develop at all conditions of potential and acid concentration over long periods of time in aggressive plant conditions.[20] The diagram also may be affected by the degree of sensitization. Nevertheless, the possibility exists that mildly sensitizing treatments or less expensive sensitizable alloys could be tolerated in some less aggressive conditions.

9.2 Intergranular Corrosion of Other Alloys

9.2.1 Ferritic and Duplex Stainless Steels

The ferritic iron-chromium stainless steels without nickel also sensitize to intergranular corrosion (IGC) by a chromium-depletion mechanism.[11] The testing procedures for the ferritic and austenitic stainless steels are generally the same, as summarized in Table 9.2. However, compositions, times, and temperatures of sensitiza-

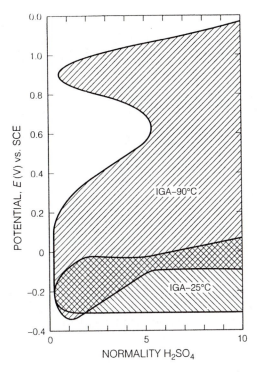

FIGURE 9.11   Conditions of potential and acid concentration favoring intergranular corrosion of sensitized Type 304 stainless steel. *(From W. D. France and N. D. Greene, Corros. Sci., Vol. 8, p. 9, 1968. Reprinted by permission, Pergamon Press.)*

tion are significantly different because of the differing properties of ferrite, as compared to austenite. Nitrogen has little effect in austenitic stainless steels because of high solubility. However, nitrogen has much lower solubility in ferrite and must be considered in the ferritic stainless steels. Ferritic stainless steels sensitize only after heating above 925°C, where solubility of carbon and nitrogen become significant in ferrite. Because of the low solubility of interstitials in ferrite, the ferritic stainless steels sensitize much more rapidly and at lower temperatures, as shown in Figure 9.12.[17] IGC of ferritic stainless steel cannot be prevented by solution anneal and water quench or by reducing carbon below 0.03%. Instead, they must be soaked at 800°C or slow cooled in the region 700° to 900°C to replenish chromium-depleted grain boundaries by diffusion from the surrounding grains.

Duplex stainless steels containing intermediate levels of nickel have a dual-phase structure with metastable ferrite present within the austenite grains. These steels are remarkably resistant to IGC. Chromium is rejected (diffuses) faster from ferrite than from austenite at the usual sensitizing temperatures, and carbides form preferentially at the ferrite-austenite boundaries.[21] Sensitization to IGC is prevented because insufficient carbon remains to precipitate carbides at the austenite grain boundaries. Chromium depletion at the ferrite-austenite boundaries is widely distributed within the austenite grains and can be replenished by diffusion during stress-relief annealing.

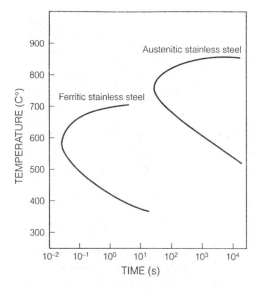

FIGURE 9.12   Time-temperature sensitization curves for intergranular corrosion of austenitic as compared to ferritic stainless steels. *(From R. L. Cowan and C. S. Tedmon, Advances in Corrosion Science and Technology, Vol. 3, M. G. Fontana and R. W. Staehle, eds., Plenum, New York, p. 293, 1973. Reprinted by permission, Plenum Press.)*

9.2.2 Nickel-Chromium Alloys

The nickel Alloy C, containing Cr, Mo, and Fe, can be subject to intergranular corrosion (IGC) due to grain boundary precipitation of molybdenum carbides or a molybdenum-rich intermetallic (μ-phase). Depletion of Mo increases attack at the grain boundaries in hot reducing acids such as HCl and H_2SO_4. Oxidizing acids such as HNO_3 attack the precipitates directly.[22] Reductions of carbon and silicon impurities resulted in the Alloy C-276 modification, which has improved resistance to sensitization and consequent IGC, especially in weldments.

The Ni-Cr-Fe Alloy 600 fails by IGC and intergranular stress corrosion cracking (IGSCC) in the caustic (alkaline) solutions that concentrate in the crevices between Alloy 600 heat-exchanger tubing and the carbon-steel tube sheet of pressurized water nuclear reactor (PWR) steam generators.[23] An example of the attack is shown in Figure 9.13. Boiling due to restricted heat transfer causes concentration of aggressive species in the crevice. The chemical agents causing the attack are uncertain but may include the caustics, NaOH and KOH, sulfur-bearing species including sulfates, and alkaline carbonates. Grain boundary carbide precipitates improve resistance to IGC whether or not chromium depletion is present. Grain boundary carbides may dissolve causitive elements (e.g., P, S, Si, N), which would normally segregate to the grain boundaries in solution or mill-annealed Alloy 600. Also, it has been suggested that enriched nickel at the grain boundaries is more resistant to caustic solutions than the alloy solid solutions of Cr and Fe in nickel.[24] That is, IGC may proceed simply by the dealloying of chromium at the grain boundaries. If present, tensile stress caus-

FIGURE 9.13   Intergranular corrosion in Alloy 600 pressurized-water nuclear-reactor steam-generator tube from tubesheet crevice. *(Photograph by courtesy of A. Agrawal, Battelle Columbus Laboratories.)*

es localized IGC penetrations and IGSCC, in addition to the usual uniform IGC (Figure 9.13).

9.2.3 Aluminum Alloys

Because aluminum and aluminum intermetallic compounds are very active in the emf series, aluminum alloys are highly susceptible to intergranular corrosion (IGC), when precipitates segregate at the grain boundaries. In Al-Mg (5xxx) alloys, Mg_2Al_8 is active to the aluminum matrix and is corroded preferentially at the grain boundaries. The xxx refers to numerical digits which specify the compositions within each alloy series, as listed in the back flyleaf. In Al-Mg-Zn (e.g., 7030) alloys, $MgZn_2$ is attacked anodically. In copper-containing alloys, such as Al-Cu (2024) and Al-Zn-Mg-Cu (7075), the copper-depleted zone adjacent to grain-boundary precipitates is attacked anodically.

Exfoliation corrosion, a particular form of IGC in aluminum, results from exposure to various industrial and marine atmospheres. Figure 9.14[25] indicates that polluted seacoast atmospheres are most aggressive followed by inland industrial atmospheres. Attack occurs at the boundaries of grains elongated in the rolling direction. Corrosion products have greater volume than the parent metal, and blocks of grains may be lifted from the metal surface, as shown in Figure 9.15.[26,27] The result is a fibrous mode of attack illustrated in the laboratory-tested specimens of Figure

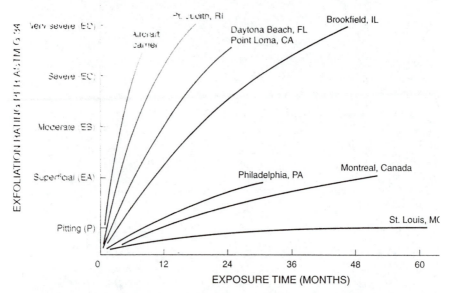

FIGURE 9.14   Exfoliation corrosion of aluminum Alloy 2124 in various atmospheres. *(From S. J. Ketcham and E. J. Jankowsky, ASTM STP 866, ASTM, Philadelphia, p. 14, 1985. Reprinted by permission, ASM International.)*

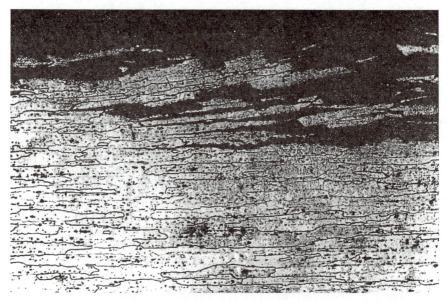

FIGURE 9.15   Grain lifting during exfoliation corrosion. *(From Standard Test Method 634, Annual Book of ASTM Standards, ASTM, Philadelphia. Photograph by courtesy of B. M. Lifka, Aluminum Co. of America. Reprinted by permission, American Society for Testing and Materials.)*

FIGURE 9.16 Fibrous nature of exfoliation corrosion after exposure to solution of 4 *M* NaCl, 0.5 *M* KNO$_3$, 0.1 *M* HNO$_3$ at ambient temperature. *(From Standard Test Method G34, Annual Book of ASTM Standards, ASTM, Philadelphia. Photograph by courtesy of D. O. Sprowls. Reprinted by permission, American Society for Testing and Materials.)*

9.16.[27] The high-strength 2xxx and 7xxx alloys are most susceptible, along with certain cold-worked 5xxx alloys.

Laboratory testing is needed for alloy development and quality assurance, but agreement of the results with subsequent service performance is not assured. Various standard acidified salt spray tests[28] are frequently utilized. The salt spray cabinets are expensive to operate, and testing generally requires 1 to 2 weeks for measurable attack. Thus, shorter, simpler continuous-immersion procedures recently have been developed, which also produce a mode of attack comparable to that observed in service. For the 2xxx and 7xxx alloys, the procedure consists briefly of continuous immersion for 1 or 2 days in a solution of 4M NaCl and 0.5M KNO$_3$, 0.1M HNO$_3$ at 25°C.[27] The extent of attack is rated by visual comparison with a set of standard photographs in ASTM method G34. Ratings progress from superficial to very severe, as indicated in Figure 9.14. The standard photograph for moderate attack is reproduced in Figure 9.16.

Removal of certain atmospheric pollutants, such as H$_2$S and SO$_2$ may sometimes be possible to prevent IGC and exfoliation. However, selection of resistant, lower-strength alloys, such as the 6xxx series, is often necessary. Tempering past maximum age hardening improves resistance, but susceptibility to stress corrosion cracking often remains.

9.3 Factors Affecting Weldment Corrosion

Welding is a physical process that melts the adjoining zones of metal parts to bond them together. This rather drastic heating, melting, mixing, freezing, and cooling cycle affects the metallurgical and mechanical constitution of the weldment. The weldment is defined as the melted and solidified weld metal and adjoining zones of the base metal alloy affected metallurgically by the heating and cooling cycle. The weldment often has substantially lower corrosion resistance and mechanical properties than the base metal parts of nominally identical composition.

The weldment structure is complex, as illustrated in Figure 9.17.[29] The weld decay region of intergranular corrosion, within the heat-affected zone of stainless steels, was discussed in Section 9.1.2. The basic weldment structure is affected by composition and temperature of the molten weld pool, thickness and geometry of the welded parts, and rates of heating and cooling.

The method of welding has a substantial effect on weldment properties. The heat necessary for welding usually is provided by either an electric arc or combustion of a flammable gas. The arc is formed between the workpiece (parts to be welded) and an electrode that may be consumable or inert. A consumable electrode is melted continuously and provides filler metal for the weld, as shown in Figure 9.18. In the gas metal arc (GMA) process, the electrode is in the form of wire, and the weld pool is protected with an inert gas cover stream. In shielded metal arc (SMA) welding, a rod electrode is used, with a coating to provide a protective flux. In gas tungsten arc (GTA) welding, tungsten is used as an inert electrode, which does not melt in the arc, and filler is fed to the weld pool separately. An inert gas and/or a flux are used in most welding processes to prevent contamination and oxidation of the weldment, especially for metals such as stainless steel, titanium, and aluminum that oxidize readily at elevated temperature.

In gas welding, the flame from a combustible gas (often acetylene, C_2H_2) burned in air or oxygen melts a rod of filler metal and the adjoining zones of the parts to be joined.

Microbiologically influenced corrosion (MIC), which is discussed more generally in Section 11.3, is often observed at welds in carbon and stainless steel.[36] Figure

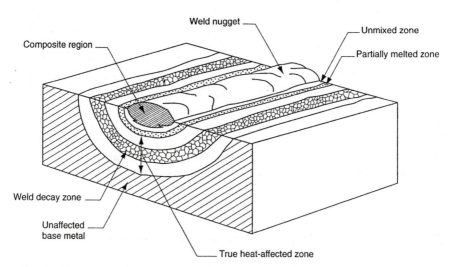

FIGURE 9.17   Schematic weldment structure. *(From W. F. Savage,* Welding Design and Engineering, *Dec. 1969.)*

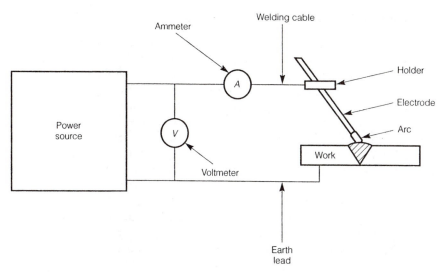

FIGURE 9.18   Consumable arc welding process.

9.27[36,42] shows selective attack at the weld metal in a Type 304 stainless-steel weldment. MIC may also be concentrated at the heat-affected zone in other cases. The attack typically occurs in slow moving or stagnant natural waters, which are the breeding grounds for a myriad of bacterial organisms. Metallurgical and biological parameters affecting MIC have not been firmly established, and the mechanism of attack at welds is uncertain at present for any given conditions.[43]

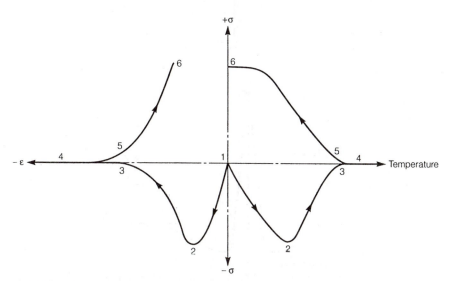

FIGURE 9.19   Relationships between temperature, stress σ and strain ε during welding. *(From K. Easterling,* Introduction to the Physical Metallurgy of Welding, *Butterworths, London, p. 35, 1983. Reprinted by permission, Butterworths)*

9.3.1 Residual Stresses and Stress Concentrations

Tensile components of residual stress in weldments are important because they may cause stress corrosion cracking (SCC) and hydrogen induced cracking (HIC) (Chapter 8). Welding produces residual stresses as follows. Figure 9.19[30] traces stress as functions of temperature and strain during heat up, melting, freezing, and cool-down in the weld. Beginning at 1, initial heating induces thermal expansion and compressive stresses in parts that are physically constrained in their required positions near one another. As temperature rises at 2, the metal softens and creeps to release compressive stresses back to zero. Temperature increases and thermal expansion continue through 3 and 4 in the softened and melted metal, respectively. After the molten zone passes, temperature falls and the weld metal freezes at 5. Freezing produces a large volume contraction from molten to solid weld metal with additional thermal shrinkage upon subsequent cooling. In constrained base metal parts, these contractions cause large, ambient-temperature, residual, tensile stress, σ_r, and strain, ε_r, at 6. Distortions may result from shrinkage in weldments which are not rigidly constrained.

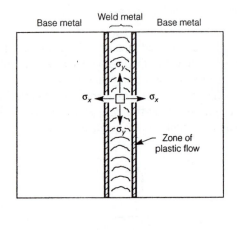

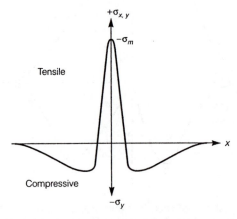

FIGURE 9.20   Distribution of stress across a butt weld.

The distribution of stress across a completed butt weld is shown in Figure 9.20. Within the weld metal, high tensile stresses, σ_x and σ_y, are apparent, with significant compressive stresses in the surrounding base metal. Thermal expansion and softening during heatup is sufficient to cause plastic deformation in the base metal near the molten zone. Subsequent freezing and thermal contraction generate tensile stresses up to yield in and near the frozen weld metal. Compressive stresses in the surrounding base metal are necessary to resolve the total stress state to zero. Other weld geometries result in still more complex stress fields around the welds.[31] It is important to appreciate that welding produces residual tensile stresses, which can cause SCC and HIC in planes transverse and longitudinal to the weldment. Figure 9.21[32] shows SCC due to polythionic acids in a Type 304 stainless-steel furnace tube near a weld to a carbon-steel tube. Note cracking both parallel and perpendicular to the weld.

Welding produces a geometric discontinuity in the cross section, which concentrates any applied or residual stresses. As a result, corrosion fatigue cracking (CFC), as well as SCC and HIC failures, often originate at the toe of a weld, as shown in Figure 9.22.[33]

Although equipment is available for continuous, automatic production, welding is often a field or shop process. Quality is very much dependent on the skill of the welder and quality of the materials and equipment in use. As a result, defects occur rather frequently; some examples are illustrated in Figure 9.23. Such defects will

FIGURE 9.21 Stress corrosion cracking of Type 304 stainless steel in polythionic acid showing cracks parallel and perpendicular to weld due to residual stresses. *(From J. Gutzeit, R. D. Merrick, and L. R. Sharfstein, Metals Handbook, 9th ed., Vol. 13, Corrosion, ASM, Metals Park, OH, p. 1262, 1986. Reprinted by permission, ASM International.)*

FIGURE 9.22 Corrosion fatigue crack originating from weld toe in carbon steel. *(From T. G. Gooch,* Process Industries Corrosion, *B. J. Moniz and W. J. Pollock, eds., NACE, Houston, p. 739, 1986. Reprinted by permission, National Association of Corrosion Engineers.)*

further aggravate stress concentrations, which lead to mechanical and environmentally induced cracking.

High residual stresses and stress concentrations make weldments the weak point for all forms of environmentally induced cracking, including CFC, SCC, and HIC. Thus, any combination of stress, alloy, and environment that would cause cracking in the base metal will often cause cracking preferentially at welds.

9.3.2 Weld Metal Composition

The filler metal, often added to the molten weld metal during the welding process, is usually of somewhat different composition than the base metal. This allows flexibility in designing weld metal properties to compensate for residual stresses and reduced corrosion resistance in the weldment. Filler alloys for the common austenitic and ferritic stainless steels are summarized in Table 9.3, taken from Nippes.[34]

In all cases, the weld metal composition must be adjusted to make the corrosion potential noble to the base metal being welded. In this way, the weld is cathodically protected in the galvanic couple between weld and base metal (Section 6.1.2). Detrimental effects of anodic dissolution at the base metal usually are minimal because the anodic current is spread over a relatively large surface area.

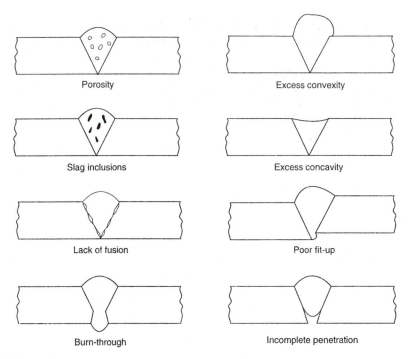

| Porosity | Excess convexity |

| Slag inclusions | Excess concavity |

| Lack of fusion | Poor fit-up |

| Burn-through | Incomplete penetration |

FIGURE 9.23   Schematic diagram of typical weld defects.

TABLE 9.3   Filler Alloys for Stainless Steels[a]

Type	Base Metal Alloy	Filler Alloy
Austenitic	301, 302, 304, 308	308
	302B	309
	304L	308L, 347
	309, 309S	309
	310, 310S	310
	316	316
	316L	318, 316L
	318, 316Cb	318
	321	347
	347	347
	348	347
Ferritic	405	405Cb, 430
	405, 430	308, 309, 310
	446	308, 446

From E. F. Nippes, Metals Handbook, *Desk Edition, ASM , Metals Park, OH, pp. 30–34, 1985. Reprinted with permission, ASM International.*

[a]More complete information, including precautionary notes, should be studied in the source reference[34] before final selection of filler alloy and welding procedures.

Highly alloyed stainless steels may be subject to localized anodic pitting if the alloying elements segregate or are diluted in the weld metal, as shown in Figure 9.24.[33]

When welding high-strength alloys, filler rods must be kept stringently free of moisture. Water will decompose at welding temperatures to contaminate welds with hydrogen, leading of course to hydrogen induced cracking (HIC). By the same token, low-hydrogen electrodes, which do not use organics as binders or flux additions, are often required.

Filler alloys often must be richer in alloying elements than the corresponding base metals. Alloying elements may be lost by oxidation and/or volatilization. They will inevitably be diluted in the weld metal by mixing with the base metal. Surface oxidation (heat tinting) of the base metal can deplete the surface of chromium or other protective alloying elements and thereby reduce the corrosion resistance.[34]

Wire brushing to remove slag and oxides can lead to enhanced corrosion. Figure 9.25[35,36] shows surface pitting induced by stainless wire brushing of a stainless-steel weldment. Such effects can be caused by incomplete removal of depleted surface layers or by deposit of lower-grade alloy material from the wire brush. Pickling or glass bead blasting is sometimes recommended[36] to avoid surface contamination from wire brushing.

In cases of severe corrosivity, it is not unusual to specify a filler of the next alloy higher in corrosion resistance. For example, Type 347 (niobium-stabilized) stainless steel is often used as filler for Type 304L stainless steel (Table 9.3). Type 321 stabilized with Ti is never specified as a filler alloy because the Ti can be lost by volatilization of TiO_2. Also, some of the high nickel-chromium-molybdenum (Hastelloy) alloys are sometimes specified as filler in welding the duplex stainless steels.[36]

9.3.3 Weldment Metallurgy

The metallurgical effects of welding are often detrimental to corrosion resistance. Nonequilibrium cooling of the weld metal may produce grain boundary segregation (coring) and primary phase substructures. The unmixed zone in Figure 9.17 is a region always present in the base metal, which is not changed in composition by mixing with the molten weld metal. Rapid nonequilibrium cooling in the unmixed zone can potentially form phases and precipitates that are detrimental to mechanical and corrosion properties of the weldment. Hence, the weldment is metallurgically inhomogeneous compared to wrought base metal.

9.4 Weldment Corrosion of Various Alloy Systems

9.4.1 Austenitic Stainless Steels

Sensitization to intergranular corrosion (IGC) by grain boundary depletion of chromium and consequent weld decay in the heat affected zone are described in

FIGURE 9.24   Chloride pitting in GTA weld metal. 18Cr-10Ni-2.5Mo. *(From T. G. Gooch,* Process Industries Corrosion, *B. J. Moniz and W. J. Pollock, eds., NACE, Houston, p. 739, 1986. Reprinted by permission, National Association of Corrosion Engineers.)*

Sections 9.1.1 and 9.1.2 for austenitic stainless steels. Sensitized weld structures are also susceptible to intergranular stress corrosion cracking (IGSCC) in reactor cooling water (Figure 9.26[37]) and polythionic acid refinery environments (Section 11.2.3). Weldments may be preferentially attacked by microbiological organisms, as well (Figure 9.27). The mechanism of such attack is not well understood (Section 11.3.4).

The filler alloy for austenitic stainless-steel weldments is usually of different composition to maximize resistance to IGC and IGSCC. For example, Type 308, usually specified as the filler alloy for Type 304 stainless steel, has somewhat higher chromium and nickel, which confers higher corrosion resistance to the weld metal. Even more significant is the fact that higher chromium stabilizes the ferrite phase, causing Type 308 weldments to freeze in a duplex, austenite plus ferrite, structure. Primary ferrite in the weld metal is desirable to prevent microcracking

FIGURE 9.25   Pitting of a stainless steel weldment caused by stainless wire brushing. *(From A. Garner,* Metals Prog., *Vol. 127 (No. 5), p. 31, April 1985. Photograph by courtesy of K. F. Krysiak. Reprinted by permission, ASM International.)*

(hot cracking) at interdendritic boundaries[34] during weld metal cooling and solidification. Also, the duplex structure is resistant to intergranular corrosion because carbides segregate at the dispersed austenite-ferrite boundaries and cannot easily precipitate and produce chromium depletion at the austenite grain boundaries (Section 9.2.1). Furthermore, stress corrosion cracks cannot easily propagate through the resistant ferrite second phase, and the duplex alloys are resistant to stress corrosion cracking.

9.4.2 Nickel Alloys

Weldments in nickel alloys are subject to sensitization, as are the austenitic stainless steels described in Section 9.4.1. Weld decay and knifeline attack may result from grain boundary precipitation of carbides in the nickel alloys, but also from precipitation of intermetallic phases of chromium, molybdenum, tungsten, and niobium. For example, the nickel-molybdenum Alloy B, used for service in reducing acids such as HCl, is subject to both weld decay and knifeline attack. Postweld annealing eliminates the attack but may be impractical in large welded structures. The low-carbon Alloy B-2 exhibits much greater resistance to both forms of attack.

Nickel-chromium-molybdenum (Hastelloy) Alloy C is also subject to intergranular corrosion by both molybdenum carbides and the intermetallic μ-phase (Ni_7Mo_6), which are rich in chromium, tungsten, and molybdenum. Silicon is also involved, apparently, because reduced carbon and silicon in Alloy C-276 reduce the sensitiza-

FIGURE 9.26 Intergranular stress corrosion cracking in the heat affected zone of Type 304 stainless steel pipe weldment. *(From B. M. Gordon and G. M. Gordon, Metals Handbook, 9th ed., Vol. 13, Corrosion, ASM, Metals Park, OH, p. 927, 1986. Photograph by courtesy of B. M. Gordon, General Electric Co. Reprinted by permission, ASM International.)*

tion kinetics considerably and permit service in the as-welded condition without postweld annealing. However, sensitization at longer times is present from μ-phase precipitation, and further reductions in tungsten, iron, and cobalt give still better resistance for Alloy C-4 to intergranular corrosion. A better balance of tungsten in Alloy C-22 gives improved resistance to pitting corrosion while retaining resistance to intergranular corrosion.

9.4.3 Ferritic Stainless Steels

The ferritic stainless steels are more difficult to weld. They are susceptible to grain growth and coarsening at welding temperatures, which may result in losses of low-temperature toughness. Interstitial contamination from carbon, nitrogen, oxygen, and hydrogen sensitize the alloys to intergranular corrosion. Therefore, extreme cleanliness and care with gas-shielded tungsten arc processes are recommended to obtain acceptable as-welded properties.[34] Variations within specified composition limits often result in the formation of austenite, which embrittles the weldment when transforming to martensite. Postweld annealing, which trans-

FIGURE 9.27 Selective corrosion of weld metal by microbiologically induced corrosion in austenitic stainless steel. *(From S. W. Borenstein,* Microbiologically Influenced Corrosion Handbook, *Woodhead Publishing Ltd., p. 169, 1994. Reprinted by permission, Woodhead Publishing Ltd., Cambridge, England.)*

forms the martensite to more ductile ferrite, is often required but is expensive. Aluminum at 0.20% is added to Type 405 ferritic stainless steel to promote ferrite formation.

Ferritic stainless-steel filler alloys have the advantage of similar thermal expansion and corrosion properties. However, austenitic filler alloys (Types 308 and 309) are often used with the ferritic stainless steels to improve weldment ductility (Table 9.3). However, dissimilar metal galvanic cells may form in some environments.

9.4.4 Duplex Stainless Steels

Duplex stainless steels must be welded with special procedures and filler metal to retain a good balance between stable austenite and metastable ferrite (Section 9.2.1). Welds with >60% ferrite become brittle, and those with <25% ferrite lose resistance to chloride stress corrosion cracking. Autogenous welds without any filler can produce up to 80% ferrite in the weld metal, and postweld annealing is often required to convert some of the ferrite to austenite. Filler metals of enriched nickel are often recommended to decrease ferrite stability. Although higher heat input and lower cooling rates give more time for conversion of ferrite to austenite in base and weld metal, adverse effects may result from grain growth and precipitation of other embrittling phases, as well as dilution of nickel in the weld metal. A proper balance of nickel-enriched filler metal with minimum heat input and temperature is usually recommended to ensure the optimal duplex microstructure in the weld metal.[36]

Duplex weldments may be attacked preferentially in very aggressive media. For example, Figure 9.28[36] shows preferential corrosion of the continuous austenite phase in the weld metal of an autogenous weldment on Ferralium 255 exposed to

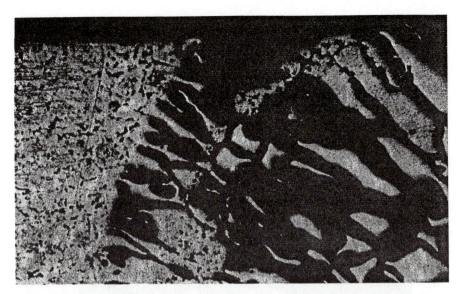

FIGURE 9.28   Preferential corrosion of austenite in autogenous weld metal in Ferralium 255 duplex stainless steel. *(From K. F. Krysiak,* Metals Handbook, *9th ed., Vol. 13,* Corrosion, *ASM, Metals Park, OH, p. 344, 1986. Photograph by courtesy of K. F. Krysiak. Reprinted by permission, ASM International.)*

synthetic seawater at 100°C, according to ASTM D1141. In such instances it may be necessary to specify a higher nickel alloy, such as one of the Hastelloys, for the filler.

9.4.5 Carbon Steels

Rapid cooling of various carbon and low-alloy steels in the partially melted unmixed zone of the weldment (Figure 9.17) produces martensite structures that are susceptible to hydrogen induced cracking (HIC). Figure 9.29[38] shows HIC through a martensitic band in a weldment from a carbon steel storage tank for anhydrous hydrofluoric acid. Such failures are commonly referred to as *underbead cracking*.

Prevention requires careful selection of base- and weld-metal hardenability, cooling rate, and postweld heat treatment to keep hardness below specified levels. The equivalent carbon content (ECC) relates composition to hardenability:

$$\text{ECC} = \%\text{C} + \frac{\%\text{Mn}}{6} + \frac{\%\text{Ni}}{15} + \frac{\%\text{Cr}}{5} + \frac{\%\text{Cu}}{13} + \frac{\%\text{Mo}}{4}.$$

Thus, Mn has one-sixth the effect of carbon on hardenability, Ni one-fifteenth, Cr one-fifth, Cu one-thirteenth, and Mo one-fourth. The general rule is that hydrogen underbead cracking is unlikely if ECC < 0.40%. By keeping ECC low, hardenability is minimized, weldment hardness is lower, martensite is less easily formed by slow cooling, and HIC is less likely. If the ECC is between 0.40 and 0.65%, pre-

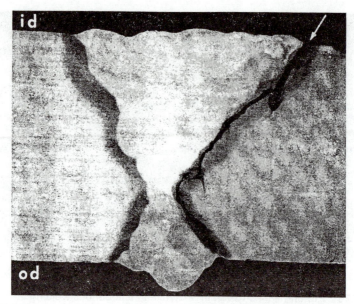

FIGURE 9.29   Hydrogen induced cracking (underbead cracking) in the heat affected zone of a carbon steel weldment. *(From D. Warren, Process Industries Corrosion, B. J. Moniz and W. J. Pollock, eds., NACE, Houston, p. 31, 1986. Reprinted by permission, National Association of Corrosion Engineers.)*

heating the weld is recommended to increase cooling time and permit more conversion of martensite to ferrite. If ECC >0.65, postweld heat treatment is required, as well.

Electric resistance welded (ERW) carbon-steel pipe has failed by selective (grooving) corrosion at the longitudinal seam weld. Continuous ERW welds are autogenous and are produced by passing a high-frequency alternating current through the contacting edges of plate as it is formed into pipe. The bond at the seam is formed by a combination of melting and upset forging as the edges are pressed together in the presence of the electric current. A recent failure of buried pipe-carrying liquid automotive fuel led to considerable damage and loss of life in the resulting fire[39] (Section 1.2.2). Figure 9.30[39] shows an ERW weld with grooving attack and alteration of structural morphology at the weld. It has been suggested[40] that the welding process impairs corrosion resistance by exposing the ends of MnS stringers at the surface of the welded pipe seam. Anodic polarization in 3.5% salt solution successfully simulated the grooving attack, which was moderated by lower sulfur content and additions of copper and calcium to deactivate the sulfur.[41]

9.4.6 Aluminum

Galvanic reactions in the weld metal constitute the major concern for corrosion resistance of aluminum alloy weldments. Again, the weld metal should be noble to the base metal to prevent preferential attack. Nonequilibrium precipitates can lead to

FIGURE 9.30 Grooving corrosion of electric resistance welded line-pipe steel. *(From R. J. Eiber and G. O. Davis,* Investigation of Williams Pipeline Company Mounds View, MN, Pipeline Rupture. *Final Report to Transportation Systems Center, U. S. Dept. of Transportation, Battelle-Columbus Lab., October 14, 1987. Photograph by courtesy of R. J. Eiber, Battelle Columbus Laboratories.)*

serious localized galvanic corrosion in the weldment. Figure 9.31[44] shows the effects of postweld annealing to even out potential differences caused by welding. Silicon is a frequent alloying element in filler metals for aluminum, because it ennobles the corrosion potential and produces lower-melting, more fluid molten metal in the weld pool. Silicon cannot be used, however, for alloys containing significant amounts of magnesium, due to formation of a brittle MgSi precipitate. Charts for filler metal recommendations are given by Hatch.[44]

9.5 Dealloying and Dezincification

Dealloying occurs when one or more components of an alloy are more susceptible to corrosion than the rest and, as a result, are preferentially dissolved. The susceptible alloying elements are generally more active electrochemically and are anodically dissolved in galvanic contact with the more noble components. The most important example of dealloying is the selective removal of zinc from brass (dezincification). The other common example is graphitic corrosion of cast iron, often during long-term burial in soil. Both dezincification and graphitic corrosion were introduced in Section 1.5.8 and will be further amplified in this section. Other instances of dealloying are observed on occasion, and some are listed in Table 9.4.[45]

Many cases in Table 9.4 involve copper, a very noble metal, combined with active alloying elements such as zinc, aluminum, and silicon, which are removed by dealloying. Dealloying involving elements closer in the galvanic series is limited by kinetics (rates) at relatively low temperature. Other cases involve oxidation in hot gases, discussed further in Chapter 11, in which kinetics match the thermodynamics more closely.

9.5.1 Graphitic Corrosion

Graphitic corrosion occurs exclusively in gray cast iron, which has a continuous graphite network in its microstructure. The graphite, acting as cathode, accelerates anodic dissolution of nearby iron, leaving behind the graphite network, which maintains structural shape but loses mechanical strength. Ductile or malleable cast irons do not have a continuous graphite network. Although they corrode by uniform penetration, graphitic corrosion is not present. Graphitic corrosion is observed in buried

(a) AS WELDED   (b) ANNEALED

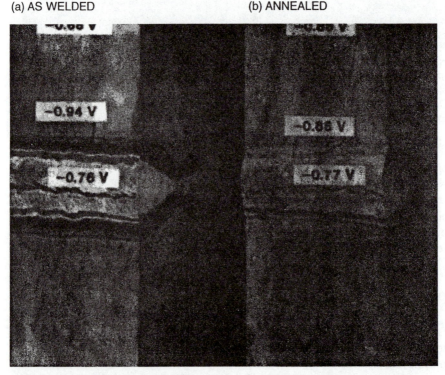

FIGURE 9.31 Effects of postweld annealing to eliminate corrosion in the heat-affected zone in aluminum weldments. *(From J. E. Hatch,* Aluminum: Properties and Physical Metallurgy, *ASM, Metals Park, OH, pp. 285–294, 1984. Photograph by courtesy of K. F. Krysiak. Reprinted by permission, ASM International.)*

TABLE 9.4   Environments Causing Dealloying for Various Alloys

Alloy	Environment	Element Removed
Brasses	Many waters, especially under stagnant conditions	Zinc (dezincification)
Gray Cast Iron	Soils, many waters	Iron (graphitic corrosion)
Aluminum Bronzes	Hydrofluoric acid, acid chloride solutions	Aluminum (dealuminification)
Silicon Bronzes	High-temperature steam, acidic species	Silicon (desiliconification)
Tin Bronzes	Hot brine or steam	Tin (destannification)
Copper-Nickels	High heat flux and low water velocity (in refinery condenser tubes)	Nickel (denickelification)
Monels (Ni-Cu)	Hydrofluoric and other acids	Copper in some acids, nickel in others
Gold Alloys	Sulfide solutions, human saliva	Copper, Silver
High-Nickel Alloys	Molten salts	Chromium, Iron, Molybdenum, Tungsten
Medium, High-Carbon Steels	Oxidizing atmospheres, hydrogen at high temperatures	Carbon (decarburization)
Iron-Chromium Alloys	High-temperature oxidizing atmospheres	Chromium which forms a protective film
Nickel-Molybdenum Alloys	Oxygen at high temperature	Molybdenum

Compiled from Polan.[45]

cast iron pipe after many years' exposure in soil. Old pipe carrying hazardous materials may cause explosions or contamination when disturbed. Cast iron fittings and pipe carrying water may also be affected (Section 1.5.8), along with pump housings and impellers, in which erosion of the outer graphite layers hastens failure.[46] Coatings and cathodic protection will arrest the attack on buried surfaces.

Replacement with a more resistant material, such as other grades of cast iron, may be possible.

Graphitic corrosion should not be confused with graphitization, which is a long-term high-temperature metallurgical conversion of carbides to graphite not involving corrosion.

9.5.2 Dezincification: Description and Testing

Selective dissolution of zinc occurs in brasses (Cu-Zn alloys) of >15% Zn in prolonged exposures to aerated water high in CO_2 and/or chlorides.[45] Two-phase, $\alpha + \beta$, alloys are more susceptible, especially if the high-zinc β phase is continuous. Slow-moving or stagnant solutions favor the attack. Removal of zinc leaves behind a porous, weak layer of copper and copper-oxide corrosion product. The weak, brittle dezincified layer on a brass bolt is illustrated in Figure 1.14. Dezincification continues beneath the previously formed dealloyed layer, as shown by the deep penetration in Figure 9.32a. Uniform dezincification is favored by acid-chloride solutions or slightly acidic pure water at room temperature. Elevated temperature alkaline chloride solutions favor localized "plug" dezincification (Figure 9.32b), which often results in rapid penetration rates.

Solutions containing $CuCl_2$ have been used often to accelerate dezincification.[47,48] Most recently, the average penetration by dezincification in a test solution of 1 % $CuCl_2$ held at 75°C has been found[49] to correlate with field tests in Australia. Lucey[47] exposed brass specimens to a saturated solution of $CuCl_2$ with small $ZnCl_2$ additions and measured the potential between the specimen and a pure copper electrode. A large potential difference indicated dezincification. Langenegger and Callaghan[50] demonstrated that the rate of dezincification increases with the oxidizing power of the solution, using additions of $CuCl_2$, $FeCl_3$, and $Na_2Cr_2O_7$. Furthermore, the depth of dezincification was the same whether the "potential shift" or polarization was supplied by the chemical oxidizers or by a galvanostat (Section 3.5.1), as shown in Figure 9.33. Potentiostatic currents in controlled potential tests have also been observed to correlate with the extent of dezincification.[49]

The described procedures have been applied primarily to two-phase, $\alpha + \beta$, alloys. Their applicability to the more resistant single phase α alloys is yet unclear. There has been greater interest in testing for dezincification in Australia and South Africa, where the common domestic waters are high in chloride and dissolved solids.

9.5.3 Prevention of Dezincification

Dezincification is relatively slow in pure, neutral water of low ionic content. However, purification of the process stream is usually not economically feasible, and substitution of a more resistant alloy provides the best means of prevention. Low-zinc red brass (<15% Zn) is generally immune to dezincification, but tube fittings are expensive, because the alloy cannot be economically die cast or forged.[45] Tin additions to $\alpha + \beta$ castings inhibit attack in the α phase and produce reasonably good resistance to dezincification in marine environments. However, dezincification of the high-zinc β phase cannot be inhibited metallurgically. Finer distribution of β

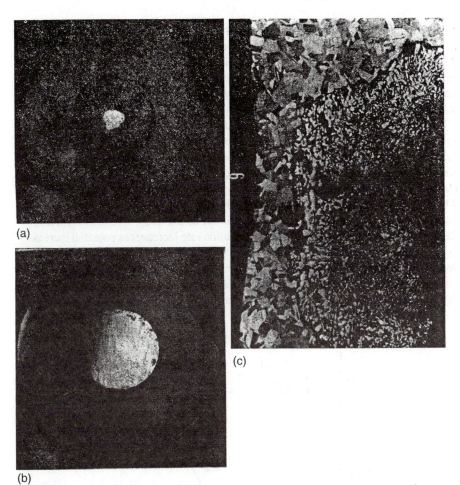

(a)

(b)

(c)

FIGURE 9.32   (a) Deep-layer dezincification of a yellow brass bolt compared to
(b) another with little attack. *(Specimens provided by courtesy of S. L. Pohlman,
Kennecott Corp.)*   (c) Metallographic cross section of plug dezincification in yellow
brass. *(Photograph by courtesy of A. Cohen, Copper Development Institute.)*

has been reported[49] to improve the resistance of two-phase alloys. Dezincification of
single-phase α admiralty brass (Cu-28Zn-1Sn) is inhibited by small additions of P,
As, or Sb. Cupronickels containing 10 to 30% Ni in copper are recommended for
severely corrosive environments.

9.5.4 Mechanism of Dezincification

Dezincification may occur by two possible processes.[51] The most straightforward
suggests that the more active alloying element, zinc, selectively dissolves or "leaches" out of the brass, leaving behind a porous, weak copper structure. The alternative
requires that both zinc and copper dissolve, and the more noble copper then rede-

posits as a porous layer. More recent studies[52,53] indicate that both processes occur in separate, but overlapping potential regimes. Selective dissolution requires solid diffusion of zinc, which seems too slow to account for the relatively rapid rates of penetration observed experimentally. Therefore, it seems probable that dezincification occurs by simultaneous dissolution and redeposition, with selective dissolution also present but not rate controlling.

The potential ranges of zinc and copper dissolution and copper redeposition are summarized in Figure 9.34, which is based on potentiostatic data in 0.1 M chloride solution.[52] Thermodynamic half-cell electrode potentials for significant electrode reactions are also marked for reference. Below 0.0 V versus the standard hydrogen electrode (SHE), copper metal is thermodynamically stable and cannot dissolve. Above −0.90 V, zinc dissolves by selective dissolution, but dezincification is slow[52] and may be limited by rate of zinc diffusion out of the alloy. Above 0.0 V versus SHE, copper begins to dissolve and its dissolution rate increases as potential increases. Simultaneously, the rate of dezincification increases. Between 0.0 and +0.2 V versus SHE, Cu^{2+} accumulated in solution can redeposit on the surface as copper metal, especially in unstirred solutions or solutions entrapped in occluded regions. At more noble potentials, copper and zinc dissolve at high, but equivalent rates with

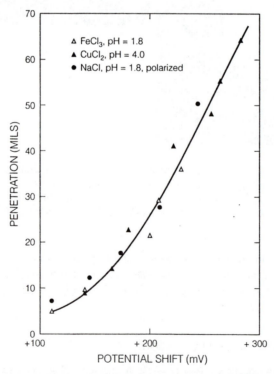

FIGURE 9.33   Dezincification caused by polarization from chemical oxidizers and applied galvanostatic current. *(From E. E. Langenegger and B. G. Callaghan,* Corrosion, *Vol. 28, p. 245, 1972. Reprinted by permission, National Association of Corrosion Engineers.)*

no dezincification. The potential above which dezincification ceases is uncertain and depends on the concentration of dissolved copper and chloride, the latter forming complexes with copper.

The following listed reactions clarify the empirical data in Figure 9.34. Nernst equations (Section 2.1.3) for each reaction are included, and half-cell electrode potentials, e, are calculated as volts versus SHE, assuming that activities are $(Cl^-) = 0.1$ and all other dissolved ions are 10^{-6} (Section 2.2.3), unless otherwise indicated.

$$Zn = Zn^{2+} + 2e^- \tag{1}$$
$$e_{Zn/Zn^{2+}} = -0.763 + 0.0295 \log(Zn^{2+}) = -0.90 \text{ V};$$
$$Cu + 2Cl^- = CuCl_2^- + e^- \tag{2}$$
$$e_{Cu/CuCl_2^-} = 0.208 + 0.0591 \log(CuCl_2^-) - 0.1182 \log(Cl^-) = -0.03 \text{ V};$$
$$CuCl_2^- = Cu^{2+} + 2Cl^- + e^- \tag{3}$$
$$e_{CuCl_2^-/Cu^+} = 0.465 + 0.0591 \log(Cu^{2+})/(CuCl_2^-) + 0.1182 \log(Cl^-)$$
$$= 0.10 \text{ V for } (CuCl_2^-) = 0.01;$$
$$Cu = Cu^{2+} + 2e^- \tag{4}$$
$$e_{Cu/Cu^{2+}} = 0.337 + 0.0295 \log(Cu^{2+}) = +0.16 \text{ V};$$

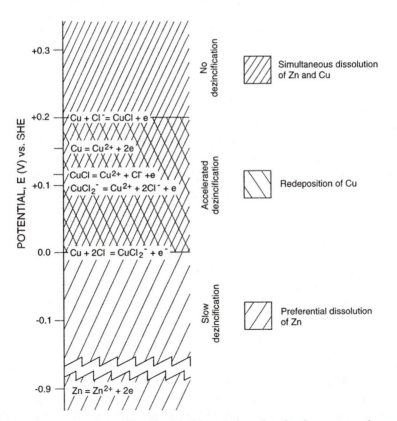

FIGURE 9.34 Potential regions in chloride solutions for simultaneous and separate dissolution of copper and zinc from α brass, and redeposition of copper in chloride solutions. *(From data of Heidersbach and Verink.[55])*

$$CuCl = Cu^{2+} + Cl^- + e^- \tag{5}$$
$$e_{CuCl/Cu^{2+}} = 0.537 + 0.0591 \log(Cu^{2+}) + 0.0591 \log(Cl^-) = +0.12 \text{ V};$$
$$Cu + Cl^- = CuCl + e^- \tag{6}$$
$$e_{Cu/CuCl} = 0.137 - 0.0591 \log(Cl^-) = +0.20 \text{ V}.$$

At active potentials above -0.90 V, (1) proceeds by selective dissolution of zinc from brass, leaving behind a copper-enriched (dezincified) surface. Reaction (2) is thought to initiate the dissolution of copper as potential becomes more positive (noble) than about -0.03 V, which is not far removed from the experimental value of 0.0 V[53] shown in Figure 9.34.

As the complex ion, $CuCl_2^-$, accumulates in solution near the surface, Cu^{2+} may be generated by (5). The half-cell electrode potential of 0.1 V is calculated assuming that $(CuCl_2^-)$ has reached an activity of 0.01. Oxidation to Cu^{2+} is favored at any potential noble to the half-cell electrode potential of (3). The Cu^{2+} so formed may be reduced immediately to Cu directly by (4), or by (6) consecutively with (5). If Cu^{2+} concentration increases, the half-cell electrode potentials of (3), (4), and (5) are elevated to more noble values. Whether or not Cu^{2+} can accumulate at the surface without immediate conversion to Cu depends on the relative reaction rates of (3), (4), (5), and (6), which cannot be predicted from the present thermodynamic considerations. Nevertheless, assuming increased (Cu^{2+}), CuCl becomes stable when the half-cell electrode potential of (4) becomes noble to that of (6), which is independent of (Cu^{2+}) and pH at $e = +0.2$ V for $(Cl^-) = 0.1$. When these potentials are maintained potentiostatically, the expected deposition of metallic copper out of solution by (6) has been verified below $+0.2$ V in solutions containing excess Cu^{2+} [52,53], as indicated in Figure 9.34.

Selective dissolution from α brass, which requires solid-state diffusion of zinc, is too slow to account for the usual penetration rates observed in service and laboratory dezincification. Redeposition seems necessary to account for accelerated dezincification in the potential region between 0.0 and $+0.2$ V, where copper can be deposited. Heidersbach and Verink[52] observed accelerated dezincification of α brass only in unstirred solutions or in occluded regions where dissolved copper could accumulate, leading to copper deposition by either of the reduction reactions (6) or (4) below $+0.2$ V. In another example, a cold-worked structure was not propagated into the dezincified copper layer,[54] suggesting that copper had redeposited. Yet, depletion of zinc by electron microprobe analysis in the adjacent brass was evident, suggesting further that selective dissolution was also operative but not controlling.

When copper begins to dissolve at the calculated potential of -0.03 V, the rate of copper removal is still much below that of zinc, because the anodic overvoltage applied on (1) is much greater than on (2). However, the relative rate of copper dissolution is expected to approach that of zinc as potential increases, because of the lower concentration of zinc in the brass alloy. Pickering and Byrne[55] report that copper and zinc dissolve simultaneously in the same proportions as in the alloy above $+0.2$ V. However, this potential is apparently lower in the presence of chloride,[52] perhaps because of reduced overvoltage for anodic copper dissolution by formation of the chloride complex ion in (2).

The research described in this section applies primarily to single-phase α brass. However, the same principles would be expected to apply to the β phase. The dissolution potential of copper from (2) may be somewhat lower than 0.03 V because of the higher zinc content, which could reduce copper activity in the alloy below unity in the Nernst equation.

References

1. A. Joshi and D. F. Stein, *Corrosion*, Vol. 28, p. 321, 1972.
2. G. S. Was, R. G. Ballinger, R. M. Latanision, and R. M. Pelloux, *2nd Semiannual Progress Report*, Research Project 1166-3, Electric Power Research Institute, Palo Alto, CA, 1977.
3. K. Osozawa and H. J. Engell, *Corros. Sci.*, Vol. 6, p. 389, 1967.
4. R. M. Davison, T. DeBold, and M. J. Johnson, *Metals Handbook*, Vol. 13, *Corrosion*, 9th ed., ASM International, Metals Park, OH, p. 547, 1986.
5. M. G. Fontana, *Corrosion Engineering*, 3rd ed., McGraw-Hill, New York, p. 78, 1986.
6. H. H. Uhlig and R. W. Revie, *Corrosion and Corrosion Control*, 3rd ed., Wiley, New York, p. 307, 1985.
7. Welding Research Council Bulletin 138, February 1969.
8. B. M. Gordon and G. M. Gordon, *Metals Handbook*, Vol. 13, *Corrosion*, 9th ed., ASM International, Metals Park, OH, p. 927, 1987.
9. ASTM Standard Practice A262, *Annual Book of ASTM Standards*, Vol 3.02, p. 4, 1988.
10. M. A. Streicher, *ASTM STP 656*, p. 3, 1978. Reprinted in *Process Industries Corrosion*, B. J. Moniz and W. I. Pollock, eds., NACE, Houston, p. 123, 1986.
11. R. A. Corbett and B. J. Saldanha, *Metals Handbook*, Vol. 13, *Corrosion*, 9th ed., ASM International, Metals Park, OH, p. 239, 1987.
12. A. J. Sedricks, *Corrosion of Stainless Steels*, Wiley, p. 110–36, 1979.
13. K. Osozawa, K. Bohnenkamp, and H. J. Engell, *Corros. Sci.*, Vol. 6, p. 421, 1966.
14. W. L. Clarke, V. M. Romero, and J. C. Danko, CORROSION/77, Paper 180, NACE, Houston, 1977.
15. V. Cihal, A. Desestret, M. Froment and G. H. Wagner, *Proc. Conf. European Federation on Corrosion*, Paris, p. 249, 1973.
16. W. L. Clarke, R. L. Cowan, and W. L. Walker, *ASTM STP 656*, ASTM, Philadelphia, p. 99, 1978.
17. R. L. Cowan and C. S. Tedmon, *Advances in Corrosion Science and Technology*, Vol. 3, M. G. Fontana and R. W. Staehle, eds., Plenum, New York, p. 293, 1973.
18. W. L. Clarke and D. C. Carlson, *Mater. Perform.*, Vol. 19., No. 3, p. 16, Mar. 1980.
19. W. D. France and N. D. Greene, *Corros. Sci.*, Vol. 8, p. 9, 1968.
20. M. A. Streicher, *Corros. Sci.*, Vol. 9, p. 53, 1969; Vol. 11, p. 275, 1971.
21. T. M. Devine, *J. Electrochem. Soc.*, Vol. 126, p. 374, 1979; *Metall. Trans. A*, Vol. 11A, p. 791, 1980.
22. M. A. Streicher, *Corrosion*, Vol. 32, p. 79, 1976.
23. S. J. Green, *Metals Handbook*, Vol. 13, *Corrosion*, 9th ed., ASM International, Metals Park, OH, p. 937, 1987.
24. J. B. Lumsden, CORROSION/88, Paper 252, NACE, Houston, 1988.
25. S. J. Ketcham and E. J. Jankowsky, *ASTM STP 866*, ASTM, Philadelphia, p. 14, 1985.

26. E. H. Hollingsworth and H. Y. Hunsicker, *Metals Handbook*, Vol. 13, *Corrosion*, 9th ed., ASM International, Metals Park, OH, p. 583, 1987.
27. Standard Test Method G34, *Annual Book of ASTM Standards*, ASTM, Philadelphia.
28. Standard Test Method G85, *Annual Book of ASTM Standards*, ASTM, Philadelphia.
29. W. F. Savage, Welding Design and Engineering, Dec. 1969.
30. K. Easterling, *Introduction to the Physical Metallurgy of Welding*, Butterworths, London, p. 35, 1983.
31. K. Masubichi, *Metals Handbook*, Vol. 6, *Welding, Brazing and Soldering*, 9th ed., ASM International, p. 856, 1983.
32. J. Gutzeit, R. D. Merrick, and L. R. Sharfstein, *Metals Handbook*, Vol. 13, *Corrosion*, 9th ed., ASM International, Metals Park, OH., p. 1262, 1987.
33. T. G. Gooch, *Process Industries Corrosion*, B. J. Moniz and W. J. Pollock, eds., NACE, Houston, p. 739, 1986.
34. E. F. Nippes, *Metals Handbook, Desk Edition*, ASM International, Metals Park, OH, p. 30–34, 1985.
35. A. Garner, *Metals Prog.*, Vol. 127 (No. 5), p. 31, April 1985.
36. K. F. Krysiak, *Metals Handbook*, Vol. 13, *Corrosion*, 9th ed., ASM International, Metals Park, OH, p. 344, 1987.
37. B. M. Gordon and G. M. Gordon, *Metals Handbook*, Vol. 13, *Corrosion*, 9th ed., ASM International, Metals Park, OH, p. 927, 1987.
38. D. Warren, *Process Industries Corrosion*, B. J. Moniz and W. J. Pollock, eds., NACE, Houston, p. 31, 1986.
39. R. J. Eiber and G. O. Davis, *Investigation of Williams Pipeline Company Mounds View, MN., Pipeline Rupture*. Final Report to Transportation Systems Center, U. S. Dept. of Transportation, Battelle-Columbus Lab., October 14, 1987.
40. C. Kato, Y. Otoguro, S. Kado, and Y. Hisamatsu, *Corrosion Science*, Vol. 18, p. 61, 1978.
41. K. Masamura and I. Matsushima, Paper No. 75, CORROSION/81, NACE, Houston, 1981.
42. G. Kobrin, *Mater. Perform.*, Vol. 15, No. 7, 1976.
43. S. C. Dexter, *Metals Handbook*, Vol. 13, *Corrosion*, 9th ed., ASM International, Metals Park, OH, p. 114, 1987.
44. J. E. Hatch, *Aluminum: Properties and Physical Metallurgy*, ASM International, Metals Park, OH, pp. 285–94, 1984.
45. N. W. Polan, *Metals Handbook*, Vol. 13, *Corrosion*, 9th ed., ASM International, Metals Park, OH, p. 614, 1987.
46. J. J. Snyder, *Metals Handbook*, Vol. 11, *Failure Analysis*, 9th ed., ASM International, Metals Park, OH, p. 372, 1986.
47. V. F. Lucey, *Br. Corros. J.*, Vol. 1, p. 7, 1965; Vol. 2, p. 53, 1966.
48. E. E. Langenegger and F. P. A. Robinson, *Corrosion*, Vol. 25, p. 59, 1968.
49. B. I. Dillon, *Proc. 7th Int. Cong. Metallic Corrosion*, Rio de Janeiro, 1978.
50. E. E. Langenegger and B. G. Callaghan, *Corrosion*, Vol. 28, p. 245, 1972.
51. H. H. Uhlig and R. W. Revie, *Corrosion and Corrosion Control*, 3rd ed., Wiley, New York, p. 334, 1985.
52. R. H. Heidersbach and E. D. Verink, *Corrosion*, Vol. 28, p. 397, 1972.
53. E. D. Verink and R. H. Heidersbach, *ASTM STP 516*, ASTM, Philadelphia, p. 303, 1972.
54. R. Natarajan, P. C. Angelo, and N. F. George, *Corrosion*, Vol. 31, p. 302, 1975.
55. H. W. Pickering and P. J. Byrne, *J. Electrochem. Soc.*, Vol. 116, p. 142, 1969.

Exercises

9-1. Type 304L stainless steel containing 0.02% C shows intergranular corrosion (IGC) after exposure to hot corrosive gases at 600°C for 7 days. How can this be, since the alloy is normally resistant to IGC because of its low carbon content? How could this attack be prevented, still using austenitic stainless steel?

9-2. An accidental arc strike from a welding tool causes surface melting and refreezing at a spot on the surface. Would this likely cause intergranular corrosion problems in the spot if the alloy were Type 304 stainless steel? Type 347 stainless steel?

9-3. Newly manufactured Type 347 stainless steel tanks will be used to hold fuming nitric acid at temperatures ranging from 0 to 120°F in a military desert location. It has been discovered that some sheets used to fabricate the tanks were accidentally heated to 2400°F and water cooled before tank fabrication by welding. Knowing that nitric acid in these conditions will cause intergranular corrosion and knifeline attack on sensitized austenitic stainless steel, **(a)** would you predict that either IGC or KLA could be a danger if the tanks are put into service? **(b)** If so, what are the possible remedies? **(c)** How could the problem have been prevented?

9-4. What testing procedure(s) would you require to assure that incoming shipments of Type 347 stainless steel will be resistant to sensitization during fabrication of the tanks in Problem 9-3? Who should be responsible for the testing—supplier, fabricator, or user?

9-5. Calculate the degree of sensitization for a Type 304 stainless steel which shows the following results from the EPR test: reactivation charge = 1.71 coul, ASTM grain size = 5, surface area = 2.12 cm^2. How would you classify this material according the sensitization guidelines in Section 9.1.5?

9-6. Calculate the equivalent carbon content for ASTM A588 Grade B steel of the following composition. Recommend pre- or postweld heat treatments if necessary. Does this composition conform to the specifications for this steel?

%C: 0.18, %Mn: 1.13, %Cr: 0.22, %Ni: 0.47, %Mo: 0.06, %Cu: 0.31

9-7. A friend who is an engineer for a recreational vehicle manufacturer calls you and describes corrosion at the interface between the Type 321 stainless steel walls and the Type 347 stainless steel weld metal in a 15-gallon cylindrical tank for potable water. Wall thickness is 0.30 in., and the water is often liberally laced with chlorine for sanitation. What do you tell your friend? Is this a good design? What are the possible causes of the corrosion? There is no opportunity to examine the tank.

Corrosive Damage by Hydrogen, Erosion, and Wear

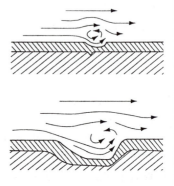

Aqueous corrosion is primarily an electrochemical reaction between a metal and the surrounding environment. The reaction can be profoundly affected, however, by physical factors or processes present before or during corrosion. Three such factors—hydrogen, erosion, and wear—are discussed in this chapter.

Hydrogen produced by corrosion readily dissolves and diffuses in metal crystals and can have dramatic effects on mechanical strength and ductility. The mechanism of embrittlement by hydrogen is discussed in this chapter, along with other irreversible forms of hydrogen damage.

Physical erosion by the passing corrosive solution and wear from a contacting solid are physical processes that have significant effects on simultaneous corrosion reactions. Both erosion and wear increase corrosion rates by disrupting otherwise protective surface films and coatings. Conversely, corrosion and its products can also enhance erosion and wear damage, as will be elaborated later in the chapter.

10.1 Hydrogen Damage

Hydrogen with only one proton in the nucleus is so small that it migrates readily through the crystal structure of most metals and alloys. Hydrogen damage includes a number of detrimental effects on metallurgical and mechanical properties, resulting from hydrogen dissolved in the crystal lattice. Initial hydrogen entry can result in loss of ductility and brittle cracking, as discussed in Section 8.3.1. Additional dissolved hydrogen may nucleate hydrogen gas, which forms internal voids and surface blisters. Some reactive metals, such as titanium and niobium, can form brittle hydrides from dissolved hydrogen. Hydrogen also can react with carbon in steel to the detriment of strength and ductility.

10.1.1 Sources of Hydrogen

Hydrogen may be made available to a metal surface from various sources, including the cathodic reduction of hydrogen or water:

$$2H^+ + 2e^- \rightarrow H_2 \tag{1}$$

$$2H_2O + 2e^- \rightarrow H_2 + 2OH^-. \tag{2}$$

These cathodic reactions may be present during corrosion, cathodic protection, pickling, and other cleaning operations. Hydrogen enters the lattice as nascent or atomic hydrogen, which is an intermediate in the formation of molecular H_2 on the surface by (1) or (2). Processes involving cathodic polarization, such as cathodic protection and electroplating, accelerate hydrogen formation by (1) or (2), and alloys must be chosen and operations selected judiciously to prevent hydrogen damage.

Dissolved "poisons," which retard the formation of molecular hydrogen and increase the residence time of nascent hydrogen on the surface, enhance hydrogen entry, and increase hydrogen damage. Commonly known poisons include phosphorous, arsenic, antimony, sulfur, selenium, tellurium, and cyanide ions. The most technically important poison is sulfide, which is commonly present in geologic fluids such as petroleum, natural gas, well waters, and geothermal steam and condensates. Hydrogen sulfide, H_2S, is especially aggressive in promoting hydrogen damage because it provides not only the ionic sulfide poison, but also H^+ as well, for reduction to H_2 by (1).

Hydrogen may also enter the metal from hydrogen-bearing atmospheres during heat treating, welding, or other manufacturing processes. Water vapor and steam may be decomposed to hydrogen at hot surfaces during welding or heat treating (Section 9.3.2). Damage often occurs during exposure to a hydrogen bearing environment, but conditions may conspire to produce delayed failure from previous hydrogen exposures as well.

10.1.2 Hydrogen Embrittlement

The first traces of hydrogen dissolved in the crystal lattice produce effects generally classified as hydrogen embrittlement. The major form of hydrogen embrittlement discussed here is hydrogen induced cracking (HIC) of various high-strength alloys, ferrous and nonferrous, under sustained tensile loads in the presence of hydrogen or hydrogen-bearing environments. Lower-strength stainless steels and nonferrous alloys of high ductility may also experience a loss in tensile ductility, but still retain a ductile fracture mode in the presence of corrosion or high-pressure hydrogen. Hydrogen embrittlement is also sometimes used to describe detrimental effects of hydrogen absorbed in the molten metal and later precipitated in the metallurgical structure during solidification and cooling. HIC is used here to avoid confusion with these or other effects on mechanical properties at relatively low levels of dissolved hydrogen, although hydrogen embrittlement and HIC are synonymous in many usages elsewhere. HIC in low-alloy and carbon steel weldments is called underbead cracking (Section 9.4.5).

HIC is usually classified as one form of environmentally induced cracking, the subject of Chapter 8. Therefore, HIC was initially discussed in Sections 8.1.3, 8.3.1, 8.7.2, and 8.8.1, in conjunction with stress corrosion cracking (SCC) and corrosion fatigue cracking. A frequent alternative name for HIC is hydrogen stress cracking, which is not used here or in Chapter 8, to avoid possible confusion with SCC processes, which do not involve hydrogen in the cracking mechanism. HIC accelerated by sulfide, sometimes called sulfide stress cracking, is discussed further in Section 11.2.1. Testing methods for HIC are held in common with SCC and discussed in Sections 8.6 and 8.7.

HIC can be reversed if the hydrogen is removed by baking at elevated temperatures (Figure 8.11), and HIC is often less critical at higher service temperatures where hydrogen is more mobile (Section 8.3.1). The temperature dependence of hydrogen damage in steels is complicated by the high-temperature face-centered cubic austenite phase, which has relatively high hydrogen solubility. Up to the transformation temperature (723°C), hydrogen solubility decreases in body centered cubic ferrite. Thus, hydrogen can be baked out of steels at moderate temperatures but will be rapidly absorbed at higher temperatures, where austenite is stable.[1] Subsequent cooling causes hydrogen damage when hydrogen previously dissolved in austenite is rejected from the low temperature ferrite.

The cracks from HIC may be transgranular or intergranular and usually have sharp tips with minor branching and cleavage-like morphology of the crack surfaces. Various mechanisms have been proposed for HIC, including hydrogen pressure, surface adsorption, decohesion, and enhanced plastic flow. Hydrogen pressure has long been thought to build up internally to weaken the structure at crystalline and metallurgical defects, such as dislocations and second-phase interfaces, respectively.[2] Surface adsorption of hydrogen has been suggested to occur on free surfaces in the volume ahead of the crack tip, reducing the energy necessary to form brittle crack surfaces.[3] Neither of these mechanisms has been widely accepted for HIC or predominantly supported by experimental data. The pressure theory is generally accepted, however, to account for hydrogen blistering, discussed in Section 10.1.4. A decohesion model has been postulated in which dissolved hydrogen weakens the interatomic bonding forces.[4,5] Direct evidence of bond weakening by dissolved hydrogen has not been forthcoming, but decohesion has not been ruled out entirely.

Most recently, considerable evidence has accumulated to support localized enhanced plastic flow to account for HIC.[6,7] Based on careful fractographic studies of crack surfaces, which showed microscopic evidence of local plastic tearing, Beachem[8] first suggested that HIC may occur by localized enhanced plasticity.

Later, Matsui et al.[9] observed a decrease in the flow stress of pure iron caused by either precharged or dynamically charged hydrogen. Transmission electron microscopy showed increased dislocation velocity and resultant plasticity in the presence of dissolved hydrogen.[10] The mechanism of enhanced plasticity has not been defined. However, some form of bond weakening or disruption (decohesion) in the strain field around a dislocation is implied to account for easier movement.

A mechanism to account for embrittlement by enhanced plasticity is described in Section 8.8.6. Enhanced plasticity and attenuation of strain hardening by anodic dissolution and cyclic stress can also account for brittle cracking by stress corrosion and corrosion fatigue cracking, respectively. At an emerging surface slip band, hydrogen penetration and consequent local plasticity may be enhanced. Due to the constraints imposed by the surrounding undeformed material, a localized plane strain (hydrostatic or triaxial) stress state is formed in the slip band, suppressing further plasticity (Section 8.8.6). The same bond-weakening processes that account for enhanced plasticity in a biaxial stress state could also induce a reduced fracture stress in a triaxial stress state. Thus, a brittle crack may be expected to initiate eventually at the slip band.

10.1.3 Hydrogen Trapping

Hydrogen diffusing through a metal lattice accumulates at metallurgical inhomogeneities or "traps." The accumulation causes a lag in the hydrogen flux through a sheet specimen of the metal, as shown in Figure 10.1.[11] The difference between consecutive transients shows that hydrogen penetrates in a much shorter time after the traps have been filled. Further, the addition of 1.5% Ti creates additional traps which prolong the initial penetration of hydrogen. The rate of steady-state hydrogen diffu-

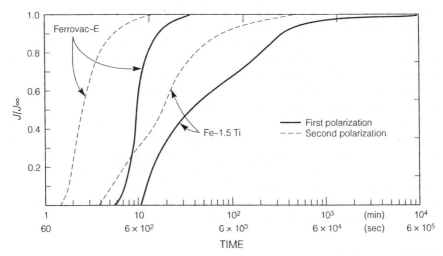

FIGURE 10.1   Hydrogen transients through iron and iron-titanium sheet. *(From I. M. Bernstein,* Environmental Sensitive Fracture of Metals and Alloys, *Proc. Office of Naval Research Workshop, Washington, D. C., June 3–4, 1985, Office of Naval Research, Arlington, VA, p. 89, 1987.)*

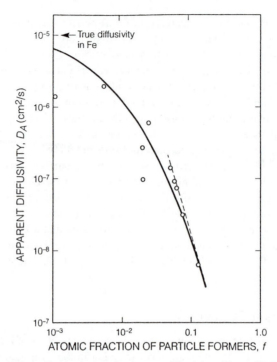

FIGURE 10.2 Effect of trapping on hydrogen diffusivity. *(From W. W. Gerberich, Hydrogen in Metals, I. M. Bernstein and A. W. Thompson, eds., ASM, Metals Park, OH, p. 115, 1974. Reprinted by permission, ASM International.)*

sion, as measured by diffusivity, is reduced by the presence of traps, as well (Figure 10.2[12]), indicating that filled traps have an attraction for the passing hydrogen flux.

A classification of traps is shown in Table 10.1.[13] It is apparent that traps may result from solute atoms, dislocations, particle-matrix interfaces, grain boundaries, and internal voids and cracks. Traps may be reversible or irreversible, depending on whether the trapped hydrogen is easily released or tightly bound, as measured by the interaction energies listed in Table 10.1. They may be mobile (dislocations) or stationary (solute atoms, particles, grain boundaries).

Because traps are associated with various microstructural features, alloying and metallurgical treatments to modify trapping properties offer a promising avenue to the development of alloys with maximum resistance to hydrogen damage. Consistent with a hydrogen trapping theory of HIC,[14] deep or irreversible trapping reduces the population of hydrogen at the crack tip and often increases resistance to HIC. Conversely, shallow or more reversible traps permit more rapid hydrogen transport, which allows some traps to reach a critical concentration necessary to initiate cracking.[11]

10.1.4 Blistering

When sufficient hydrogen builds up at crystalline or metallurgical inhomogeneities (traps), atomic or nascent hydrogen will recombine to form molecular H_2.

TABLE 10.1   Classification of Hydrogen Traps

Trap class	Example of trap — Elements at the left of iron	Example of trap — Elements with a negative $\varepsilon^i_H(b)$	Interaction energy(a), eV	Character if known	Influence diameter, D_i
Point	. . .	Ni	(0.083)	Most probably reversible	a few inter-atomic spacings
	MN	MN	(0.09)		
	Cr	Cr	(0.10)		
	V	V	(0.16)		
	. . .	Ce	(0.16)	———	
	. . .	Nb	(0.16)		
	Ti	Ti	0.27	Reversible	———
		(vacancy)			
	Sc	O	(0.71)		
	Ca	Ta	(0.98)	Getting more irreversible	
	K	Ia	(0.98)		
	. . .	Nd	(1.34)		
Linear	Dislocations		0.31 / 0.25 (average values)	Reversible / Reversible	3 nm for an edge dislocation
	Intersection of three grain boundaries		. . .	Depends on coherency	. . .
Planar or bidimensional	Particle/matrix interfaces TIC (incoherent) Fe$_3$C MnS		0.98 / 0.8–0.98	Irreversible, gets more reversible as the particle is more coherent	Diameter of the particle, or a little more as coherency increases
	Grain boundaries		0.27 / Average value 0.55–0.61 (high angle)	Reversible / Reversible or irreversible	Same as dislocation
	Twins		. . .	Reversible	A few interatomic spacings
	Internal surfaces (voids)		. . .	. . .	. . .
Volume	Voids		>0.22	. . .	Dimension of the defect
	Cracks		. . .	. . .	
	Particles		Depends on exothermicity of the dissolution of H by the particle	. . .	

From G. M. Pressouyre, Metall. Trans. A, Vol. 10A, p. 1571, 1979. Reprinted by permission, ASM International.

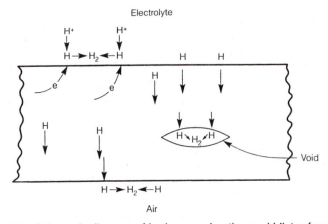

FIGURE 10.3   Schematic diagram of hydrogen migration and blister formation.

Accumulated molecules nucleate the gas phase, which develops very high pressures sufficient to rupture interatomic bonds, forming microscopic voids and macroscopic blisters (Figure 10.3). Thus, the hydrogen pressure theory, mentioned initially in the preceding section, is accepted as valid for blister formation. The blisters will embrittle the lattice and generally degrade mechanical properties. Figure 10.4[15] shows an example of an internal microscopic blister nucleated at an inclusion. Blisters erupting at the surface form surface cracks. Examples of surface hydrogen blisters are shown in Figure 1.12. Steels high in oxygen and sulfur, which form intermetallic compounds with iron and other alloying elements, decrease resistance to hydrogen blistering. Blistering is especially prevalent in petroleum drilling and refining equipment, because of the ubiquity of sulfides and H_2S in the associated environments.[16]

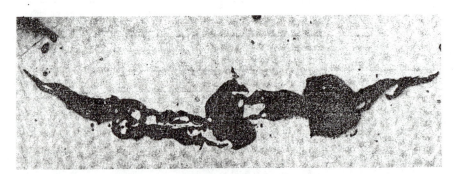

FIGURE 10.4   Hydrogen blister nucleating at a metallurgical inclusion. *(From C. D. Kim, Metals Handbook, Vol. 11, Failure Analysis, 9th ed., ASM International, Metals Park, OH, p. 245, 1986. Reprinted by permission, ASM International.)*

10.1.5 Internal Hydrogen Precipitation

Hydrogen has relatively high solubility in the face-centered cubic austenite phase, as compared to body-centered cubic ferrite. Considerable hydrogen can be absorbed at high temperature from moisture in the atmosphere and in additives to the melt. Upon forming body-centered cubic ferrite below the transformation temperature at 723C, insoluble hydrogen may be released, especially in heavy forgings. Regions around inclusions, precipitates, and laminations become embrittled by supersaturated hydrogen, as temperature drops from the austenite region during forging operations. Nucleation of hydrogen gas in these regions causes localized brittle ruptures or "flakes" in the structure.

"Fisheyes" are sometimes evident on the brittle fracture surfaces of tensile specimens taken from plate or forgings of high hydrogen content. Fisheyes are often associated with microscopic pores or inclusions on the fracture surfaces (Figure 8.13[17]). Baking or long-term room-temperature aging will sometimes eliminate fisheyes and restore normal ductility.

10.1.6 Hydrogen Attack

Hydrogen in steels at elevated temperature can react with carbides to form decarburized structures and methane gas, which cannot easily diffuse out of the structure because of its size. Methane bubbles form and eventually grow into fissures at grain boundaries. Decarburization, along with the fissures, may produce sudden reductions in strength and ductility, especially in operating petroleum refinery equipment. High-temperature failures by hydrogen attack should be contrasted with other forms of hydrogen damage which occur primarily at temperatures below 200°C.

10.1.7 Hydride Formation

Titanium, tantalum, niobium, zirconium, and uranium are reactive metals that form brittle hydrides with dissolved hydrogen. Hydrogen diffuses readily in these metals above 250°C and precipitates as insoluble brittle hydrides. The hydrides increase tensile strength and decrease ductility. The integrity of the surface oxide film on titanium has a great effect on hydrogen pickup. Surface contamination by iron promotes localized hydrogen damage of titanium.

10.1.8 Prevention of Hydrogen Damage

Hydrogen damage can be prevented by modifications of the environment or selection of materials more resistant to embrittlement. Elimination of cathodic protection systems, stray currents, or galvanic couples (including galvanized coatings) that apply cathodic currents to a sensitive material may be very helpful. Removal of sulfides to below 50 ppm will improve the resistance of most carbon and low-alloy steels. Cyanides may also be critical in some oil-field and petroleum refining operations. Hydrogen induced cracking (HIC) with sulfides is reduced markedly at pH above 8.[15] Removal of moisture from an H_2S gas stream will greatly reduce embrittlement. Operating temperatures above ambient will reduce HIC of steels; above 200°C, hydrogen attack may be a problem, however. Inhibitor additions (Section 14.4) which reduce the general corrosion rate also reduce the rate of hydrogen generation on the surface and thereby mitigate hydrogen damage.

Tensile strength is inversely proportional to HIC resistance. Appropriate reductions in strength level by annealing or alloy substitutions will reduce or eliminate failures. Steels with tensile strength below 690 MPa (100 ksi) are resistant to HIC in the absence of sulfides. This threshold may be elevated somewhat for appropriately treated low-alloy high-strength steels.[15] Alloying elements at low levels have an indeterminate effect, but nickel above 1% is detrimental to HIC resistance with sulfides. Because HIC is reversible, baking at temperatures between 100° and 650°C will restore ductility to carbon and low-alloy steels exposed to plating and pickling operations. Untempered martensite is especially susceptible to HIC, and stress-relief treatments to reduce internal stress and form tempered martensite and bainitic structures in welds are generally recommended. Baking at 230°C for 1 hour per 25 mm (inch) of thickness has been recommended[16] to restore ductility after welding.

Clean, high-quality steels free of inclusions and other inhomogeneities have increased resistance to hydrogen blistering. Killed steels with more inclusions are often more susceptible than semikilled steels,[18] but susceptibility is sensitive to inclusion morphology. A major challenge facing modern steelmakers is the production of clean, high-quality steels which are resistant to hydrogen damage in sour (H_2S-bearing) environments common in oil fields, refineries, natural gas wells, and geothermal systems.

Water and temperature enhance hydrogen attack; maintaining temperatures below 700°C (1290°F) or dewpoint below −45°C (−50°F) will prevent decarburization of hypoeutectoid steels.[15] Nelson diagrams define the limits of temperature and hydrogen partial pressure for resistance to hydrogen attack for various alloys measured in

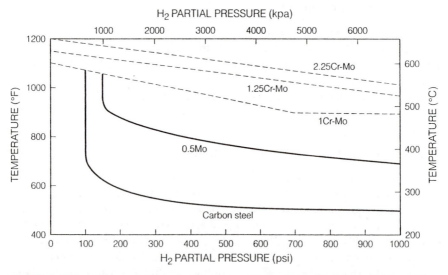

FIGURE 10.5 Nelson diagram for hydrogen attack. *(From R. Chiba, K. Omnishi, K. Ishii, and K. Maeda, Corrosion, Vol. 41, p. 415, 1985. Reprinted by permission, National Association of Corrosion Engineers.)*

service. An example in Figure 10.5[19] also shows that molybdenum and chromium additions enhance resistance to hydrogen attack and decarburization.

Anodizing in 10% ammonium sulfate solution removes surface contamination and thickens the usual surface oxide on titanium and its alloys for increased resistance to hydrogen entry.[15] Oxide surface films are generally resistant to hydrogen penetration and provide a degree of protection for most alloy systems. Welding and heat treating of titanium requires inert gases for a protective covering atmosphere to prevent embrittlement by hydrogen. Hydrides will decompose to some extent, and baking at elevated temperature in a vacuum may be effective in restoring ductility to titanium in some cases.

10.2 Erosion-Corrosion

10.2.1 Description and Causes

Rapidly flowing solutions can often disrupt adherent surface films and deposits that would otherwise offer protection against corrosion. Thinning or removal of surface films by erosion from the flowing stream results in accelerated corrosion, called *erosion-corrosion* or *impingement attack*. The attack is accelerated at elbows, turbines, pumps, tube constrictions, and other structural features that alter flow direction or velocity and increase turbulence. Erosion-corrosion often occurs when the corrodent is in the liquid phase. Suspended solids further aggravate the erosion of surface films and increase erosion-corrosion. Attack may be especially serious in two-phase flow, in which steam and water vapor condensate droplets are present together.

The lower-strength, less-corrosion-resistant alloys, such as carbon steel, copper, and aluminum, are especially susceptible to erosion-corrosion. However, strength does not always control resistance. Stainless steels are resistant in conditions where carbon steels of the same or higher hardness fail by erosion-corrosion. The stainless steels, nickel alloys, and titanium are usually resistant, because of their tenacious and durable passive surface films. Even these may in some cases show erosion-corrosion if corrosive conditions are sufficiently severe. A few, but by no means all, alloy-corrodent combinations that have produced erosion-corrosion are summarized in Table 10.2.

Erosion-corrosion takes the form of grooves, waves, gullies, teardrop-shaped pits, and horseshoe-shaped depressions in the surface. The effect of hydrodynamics is not well understood. For example, undercutting may occur in either the upstream or the downstream directions.[20] Upstream undercutting was shown in Figure 1.16; downstream undercutting appears in Figure 10.6[21] for erosion-corrosion of brass condenser tubing. Commonly observed horseshoe-shaped pits (Figure 1.16) are undercut at the upstream end, but teardrop-shaped pits are usually elongated and undercut in the downstream direction.

Figure 10.7[22] shows teardrop-shaped erosion-corrosion pits on the surface of a water-cooled aluminum-clad nuclear fuel element. Turbulent eddies probably thin the protective film locally to account for downstream undercutting, as described schematically in Figure 10.8.[23] The corrosivity of the flowing corrodent had a significant effect in this example. The attack was seasonal and unpredictable and could be simulated only with water containing a combination of dissolved bicarbonate and silicate.[22]

TABLE 10.2 Conditions Causing Erosion-Corrosion

Alloy Component	Corrosive Conditions	Ref.
Brass Condenser Tubes	Brackish water, seawater, and polluted cooling waters.	20
Aluminum Heat Transfer Surfaces	Filtered river water containing dissolved silicates and bicarbonates.	22
Carbon Steel Pipe	Exhausted steam from turbines containing mixed vapor/liquid.	24
Carbon Steel Petroleum Refinery Equipment	Liquid and liquid/vapor process streams containing H_2S.	16
Carbon Steel Pipe and Storage Tanks	Sulfuric acid, 65-100%, >0.9 m/s. Hydrogen bubbles evolved by corrosion in sulfuric acid.	25
Cast Austenitic Stainless Steel Pump Parts	Acid process stream, reducing conditions	27

Vapor-phase steam-carrying water condensate droplets, can be particularly damaging to steel. Steel pipe surfaces normally develop protective oxide scales, but the momentum of impinging water droplets erodes and thins the oxide and accelerates the reoxidation and corrosion of the underlying surface. Erosion-corrosion caused sudden rupture of a steam return line in a nuclear power plant [24] and several fatalities in a maintenance crew working nearby (Section 1.2.2). Similar attack occurs in the return lines from distillation units in petroleum refineries.[16] In this case, impinging liquid droplets erode iron sulfide scales in liquid-vapor streams containing H_2S. Figure 10.9 shows an example of the broad lake-like depressions causing perforation of a petroleum refinery pipe.[16]

Sulfate surface scales can be eroded by sulfuric acid at room temperature if flow velocities exceed 0.9 m/s or if turbulence is present.[25] Hydrogen bubbles rising over the surfaces of carbon-steel pipe and storage vessels cause similar effects in sulfuric acid. Under quiescent conditions, hydrogen bubbles from corrosion follow the same

FIGURE 10.6 Erosion-corrosion of brass condenser tubing showing individual tear-drop shaped pits with undercutting in the downstream direction. *(From H. H. Uhlig and R. W. Revie,* Corrosion and Corrosion Control, *3rd. ed., Wiley, p. 328, 1985. Reprinted by permission, John Wiley & Sons.)*

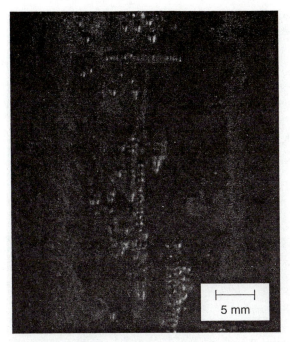

FIGURE 10.7 Erosion-corrosion of aluminum in cooling water. Flow direction: bottom to top. *(From D. A. Jones,* Corrosion, *Vol. 37, p. 563, 1981. Reprinted by permission, National Association of Corrosion Engineers.)*

tracks up the surface, erode the surface sulfate deposit, and produce attack known as hydrogen grooving (Figure 10.10[25]). The mechanism of surface erosion is rather similar to conventional erosion-corrosion. Periodic movement or stirring will disrupt bubble flow and control hydrogen grooving.

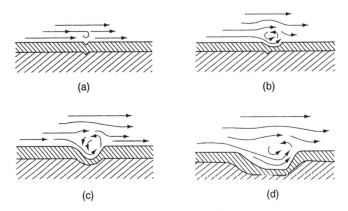

FIGURE 10.8 Turbulent eddy mechanism for downstream undercutting of erosion-corrosion pits. *(From D. A. Jones,* Corrosion Processes, *R. N. Parkins, ed.,* Applied Science, *p. 177, 1982. Reprinted by permission, Applied Science Publishers.)*

FIGURE 10.9 High-temperature erosion-corrosion of 500-mm carbon steel pipe by water-droplet impingement in a high-velocity petroleum-refinery vapor stream containing sulfides. *(From J. Gutzeit, R. D. Merick, and L. R. Scharfstein,* Metals Handbook, *Vol. 13,* Corrosion, *9th ed., ASM International, Metals Park, OH, p. 1265, 1987. Photograph by courtesy of J. Gutzeit, Amoco Corp. Reprinted by permission, ASM International.)*

Stainless steel is normally resistant to erosion-corrosion with an adherent passive film. However, a cast austenitic stainless steel pump impeller, which had served for several years, failed in only a few months,[26] when oxidizers were removed from solution, destabilizing the passive film. A similar impeller showing the effects of erosion-corrosion in hot acid with suspended solids appears in Figure 10.11.[27]

10.2.2 Prevention

Erosion-corrosion can be prevented by either design or materials selection. Redesign of equipment can reduce surface velocity and turbulence of the impinging process stream. Larger radius elbows, larger pipe diameter, and gradual rather than abrupt changes in flow channel dimensions are examples of measures to lessen erosion-corrosion. Subtle changes in solution corrosivity (e.g., pH, dissolved oxygen content) can have significant effects. Inhibitor or passivator additions can also control

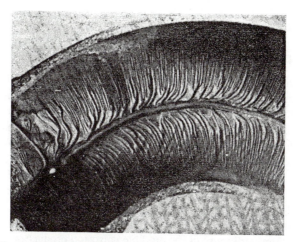

FIGURE 10.10   Grooving in a carbon-steel pipe, containing quiescent concentrated sulfuric acid. *(From S. K. Brubaker, Process Industries Corrosion, B. J. Moniz and W. J. Pollock, eds., NACE, Houston, p. 243, 1986. Reprinted by permission, National Association of Corrosion Engineers.)*

erosion-corrosion. Design for greater section thickness or easy replacement of susceptible parts is sometimes the most economical measure. When design or process changes are unreasonable, selection of a slightly more corrosion-resistant alloy will often improve the stability of surface films sufficiently to stop the attack.

FIGURE 10.11   Erosion-corrosion of an austenitic stainless steel pump impeller. *(From G. A. Minick and D. L. Olson, Metals Handbook, 9th ed., Vol 13, Corrosion, ASM International, Metals Park, OH, p. 1293, 1987. Photograph by courtesy of G. A. Minick. Reprinted by permission, ASM International.)*

10.3 Cavitation

10.3.1 Description and Causes

Cavitation is a form of erosive attack similar to erosion-corrosion. While corrosion often enhances cavitation, it is not essential, as in conventional erosion-corrosion. Cavitation results from collapsing bubbles created by pressure changes across surfaces exposed to high-velocity liquid flow. Flow across a curved surface produces a pressure drop that causes local boiling when the pressure is reduced below the vapor pressure of the liquid. The pressure drop is similar to that causing lift during flow across an air foil. When the bubbles land on the surface and collapse or *implode*, the repeated pressure impacts are sufficient to erode or cavitate the surface, especially in the presence of friable corrosion product scales. As in erosion-corrosion, removal or thinning of protective surface films allows more rapid corrosion of the underlying metal, which forms still more of the erodible corrosion product on the surface. However, implosion pressure surges are sufficient to cavitate the surface of the metal itself in many cases.

Cavitation is not uncommon on ship propellers, pump impellers, and turbine runners, which are subject to high-velocity liquid flow. Figure 1.17 shows advanced cavitation of a cast-iron pump inlet. In a second example, Figure 10.12 shows cavitation attack of a cast steel turbine runner from the hydroelectric power plant at Hoover Dam on the lower Colorado River. Turbulence due to the roughened cavitated surface reduces turbine efficiency. The added turbulence further enhances cavitation, which can eventually impair the structural integrity of the turbine runners. Turbine runners have been repaired routinely with overlay welding, which is apparent in Figure 10.12. Most have been replaced at Hoover Dam with cast stainless-steel runners: for example, grade CA-15 containing 11.5 through 14% Cr and 1 through 2% Ni. Even this stainless steel will cavitate with little or no corrosion, unless the runners are carefully designed to minimize pressure drops across the surfaces.

Pure cavitation of ductile alloys without corrosion probably occurs by a mechanism of progressive surface work hardening, fatigue, embrittlement, and microfracture by repeated cavitation blows. An incubation period is necessary before attack begins, and harder alloys are more resistant.

10.3.2 Prevention

Cathodic protection is sometimes beneficial, not because of the reduced corrosion rate but because of the cushioning effect of hydrogen evolved on the surface. Removal of dissolved air is often beneficial because dissolved gases more easily nucleate cavitating bubbles at lesser pressure reductions. As already mentioned, more corrosion-resistant alloys resist cavitation but are not immune. Careful design is required to minimize pressure drops across surfaces subject to high-velocity liquid flow. Proper operation of pumps and other equipment is necessary. For example, a pump should not be operated on plugged or impaired flow lines. The resulting low inlet (suction) pressure decreases ambient pressure on the liquid and nucleates cavitating bubbles within the pump.

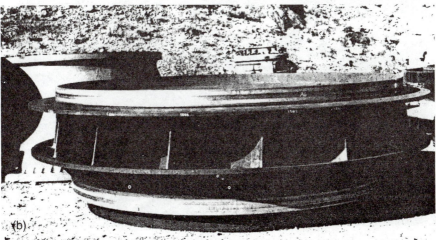

FIGURE 10.12   Cavitation (a) on the surfaces of a Hoover-Dam turbine runner (b). *(Photographed by permission, U. S. Dept. of the Interior.)*

10.4 Fretting

10.4.1 Description and Causes

Fretting is a wear process enhanced by corrosion. On the other hand, erosion-corrosion and its variations are the reverse—corrosion processes enhanced by wear or erosion. Fretting involves wear of an alloy contacting another solid in dry or humid air, in contrast to erosion-corrosion, in which an alloy is eroded by a flowing liquid. Fretting results from abrasive wear of surface oxide films, which form on contacting surfaces under load in air. Slight motion, for example by vibration, of only a few tenths of a millimeter wears away the surface oxide and underlying metal. Metallic wear particles are rapidly oxidized to hard oxides that act as an additional abrasive. Further motion grinds the oxide particles down to a fine rouge, which further enhances the wear process. Figure 10.13[28] shows evidence of wear and oxide debris on a fretted surface in the scanning electron microscope. Fretting corrosion of ferrous alloys is characterized by a fine red powder of Fe_2O_3 issuing from between the fretting surfaces. Humid air may accelerate fretting corrosion by increasing the corrosion product formation. However, excess liquid condensed between the fretting surfaces may prevent attack by acting as a lubricant. Damage increases with normal load on the contacting surfaces and with the amplitude of motion.[29] There is no known minimum amplitude below which fretting stops.

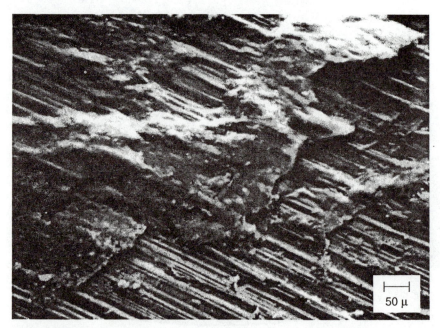

FIGURE 10.13   Fretted steel surface. ×480. *(From L. Engel and H. Klingele,* Atlas of Metal Damage, *Prentice-Hall, p. 180, 1981. Reprinted by permission, Springer Verlag, Munich.)*

Fretting attack may appear anyplace where parts are in dry contact and subject to relative motion. The necessary conditions are present in many surprising and unusual circumstances. In operating equipment, fretting will loosen wheels and turbines from attached shafts. On a microscale, fretting can ruin the electrical contact between gold and copper parts in electronic assemblies. Fretting often appears in bearings and other contact points during shipping of components and equipment. Figure 10.14[30] shows fretting in a stainless-steel spacer used to mount a control rudder on the space shuttle orbiter. Impacted oxide and wear debris are shown on the surface in the metallurgical section. Only minor surface marring was present on the matching nickel alloy bearing.

10.4.2 Prevention

Fretting may be mitigated by any measure that interrupts the required surface contact and motion. Several such methods that have seen at least partial success are as follows:

1. Lubrication of contacting surfaces is common and often prevents the wearing action in fretting.
2. Insertion of an insulating and cushioning material between surfaces as a sealant, coating, or loose gasket is often helpful.
3. Reduced load between fretting surfaces allows greater motion but reduces the wearing action between them.
4. Greater load may be helpful to reduce relative motion between the surfaces but must be used with caution because only very slight motion is often sufficient to produce fretting attack.
5. Increased friction by surface roughening may reduce relative surface movements in the same way but may not stop fretting entirely.
6. Substituting more wear-resistant materials is often helpful but depends also on the hardness and abrasiveness of wear debris in a particular atmosphere.

Substituting more corrosion-resistant materials is noticeably absent from the list above. Fretting is controlled by the abrasiveness of wear debris and wear resistance of materials. Conventionally corrosion-resistant alloys can still oxidize to form hard abrasive wear debris and thus be highly susceptible to fretting. In a similar manner, reducing the corrosivity of the atmosphere is often inadequate to prevent fretting. Fretting has been reported even for nonreactive gold surfaces and for other alloys in vacuum and outer space.[29]

10.5 Erosive and Corrosive Wear

10.5.1 Description and Causes

As erosive intensity increases, a metal can be eroded directly with no necessary contribution from corrosion. Cavitation, described earlier, is an example, in which

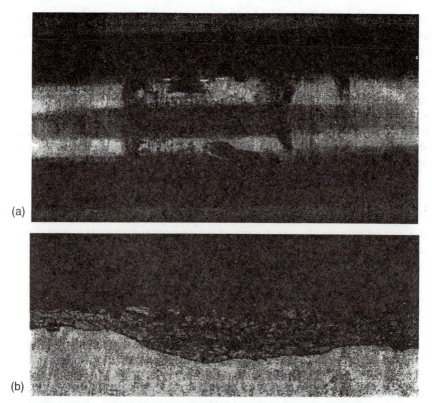

FIGURE 10.14   Fretting in a precipitation-hardened stainless steel mounting assembly for the space shuttle orbiter: (a) wear surface: (b) surface oxide. *(From L. J. Korb, Metals Handbook, 9th ed., Vol. 13, Corrosion, ASM International, Metals Park, OH, p. 1058, 1986. Photograph by courtesy of L. J. Korb, Rockwell International. Reprinted by permission, ASM International.)*

attack may be present without corrosion, when the shock impact of collapsing bubbles is sufficient to remove metal directly. Erosive intensity may also be increased by solid particles suspended in a liquid slurry or by solid particles or water droplets impacting at high velocity from a vapor or gas such as air. Grinding media, balls or rods, tumbled in grinding mills, experience wear during comminution of ores and other materials charged into the mill. In all these cases, wear is present without corrosion, but the addition of water and other corrosive atmospheres may increase the effects of wear.

Erosion in slurries may be very intense in pipe, valves, and pumps. Figure 10.15[29] shows erosion of a valve plug used to control slurry flow by throttling through a narrow orifice. Water-drop impingement is a particular problem in steam turbines, helicopter rotor blades, and aircraft propellers. The hydrodynamic intensity is much greater in these instances than in erosion-corrosion of liquid-vapor piping systems described in Section 10.2.1. The rotating surfaces experience high-impact blows from water droplets when intersecting a liquid-vapor stream. These impacts are suf-

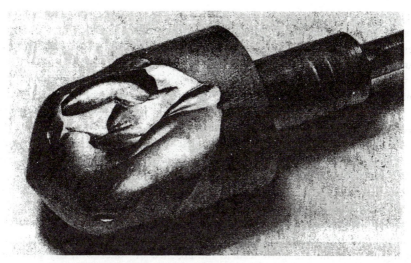

FIGURE 10.15 Erosion of surface hardened stainless steel valve plug in slurry flow. *(From W. Glaeser and I. G. Wright,* Metals Handbook, *9th ed., Vol. 13,* Corrosion, *ASM International, Metals Park, OH, p. 136, 1986. Photograph by courtesy of W. A. Glaeser, Battelle Columbus Laboratories. Reprinted by permission, ASM International.)*

ficiently intense to erode the substrate alloy directly, without benefit of surface corrosion products. Liquid impingement erosion has a similar appearance to that of cavitation, and some theories suggest that bubble implosions produce a local jet impingement which causes erosion during cavitation.[31]

FIGURE 10.16   Internal view of a ball mill.

Corrosive wear of grinding media, balls and rods, in coal and ore grinding systems is a significant economic liability.[32] The most common grinding process utilizes large rotating cylinders (mills) several feet in diameter, which contain the ore charge along with the grinding media in the form of steel balls or rods. Figure 10.16 shows the ball charge in a cylindrical ball mill. During rotation, the grinding media tumble with the ore charge, and the ore fragments are fractured by impact first with one another, but primarily with the grinding media. The grinding process may be conducted either dry or wet, but wet grinding is more usual in practice, because the grinding efficiency is much higher. Unfortunately, the benefits of wet grinding are accompanied by profound increases in consumption of the grinding media by wear.

The dual nature of corrosive wear is shown in Figure 10.17 by the results of laboratory ball mill tests.[33] Increase in pH, which stabilizes the passive film and increases corrosion resistance, also reduced metal losses in the grinding balls. At the same time, higher ball hardness reduced metal losses still further even under the most favorable circumstances of aqueous corrosion. However, these losses are still higher than those experienced in the absence of water with dry grinding. Improved grinding efficiency in the presence of water has yet to be explained. The

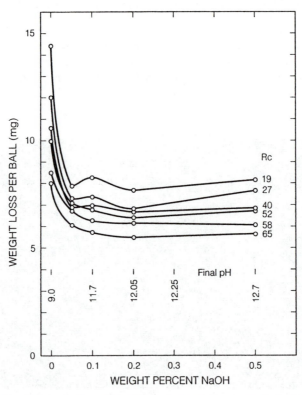

FIGURE 10.17   Effect of pH and hardness on corrosive wear of steel in a laboratory ball mill. *(From R. J. Brigham,* Canadian Met. Quarterly, *Vol. 15, p. 389, 1976. Reprinted by permission, Pergamon Press.)*

mechanisms of rock fracture and metal wear in the presence of a corrosive atmosphere are probably related to those of environmentally induced cracking, discussed in Chapter 8. All are uncertain and subject to continuing investigation and study.

10.5.2 Prevention

Erosive and corrosive wear can be reduced by any means that decrease hydrodynamic intensity. These include reduced velocity, slurry solids content, and water-drop size and concentration. Larger pipe sections, larger radius of curvature, and avoidance of small flow orifices will all reduce impingement and velocity and reduce attack. Reduced corrosion rate by control of process parameters, inhibitor additions, or alloy substitutions will reduce erosion and corrosive wear, but may not stop them entirely, because both have a strong wear component. Harder, more wear-resistant alloys will usually reduce the attack.

References

1. G. Kobrin, *Metals Handbook*, Vol. 13, *Corrosion*, 9th ed., ASM International, Metals Park, OH, p. 331, 1987.
2. C. Zapffe and C. Sims, *Trans. AIME*, Vol. 145, p. 225, 1941.
3. N. J. Petch and P. Stables, *Nature*, Vol. 169, p. 852, 1952.
4. A. R. Troiano, *Trans. ASM*, Vol. 52, p.54, 1960.
5. R. A. Oriani and P. H. Josephic, *Acta Metall.*, Vol. 22, p. 1065, 1974.
6. J. P. Hirth, *Environment Sensitive Fracture of Metals and Alloys*, Proc. Office of Naval Research Workshop, Washington, D. C., June 3–4, 1985, Office of Naval Research, Arlington, VA, p. 79, 1987.
7. H. K. Birnbaum, *Environment Sensitive Fracture of Metals and Alloys*, Proc. Office of Naval Research Workshop, Washington, D. C., June 3–4, 1985, Office of Naval Research, Arlington, VA, p. 105, 1987.
8. C. D. Beachem, *Metall. Trans.*, Vol. 3, p. 437, 1972.
9. H. Matsui, H. Kimura and S. Moriya, *Mater. Sci. Eng.*, Vol. 40, pp. 207, 217, 1979.
10. T. Tabata and H. K. Birnbaum, *Scr. Metall.*, Vol. 17, p. 947, 1983.
11. I. M. Bernstein, *Environment Sensitive Fracture of Metals and Alloys*, Proc. Office of Naval Research Workshop, Washington, D. C., June 3–4, 1985, Office of Naval Research, Arlington, VA, p. 89, 1987.
12. W. W. Gerberich, *Hydrogen in Metals*, I. M. Bernstein and A. W. Thompson, eds., ASM International, Metals Park, OH, p. 115, 1974.
13. G. M. Pressouyre, *Metall. Trans. A*, Vol. 10A, p. 1571, 1979.
14. G. M. Pressouyre and I. M. Bernstein, *Metall. Trans. A*, Vol. 12A, p. 835, 1981.
15. C. D. Kim, *Metals Handbook*, Vol. 11, *Failure Analysis*, 9th ed., ASM International, Metals Park, OH, p. 245, 1986.
16. J. Gutzeit, R. D. Merrick, and L. R. Scharfstein, *Metals Handbook*, Vol. 13, *Corrosion*, 9th ed., ASM International, Metals Park, OH, p. 1265, 1987.
17. L. Engel and H. Klingele, *Atlas of Metal Damage*, Prentice Hall, Englewood Cliffs, NJ, pp. 125–9, 1981.

18. G. Kobrin, *Metals Handbook*, Vol. 13, *Corrosion,* 9th ed., ASM International, Metals Park, OH, p. 321, 1987.

19. R. Chiba, K. Omnishi, K. Ishii, and K. Maeda, *Corrosion*, Vol. 41, p. 415, 1985.

20. N. W. Polan, *Metals Handbook*, Vol. 13, *Corrosion*, 9th ed., ASM International, Metals Park, OH, p. 610, 1986.

21. H. H. Uhlig and R. W. Revie, *Corrosion and Corrosion Control*, 3rd ed., Wiley, New York, p. 328, 1985.

22. D. A. Jones, Corrosion, Vol. 37, p. 563, 1981.

23. D. A. Jones, *Corrosion Processes*, R. N. Parkins, ed., Applied Science, p. 177, 1982.

24. N. S. Hirota, *Metals Handbook*, Vol. 13, *Corrosion*, 9th ed., ASM International, Metals Park, OH, p. 964, 1986.

25. S. K. Brubaker, *Process Industries Corrosion*, B. J. Moniz and W. J. Pollock, eds., NACE, Houston, p. 243, 1986.

26. M. G. Fontana, *Corrosion Engineering*, 3rd ed., McGraw-Hill, New York, p. 92, 1986.

27. G. A. Minick and D. L. Olson, *Metals Handbook*, Vol. 13, *Corrosion*, 9th ed., ASM International, Metals Park, OH, p. 1293, 1987.

28. L. Engel and H. Klingele, *Atlas of Metal Damage*, Prentice-Hall, Englewood Cliffs, NJ, p. 180, 1981.

29. W. Glaeser and I. G. Wright, *Metals Handbook*, Vol. 13, *Corrosion*, 9th ed., ASM International, Metals Park, OH, p. 136, 1986.

30. L. J. Korb, *Metals Handbook*, 9th ed., Vol. 13, *Corrosion*, ASM International, Metals Park, OH, p. 1058, 1986.

31. F. G. Hammitt and F. J. Heymann, *Metals Handbook*, Vol. 11, *Failure Analysis*, 9th ed., ASM International, Metals Park, OH, p. 163, 1986.

32. D. A. Jones, *J. Met.*, Vol. 37, p. 20, June 1985.

33. R. J. Brigham, *Can. Metall. Q.*, Vol. 15, p. 389, 1976.

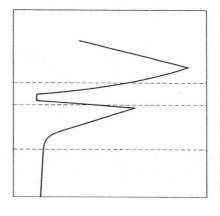

11

Corrosion in Selected Corrosive Environments

In this chapter, the corrosive properties of the most common and significant natural and industrial environments are reviewed, and the parameters most affecting corrosion are discussed. Some, but not all, of these environments have been mentioned in previous chapters. The purpose of this chapter is to provide guidelines for materials selection in these environments. The corrosion of wrought carbon steel is the common example in this chapter due to its widespread use in numerous weakly to moderately corrosive applications. However, other alloys are mentioned when appropriate, especially in later sections dealing with more aggressive environments.

11.1 Water and Aqueous Solutions

11.1.1 Effect of pH

The effect of pH on the corrosion of iron in aerated water is shown in Figure 11.1.[1] The anodic reaction is

$$Fe \rightarrow Fe^{2+} + 2e^-$$
(1)

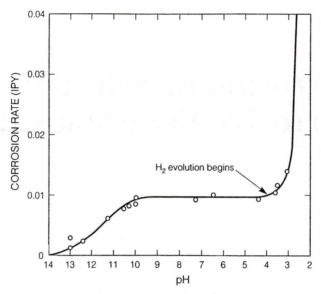

FIGURE 11.1   Effect of pH on corrosion of iron, using HCl and NaOH to control pH in water, containing dissolved oxygen. *(From W. Whitman, R. Russell, and V. Altieri, Ind. Eng. Chem., Vol. 16, p. 665, 1924. Reprinted by permission, The American Chemical Society.)*

at all pH values, but the corrosion rate varies due to changes in the cathodic reduction reaction. In the intermediate pH 4 through 10 range, a loose, porous, ferrous-oxide deposit shelters the surface and maintains the pH at about 9.5 beneath the deposit.[2] The corrosion rate is nearly constant and is determined by uniform diffusion of dissolved oxygen through the deposit in this intermediate range of pH. At the metal surface under the deposit, oxygen is reduced cathodically by

$$O_2 + 2H_2O + 4e^- \rightarrow 4OH^-. \tag{2}$$

In more acidic solutions below pH 4, the oxide is soluble and corrosion increases, due to availability of H^+ for reduction by

$$2H^+ + 2e^- \rightarrow H_2. \tag{3}$$

The absence of the surface deposit also enhances access of dissolved oxygen, which, if present, further increases corrosion rate. Dissolved oxygen is cathodically reduced in acid according to

$$O_2 + 4H^+ + 4e^- \rightarrow 2H_2O, \tag{4}$$

which is the equivalent of (2) in acid solutions (Section 2.2.2). Reactions (3) and (4) occur simultaneously in acid solutions with dissolved oxygen.

At pH above 10, corrosion rate is low, due to formation of the passive ferric oxide film in the presence of dissolved oxygen. At pH above about 14 without dissolved oxygen, corrosion rate may increase when the soluble ferrite ion, $HFeO_2^-$, forms (Figure 2.10).

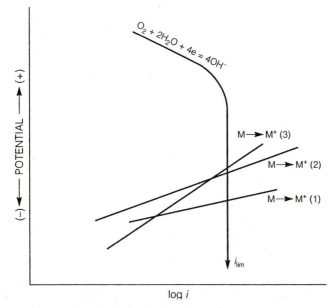

FIGURE 11.2 Schematic polarization diagram showing that metallurgical effects on the anodic dissolution process have no effect on the corrosion of carbon steel or iron, controlled by diffusion of dissolved oxygen at i_L.

Diffusion of dissolved oxygen controls the corrosion rate at a constant level in the pH range 4 through 10. Thus, metallurgical variables affecting the anodic reaction (1) have no effect on the corrosion rate, as shown in Figure 11.2. Such is not the case for acid pH < 4, where the cathodic reaction (3) is under activation control. The carbide phase shows low overvoltage (higher rate) for reduction of H^+. Thus, high-carbon steels have a higher corrosion rate in acid solutions than that of the low-carbon steels.[3]

For aluminum, the corrosion rate is low at intermediate pH between about 3 and 10, where the oxide film is stable, as shown in Figure 2.9, and predicted by the Pourbaix diagram. Like aluminum, many of the common constructional metals have increased corrosion in low pH solutions (Figure 2.11) due to the higher availability of oxidizing H^+ and solubility of oxides. Many are also independent of pH in nearly neutral solutions, where the oxides are stable. Other common constructional metals, such as copper, iron, and nickel, also show increased corrosion at very alkaline pH values, but usually to a lesser degree than aluminum.

11.1.2 Dissolved Oxygen and Other Dissolved Gases

Appreciable corrosion of iron or steel at near ambient temperature requires *dissolved oxygen* in neutral and alkaline solutions, as shown in Figure 11.3.[4] A protective magnetite film is stable in the absence of dissolved oxygen. Any factors affecting dissolved oxygen thus affect the corrosion of steel proportionally. Solution agitation or stirring increases transport of dissolved oxygen and increases corrosion rate (Figure 3.13). An increasing temperature initially increases corrosion rates, as expected, but

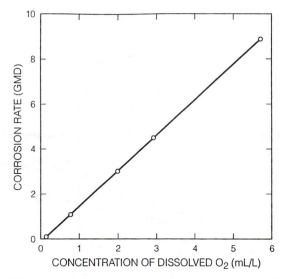

FIGURE 11.3 Effect of dissolved oxygen on the corrosion of iron in water containing 165 ppm CaCl₂. *(From H. H. Uhlig, D. N. Triadis, and M. Stern,* J. Electrochem. Soc., *Vol 102, p. 59, 1955. Reprinted by permission, The Electrochemical Society.)*

decreasing solubility of dissolved oxygen reduces corrosion rate above about 80°C (Section 11.1.4). Dissolved solutes also decrease the solubility of dissolved oxygen and the consequent corrosion rate.

Differences in dissolved oxygen transport readily create differential aeration cells (Section 6.5), which result in localized corrosion on iron or steel surfaces at ambient temperature. Dissolved oxygen often has variable access to different points on a larger surface. Those of higher access become cathodes at the expense of nearby anodes which corrode preferentially. Lower pH develops under oxide rust deposits at the anodes, which are surrounded by high-pH cathodes, resulting from reduction of dissolved oxygen by reaction (2). Dissolved oxygen is also critical in corrosion in pressurized cooling waters, and its control in boilers and steam generators is discussed in Section 11.1.4.

Chlorine may be dissolved intentionally in water as a biocide. It has little effect on corrosion of steel if pH is maintained above 7 to suppress formation of acid hydrolysis products by

$$Cl_2 + H_2O \rightarrow HClO + HCl.$$

However, chlorine will attack copper alloys, even at higher pH, apparently by reaction with the Cu_2O surface film.[5]

Ammonia results from intentional additions of NH_4OH as an acid neutralizer (Section 11.1.4), decay of intrinsic organic humic matter, or thermal degradation of nitrogen-bearing inhibitors and conditioners. It has little effect on iron and steel but has a strong effect on copper alloys. Ammonia forms complexes with copper that can cause rapid general corrosion and/or stress corrosion cracking (Section 8.4.4) of copper alloys.

Carbon dioxide, dissolved from air and from chemical processes, has a myriad of effects. Carbon dioxide from ambient air has a considerable effect on pH and forms insoluble scales on the surface, as discussed in the following section. Carbon dioxide is dissolved in some chemical process waters and is injected into waters associated with geologic petroleum reservoirs, where it enhances recovery by decreasing the viscosity of crude oil. Carbonic acid, formed from dissolved CO_2, is mildly corrosive, but corrosion product $FeCO_3$ surface films are normally protective. Erosion-corrosion (Section 10.2.1) may occur if turbulent flow disrupts the surface film. Scaling problems may result when the acid dissolves geologic calcium carbonate in wells,[6] which redeposits later on pipe walls. Dissolved carbon dioxide at elevated temperature must also be controlled to prevent excessive corrosion in boiler waters (Section 11.1.4).

11.1.3 Hardness

Hard waters contain dissolved calcium and magnesium cations and are less corrosive because a protective carbonate (carbonaceous) film is formed on the surface. Carbon dioxide dissolved in water forms carbonic acid, H_2CO_3, and reduces pH by dissociation to H^+ and the bicarbonate ion, HCO_3^-:

$$CO_2 + H_2O \rightarrow H_2CO_3 \rightarrow H^+ + HCO_3^-. \tag{5}$$

Equilibrium in (5) favors concentration of the bicarbonate as pH is increased. The bicarbonate ion forms insoluble calcium carbonate, $CaCO_3$, surface films in alkaline solutions by

$$Ca^{2+} + 2HCO_3^- \rightarrow Ca(HCO_3)_2 \rightarrow CaCO_3 + CO_2 + H_2O.$$

Similar reactions in seawater form calcareous scales in the alkaline solutions resulting from cathodic protection (Section 13.1.5). Such scales reduce the corrosion of steel surfaces by providing a diffusion barrier that decreases the rate of dissolved oxygen reduction (Figure 13.8).

The tendency for calcium carbonate to precipitate and provide corrosion resistance in fresh water is measured by the saturation index (SI),[7] which is defined by

$$SI = pH - pH_s,$$

where pH is experimentally measured, and pH_s is the pH value at which the water is in equilibrium with solid $CaCO_3$. A positive SI indicates that the water has sufficient alkalinity that $CaCO_3$ will precipitate and reduce corrosion. pH_s is given by

where
$$
\begin{aligned}
pH_s &= (pK_2' - pK_s') + pCa + pAlk \\
pK_2' &= -\log_{10}(H^+)(CO_3^{2-})/(HCO_3^-) \\
pK_s' &= -\log_{10}(Ca^{2+})(CO_3^{2-}) \\
pCa &= -\log_{10}(Ca^{2+}) \\
pAlk &= -\log_{10}(\text{total alkalinity}).
\end{aligned}
$$

The total alkalinity is the concentration (moles/liter) of acid, $(H^+)_{tit}$, needed to neutralize dissolved alkaline ions by titration according to

$$\text{total alkalinity} = (H^+)_{tit}$$
$$= 2(CO_3^{2-}) + (HCO_3^-) + (OH^-).$$

The SI in a domestic water system is maintained at a positive value to minimize corrosion. This can be done by adding lime, $Ca(OH)_2$, or soda ash, Na_2CO_3, to decrease pCa and pAlk. The SI is essentially a thermodynamic quantity and indicates only tendencies and not rates. Thus, SI will not account for unpredictable supersaturation of $CaCO_3$ or the time required to precipitate a protective film. Other dissolved ions may codeposit and alter the capacity and rate of $CaCO_3$ deposition as well.

Negative SI may be desirable in condensers and other heat exchangers where scale deposition retards heat transfer. Often, scale inhibitors such as polyphosphates and organic polymers are added which retard deposition or act as colloidal dispersants to control deposition in water cooling systems.[5]

11.1.4 Elevated Temperature

For steel, heating above room temperature initially increases corrosion rate, as expected, but also reduces solubility of dissolved oxygen. Thus, above 80°C, corrosion rate decreases in an open system, which allows oxygen to escape (Figure 11.4[8]). However, in a closed system that retains the dissolved oxygen, the corrosion increases, even above 80°C, as would be expected for most chemical reactions. Stringent control of dissolved oxygen to absolute minimum levels is an obvious requirement to control corrosion in closed high-temperature boiler systems.

Steam condensate in contact with air absorbs oxygen and heavily corrodes steel vessels and piping by formation of highly aggressive differential aeration cells (Section 6.5.1) above about 50°C. Corrosion takes the form of irregular, deep pitting, as shown in Figure 11.5.[9] The bottoms of active pits contain black magnetite,

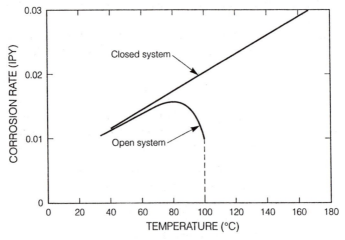

FIGURE 11.4 Effect of temperature on corrosion of iron in water containing dissolved oxygen. *(From F. N. Speller,* Corrosion, Causes and Prevention, *3rd ed., McGraw-Hill, p. 168, 1951. Reprinted by permission, McGraw-Hill Book Company.)*

FIGURE 11.5   Pitting corrosion of carbon steel by steam condensate with oxygen contamination. *(From J. J. Maguire, ed., Handbook of Industrial Water Conditioning, 8th ed., Betz Laboratories, Inc. Trevose, PA, pp. 145–146, 1980. Reprinted by permission, Betz Laboratories Inc.)*

Fe_3O_4, with iron partially in the reduced Fe^{2+} state, while the surrounding surfaces have the characteristic red rust color of Fe_2O_3 with iron in the oxidized Fe^{3+} state. In the absence of dissolved oxygen, the magnetite surface film is protective; corrosion rates are uniform and low.

Dissolved oxygen may be removed mechanically or chemically. Mechanical deaeration involves either heating to reduce solubility and drive off oxygen near the boiling point of the water, or purging with a countercurrent flow of gas to strip the oxygen from the water. Both heating and stripping may be accomplished with a countercurrent flow of steam,[10] and the heated water is useful as feed for boilers. Gas stripping is suitable if the water must be cold and evacuation will hasten the process.[11]

Chemical scavenging is usually necessary to further reduce dissolved oxygen to levels acceptable for many applications. Sodium sulfite and hydrazine react with dissolved oxygen:

$$Na_2SO_3 + \tfrac{1}{2}O_2 \rightarrow Na_2SO_4 \tag{6}$$
$$N_2H_4 + O_2 \rightarrow N_2 + 2H_2O. \tag{7}$$

Sulfite in 1 to 2 ppm stoichiometric excess of the dissolved oxygen (about 8 ppm sulfite/ppm O_2 with 0.1 to 0.2 ppm $CoCl_2$ as a catalyst) will deaerate water completely in a few minutes by reaction (6).[11] However, sulfite is unstable at higher temperature, and the accompanying alkali metal cation Na^+ results in harmful concentration of caustic NaOH on boiling surfaces in boilers and steam generators. Sulfur-containing degradation products at high temperature may also be harmful. Hydrazine, on the other hand, creates only volatile products, that do not accumulate in the system. Hydrazine is ineffective for low-temperature deaeration, because reaction (7) proceeds at an adequate rate only in boiler waters above about 300°C.

Hydrogen added to the pressurized water of boiling-water nuclear reactors combines with dissolved oxygen and reduces electrochemical potentials to prevent intergranular stress corrosion cracking in the heat-affected zones of weldments.[12]

Carbon dioxide that dissolves according to

$$CO_2 + H_2O \rightarrow H_2CO_3 \rightarrow H^+ + HCO_3^- \tag{5}$$

also increases corrosion in steam-condensate systems by lowering the pH. Various amines of the form $R—NH_2$ are routinely added to steam and condensate waters to neutralize acidity caused by carbon dioxide. They hydrolyze to OH^- by

$$R—NH_2 + H_2O \rightarrow R—NH_3^+ + OH^-,$$

which neutralizes the acidity produced by reaction (5). The effectiveness of a particular amine, incorporating a specific radical R, depends on neutralization capacity (basicity), stability at temperature, and distribution ratio. The distribution ratio, which measures volatility of the amine, is the ratio of the amine content of the steam and liquid phases at temperature. The distribution ratios for some common amines are as follows:[9]

morpholine	0.4
diethylaminoethanol	1.7
dimethyllisopropanolamine	1.7
ammonia	10.0

A low distribution ratio is desirable to prevent losses from steam venting, but a high ratio is necessary to carry the amine from the feedwater through the steam phase to the condensate.

The first attempt to control corrosion in boilers was by NaOH additions to adjust pH to alkaline conditions, where ferrous alloys are normally passive at ambient and near ambient temperatures. However, boiling concentrates even the most dilute solutes on heat transfer surfaces, and concentrated NaOH causes much higher corrosion than expected at lower temperatures. Figure 11.6[10,13] shows corrosion of carbon steel at a boiler temperature of 310°C as a function of HCl and NaOH additions to control room temperature pH. Phosphate additions such as Na_3PO_4 have been added to control pH by buffering. However, in crevices, the concentration may exceed the buffering capacity of the compound, and excessive alkaline concentration can result in general corrosion and stress corrosion cracking of steel.

Boiling at steel surfaces concentrates chloride and reduces pH, when ferrous-chloride corrosion products hydrolyze (Section 7.3.2). A recent example is accelerated corrosion in the crevice between nickel Alloy 600 tubing and carbon-steel support plates. Boiling in this restricted volume results in chloride concentration, exces-

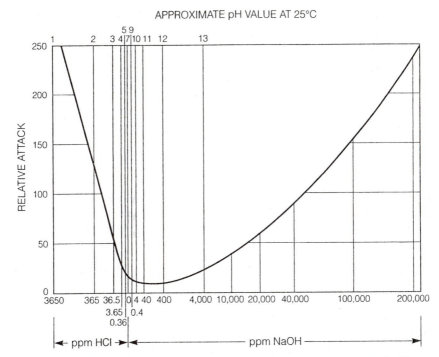

FIGURE 11.6   Effect of room-temperature pH on steel corrosion at 310°C. *(From H. H. Uhlig and R. W. Revie,* Corrosion and Corrosion Control, *3rd ed., Wiley, p. 278–290, 1985. Reprinted by permission, John Wiley & Sons.)*

sive corrosion, and deposition of insoluble expanded oxide corrosion products, which locally collapse or "dent" the tubing at the crevice.

A major source of undesirable impurities in boiler water is contamination by cooling water through condenser leakage. Seawater and brackish water supply chlorides, which concentrate to acid solutions. Many inland waters (e.g., from the Great Lakes) are high in alkali cations, which concentrate to corrosive caustic solutions by boiling.

11.1.5 Sodium Chloride and Seawater

The salts discussed in this section are those that do not appreciably alter the pH when dissolved in water. The major example is sodium chloride, NaCl, which is plentiful in seawater, brackish waters, many chemical process waters, and mammalian body fluids. The schematic effect of NaCl concentration on corrosion rate of iron in aerated room-temperature solutions is shown in Figure 11.7. The initial increase in corrosion rate is due to enhanced solution conductivity. Low conductivity allows only closely spaced anodes and cathodes, and anodic reaction products tend to limit cathodic oxygen reduction. Higher conductivity permits lower polarization with higher corrosion currents between adjoining anodes and cathodes. However, still higher dissolved salt decreases the solubility of dissolved oxygen, and corrosion rate steadily decreases beyond the maximum, at about 3% NaCl. Other alkali metal salts,

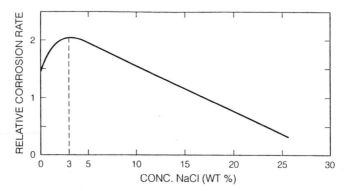

FIGURE 11.7 Effect of NaCl concentration on corrosion of iron in aerated solutions. *(From H. H. Uhlig and R. W. Revie,* Corrosion and Corrosion Control, *3rd ed., Wiley, pp. 96–111, 1985. Reprinted by permission, John Wiley & Sons.).*

such as KCl, LiCl, Na_2SO_4, KI, NaBr, and so on, produce approximately the same effects as NaCl.

An NaCl solution of about 3.5% is sometimes used to simulate seawater in the laboratory. However, such solutions are often more aggressive than natural seawater, especially toward carbon steel.[14] This is probably due at least in part to significant concentrations of Ca^{2+} and Mg^{2+}. Cathodic reduction of dissolved oxygen by (2) produces slightly alkaline surface conditions which precipitate $CaCO_3$ and $Mg(OH)_2$ by (Section 13.1.5)

$$Ca^{2+} + HCO_3^- + OH^- \rightarrow H_2O + CaCO_3,$$
$$Mg^{2+} + 2OH^- \rightarrow Mg(OH)_2.$$

These precipitates in turn inhibit further cathodic reduction and corrosion. However, variable temperature, dissolved oxygen content, currents, biological organisms, and pollutants can increase corrosion rates in coastal and ocean seawaters. Biological activity in seawater also affects its corrosivity by producing crevice conditions, acids, and sulfides, as discussed further in Section 11.3.6. Development of a thick, uniform biofouling surface layer in seawater is sometimes beneficial for stationary equipment, limiting access by dissolved oxygen.

Thus, seawater is a complex chemical system affected by concentration and access of dissolved oxygen, salinity, concentration of minor ions, biological activity, and pollutants.[15] These are in turn affected by temperature, depth, and ocean currents. Samples of seawater upon standing are known to change in corrosivity from the bulk seawater mass from which they were taken.[15] A synthetic seawater made up from laboratory chemicals is available as a standard,[16] but corrosivity is not likely to match any particular seawater site quantitatively.

Figure 11.8[17] shows the corrosion rate of carbon steel as a function of distance from the seawater surface. Corrosion is maximized in the splash zone, where dissolved oxygen has easy access and chloride may be somewhat concentrated by drying of moist films formed by spray and condensation. Corrosion is often much lower below the mudline, where dissolved oxygen has least access. However, sulfate-

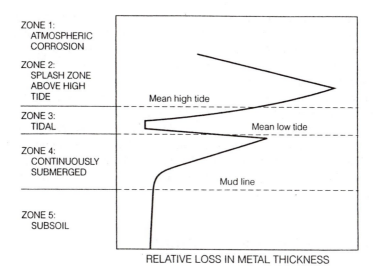

ZONE 1: ATMOSPHERIC CORROSION

ZONE 2: SPLASH ZONE ABOVE HIGH TIDE

ZONE 3: TIDAL

ZONE 4: CONTINUOUSLY SUBMERGED

ZONE 5: SUBSOIL

Mean high tide

Mean low tide

Mud line

RELATIVE LOSS IN METAL THICKNESS

FIGURE 11.8   Effect of seawater depth on the corrosion of steel *(From F. L. LaQue, Marine Corrosion, Wiley-Interscience, 1975. Reprinted by permission, John Wiley & Sons.)*

reducing bacteria (Section 11.3.2), which are often present in coastal muds, can increase corrosion rates.

11.1.6 Other Dissolved Salts

Certain salts, including $AlCl_3$, $NiSO_4$, $MnCl_2$, and $FeCl_2$, hydrolize to form acid products, when dissolved in water. A typical hydrolysis reaction is

$$AlCl_3 + 3H_2O \rightarrow Al(OH)_3 + 3HCl.$$

The HCl is a strong, highly dissociated acid, while the hydroxide is an undissociated weak base. Thus, the corrosion of most metals increases in correspondence to the acid produced. Others, such as $FeCl_3$, $CuCl_2$, and $HgCl_2$, produce not only acid but also a redox couple (e.g., $Fe^{3+} + e^- \rightarrow Fe^{+2}$), which further accelerates corrosion by providing additional oxidizing power. These salts are often the corrosion products which account for localized crevice and pitting corrosion within small occluded cells (Sections 7.2 and 7.3).

Ammonium salts (e.g., NH_4Cl) are also acidic, but the corrosion of both steel and copper alloys are much higher than can be accounted for by the acid produced. NH_4^+ forms ammonia complexes with both iron and copper, which effectively reduce the activity of corrosion product Fe^{2+} and Cu^{2+} and increase the anodic dissolution rate by reducing the anodic polarization.

Nitrate acts as an oxidizer in acid solutions through the reaction (Table 2.1)

$$NO_3^- + 4H^+ + 3e^- \rightarrow NO + 2H_2O.$$

Therefore, nitric acid is known as an oxidizing acid and is more corrosive than the other strong (reducing) acids, hydrochloric and sulfuric. In sufficient concentration, nitric acid will have enough oxidizing power (sufficiently noble half-cell potential) to passivate pure iron (Section 4.1.1). Few redox reactions are sufficiently oxidizing to accomplish this feat. Ammonium nitrate (NH_4NO_3), present in many fertilizer solutions, hydrolyzes to form dilute nitric acid and is consequently more corrosive than either the chloride or sulfate salts of ammonium.

Salts that hydrolyze to alkaline products may act as inhibitors for steel. These include trisodium phosphate, Na_3PO_4; sodium tetraborate, $Na_2B_2O_7$; sodium silicate, Na_2SiO_3; and sodium carbonate, Na_2CO_3. All form a strong base, NaOH, and a weak acid with the anion, when reacting with water. In addition to passivating steel in aerated solutions in the manner of NaOH, they can also form insoluble ferrous and ferric surface films (e.g., phosphates and silicates), which are more effective diffusion barriers than the usual oxides.[2]

11.2 Sulfur-Bearing Solutions

Sulfur can exist in oxidation states from −2 to +6 in aqueous electrolytes and is widely dispersed in mineral, organic, and biologic materials. Therefore, many chemical forms of sulfur are very important in corrosion, because of sulfur's multidimensional chemical reactivity. Furthermore, sulfur is a frequent product of biological decay and is often found in waters from domestic and industrial wells, geothermal energy recovery systems, and petroleum production and refining plants.

Table 11.1 lists the common anionic sulfur species, together with their structures and oxidation numbers. Sulfate, elemental sulfur, sulfide S^{2-}, hydrosulfide HS^-, bisulfate HSO_4^-, and hydrogen sulfide, H_2S, are stable and appear on the Pourbaix diagram for sulfur and water (Figure 11.9[18]). Reduced sulfur is metastable in oxidation states between +2 and +5 but is the cause of some unique and very damaging corrosive effects.

The sulfate ion, SO_4^{2-}, is relatively unreactive, with sulfur in its fully oxidized state at +6, and behaves much like many other anions in alkali metal salts (Section 11.1.5). The industrial importance of sulfuric acid merits separate treatment in Section 11.6.1. Corrosivity of dissolved sulfur in reduced states of oxidation is discussed in the following sections.

11.2.1 Sulfides and Hydrogen Sulfide

Hydrogen sulfide, H_2S, is one of the most aggressive of the common industrial aqueous corrosive solutes. Although not classified as a strong acid, H_2S dissociates sufficiently to yield a minimum pH of about 4 in saturated solutions. The sulfide anion further acts as a "poison," which retards recombination of nascent H-atoms on corroding surfaces, increases the residence time of nascent H, and enhances hydrogen penetration into the metal lattice. Thus, all modes of hydrogen damage (Section 10.1), including hydrogen induced cracking, hydrogen blistering, hydrogen attack, and hydride formation are accelerated in susceptible alloys by the presence of H_2S.

TABLE 11.1   Aqueous Sulfur Species

Formula	Name	Structure	Sulfur Oxidation Number
H_2S or S^{2-}	Sulfide		-2
H_2S_2 or S_2^{2-}	Polysulfides	$[S-S]^{2-}$	-1
H_2S_3 or S_3^{2-}		$[S-S-S]^{2-}$	-2/3
H_2S_n		$[S-S \cdots]^{2-}$	-2/n
S	Sulfur	S_8 rings	0
$S_2O_3^{2-}$	Thiosulfate	$\begin{bmatrix} & O & \\ & \| & \\ O- & S & -S \\ & \| & \\ & O & \end{bmatrix}^{2-}$	+2
$S_4O_6^{2-}$	Tetrathionate	$\begin{bmatrix} O & & & O \\ \| & & & \| \\ O-S & -S-S- & S-O \\ \| & & & \| \\ O & & & O \end{bmatrix}^{2-}$	+2.5
SO_3^{2-} (H_2SO_3)	Sulfite (sulfurous acid)	$\begin{bmatrix} & O & \\ & \| & \\ O- & S & -O \end{bmatrix}^{2-}$	+4
SO_2	Sulfur dioxide	Gas	+4
$S_2O_6^{2-}$	Dithionate	$\begin{bmatrix} O & O \\ \| & \| \\ O-S & -S-O \\ \| & \| \\ O & O \end{bmatrix}^{2-}$	+5
SO_4^{2-}	Sulfate	$\begin{bmatrix} & O & \\ & \| & \\ O- & S & -O \\ & \| & \\ & O & \end{bmatrix}^{2-}$	+6

Furthermore, S^{2-} is reactive in itself, and H_2S solutions will corrode carbon-steel and copper alloys, forming soluble sulfide corrosion products and insoluble sulfide deposits.

In the petroleum refining industry, various process waters derive their sour taste from the dissolved weak acid, H_2S. Such solutions, also containing ammonia, cyanide, and various dissolved organics, are corrosive to steel and cause brittle cracking of stressed high-strength steels. Such brittle failures are sometimes referred to as hydrogen stress cracking or sulfide stress cracking. Both are misnomers because the failures do not involve the anodic mechanism of embrittlement required

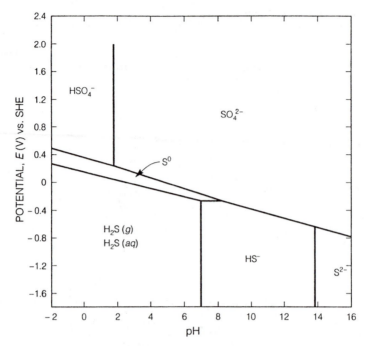

FIGURE 11.9 Pourbaix diagram for sulfur and water. Sulfur and all dissolved ions assumed to be at unit activity. *(Adapted from Pourbaix.[18])*

of usual stress corrosion cracking (Section 8.1.1). Hydrogen induced cracking is the term used for these failures in this book; mechanism and methods of prevention are discussed in sections 10.1.2 and 10.1.8, respectively.

Sour waters high in ammonia also form ammonium bisulfide solutions, which aggressively attack carbon-steel and admiralty metal tubing and other structural components in petroleum refineries.[20] H_2S solutions neutralized to pH 8 or above do not normally attack steel. However, cyanides in petroleum refinery waters complex the protective FeS film as soluble $Fe(CN)_6^{4-}$, and the unprotected steel corrodes rapidly.[19]

Copper alloys are subject to attack by seawater and brackish waters polluted with sulfides and hydrogen sulfide from the decay of organic matter in tidal marshes, bays, and estuaries. CuS films are less protective than the usual oxide films, and copper-alloy condensers are especially susceptible to erosion-corrosion in the presence of sulfides (Section 10.2.1). Figure 11.10 shows the combined effects of sulfide level and seawater velocity.[20] Sulfides in coastal waters are often generated by the action of sulfate-reducing bacteria in the decaying biomass of silts and muds. In surface slimes on iron and copper, bacterial action creates a local concentration of sulfide which accelerates corrosion. It is notable that copper corrosion products are toxic to large fouling organisms (Section 11.3.6), but certain strains of sulfate-reducing bacteria can be sufficiently copper resistant to grow on copper alloy surfaces (Section 11.3.1).

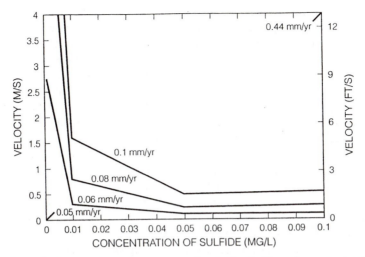

FIGURE 11.10 Effect of sulfide concentration and velocity on isocorrosion lines for Cu-Ni alloy C070600 in seawater. *(From N. W. Polan,* Metals Handbook, Vol 13, Corrosion, *9th ed., ASM, p. 625, 1987. Reprinted by permission, ASM International.)*

11.2.2 Sulfite Liquors

Solutions of SO_2 in calcium, sodium, magnesium, or ammonium hydroxide are used to convert wood chips to fiber pulp for papermaking. The resultant liquor contains a mixture of bisulfite (HSO_3^-) and sulfite (SO_3^{2-}), together with dissolved SO_2. Sulfite and bisulfite are generally noncorrosive to the austenitic stainless steels used in the sulfite pulping process. SO_2 acts as an oxidizer which helps maintain passivity of stainless steel, and insufficient SO_2 may thus result in depassivation. However, SO_2 also contributes to corrosivity by aiding in partial decomposition of the liquor to sulfuric acid,[21] producing a special case of sulfuric acid corrosion (Section 11.6.1).

The most unique aspect of corrosion in sulfite pulping liquors is thiosulfate pitting. Thiosulfate ($S_2O_3^{2-}$) is formed spontaneously in sulfite liquors. During pitting, it is reduced to a sulfur monolayer on the surface in the narrow potential range (Figure 11.9) of elemental sulfur stability.[21] Surface-adsorbed sulfur apparently accelerates the local anodic dissolution of the metal, producing an acidified pit.

11.2.3 Polythionic Acids

Polythionic acids of the form $H_2S_nO_6$ are formed when oxygen and water react with iron-chromium sulfides. The sulfides form on the stainless-steel surfaces of furnaces, heat exchangers, and vessels exposed to high-temperature sulfide gases in petroleum refineries.[19] During shutdowns, the surface sulfides react with oxygen and water or water vapor, forming polythionic acids, which cause rapid intergranular stress corrosion cracking (SCC) (Section 8.4.1) near welds, where a microstructure sensitized to intergranular corrosion (Section 9.1.2) is present. SCC near a Type 304 stainless steel weld exposed to polythionic acids appears in Figure 9.21.

The exact form(s) and oxidation state(s) of dissolved sulfur responsible for poly-thionic acid SCC are uncertain. Sulfurous acid (H_2SO_3) is included as a possible agent for SCC in refinery systems.[19] Furthermore, thiosulfate ($S_2O_3^{2-}$) at ppm levels is known to cause SCC of sensitized austenitic stainless steel[22] in water at ambient and slightly elevated temperatures. In a similar manner, both thiosulfate and tetrathionate ($S_4O_6^{2-}$) have produced SCC of sensitized Alloy 600.[23] SCC occurs in a narrow range of potential, where active dissolution of Ni and stability of elemen-tal sulfur coincide. Therefore, dissolved metastable reduced sulfur species apparent-ly are cathodically converted to adsorbed sulfur on the surface during SCC.

SCC by polythionic acid or reduced sulfur species can occur in unexpected cir-cumstances. Following the well-known 1981 nuclear reactor failure at Three Mile Island in Pennsylvania, steam generators not directly involved in the accident were repressurized after a year-long shutdown. Extensive leaking from the primary (reac-tor) side to the secondary (steam) side of the generators aborted the startup.[24] Extensive subsequent investigation revealed[25] intergranular cracking in the Alloy 600 heat exchanger tubing, which had been sensitized by stress-relief annealing dur-ing manufacture. Elemental sulfur was detected on the crack surfaces and on the pri-mary side tube surfaces. Levels of 0.7 ppm thiosulfate were measured in the prima-ry water, probably due to accidental leakage of water containing thiosulfate from an auxiliary safety cooling system. It was concluded that the failures were due to sul-fur-induced SCC.[25]

11.3 Biologically Influenced Corrosion

Biological organisms have long been known to contribute to corrosion,[26] but wide-spread recognition has been delayed until recently. Most of this section is devoted to the subtle effects of bacteria present in groundwater, seawater, and domestic and industrial fresh waters. The effects of larger macrofouling organisms in seawater are discussed in Section 11.3.6.

11.3.1 Recognition and Prevention

Microbiologically influenced corrosion (MIC) is a special danger when nearly neu-tral water, pH 4 to 9, 10° to 50°C, is in constant contact, especially stagnant, with carbon steel, stainless steel, and alloys of aluminum and copper. Specific strains of bacteria can widen these limits still further. Industrial instances where MIC has been recognized are summarized in Table 11.2.[27] Other industries may be affected unknowingly as well, because the accelerating effects of biological organisms on corrosion are not widely appreciated or easily identified. Table 11.3[27] summarizes the most significant microorganisms recognized to affect corrosion.

The first sign of MIC is often unexpected, severe corrosion of any of the metals noted above in nearly neutral ambient temperature water or dilute solutions where

TABLE 11.2   Industries Affected by Microbiologically Influenced Corrosion

Industry	Problem areas
Chemical-processing industries	Stainless steel tanks, pipelines, and flanged joints, particularly in welded areas after hydrotesting with natural river or well waters
Nuclear power generation	Carbon and stainless steel piping and tanks; copper-nickel, stainless, brass and aluminum-bronze cooling water pipes and tubes, especially during construction, hydrotest, and outage periods
Onshore and offshore oil and gas industries.	Mothballed and waterflood systems: oil and gas handling systems, particularly in those environments soured by sulfate reducing bacteria-produced sulfides
Underground pipeline industry.	Water-saturated clay-type soils of near-neutral pH with decaying organic matter and a source of SRB
Water treatment industry	Heat exchangers and piping
Sewage handling and treatment industry	Concrete and reinforced concrete structures
Highway maintenance industry	Culvert piping
Aviation industry.	Aluminum integral wing tanks and fuel storage tanks
Metalworking industry	Increased wear from breakdown of machining oils and emulsions

From S. C. Dexter, Metals Handbook, Vol. 13, Corrosion, 9th ed., ASM International, p. 118, 1987. Reprinted by permission, ASM International.

corrosion rates are normally low. Excessive deposits or tubercles are characteristic. Breaking the deposits often reveals biologic slimes with black magnetite and iron sulfide deposits having the characteristic rotten-egg odor of H_2S. Treatment of the deposits with dilute hydrochloric acid will also generate the H_2S smell. Pit surfaces beneath the deposits are usually bright but rust rapidly on exposure to ambient air.

Soils capable of producing MIC are characteristically waterlogged and contain a large fraction of clay, which inhibits drainage. Soft, black, slime deposits on the metal surface and in the surrounding soil are usually present along with the H_2S odor. In buried gray cast-iron structures, graphitic corrosion (Section 9.5.1), in which iron is leached out of the metallurgical graphite network, is a frequent result of MIC in soil.[26]

Prevention of MIC requires frequent mechanical surface cleaning and treatment with biocides to control populations of bacteria. Biocide treatments without cleaning may not be effective because organisms sheltered beneath deposits may not be reached by the injected chemicals. During storage and layup or after hydrotesting,

TABLE 11.3 Bacteria Known to Cause Microbiologically Influenced Corrosion

Genus or Species	pH Range	Temperature Range °C	Oxygen Requirement	Metals Affected	Action
Bacteria					
Desulfovibrio					
Best known: *D. desulfuricans*	4–8	10–40	Anaerobic	Iron and steel, stainless steels, aluminum zinc, copper alloys	Utilize hydrogen in reducing SO_4^{2-} to S^{2-} and H_2S: promote formation of sulfide films
Desulfotomaculum					
Best known: *D. nigrificans* (also known as *Clostridium*)	6–8	10–40 (some 45–75)	Anaerobic	Iron and steel, stainless steels	Reduce SO_4^{2-} to S^{2-} and H_2S (spore formers)
Desulfomonas	...	10–40	Anaerobic	Iron and steel	Reduce SO_4^{2-} to S^{2-} and H_2S
Thiobacillus thiooxidans	0.5–8	10–40	Aerobic	Iron and steel, copper alloys, concrete	Oxidizes sulfur and sulfides to form H_2SO_4; damages protective coatings
Thiobacillus ferrooxidans	1–7	10–40	Aerobic	Iron and steel	Oxidizes ferrous (Fe^{2+}) to ferric (Fe^{3+})

TABLE 11.3 (Cont.)

Genus or Species	pH Range	Temperature Range °C	Oxygen Requirement	Metals Affected	Action
Gallionella	7–10	20–40	Aerobic	Iron and steel, stainless steels	Oxidizes ferrous (and manganous) to ferric (and manganic); promotes tubercule formation
Sphaerotilus	7–10	20–40	Aerobic	Iron and steel, stainless steels	Oxidizes ferrous (and manganous) to ferric (and manganic); promotes tubercule formation
S. natans	...	...	...	Aluminum alloys	
Pseudomonas	4–9	20–40	Aerobic	Iron and steel, stainless steels	Some strains can reduce Fe^{3+} to Fe^{2+}
P. aeruginosa	4–8	20–40	Aerobic	Aluminum alloys	
Fungi *Cladosporium resinae*	3–7	10–45 (best at 30–35)	...	Aluminum alloys	Produces organic acids in metabilizing certain fuel constituents

From S. C. Dexter, Metals Handbook, Vol. 13, Corrosion, 9th ed., ASM International, p. 114, 1987. Reprinted by permission, ASM International.

water should not be permitted to stand for periods of more than a few days. Complete drainage and, where possible, mop drying are recommended to avoid possible MIC. Copper alloys are more resistant because of natural toxicity of the corrosion products, but some strains of bacteria will attack copper alloys, especially those alloys containing iron and manganese.

Cathodic protection can prevent or arrest MIC.[28] However, MIC may persist where cathodic currents are low on surfaces remote from the anodes (Section 13.2.3) or under already established deposits. Thus, cathodic protection systems should be carefully designed and applied with added current capacity to new or freshly cleaned surfaces, when biological activity is known or suspected.

11.3.2 Effects of Anaerobic Bacteria on Iron

Iron and carbon steel usually have very low corrosion rates in deaerated neutral water and dilute salt solutions because the only available cathodic reduction reaction,

$$2H_2O + 2e^- \rightarrow H_2 + 2OH^-,$$

occurs only very slowly. However, saturated soils and deaerated cooling waters may support relatively high corrosion rates attributed to anaerobic bacteria, that do not require oxygen for growth. As indicated in Table 11.3, the responsible anaerobic bacteria reduce sulfate, forming corrosive hydrogen sulfide with water. These sulfate-reducing bacteria (SRB) are easily the most notorious and harmful of the microorganisms known to enhance corrosion. Although SRB require anaerobic conditions for growth, many strains can stay alive for long periods in aerobic conditions, ready for activation when conditions become favorable.

The mechanism by which anaerobic SRB accomplish an increase in corrosion of iron and carbon steel is uncertain, but a recent review[26] highlights some important facts. In the "classic" theory, the cathodic reduction reaction,

$$2H^+ + 2e^- \rightarrow 2H \rightarrow H_2, \tag{3}$$

is accelerated (depolarized) when nascent hydrogen, H, reacts with oxygen, O, made available from sulfate reduction,

$$SO_4^{2-} \rightarrow S^{2-} + 4O.$$

The hydrogenase enzyme, possessed by certain strains of SRB, has been shown to increase cathodic currents on steel electrodes, and corrosion rate must increase correspondingly. However, in other studies, hydrogenase-negative SRB were found equally aggressive. Furthermore, anaerobic, sulfate-rich conditions do not invariably result in severe corrosion. More recently, the corrosion product sulfides are thought to be important since both dissolved and solid sulfides are known to accelerate corrosion. However, FeS films are protective in many neutral sulfide process waters (Section 11.2.1). Hydrogenase may assist hydrogen removal on or from within the sulfide film. Cyclic anearobic, followed by aerobic, conditions have sometimes shown increased corrosive activity. SRB may maintain the activity of dissolved and deposited sulfides as agents for cathodic reduction of dissolved oxygen and other oxidizers.

11.3.3 Effects of Aerobic Bacteria on Iron

From the standpoint of corrosion, oxygen-consuming aerobic bacteria can serve one or more of several functions, such as (a) formation of slime, (b) oxidation of sulfide, (c) oxidation of iron, and (d) generation of acidic byproducts (metabolites). Slime is a polymer excreted by microorganisms to enhance the growth environment and adherence to surfaces. A massive sample of slime from a paper mill is shown in Figure 11.11.[29] Many organisms are slime formers, including filamentous fungi, algae, protozoa, and diatoms, as well as bacteria. Other aerobic bacteria (Table 11.3), which oxidize iron or sulfur, also produce slimes. Hydrated biofouling slimes shield the surface, create differential aeration cells, and eventually provide a sheltered environment for anaerobic bacteria.

Sulfide-oxidizing bacteria produce sulfuric acid, which is potentially corrosive. However, in the presence of sulfate, the more important function of the sulfide oxidizers may be to produce slimes that provide a localized anaerobic environment and nourish sulfate-reducing bacteria.[27]

Iron-oxidizing bacteria (Table 11.3) oxidize soluble ferrous ions to less soluble ferric, Fe^{3+}, ions. The lower Fe^{2+} activity increases the rate of the anodic reaction,

$$Fe \rightarrow Fe^{2+} + 2e^-, \tag{1}$$

and the iron oxidizers convert significant amounts of Fe^{2+} to Fe^{3+} at the anode sites. As a result, insoluble tubercles, consisting of hydrated ferric oxides and excreted biological slimes, grow on the surface.[26] Figure 11.12[30] shows a tubercle that grew

FIGURE 11.11   Biological slime accumulations in paper-mill process water. *(From J. J. Maguire, ed., Handbook of Industrial Water Conditioning, 8th ed., Betz Laboratories, Inc., Trevose, PA, p. 296, 1980. Reprinted by permission, Betz Laboratories, Inc.)*

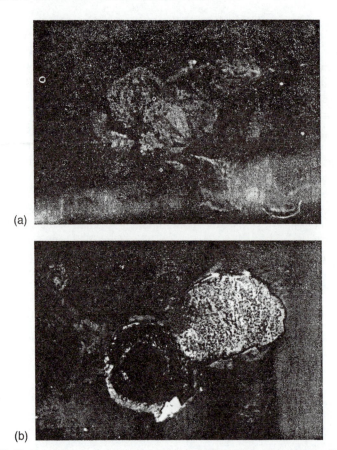

(a)

(b)

FIGURE 11.12   Tubercule (a) and resulting pit (b) in the wall of epoxy lined steel pipe carrying water. *(From M. Bibb,* Biologically Induced Corrosion, *S. C. Dexter, ed., NACE, Houston, p. 96, 1986. Reprinted by permission, National Association of Corrosion Engineers.)*

at a defect in the epoxy lining for steel water pipe. It is apparent that penetration was deep and rapid. Tubercles can grow to massive dimensions, enough even to block pipe flow in the example of Figure 11.13.[31]

The sheltered areas beneath deposits become the anode in a differential aeration cell, with oxygen being reduced at the surrounding metal and on the electrically conductive deposit surfaces by

$$O_2 + 2H_2O + 4e^- \rightarrow 4OH^-. \tag{2}$$

Increased surface concentration of OH^- promotes further precipitation of $Fe(OH)_3$ or possibly $Fe_2(CO_3)_3$. The tubercules serve as harbors for anaerobic SRB, which probably feed on the decaying biomass under the deposits and produce acidic H_2S. If present in the bulk water, chlorides may migrate into the anode area and hydrolize to further acidify the anode volume under deposits. Figure 11.14[27] summarizes the electrochemical and biological processes that are possible. Not all are likely to occur

FIGURE 11.13   Tubercules causing flow blockage in carbon steel pipe. *(From W. P. Iverson, ASTM STP 741, p. 40, 1981. Photograph by courtesy of E. Escalante and W. P. Iverson, National Institute of Standards and Technology. Reprinted by permission, American Society for Testing and Materials.)*

in any given situation. The similarity to underdeposit pitting described in Section 7.5 (Figure 7.29) is notable, and the major role of aerobic bacteria indeed may be to enhance this mechanism for iron and steel.

11.3.4 Biological Corrosion of Stainless Steels

Slimes formed by bacteria can create sites for initiation of pitting corrosion of stainless steels in the presence of seawater or fresh water. There is evidence to indicate that aerobically produced biofilms catalyze the cathodic reduction of dissolved oxygen,[32,33] raising the corrosion potential of stainless steels above the critical potential for pitting when chloride is present.[34] Once underway, pitting or crevice corrosion can propagate in the occluded pit interior anode by the acid hydrolysis mechanism described in Section 7.2.2. Biological slimes can enhance formation of deposits at the pit or crevice opening. When available, sulfate reducing bacteria (SRB) may flourish in the interior and further accelerate anodic corrosion in the pit or crevice. Figure 11.15[35] shows pitting attack on stainless steel by microbiologically induced corrosion (MIC) in a paper mill.

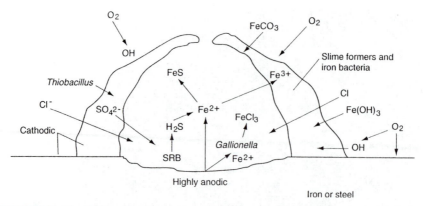

FIGURE 11.14 Schematic diagram of possible processes which may occur during pitting of iron or steel with biological activity. *(From S. C. Dexter, Metals Handbook, Vol. 13, Corrosion, 9th ed., ASM International, p. 114, 1987. Reprinted by permission, ASM International.)*

MIC on stainless steels has often been observed at weldments.[28,36] Attack may be higher in the weld metal itself or in the heat affected zone (HAZ) near the weld. An example of preferential attack of weld metal in stainless steel appears in Figure 9.27. The mechanism of preferential attack at welds is presently uncertain. There is no good correlation with metallurgical structure; either ferrite or austenite may be

FIGURE 11.15 Pitting of austenitic stainless steel in paper-mill process water due to microbiologically influenced corrosion. *(From J. J. Maguire, ed., Handbook of Industrial Water Conditioning, 8th ed., Betz Laboratories, Inc., Trevose, PA, p. 306, 1980. Reprinted by permission, Betz Laboratories, Inc.)*

attacked preferentially.[34] Cubicciotti[36] has suggested that biofilms elevate the electrode potential of the metal in the weld bead or the HAZ to values above the pitting potential. Figure 9.27 and other case histories[28] suggest that MIC may alter corrosion potentials in the weldment, making the weld metal or HAZ anodic to the surrounding alloy and inducing galvanic corrosion. Reported examples of MIC at welds are generally under aerated conditions, but the biomechanisms and microbe types responsible for MIC at welds are not well documented.

11.3.5 Biological Corrosion of Other Metals

Aluminum corrosion can also be accelerated by microorganisms in neutral waters, as indicated in Table 11.3. One fungus feeds on certain hydrocarbons in aircraft fuel, producing acids that dissolve in the water-contaminant phase and attack aluminum fuel tanks.[15] Copper, although more resistant due to its known toxicity, can still be affected by certain bacteria having a high tolerance for Cu^{2+}. For example, *thiobacillus thiooxidans* can tolerate copper concentrations up to 2%.[15]

11.3.6 Macrofouling Organisms

The largest organisms affecting corrosion are readily visible and generally familiar to even the casual observer of seagoing vessels and seacoast structures, such as docks and piers. Barnacles, mollusks, and similar creatures increase the drag on ship hulls, reduce the flow of seawater in pipe and tubing, and eventually destroy wooden structures. Fouling characteristics are often more important than the attendant corrosion. Nevertheless, the by-products of organism growth (metabolites) are often acidic, accelerating the corrosion of metal substrates. Furthermore, these organisms shelter the underlying metal from access by dissolved oxygen and create differential aeration cells (Section 6.5.1), which also accelerate corrosion. The oxygen-free (anaerobic) environment beneath macroorganisms can further host harmful sulfate-reducing bacteria, which are discussed more fully in Section 11.3.2.

Macrofouling organisms in seawater create crevices, which are sites for traditional stainless-steel crevice corrosion described in Section 7.3. Crevice corrosion under barnacles, cleaned from the surface of stainless steel, is shown in Figure 11.16.[37]

Seawater fouling of closed systems is mitigated by periodic mechanical cleaning and continuous, or at least frequent, injection of dissolved chlorine (chlorination). Copper alloys are resistant (but not always immune), because cupric corrosion products are toxic to fouling organisms, and corrosion products slough off the surface, preventing adhesion of the organisms. Antifouling paints are also impregnated with toxic chemicals. Seagoing vessels are subject to periodic dry docking for cleaning and repainting. Exposure to fresh-water river estuaries will often kill fouling organisms on ship hulls, but the encrustations remain and still require removal. Cathodic protection, unfortunately, will not stop the growth of large fouling organisms.

FIGURE 11.16 Cleaned surface of stainless steel showing crevice corrosion under barnacles growing in seawater. *(From K. D. Efird,* Corrosion/75, *Paper No. 124. Reprinted by permission, National Association of Corrosion Engineers.)*

11.4 Soils

11.4.1 Significance

Millions of miles of buried steel pipelines and more millions of buried steel tanks are used throughout the world for transport and storage of natural gas, fuel, chemicals, and water. Because the soil environment is relatively benign, it is easy to ignore the corrosive properties of soil. However, the shortsighted view places a heavy economic burden on subsequent generations who bear the expense of repairing and replacing so many such structures. Long-term corrosion over decades results in loss of product, contamination of the soil environment, and accidents that cause loss of property and life. The necessity of replacing or retrofitting most of the present automotive service-station fuel-storage tanks in the United States was cited in Section 1.2.1 as an example of the economic liabilities resulting from corrosion in soil.

Extensive, long-term soil-testing programs conducted from 1910 to 1955 by the National Bureau of Standards (NBS) provide the foundation for data on soil corrosion in the United States.[38] Table 11.4, derived[39] from the NBS data, shows corrosion rates in typical U.S. soils. Corrosion of most metals in soil, including steel, is controlled by diffusion of dissolved oxygen in the water entrapped in the soil. Thus, the composition of carbon steel has little effect on corrosion resistance in a given soil, as indicated in Table 11.4. Averaged corrosion rate of zinc in Table 11.4 is about the same as for the steels, suggesting common control by diffusion of dissolved oxy-

TABLE 11.4 Corrosion Rates in Soils

Maximum penetration in mils (1 mil = 0.001 in. = 0.025 mm) for total exposure period
Average corrosion rates in g m^{-2}d^{-1}(gmd)

Soil	Open Hearth Iron 12-Year Exposure		Wrought Iron 12-Year Exposure		Bessemer Steel 12-Year Exposure		Copper 8-Year Exposure		Lead 12-Year Exposure		Zinc 11-Year Exposure	
	gmd	mils	gmd	mils	gmd	mils	gmd	mils	gmd	mils	gmd	mils
Average of several soils	0.45 (44 soils)	70	0.47 (44 soils)	59	0.45 (44 soils)	61	0.07 (29 soils)	<6	0.052 (21 soils)	>32	0.3 (12 soils)	>53
Tidal marsh, Elizabeth, NJ	1.08	90	1.16	80	1.95	100	0.53	<6	0.02	13	0.19	36
Montezuma clay Adobe, San Diego, CA	1.37	>145	1.34	>132	1.43	>137	0.07	<6	0.06 (9.6 years)	10	—	—
Merrimac gravelly sandy loam, Norwood,MA	0.09	28	0.10	23	0.10	21	0.02 (13.2 years)	<6	0.013	19	—	—

From H. H. Uhlig and R. W. Revie, Corrosion and Corrosion Control, 3rd ed., Wiley, New York, p. 182, 1985. Reprinted by permission, John Wiley & Sons.

gen in the soil water. However, copper and lead, which form barrier corrosion product films, show lower corrosion rates in the same soils.

11.4.2 Soil Corrosivity

Soils having a high moisture content, high dissolved salt concentration, and high acidity are expected to be the most corrosive. However, soil composition alone has been found to have little correlation with soil corrosivity.[40] Most of the entrained soil water or moisture is relatively noncorrosive in the short term. In the long term, the residence time of water or moisture on the metal surface will control the degree of corrosion in soil.[40] Measuring this residence time is difficult or impossible in practice. Therefore, it becomes necessary to use more easily measured soil characteristics, which have uncertain correlation with corrosivity.

Corrosion rates in marshy or swampy areas may approach or exceed those measured for fully immersed specimens in aerated waters. In fact, occasionally high corrosion rates in saturated or swampy soils of low dissolved oxygen content are contrary to the expected proportionality with dissolved oxygen content of waters. The reason may be in part attributable to microbiologically influenced corrosion by anaerobic sulfate-reducing bacteria (SRB) (Section 11.3.2) which are common residents of such soils. The redox potential measured in the field on platinum in terms of the standard hydrogen electrode gives an indication of corrosivity in water-saturated biologically-active soils:[41]

Potential, mV	Corrosivity
< 100	Severe
100–200	Moderate
200–400	Slight
> 400	Noncorrosive

The soils are more corrosive with reducing (anaerobic) potentials, which favor the presence of SRB.

Drainage, defined as the ability to allow water percolation, varies with the particle size distribution of a soil. A poorly drained soil has small particle size, small interstitial openings between soil particles, and therefore retains water more easily. Soil may be classified according to texture, which reflects the size distribution of particles in the soil. Soil components according to size are as follows:

Gravel and stones	> 2 mm
Fine gravel	1 to 2 mm
Sand	0.05 to 1 mm
Silt	0.002 to 0.05 mm
Clay	< 0.002 mm.

Combinations of sand, silt, and clay form loam, the "ideal" soil texture for most agricultural purposes. The progression from sand to silt to clay decreases the drainage of water through the soil, retains water on metal surfaces more readily, and can increase corrosivity of a soil. The inherent molecular structure of clay traps moisture and thereby further impedes drainage. Consequently, the gravelly loam in Table 11.4

shows a low corrosion rate, and the clay shows a high corrosion rate. However, drainage cannot fully characterize corrosivity. A well-drained soil in a rainy, moist climate may always contain considerable moisture and consequently be rather corrosive. Conversely, a poorly drained soil, high in clay, in a dry climate with low moisture would probably be noncorrosive.

Soil resistivity gives a composite measure of moisture content of soil and dissolved electrolytes in the soil water. It is commonly measured by the 4-pin Wenner technique (Figure 11.17), in which portable instrumentation applies an alternating current, I, between two outer pin contacts in the soil and induces a potential drop, ϕ, between two inner pin contacts. Resistivity, ρ, to an approximate depth equal to the uniform spacing, a, between pins is calculated from

$$\rho = \frac{2\pi a \phi}{I} .$$

Soil resistivity is frequently used to estimate soil corrosivity, because it is easy to measure. However, it does not measure the residence time of water on buried surfaces and consequently does not correlate well with soil corrosivity. Low soil resistivity values do, however, indicate areas of potentially high corrosivity that require further investigation in a survey.[40]

It must be concluded that there is no single easily measured soil parameter that can be used to determine soil corrosivity. Instead, a number of characteristics must be combined to estimate the corrosion that may be expected from a particular soil. Table 11.5[40] summarizes properties of soils having corrosivities ranging from very high to very low. Note that a soil of any particular classification need not have all of the characteristics listed.

11.4.3 Prevention

The engineer often has only limited control over the materials or the environment in soil corrosion. Carbon steel is often the only economic choice for large equipment such as pipelines and storage tanks. Special backfills are often too expensive over

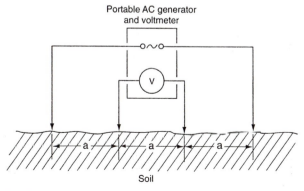

FIGURE 11.17   Schematic Wenner 4–pin apparatus for measuring soil resistivity.

Drainage, Texture and Aeration.

Soil Corrosivity, class	Total Acidity, meq/100g	Resistivity, Ω cm	Conductivity, mmho/cm	General Relationships	
				Drainage-Texture Relationship	Water-Air Permeability
Very low	<4	>10,000	<0.1	Somewhat excessive—excessively drained coarse-textured soil	Rapid to very rapid
Low	4 to 8	5000 to 10,000	0.1 to 0.2	Well-drained with moderately coarse and medium-textured control section; somewhat poorly drained with coarse-textured control section	Moderate to rapid
Moderate	8 to 12	2000 to 5000	0.2 to 0.4	Well-drained with moderately fine-textured control section; moderately well drained with medium-textured control section; somewhat poorly-drained with moderately coarse-textured control section; very poorly-drained with high, nonfluctuating water table	Moderately slow to slow
High	12 to 16	1000 to 2000	0.4 to 1.0	Well drained and moderately well-drained, fine-textured soils; moderately well-drained, moderately fine-textured soils; somewhat poorly drained; medium and moderately fine-textured control sections; poorly drained with coarse to moderately fine-textured control sections; very poorly drained soils; water table fluctuates within ft of surface;	Slow and very slow; saturated
Very high	>16	<1000	>1.0	Somewhat poorly to very poorly drained fine-textured soils; mucks, peats with a fluctuating water table	Very slow; saturated

From F. E. Miller, J. E. Foss, and D. C. Wolf, ASTM STP 741, p. 19, 1981. Reprinted by permission, American Society for Testing and Materials.

long lengths of pipeline and even ineffective in swampy, wet soils. The common prevention methods are coatings (Chapter 14) and cathodic protection (Chapter 13). Long lengths of cross-country pipeline universally employ both. Added cathodic protection capacity may be required at "hot spots" where soil conditions indicate elevated corrosivity.

Many water and waste-water lines used in domestic plumbing and municipal utility systems have been constructed of polymeric materials (e.g., polyvinyl chloride) in recent years.

11.5 Concrete

Corrosion of reinforcing steel in concrete structures is one of the most expensive corrosion problems in the United States. The structural integrity of thousands of bridges, roadbeds, overpasses, and other concrete structures has been impaired by corrosion, urgently requiring expensive repairs to ensure public safety. A recent committee report of the American Concrete Institute[42] gives a good review of the subject.

11.5.1 Corrosive Characteristics

Concrete is a composite of cured portland cement, aggregate, and, often, various admixtures. Dry portland cement is a powdered anhydrous mixture of calcium silicates and calcium aluminates. When combined with the sand and gravel aggregate, portland cement forms a workable mixture with water which can be poured into forms of the desired structural shape. The anhydrous compounds in portland cement react slowly with water and "cure" into a hardened monolithic structure of hydrated compounds having high compressive strength but low tensile strength. As a result, concrete structures must be designed so that loading is primarily compressive. The aggregate reduces the cost and increases both compressive and tensile strength. Strength is improved routinely in concrete structures by pouring the concrete mix around a network of steel reinforcing wire and/or rod. Reinforcing wire may be pre- or post-tensioned before or after the concrete cures to impose compressive stresses that must be overcome by service loads before any tensile stresses are produced in the structure.

Corrosion of normally resistant reinforcing steel often threatens the integrity of concrete structures. Corrosion products, which usually have greater volume than the parent metal, set up tensile stresses to fracture the concrete around the reinforcement. Water in pores can also fracture the concrete by alternate freezing and thawing, thereby enhancing further water penetration. Figure 11.18[43] shows a bridge support column with concrete spalled from the steel reinforcing substructure.

Calcium hydroxide, $Ca(OH)_2$, leaches out of uncured portland cement upon addition of water, forming an alkaline solution that forms an initially resistant passive film on the steel surfaces. During curing, water first reacts with the cement compounds, and the excess then evaporates out of the hardened structure, concentrating the $Ca(OH)_2$ solution in pores and voids and further increasing the corrosion resistance of the reinforcing steel.

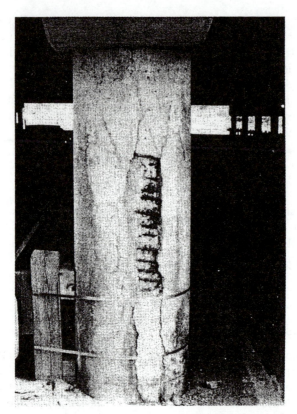

FIGURE 11.18 Degradation of a concrete bridge support column due to spalling of concrete from corroded steel reinforcement. *(From R. H. Heidersbach, Corrosion/89, Paper No. 129, NACE, Houston, 1989. Photograph by courtesy R. H. Heidersbach, California Polytechnic State University, San Luis Obispo. Reprinted by permission, National Association of Corrosion Engineers.)*

Corrosion resistance of steel in concrete would remain indefinitely in usual atmospheric exposures if the concrete cover could exclude air and water from the embedded reinforcement. When the cover is too thin or porous, corrosive damage to concrete results primarily from access of water solutions containing dissolved oxygen. No damage occurs in very dry concrete due to the absence of essential water. Nor is there much damage in continuously water-saturated concrete because oxygen has less access through the liquid phase than through air in the concrete pores. Also, the corrosion products are colloidal and uncompacted in liquid water. An unsaturated concrete, which results from alternate wetting and drying, allows more rapid transport of oxygen to the corroding steel surface. The oxygen dissolves in the pore water at the metal surface, enabling electrochemical corrosion. Periodic drying allows deposition and compaction of oxide corrosion products, which impose tensile stresses on the surrounding concrete.

Dissolved chlorides play the major role in corrosion damage to concrete. Chloride impairs passivity and increases the active corrosion rate of carbon steel in neutral and alkaline pore water solutions. The resultant ferrous corrosion products form an acid solution with chloride (Section 7.2.2) which neutralizes the alkaline concrete environment and further enhances corrosion. Chloride may be added accidentally to concrete as a contaminant in the water or aggregate or intentionally as an early setting additive in the mix water. Seacoast spray, fog, and mist, or road salts used for snow and ice control, provide dissolved chloride that penetrates the concrete cover through incipient porosity and cracks. Bridge decks and other highway structures are especially susceptible from road salt and mechanical fatigue and wear. Figure 11.19[43] shows craters in a bridge deck resulting from road salt corrosion of the reinforcing steel.

Carbon dioxide from ambient air can also increase corrosion of reinforcing steel, but at a much slower rate than similar increases due to chlorides. CO_2 from the atmosphere reacts with saturated $Ca(OH)_2$ in pore water (pH, 12.5 to 13.8) forming saturated $CaCO_3$ (pH 8 to 9). The reduced pore-water pH allows corrosion of reinforcing steel in the concrete. Again, access of CO_2 and the resulting corrosion are maximized in wet but unsaturated concretes.

11.5.2 Prevention

Good design and workmanship in mixing and pouring concrete are the keys to prevention of reinforcing-steel corrosion. The steel must be sufficiently recessed below the surface to ensure adequate covering. A minimum 2-in. (50-mm) covering is recommended but often not attained. Uniformly graded aggregate is required to prevent voids and achieve good adherence between cement and aggregate. Chloride-free water is necessary wherever available, and the water/cement ratio should be kept to a minimum to give a cured concrete of minimum porosity. Excess water above that needed for the curing reactions leaves porosity behind after evaporation. Reinforcing

FIGURE 11.19 Concrete pavement spalling on a bridge deck. *(From R. H. Heidersbach,* Corrosion/89, *Paper No. 129, NACE, Houston, 1989. Photograph by courtesy R. H. Heidersbach, California Polytechnic State University, San Luis Obispo. Reprinted by permission, National Association of Corrosion Engineers.)*

steel should have clean, scale-free surfaces to give good adherence to the surrounding concrete. However, a slightly rusted surface maximizes the bond with concrete.

Epoxy coatings have been used successfully on reinforcing steel in critical applications. Zinc-galvanized coatings have frequently been used on reinforcing steel but may not be recommended in applications where chloride may be present. Impressed-current cathodic protection systems, using conductive-polymer or mixed-oxide network anodes embedded in the concrete, have shown promise for bridge decks, but insufficient service history is available to evaluate their effectiveness. Cathodic protection must be used cautiously for prestressed tension wires due to the danger of hydrogen embrittlement.

11.6 Acid Process Streams

Various acid streams are essential in the extraction and manufacture of many drugs and chemicals. Only general guidelines and examples are given in this section for materials selection in the various important acids. The interested reader should consult references 44 and 45 for more detail.

11.6.1 Sulfuric Acid

Carbon steel can be used in concentrated, 65 to 101%, sulfuric acid due to the formation of a sulfate surface film. However, the film is easily eroded, and only static or low-velocity flow is permissible.[44] Hydrogen evolution can cause erosion and grooving corrosion, as described in Section 10.2.1. Therefore, slight stirring to randomize bubble flow is sometimes beneficial. *Cast irons* have resistance to concentrated sulfuric acid at least matching that of carbon steel and are less sensitive to erosion. Higher carbon and silicon contents may enhance resistance to sulfuric acid in the cast irons. *High-silicon cast iron* (Duriron) with about 14.5% Si is resistant to concentrated acid at temperatures up to boiling but is highly brittle and subject to cracking from accidental impact blows or rapid temperature fluctuations.[46] *Cast alloy 20* (ACI CN-7M) is frequently used over the entire concentration range. *Lead* is resistant to sulfuric acid up to about 85% at all temperatures up to 250°F but lacks strength. Figure 11.20[47] shows a composite isocorrosion (less than 20 mils/yr or 0.5 mm/yr) diagram for the common materials used for handling concentrated sulfuric acid.

Wrought *austenitic stainless steels* are not generally used for handling sulfuric acid in any concentration, because they depend on adequate oxidizers in solution to maintain corrosion resistance by passivity. Inadequate oxidizers can lead to the dangerous case of borderline passivity (Section 4.1.3), in which a preformed passive film is unstable. However, austenitic stainless steel can be used when supplemented by anodic protection. Sulfuric acid coolers are routinely protected by integral anodic protection systems (Section 4.4.1). Also, stainless steels, cast and wrought, containing 5 to 6% Si, have been reported[44,47] resistant to concentrated sulfuric acid at temperatures up to 120°C, without anodic protection.

Other alloys are used on occasion in sulfuric acid. *High-nickel alloys* often are resistant but are usually not economically competitive with carbon steel and cast

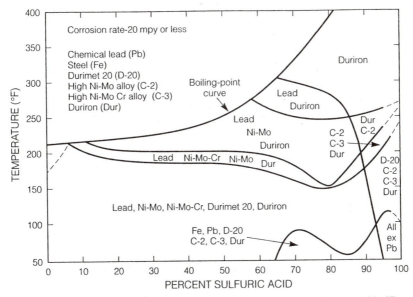

FIGURE 11.20   Isocorrosion diagram for materials to handle sulfuric acid. *(From M. G. Fontana, Corrosion Engineering, 3rd ed., McGraw-Hill, New York, pp. 319–327, 1986. Reprinted by permission, McGraw-Hill Book Company.)*

irons. However, those containing Cr and Mo may be required in the presence of halide contamination and at higher temperature.[44] Zirconium alloys are resistant in dilute (<50%) sulfuric acid up to boiling temperatures.

11.6.2 Nitric Acid

Because of its strongly oxidizing nature, nitric acid can be readily handled with *austenitic stainless steels* in both wrought and cast form in concentrations up to about 80% and at temperatures up to boiling. The isocorrosion chart for austenitic alloys is shown in Figure 11.21.[46,48] Unfortunately, the oxidizing nature of nitric acid leads to intergranualar corrosion (IGC) in welded austenitic stainless steel structures (Section 9.1.2). Low-carbon alloy modifications (Section 9.1.3) are usually the most convenient for prevention of IGC in nitric acid service. Solution annealing and stabilization with Cb or Ti alloying additions are less popular.[48] *High-silicon cast irons* and stainless steels with silicon additions have increased resistance in concentrated nitric acid. *Aluminum alloys* are used frequently at temperatures below 40°C and concentrations above 80%.

11.6.3 Hydrochloric Acid

Tantalum is about the only metal that is resistant to hydrochloric acid in most concentrations and temperatures. It is too expensive for general consideration but is required in some applications, such as steam lines, for heating the pure acid or sheathing for instrument sensor probes.[46]

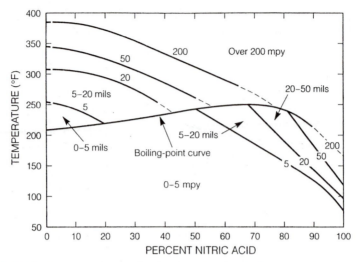

FIGURE 11.21   Isocorrosion diagram for quench-annealed austenitic 18-8 stainless alloys in nitric acid. *(From M. G. Fontana,* Corrosion Engineering, *3rd ed., McGraw-Hill, New York, pp. 319–327, 1986. Reprinted by permission, McGraw-Hill Book Company.)*

Molybdenum is a key alloying element for resistance to hydrochloric acid. *Nickel Alloy* (Hastelloy) *B-2*, containing 28% Mo, is the most resistant of the wrought alloys. The isocorrosion diagram is shown in Figure 11.22 for Hastelloy B-2. Similarly, *cast iron* containing molybdenum (Durichlor) is relatively resistant to hydrochloric acid. However, fluoride impurities, as well as oxidizing ferric and cupric ions, substantially decrease the resistance of these and most other alloys.

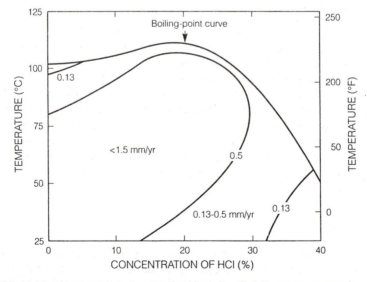

FIGURE 11.22   Isocorrosion diagram for Hastelloy B-2 in oxygen-purged hydrochloric acid. *(From T. F. Degnan,* Metals Handbook, *Vol. 13,* Corrosion, *9th ed., ASM International, p. 1160, 1987. Reprinted by permission, ASM International.)*

Normally, corrosion-resistant alloys such as the stainless steels and nickel alloys containing chromium fail by pitting and crevice corrosion (Sections 7.2 and 7.3), when exposed to hydrochloric acid, unless they contain substantial amounts of molybdenum. *Nickel Alloy C-276* and the equivalent cast alloy, Chlorimet, are resistant to hydrochloric acid with oxidizing impurities, because of the combination of chromium and molybdenum alloying elements. Nonmetallics, including glass and rubber, are often used with hydrochloric acid, when alloys do not have sufficient corrosion resistance or are too expensive.

11.6.4 Hydrofluoric Acid

Refrigerants, fluorocarbon polymers and elastomers, and certain cleaning fluids, among others, are based on hydrofluoric acid and/or anhydrous hydrogen fluoride for their manufacture. These acids cause severe, painful burns and are highly toxic as well as strongly corrosive. Hydrofluoric acid is contrary in chemical reactivity and corrosivity, compared to the other strong acids. Usually, inert glass and ceramics are attacked by hydrofluoric acid, which is used as an etchant for glass. On the other hand, usually reactive *magnesium* is resistant because of an insoluble fluoride film which forms on the surface.

Carbon and low alloy steels are suitable for handling aqueous hydrofluoric acid at concentrations from about 65 to 100%. Slow corrosion can build up dangerous hydrogen pressures in transportation tankers at lower concentrations. Hydrogen induced cracking and blistering (Section 10.1) are problems for hardened (especially alloy) steels and welds.[49] *Monels* (Ni-Cu alloys) are resistant to deaerated hydrofluoric acid solutions at all concentrations and temperatures up to 150°C. Aeration increases attack, and Monel 400 with 33% Cu is susceptible to stress corrosion cracking (SCC) in moist HF vapors with air. The critical agent apparently is Cu_2F, which is able to build up to critical levels in the thin layer of liquid phase on the surface exposed to the vapor.[49,50] *Nickel Alloy 600* is used to replace Monel in hot HF vapors and when SCC is a concern.

11.6.5 Organic Acids

The organic acids are classified as weak acids, because they are only slightly dissociated to give dissolved H^+ in aqueous solution. They all have similar corrosive properties. Truly anhydrous organic acids are often highly corrosive, and addition of a few percent water greatly increases the corrosion resistance of many alloys. Formic acid is the most strongly dissociated and therefore the most corrosive. Acetic acid is the most important and abundant in chemical processing.[51] Naphthenic acids are present in certain petroleum crude oils and cause severe corrosion at high temperatures in distillation equipment.[19]

Aluminum is suitable for handling most of the organic acids below 25°C. It is sometimes unsuitable at high temperature in dilute acids, but the difference between room-temperature and high-temperature corrosion narrows as concentration increases, as shown in Figure 11.23.[51] *Copper and bronze* are used in services similar to aluminum and are more resistant to elevated temperature. However, the copper alloys are unsuitable if air or other dissolved oxidizers are present. *Austenitic stainless*

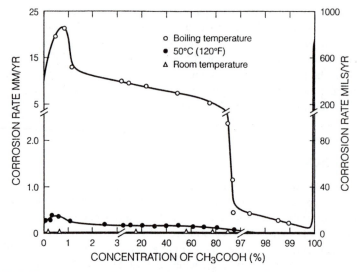

FIGURE 11.23 Corrosion of aluminum in acetic acid as affected by temperature and concentration. *(From G. B. Elder,* Metals Handbook, *Vol 13,* Corrosion, *9th ed., ASM International, p. 1157, 1987. Reprinted by permission, ASM International.)*

steels have become the materials of choice for handling and storing most organic acids; Type 316 is the most popular. The carbon steels and ferritic Type 400 stainless steels are not suitable for most organic acids.

11.7 Alkaline Process Streams

11.7.1 Caustic Hydroxides

Ordinary *carbon steel* is suitable for caustic soda (NaOH) and caustic potash (KOH) because a resistant passive surface film is stable in alkaline solutions. However, at elevated temperature, concentration of caustics can lead to accelerated general corrosion, as shown in Figure 11.6. Carbon steels are sometimes precluded from use in high-purity caustic. Agitation can disturb and redissolve the soft hydroxide surface film or contaminate the solution with suspended solids. Stress corrosion cracking (SCC) is a further danger for carbon steel, as temperature and caustic concentration increase (Figure 11.24[52]).

The resistance of steel and cast iron improve in proportion to the amount of alloyed nickel. Thus, the *austenitic stainless steels* are used for caustic service at intermediate temperatures up to 95°C and concentrations up to 50%, where carbon steel is unsuitable. However, SCC remains a problem at still higher levels of temperature and concentration, as indicated in Figure 11.24. *Nickel alloys* are used in still more severe conditions of concentration and temperature. Nickel Alloys 200, 201, 400 (Monel), and 600 (Inconel) are all recommended for caustic service at high temperature. Commercial-purity nickel (Alloy 201), low in carbon, is most resistant to

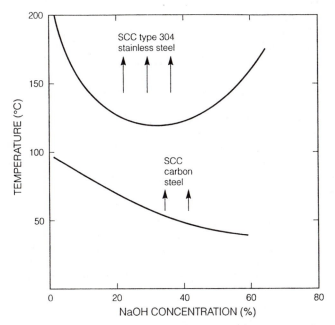

FIGURE 11.24   Approximate temperature and concentration limits for stress corrosion cracking of carbon steel and austenitic stainless steel in caustics. *(Adapted from Nelson.* [52]*)*

SCC. Alloy 600 is useful when sulfides are present in caustic.[52] Chloride apparently does not exert a major influence in caustic SCC of stainless steels and nickel alloys.[52]

11.7.2 Hypochlorites

Sodium hypochlorite is produced by chlorination of NaOH solutions and is used as a bleaching agent. Residual NaOH stabilizes the hypochlorite ion (OCl⁻) and maintains alkaline pH. The hypochlorite ion is highly corrosive due to its oxidizing nature, and only titanium shows consistent resistance. Low ambient temperatures are necessary to avoid decomposition of the hypochlorite. Titanium, commercial grade 2, is used for processing equipment, and various nonmetallic coatings and linings are required in carbon-steel tanks and pipelines for storage and transportation.[52]

References

1. W. Whitman, R. Russel, and V. Altieri, *Ind. Eng. Chem.,* Vol. 16, p. 665, 1924.
2. H. H. Uhlig and R. W. Revie, *Corrosion and Corrosion Control,* 3rd ed., Wiley, New York, p. 96-111, 1985.
3. H. J. Cleary and N. D. Greene, *Corros. Sci.,* Vol. 9, p. 3, 1969.
4. H. H. Uhlig, D. N. Triadis, and M. Stern, *J. Electrochem. Soc.,* Vol. 102, p. 59, 1955.

5. B. P. Boffardi, *Metals Handbook,* Vol. 13, *Corrosion,* 9th ed., ASM International, Metals Park, OH, p. 487, 1987.

6. H. E. Bush and J. E. Donham, *Metals Handbook,* Vol. 13, *Corrosion,* 9th ed., ASM International, Metals Park, OH, p. 1253, 1987.

7. H. H. Uhlig and R. W. Revie, *Corrosion and Corrosion Control,* 3rd ed., Wiley, New York, p. 111, 1985.

8. F. N. Speller, *Corrosion, Causes and Prevention,* 3rd ed., McGraw-Hill, New York, p. 168, 1951.

9. J. J. Maguire, ed., *Handbook of Industrial Water Conditioning,* 8th ed., Betz Laboratories, Inc., Trevose, PA, pp. 145-6, 1980.

10. H. H. Uhlig and R. W. Revie, *Corrosion and Corrosion Control,* 3rd ed., Wiley, New York, p. 278-90, 1985.

11. J. E. Donham, *Metals Handbook,* Vol. 13, *Corrosion,* 9th ed., ASM International, Metals Park, OH, p. 1244, 1987.

12. B. M. Gordon and G. M. Gordon, *Metals Handbook,* Vol. 13, *Corrosion,* 9th ed., ASM International, Metals Park, OH, p. 932, 1987.

13. E. Partridge and R. Hall, *Trans. Am. Soc. Mech. Eng.,* Vol. 61, p. 597, 1939.

14. G. Wranglen, *An Introduction to Corrosion and Protection of Metals,* Chapman and Hall, New York, p. 229, 1985.

15. S. C. Dexter, *Metals Handbook,* Vol. 13, *Corrosion,* 9th ed., ASM International, Metals Park, OH, p. 893, 1987.

16. ASTM Method D1141-86, *Annual Book of ASTM Standards,* Vol. 11.02, ASTM, Philadelphia, p. 403, 1988.

17. F. L. LaQue, *Marine Corrosion,* Wiley-Interscience, New York, 1975.

18. M. Pourbaix, *Atlas of Electrochemical Equilibria in Aqueous Solutions,* NACE, Houston, p. 551, 1974.

19. J. Gutzeit, R. D. Merrick, and L. R. Sharfstein, *Metals Handbook,* Vol. 13, *Corrosion,* 9th ed., ASM International, Metals Park, OH, p. 1262, 1987.

20. N. W. Polan, *Metals Handbook,* Vol. 13, *Corrosion,* 9th ed., ASM International, Metals Park, OH, p. 625, 1987.

21. D. A. Wensley, *Metals Handbook,* Vol. 13, *Corrosion,* 9th ed., ASM International, Metals Park, OH, p. 1186, 1987.

22. H. S. Isaacs, B. Vyas, and M. W. Kendig, *Corrosion,* Vol. 38, p. 130, 1982.

23. R. Bandy, R. Roberge and R. C. Newman, *Corrosion,* Vol. 39, p. 391, 1983.

24. *Nucl. News,* Vol. 25, p. 47 (March 1982).

25. R. L. Jones, R. L. Long and J. S. Olszewski, CORROSION/83, Paper 141, NACE Houston, 1983.

26. A. K. Tiller, *Biologically Induced Corrosion,* S. C. Dexter, ed., NACE, Houston, p. 8, 1986.

27. S. C. Dexter, *Metals Handbook,* Vol. 13, *Corrosion,* 9th ed., ASM International, Metals Park, OH, p. 114, 1987.

28. G. J. Licina, *Sourcebook for Microbiologically Influenced Corrosion in Nuclear Power Plants,* NP 5580, Electric Power Research Institute, Palo Alto, CA, 1988.

29. J. J. Maguire, ed., *Handbook of Industrial Water Conditioning,* 8th ed., Betz Laboratories, Inc., Trevose, PA, p. 296, 1980.

30. M. Bibb, *Biologically Induced Corrosion,* S. C. Dexter, ed., NACE, Houston, p. 96, 1986.

31. W. P. Iverson, *ASTM STP 741,* ASTM, Philadelphia, p. 40, 1981.

32. V. Scotto, R. Dicintio, and G. Mercenaro, *Corros. Sci.,* Vol. 25, p. 185, 1985.

33. R. Holthe, P. O. Gartland and E. Bardal, "Influence of the Micorbial Slime Layer on the Electrochemical Properties of Stainless Steel in Sea Water," *Proc. Eurocorr87.*

34. D. Cubicciotti, *Proc. 3rd Int. Symp. on Environmental Degradation of Materials in Nuclear Power Systems—Water Reactors,* TMS/AIME, Warrendale, PA, p. 653, 1988.

35. J. J. Maguire, ed., *Handbook of Industrial Water Conditioning,* 8th ed., Betz Laboratories, Inc., Trevose, PA, p. 306, 1980.

36. S. W. Borenstein, CORROSION/88, Paper 78, NACE, Houston, 1988.

37. K. D. Efird, CORROSION/75, Paper 124. Cited in D. A. Jones, *Forms of Corrosion Recognition and Prevention,* C. P. Dillon, ed., NACE, Houston, p. 35, 1982.

38. M. Romanoff, *Underground Corrosion,* NBS 579, National Bureau of Standards, 1957. Reprinted by NACE, Houston, 1989.

39. H. H. Uhlig and R. W. Revie, *Corrosion and Corrosion Control,* 3rd ed., Wiley, New York, p. 182, 1985.

40. F. E. Miller, J. E. Foss, and D. C. Wolf, *ASTM STP 741,* ASTM, Philadelphia, p. 19, 1981.

41. J. D. A. Miller and A. K. Tiller, *Microbial Aspects of Metallurgy,* J. D. A. Miller, ed., American Elsevier, New York, p. 61, 1970.

42. Committee Report ACI222R-85, *ACI Journal,* p. 3, Jan/Feb, 1985.

43. R. H. Heidersbach, CORROSION/89, Paper No. 129, NACE, Houston, 1989.

44. S. K. Brubaker, *Metals Handbook,* Vol. 13, *Corrosion,* 9th ed., ASM International, Metals Park, OH, p. 1148, 1987.

45. *Process Industries Corrosion,* B. J. Moniz and W. J. Pollock, eds., NACE, Houston, pp. 243–287, 1986.

46. M. G. Fontana, *Corrosion Engineering,* 3rd ed., McGraw-Hill, New York, pp. 319-27, 1986.

47. D. J. Chronister and D. C. Spence, *Proc. Symposium on Sulfuric Acid Corrosion,* NACE, Houston, p. 75, 1985.

48. R. D. Crooks, *Metals Handbook,* Vol. 13, *Corrosion,* 9th ed., ASM International, Metals Park, OH, p. 1154, 1987.

49. T. F. Degnan, *Metals Handbook,* Vol. 13, *Corrosion,* 9th ed., ASM International, Metals Park, OH, p. 1166, 1987.

50. L. Graf and W. Wittich, *Werkst. Korros.,* Vol. 17, p. 385, 1966.

51. G. B. Elder, *Metals Handbook,* Vol. 13, *Corrosion,* 9th ed., ASM International, Metals Park, OH, p. 1157, 1987.

52. J. K. Nelson, *Metals Handbook,* Vol. 13, *Corrosion,* 9th ed., ASM International, Metals Park, OH, p. 1174, 1987.

Exercises

11-1. It is found that in deaerated solutions of pH > 13, corrosion proceeds at a relatively high rate compared to what is shown in Figure 11.1. Explain how this can be so.

11-2. The measured saturation index of the water system in an ore concentration mill is +1 at a pH of 8. No major scaling or corrosion problems have been experienced. A new water well is brought on line to meet the requirements of expanded operations. The main initial effect on the water system chemistry is an increase in the measured carbonate activity (CO_3^{2-}) by 100 $\times$. Other water chemistry parameters are unchanged. What is the effect on the saturation index? Would you expect long-term corrosion or scaling problems?

11-3. Show with a mixed potential analysis (Chapter 3) how dissolved oxygen in acid increases corrosion rate over that expected for a deaerated acid.

11-4. Show with a mixed potential analysis (Chapter 3) how decreasing pH (Figure 11-1) below 4 can increase corrosion rate.

11-5. Sketch the cross section of a pit in carbon steel that occurs as the result of oxygen contamination of pressurized boiler water. Indicate the chemical and electrochemical reactions occurring in each region of the pit and surrounding surface area.

11-6. List possible measures for controlling corrosion of carbon steel in high temperature boiler waters and steam condensate.

11-7. In a research study, dissolved oxygen was measured in a series of soils. The corrosion rate generally increased with dissolved oxygen content for most of the soils. **(a)** Explain this behavior. **(b)** One soil, however, having very low dissolved oxygen, showed exceptionally high corrosion. Suggest a possible characteristic of this soil to account for the observed high corrosion.

11-8. Design engineers in your company are considering a high strength aluminum alloy as a substitute for steel reinforcing bar in concrete for light weight in a complex structure. The engineers tell you that aluminum has been used for electrical conduit embedded in steel with some success. Could you recommend aluminum for this application on the basis of corrosion resistance? Why is aluminum useable in such a highly alkaline environment as concrete?

11-9. List aqueous environments and conditions discussed in this chapter for which the following uncoated metals/alloys would be suitable from the standpoint of, first, corrosion resistance only and, second, corrosion resistance and cost: **(a)** tantalum, **(b)** titanium, **(c)** nickel Alloy 600, **(d)** austenitic stainless steel, **(e)** carbon steel, **(f)** aluminum.

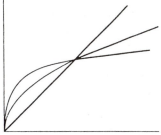

12

Atmospheric Corrosion and Elevated Temperature Oxidation

The forms of corrosion described in preceding chapters generally result from exposure to aqueous solutions. In this chapter we discuss corrosion in the vapor phase: first, at ambient and near ambient temperatures in atmospheric humid air; and second, at elevated temperatures in air and other oxidizing gases. In both, the key to corrosion resistance is a dense adherent surface film or scale that forms a protective barrier between the underlying metal and the corrosive environment. Cracked, porous, or soluble films do not protect the metal and result in relatively high corrosion rates.

12.1 Atmospheric Corrosion

Atmospheric corrosion in air is confined to temperatures and conditions resulting from exposure to the natural ambient environment, which contains variable amounts of water from rain and splashing. Accelerating effects of splashing above the high-tide line were described in Section 11.1.5. Dissolved oxygen is more readily transported through a thin layer of surface water than through bulk water during complete immersion. Thus, corrosion in the splash zone above water is higher than in full immersion. Periodic washing by rainwater or tidal fluctuation creates conditions less severe than those of continuous splashing.

General uniform attack is the normal mode of atmospheric corrosion. General atmospheric corrosion of a structural member from the Golden Gate Bridge in San Francisco, California, is shown in Figure 12.1[1] after over 50 years of exposure. Pitting and other localized forms are rare in atmospheric corrosion. However, visual evidence of galvanic attack at dissimilar-alloy contacts is often present after atmospheric exposure. Increased attack within bolted lap joints of weathering steels, exposed to humid atmospheres, is a problem described in Section 7.4.3.

12.1.1 Atmospheric Parameters Affecting Corrosion

Humidity is necessary for atmospheric corrosion. A thin layer of condensed water deposits on the surface to provide the electrolyte needed for electrochemical corrosion. Although necessary, humidity is not sufficient. Even in very humid environments, corrosion of uncontaminated surfaces is often relatively low in unpolluted atmospheres.

Pollutants or other atmospheric contaminants increase atmospheric corrosion by enhancing the electrolytic properties and stability of water films that condense from

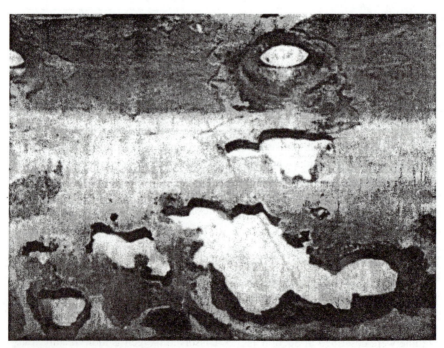

FIGURE 12.1 General atmospheric corrosion of a carbon-steel structural member from the Golden Gate Bridge after 50 years service. *(From R. H. Heidersbach, Corrosion/89, Paper 129, NACE, Houston, 1989. Photograph by courtesy or R. H. Heidersbach, California Polytechnic State University, San Luis Obispo. Reprinted by permission National Association of Corrosion Engineers.)*

the atmosphere. Sulfur dioxide, SO_2, a common industrial pollutant, which forms sulfuric acid when dissolved in the surface film, accelerates atmospheric corrosion of carbon steel significantly, as shown in Figure 12.2.[2] Corrosion product films are protective, and corrosion may remain low in the absence of SO_2. However, in the presence of SO_2, the films are nonprotective, and weight loss increases linearly with time. Thus, pollutants can provide dissolved solutes in the surface water film; SO_2, NO_2, Cl^-, and F^- are prime examples.

In atmospheres containing 0.01% SO_2, the corrosion rate of carbon steel increased rapidly above a critical humidity of 60%, as shown in Figure 12.3.[3] Attack remains low without SO_2, even near 100% relative humidity. This effect is attributed to hygroscopic corrosion product $FeSO_4$ (Section 12.1.3), which absorbs water above a critical humidity level. Hygroscopic corrosion products, and other salt particles deposited from the atmosphere, reduce the relative humidity necessary to cause water condensation. The enhanced presence of a water film increases the time of wetness and the consequent extent of corrosion.[3] Only when the relative humidity is reduced below the critical value for each such available salt is film formation suppressed and corrosion minimized. The critical relative humidity corresponds to the equilibrium humidity over a saturated solution of the salt, as shown in Figure 12.4, which shows corrosion rates as a function of humidity in the presence of numerous salts. It is evident that corrosion is minimal when relative humidity is maintained below the humidity over the saturated solution. $NaNO_2$ and Na_2CrO_4 are inhibitors (Table 14.7), and are consequently notable exceptions to the trend shown in Figure 12.4.

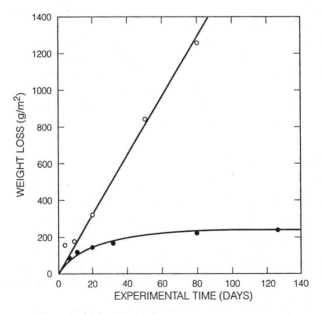

FIGURE 12.2 Comparison of atmospheric corrosion in the absence (●) and presence (○) of SO_2. *(From H. Kaesche,* Metallic Corrosion, *NACE, Houston, pp. 216–219, 1985. Reprinted by permission, National Association of Corrosion Engineers.)*

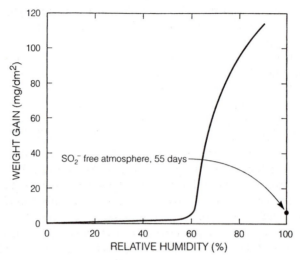

FIGURE 12.3 Effect of relative humidity on atmospheric corrosion in the presence of 0.01% SO_2. *(From P. W. Brown and L. W. Masters, Atmospheric Corrosion, W. J. Ailor, ed., Wiley, New York, p. 31, 1982. Reprinted by permission, John Wiley & Sons.)*

Temperature has a variable effect on atmospheric corrosion. Ambient atmospheric temperatures keep corrosion rates relatively low but may enhance the condensation of an aqueous surface film to increase corrosion. Exposure to sunlight raises surface temperature, but does not necessarily accelerate corrosion. Higher temperature may dry the surface and reduce corrosion. As a result, shaded surfaces often corrode more rapidly than those exposed to direct sunlight.

A combination of high humidity, high average temperature, and the presence of industrial pollutants or air-entrained sea salt increases atmospheric corrosion rates. Thus the maximum atmospheric corrosion occurs in marine tropical or semitropical environments, as shown in Table 12.1.[4] Absence of one or more of these factors may result in low corrosion. For example, despite the high ambient temperature in Phoenix, Arizona, the corrosion rate is low, due to the low humidity.

12.1.2 Weathering Steels

High-strength low-alloy (HSLA) steels, containing a few tenths percent of Cu, Cr, Ni, Si, and P, not only have higher strength but also much improved resistance to atmospheric corrosion. Table 12.2[5] shows 15.5-year thickness losses of various steel compositions in industrial and marine atmospheres. Copper seems to be the key alloying element, but others also enhance the beneficial effects of copper in all environments. Figure 12.5[5] demonstrates the improved atmospheric corrosion resistance for copper-containing steel and the further improvement afforded by additional alloying elements in HSLA steel.

Alloying elements contribute to a more compact and less porous corrosion product surface film. Figure 12.6[6] shows scanning electron micrographs of the corrosion product surface of carbon and HSLA steels. The ruptured and protective oxide sur-

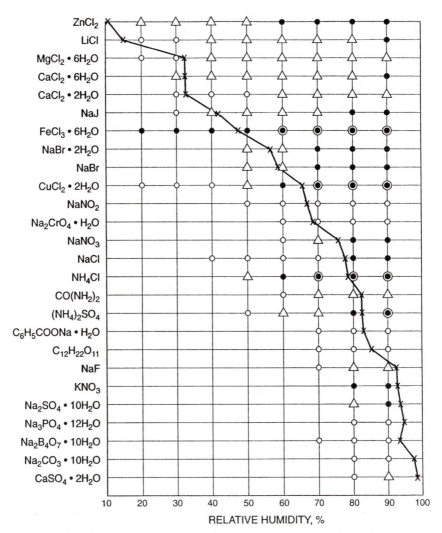

FIGURE 12.4 Effect of relative humidity over excess salt on the weight loss of steel coupons after 7-day exposure. ○ = 0-1 mg; △ = 2-5 mg; ● = 6-20 mg; ◉ = 21-100 mg. × = relative humidity above saturated solution. *(From V. Kucera and E. Mattsson, "Atmospheric Corrosion", Corrosion Mechanisms, F. Mansfeld, ed., Marcel Dekker, New York, 1987, p. 214. Reprinted by permission Marcel Dekker.)*

faces of the carbon and HSLA steels, respectively, are apparent. Adherent, protective films on HSLA steels seal the surface against further penetration of water, which does not easily wet the oxide surface.[7] Compact, cosmetically agreeable surface oxide films develop more rapidly in industrial atmospheres containing SO_2, which is probably involved in film formation, as discussed in Section 12.1.3. However, the corrosion rates of the weathering steels are not reduced in industrial atmospheres to levels lower than those in noncorrosive rural or semirural atmospheres.[8]

TABLE 12.1   Atmospheric Corrosion Comparisons (2-year weight loss from 10-cm × 15-cm steel specimens)

Phoenix, AZ	2.23	Durham, NC	13.3
Brazos River, TX	45.4	Daytona Beach, FL	144.0
Monroeville, PA	23.8	Point Reyes, CA	244.0
Pittsburgh, PA	14.9	Kure Beach, NC 250-m lot	71.0
State College, PA	11.7	Kure Beach, NC, 25-m lot	260.0
South Bend, PA	16.2	Cape Kennedy, 0.8-km from ocean	42.0
Potter County, PA	10.0	Cape Kennedy, 55-m from ocean	80.2
Bethlehem, PA	18.3	9-m elevation	
Cleveland, OH	19.0	Cape Kennedy, 55-m from ocean	215.0
Columbus, OH	16.0	ground level	
Middletown, OH	14.0	Cape Kennedy, 55-m from ocean	64.0
Waterbury, CT	11.0	18-m elevation	
Newark, NJ	24.7		
Bayonne, NJ	37.7	Galeta Point Beach, Panama	336.0
Morenci, MI	7.03	Limon Bay, Panama	30.3
Detroit, MI	7.03	Fort Amidor, Panama	7.10
East Chicago, IN	41.1	Miraflores, Panama	20.9
Manila, Philippines	26.2		
Halifax, (York Redoubt) N.S.	13.3	London (Battersea) England	23.0
Halifax (Federal Bldg.) N.S.	55.3	London (Stratford) England	54.3
Ottawa, Ontario, Canada	9.60	Melbourne, Australia	12.70
Montreal, Quebec, Canada	11.44	Widness, England	174.0
Trail, B.C., Canada	16.9	Pilsey Island, England	50.0
Norman Wells, NWT, Canada	0.73	Dungeness, England	238.0
Saskatoon, Sask. Canada	2.77		
Vancouver Island, Canada	6.50		

From Metal Corrosion in the Atmosphere, ASTM STP 435, ASTM, Philadelphia, p. 383, 1968.

Periodic drying is required for the surface film to develop its protective properties. Fully immersed HSLA steel has the same corrosion rate as that of ordinary carbon steel (Section 11.1). Apparently, iron oxides and/or sulfates must be allowed to precipitate and seal oxide pores during the drying cycle. Structures using HSLA steel for corrosion resistance must be designed to avoid features that retain water for long periods. Constantly shaded surfaces, as well as structures in northern latitudes, experience less dryout time, and corrosion rates of the HSLA steels are often increased.[7] Columns and other structural members must be painted on the lower ends exposed to constant wetting by foliage or soil. Moisture retained in bolted structural lap joints can produce voluminous corrosion products (Section 7.4.3) which distort the joints unless minimum bolt spacing guidelines are followed.[7]

The HSLA weathering steels provide an excellent base for paint coatings. It has been conservatively estimated that the life of most paint systems is doubled on weathering steel surfaces.

TABLE 12.2   Thickness Reduction of Steel in 15.5 Year Exposures

Specimen No.	Composition, wt%					Kearny, NJ (Industrial)		Thickness Reduction Kure Beach, NC, 250-m (800-ft) lot (moderate marine)	
	Cu	Ni	Cr	Si	P	μm	mils	μm	mils
1.......	0.012	...	...	...	...	731	28.8	1321	52.0
2.......	0.04	...	...	...	...	223	8.8	363	14.3
3.......	0.24	...	...	...	...	155	6.1	284	11.2
4.......	0.008	1	...	...	...	155	6.1	244	9.6
5.......	0.2	1	...	...	...	112	4.4	203	8.0
6.......	0.01	...	0.61	...	...	1059	41.7	401	15.8
7.......	0.22	...	0.63	...	...	117	4.6	229	9.0
8.......	0.01	...	...	0.22	...	373	14.7	546	21.5
9.......	0.22	...	...	0.20	...	152	6.0	251	9.9
10......	0.02	...	...	...	0.06	198	7.8	358	14.1
11......	0.21	...	...	...	0.06	124	4.9	231	9.1
12......	...	1	1.2	0.5	0.12	66	2.6	99	3.9
13......	0.21	...	1.2	0.62	0.11	48	1.9	84	3.3
14......	0.2	1	...	0.16	0.11	84	3.3	145	5.7
15......	0.18	1	1.3	...	0.09	48	1.9	97	3.8
16......	0.22	1	1.3	0.46	...	48	1.9	94	3.7
17......	0.21	1	1.2	0.48	0.06	48	1.9	84	3.3
18......	0.21	1	1.2	0.18	0.10	48	1.9	97	3.8

From C. P. Larabee and S. K. Coburn, First Int. Cong. Metallic Corrosion, *Butterworth, Sevenoaks, Kent, England, p. 276, 1962. Reprinted by permission, Butterworths Publishers Ltd.*

12.1.3 Electrochemical Mechanism

When exposed to dry or humid air, steel or iron forms a very thin oxide film composed of an inner layer of magnetite, Fe_3O_4 (FeO $\cdot$ Fe_2O_3), covered by an outer layer of FeOOH (rust). Some Fe in magnetite is in the ferrous (FeO) and the rest in the ferric (Fe_2O_3) oxidation state. The outer FeOOH layer is penetrated by fissures, from which most of the water has evaporated, allowing complete oxidation of magnetite to hydrated Fe_2O_3 or FeOOH ($Fe_2O_3 \cdot H_2O = 2FeOOH$), due to ready access of atmospheric oxygen. Pores in the magnetite are filled with condensed water and become plugged by insoluble oxide corrosion products. Thus, the oxide film remains protective in the presence of uncontaminated water vapor, as shown in Figure 12.3.

Ambient air usually contains some amount of SO_2, which reacts with water and dissolved oxygen to form sulfuric acid in the oxide pores. The acid partially dissolves the oxide, producing ferrous sulfate, $FeSO_4$, which in turn hydrolyzes to form additional acidity. The entrained acid more readily dissolves the oxide, opens pores in the inner magnetite layer, and permits easier access for the electrolyte solution to the underlying metal surface. Also, $FeSO_4$ is hygroscopic and attracts water from the

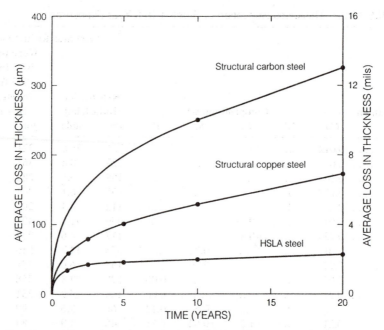

FIGURE 12.5 Effect of copper and other alloying elements on the long-term atmospheric corrosion resistance of steel. *(From C. P. Larabee and S. K. Coburn, First Int. Cong. Metallic Corrosion, Butterworth, Sevenoaks, Kent, England, p. 276, 1962. Reprinted by permission, Butterworths Publishers Ltd.)*

atmosphere, accounting for enhanced corrosion above a critical humidity in the presence of SO_2 (Figure 12.3).

The electrochemical mechanism,[9] summarized schematically in Figure 12.7, arises from the anodic dissolution of iron,

$$Fe \rightarrow Fe^{2+} + 2e^-,$$

under the inner magnetite layer. Ferrous ions in the saturated, or near-saturated, solution within the pores of the magnetite react with oxygen at the outer magnetite surfaces to form additional magnetite by

$$3Fe^{2+} + 2OH^- + \tfrac{1}{2}O_2 \rightarrow Fe_3O_4 + H_2O.$$

The cathodic reduction reaction is

$$8FeOOH + Fe^{2+} + 2e^- \rightarrow 3Fe_3O_4 + 4H_2O,$$

in which ferric iron in FeOOH (rust) is reduced to ferrous iron in Fe_3O_4 (magnetite) at the interface between the two, as indicated in Figure 12.6. Atmospheric oxygen, migrating through open fissures in the outer FeOOH layer, can then reoxidize the magnetite to rust by

$$3Fe_3O_4 + 0.75O_2 + 4.5H_2O \rightarrow 9FeOOH.$$

A certain amount of insoluble sulfate precipitates in the oxide and must be replenished by SO_2 from the atmosphere. It is important to note that diffusion of oxygen through

FIGURE 12.6   Surface oxides on (a) carbon and (b) low alloy steels showing ruptured and continuous films, respectively. *(From W. P. Gallagher, presented at the
1970 Spring Meeting, NACE, Houston, Mar. 1970.)*

either the bulk oxide or the entrapped water solution is not necessary,[2] explaining the
linear penetration rate and loss of film protectiveness shown in Figure 12.2.

Should the anodic reaction in Figure 12.7 become localized, anion enrichment
may occur locally, as it does in pitting (Section 7.2.2), and "sulfate nests"[10] may be
present in the corrosion products. However, significant macroscopic pitting is seldom a concern in atmospheric corrosion.

12.1.4 Prevention

Atmospheric corrosion can be controlled by two methods: coatings and alloy substitution. Many of the coating systems described in Chapter 14 are designed for the control of atmospheric corrosion. Often, economic analysis favors a coating system on

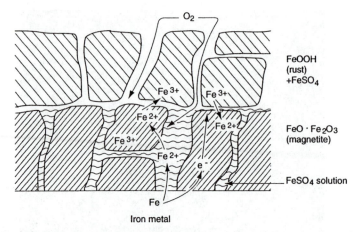

FIGURE 12.7 Schematic diagram showing electrochemical mechanism of atmospheric rusting in an SO_2-polluted environment.

inexpensive carbon steel over an uncoated, more expensive, corrosion-resistant alloy. However, if cosmetic appearance and maintenance cost are necessary considerations, many of the stainless steels and nickel alloys perform in a variety of atmospheres with almost no evidence of surface staining or corrosion. If a surface is out of public view, and appearance is not of great concern, simply increasing the section thickness to compensate for general corrosion may be the least expensive alternative.

Copper has been used for many years for roofing and drain gutters. Atmospheric exposures develop a pleasing green patina on the surface. Weathering steels that form a red-brown oxide rust film on the surface have already been mentioned in this section. Aluminum alloys are resistant to neutral atmospheres and have been used extensively for architectural trim and window and door hardware. The surfaces of aluminum alloys, like the weathering steels, should be well drained to prevent accumulation of water and debris. Stagnant water on aluminum alloys can change pH with time and cause staining and corrosion.

12.2 Oxidation at Elevated Temperature

The remaining sections of this chapter introduce the principles of high-temperature oxidation of pure metals and briefly review the selection of engineering alloys for elevated temperature service. The textbooks by Birks and Meier[11] and Kofstad[12] are recommended for additional fundamental discussions of alloy oxidation, reactions in mixed gases, formation of multiphase oxides, and hot corrosion in deposited molten salts.

12.2.1 Oxidation Reactions

A metal, M, reacts with oxygen or other gases at high temperature by initial adsorption of oxygen, chemical reaction to form the surface oxide, oxide nucleation, and lateral growth into a continuous film, that may protect the underlying metal. The

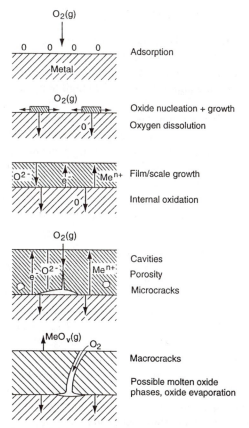

FIGURE 12.8 Film and scale formation during high-temperature metal oxidation. *(From P. Kofstad,* High Temperature Corrosion, *R. A. Rapp, ed., NACE, Houston, p. 123, 1983. Reprinted by permission, National Association of Corrosion Engineers.)*

film may also thicken into a nonprotective scale with various defects including cavities, microcracks, and porosity, as shown schematically in Figure 12.8.[13]

Oxidation in air by oxygen proceeds according to a reaction such as,

$$M + O_2 \rightarrow MO_2. \tag{1}$$

More generally,

$$xM + 1/2(yO_2) \rightarrow M_xO_y.$$

On the basis of one mole of O_2 reacted, as in (1),

$$(2x/y)M + O_2 = (2/y)M_xO_y. \tag{2}$$

A metal, M, also can be oxidized similarly by either water vapor or carbon dioxide according to

$$xM + yH_2O \rightarrow M_xO_y + yH_2 \tag{3}$$

and

$$xM + yCO_2 \rightarrow M_xO_y + yCO. \tag{4}$$

The M_xO_y oxide, as formed on the metal surface in Figure 12.8, becomes a barrier between the substrate metal and the oxidizing environment. The chemical and physical properties of the oxide film are of paramount importance in determining the rate of oxidation and the life of equipment exposed to high-temperature oxidizing environments.

Thermodynamics, as discussed in the following section, reveals whether any of reactions, (2), (3) and, (4), are possible under given conditions. Thus, it may be possible to adjust conditions to prevent oxidation during annealing, heat treatment, and other high-temperature processes. However, the rates of such reactions cannot be predicted from thermodynamics. Subsequent sections are devoted to oxide film structure, morphology, and properties that affect the ability to serve as a protective barrier and, in turn, affect the rate of high-temperature oxidation.

12.2.2 Thermodynamics of Oxidation

Each of the reactions (2), (3), and (4) for any metal is characterized thermodynamically by a standard free-energy change $\Delta G°$ which must be negative in order for the reaction to proceed spontaneously from left to right as written, with all reactants and products at standard state. Because $\Delta G° = \Delta H° - T\Delta S°$, a plot of $\Delta G°$ versus T approximates a straight line, with changes in slope where new phases form (i.e., at melting or boiling temperatures). A number of such plots for various oxidation reactions (2) at standard state ($p_{O_2} = 1$ atm) are shown in Figure 12.9,[14] which is known as the Ellingham diagram and shows the relative thermodynamic stability of the indicated oxides. The lower on the diagram, the more negative the standard free energy of formation and the more stable the oxide.

For reaction (2), a departure from standard conditions for the reactants or products gives a new free-energy change, $\Delta G_{O_2,MO}$;

$$\Delta G_{O_2/MO} = \Delta G°_{O_2/MO} + RT \ln \frac{a_{M_xO_y}{}^{2/y}}{(a_M{}^{2x/y})(a_{O_2,MO})}, \tag{5}$$

where R is the gas constant and T is absolute temperature. Because activities of pure solids in the stable form are defined as unity at all temperatures and pressures, equation (5) reduces to

$$\Delta G°_{O_2MO} = RT \ln p_{O_2MO}, \tag{6}$$

when $\Delta G_{O_2MO} = 0$ at equilibrium and $p_{O_2,MO} = a_{O_2,MO}$ by definition. Equation (6) defines the relationship between the standard-state free-energy change, $\Delta G°_{O_2MO}$, for formation of the pure oxide, MO, on the base metal, M, and the characteristic equilibrium oxygen pressure, $p_{O_2,MO}$, for dissociation of that oxide at any temperature, T.

The nonstandard state oxygen dissociation pressures leading to oxide formation or reduction on pure metals can be found from the nomogram scales shown in Figure 12.9 on the sides of the diagram. Consider copper oxidation at 900°C as an example. Extending a line from **O** on the left scale through the free energy line at 900°C for copper to the scale marked p_{O_2} at the right shows that p_{O_2,Cu_2O} for copper is

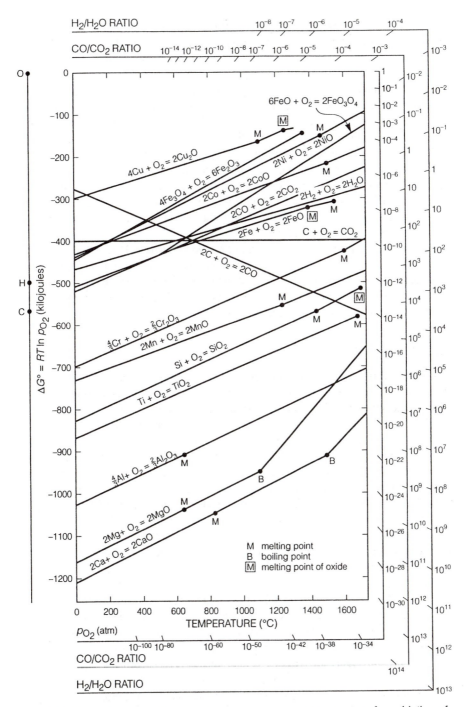

FIGURE 12.9 Ellingham diagram of free energy versus temperature for oxidation of metals. *From D. R. Gaskell,* Introduction to Metallurgical Thermodynamics, *2nd ed., McGraw-Hill, p. 287, 1983. Reprinted by permission, McGraw-Hill Book Company.)*

about 10^{-8} atmospheres. Any oxygen partial pressure above this value will oxidize pure copper; any below it will reduce pure copper oxide to pure copper at 900°C. Clearly, the oxides for most metals are rather stable in environments containing oxygen.

Similar predictions can be made for the atmospheres resulting from mixtures of water vapor and hydrogen and carbon monoxide and carbon dioxide in Figure 12.9. Again using the example of pure copper at 900°C, extending a line from the point **H** on the left scale through the free-energy line for copper at 900°C to the scale marked H_2/H_2O ratio shows an equilibrium ratio of hydrogen to water vapor of 10^{-4}. For a higher ratio, cuprous oxide is reduced to pure copper; below it, copper is oxidized. Extending a dashed line from **C** through the the the same free-energy point to the scale marked CO/CO_2 ratio predicts an equilibrium ratio of carbon monoxide to carbon dioxide of again about 10^{-4} at 900°C. Again, for a higher ratio, oxide is reduced on copper; below it, copper is oxidized.

The thermodynamic Ellingham (Figure 12.9) and Pourbaix (Figure 2.5) diagrams share a common failing: they cannot predict rates of corrosion. However, the Ellingham diagram is more useful at elevated temperatures, where the rates of most chemical reactions are much higher, than the Pourbaix diagram at room temperature, where equilibrium is usually attained very slowly.

12.2.3 Oxide Structure

The oxides formed on metal surfaces have some unique structural features that decide the oxidation behavior. A brief discussion of such properties is necessary here to understand the oxidation processes described in the next section.

Metal oxides are seldom exactly stoichiometric. They usually have either an excess or deficit of metal, equivalent to a deficit and excess of oxygen, respectively. A *metal-deficit oxide*, for example, the idealized MO shown in Figure 12.10a, contains M^{2+} cation vacancies. An example is NiO. The absence of an M^{2+} cation leaves behind a vacancy of relative charge -2. Cations diffuse in the lattice by exchange with these cation vacancies. Every cation vacancy is neutralized electrically by a pair of electron "holes," which are equivalent to M^{3+} cations, each having an electron deficit or excess charge of $+1$. The electron hole carries current by exchanging electrons with neighboring normally charged cations. Because current is carried by electron holes of positive charge, the metal deficit oxides are classified as *p*-type semiconductors.

A Li^+ solute ion present in the MO lattice in Figure 12.10b replaces a cation vacancy and maintains charge neutrality by pairing with an M^{3+} electron hole. Thus, the cation vacancy concentration is decreased and the diffusivity of M^{2+} is decreased as well. On the other hand, a Cr^{3+} solute ion in the oxide lattice, with excess positive charge shown in Figure 12.10c, is charge compensated by formation of additional cation vacancies, which increases the M^{2+} diffusivity.

A *metal-excess oxide* carries the excess as either interstitial cations or as oxygen anion vacancies. Interstitial cations, small enough to fit into interstitial spaces in the oxide lattice, as shown in Figure 12.11, jump from one interstitial position to another during diffusion. An example is ZnO. Oxides, with cations too large or too high-

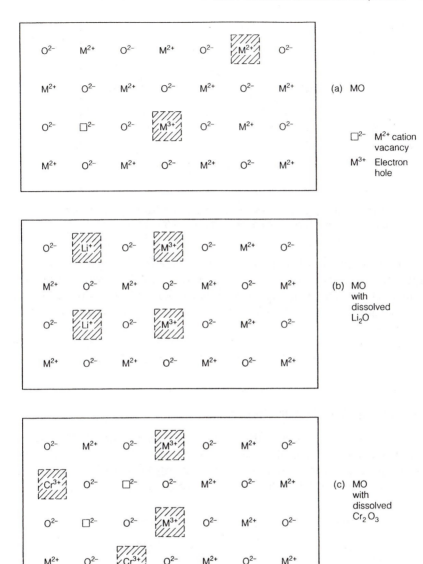

FIGURE 12.10 Schematic diagrams for (a) p-type MO metal deficit oxide semiconductor with cation vacancies; (b) decreased concentration of cation vacancies in MO by doping with Li^+; (c) increased concentration of cation vacancies in MO by doping with Cr^{3+}. NiO is an example of this type oxide.

ly charged to assume an interstitial position, require anion vacancies to form the metal-excess structure, as in Figure 12.12. For example, in ZrO_2, +2 charged oxygen anions migrate by exchange with the anion vacancies. In either case, charge compensation requires an excess of negative conduction electrons, which account for electrical conductivity, and both types of metal excess oxides are classified as n-type

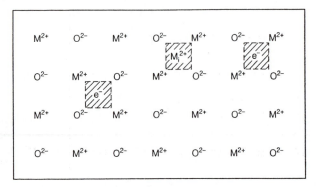

FIGURE 12.11 Metal excess *n*-type metal oxide with excess interstitial M_i^{2+} cations and free electrons, e^-. ZnO is an example of this type oxide.

semiconductors. Dopant additions have the opposite effect on metal-excess oxides compared to the metal-deficit oxides. Lower-valence additions increase either the interstitial cation or the anion vacancy concentrations. Anion diffusivity is increased correspondingly. Higher-valence additions decrease the anion diffusivity by a similar mechanism.

These descriptions of metal oxide defect structures are necessarily brief. The interested reader is referred to Birks and Meier[15] or Kofstad[12] for more complete discussions.

12.2.4 Oxide Film Growth Processes

Figure 12.12 shows the processes that occur across continuous oxide films having (a) metal excess with interstitial cations, (b) metal excess with anion vacancies, and (c) metal deficit with cation vacancies. M is arbitrarily chosen as divalent for uniformity, as in reaction (1). For all three, M atoms are oxidized to M^{2+} by

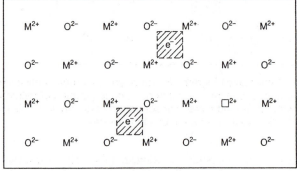

$\square^{2+}$– Anion vacancy

FIGURE 12.12 Metal excess *n*-type oxide with anion vacancies, $\square^{2+}$ and free electrons e^-. ZrO_2 is an example of this type oxide.

$$M \rightarrow M^{2+} + 2e^-, \tag{8}$$

liberating a pair of electrons, which migrate across the oxide to participate in the reduction of molecular oxygen,

$$\tfrac{1}{2}O_2 + 2e^- \rightarrow O^{2-}, \tag{9}$$

at the oxide-gas interface. The reactions (8) and (9) are, in effect, electrochemical anodic and cathodic reactions, respectively, the oxide acting as a solid-state electrolyte.

Electron transport occurs simply by movement of free electrons in n-type oxides (Figures 12.13a and 12.13b), but in the p-type oxides, electron holes migrate (Figure 12.13c) by electron transfer between neighboring M^{2+} cations with no ionic movements. They are able to do so because of the low energy difference between different oxidation states (e.g., M^{2+} and M^{3+}) for transition metals. The concentration of the dominant oxide defects (i.e., cation interstitials, anion vacancies, or cation vacancies) affect the diffusivity and, in turn, the rate of the oxidation reaction (Section 12.3.1). In these oxides, electrons or electron holes migrate much faster than any of the ionic defects.

Explanation of growth for the *n-type cation interstitial oxide* in Figure 12.13a is the most straightforward. The dominant oxide defects, interstitial cations, M_i^{2+}, are liberated at the oxide-metal interface by the anodic reaction (8) and then, by successively jumping to adjoining interstitial positions, migrate to the oxide-gas interface, where they react by reaction (9) to form O^{2-} at the surface. The O^{2-} reacts to form surface oxide by

$$O^{2-} + M_i^{2+} \rightarrow M_{ox} + O_{ox}, \tag{10}$$

where M_{ox} and O_{ox} represent metal and oxygen, respectively, incorporated into the oxide lattice. Reaction (8) also liberates electrons which enter the oxide and migrate as free electrons to the oxide/gas interface, where they participate in reaction (9).

For the *n-type anion vacancy oxide* in Figure 12.13b, O^{2-} anions migrate through the oxide lattice by exchange with anion vacancies to the oxide/metal interface where they react by

$$O^{2-} + M^{2+} \rightarrow M_{ox} + O_{ox}, \tag{11}$$

which is chemically identical to reaction (10). However, reaction (11) occurs at the oxide/metal interface, whereas reaction (10) occurs at the oxide/gas interface. Also, M^{2+} in reaction (11) migrates by vacancy exchange, whereas M_i^{2+} in reaction (10) migrates by jumping to neighboring interstitial sites. Anion vacancies, $\square_O^{2+}$, which are the dominant oxide defect in this type oxide, enable O^{2-} anion migration in the oxygen deficit oxide lattice. The lattice anions and anion vacancies must migrate in opposite directions as they exchange places.

The anodic reaction (8) provides M^{2+} at the oxide/metal interface for reaction (11) and also liberates free electrons, which migrate to the oxide/gas interface, where they participate in the reduction reaction (9), just as in Figure 12.13a for the interstitial-cation, n-type oxide. The O^{2-} anion, formed by reaction (9), is incorporated as O_{ox} in the oxide lattice by combining with an anion vacancy;

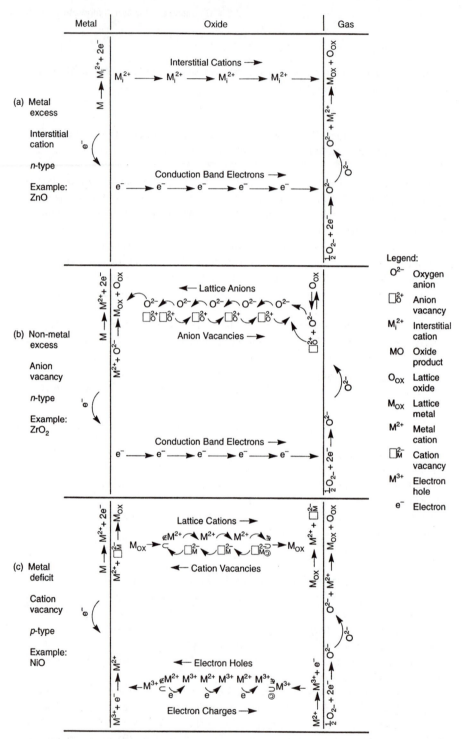

FIGURE 12.13 Processes occurring in three types of oxide surface scale during high-temperature oxidation.

$$\Box_O{}^{2+} + O^{2-} = O_{ox}. \tag{12}$$

Exchange between anion vacancies and oxide anions allows migration of O^{2-} to the oxide/metal interface, where the basic oxidation reaction (11) takes place.

The absence of the oxygen anion, O^{2-}, implies a +2 charge on the anion vacancy, i.e., $\Box_O{}^{2+}$. However, the +2 charge may attract one or two electrons, creating a vacancy of charge one, $\Box_O{}^{+}$, or zero, $\Box_O{}^{0}$, respectively. The presence of these defects may be distinguished by the effects of oxygen pressure on electrical conductivity and oxidation rate (Section 12.3.1).

For the *p-type metal-deficit oxide* in Figure 12.13c, M^{2+} cations are provided at the oxide/metal interface by reaction (8) and migrate to the oxide/gas interface by exchange with the dominant oxide defect, cation vacancies, $\Box_M{}^{2-}$. The exchange of cation vacancies and cations with metal M_{ox} in the lattice is represented by

$$M_{ox} = M^{2+} + \Box_M{}^{2-}. \tag{13}$$

At the oxide/gas interface, the M^{2+} cations form oxide ($M_{ox} + O_{ox}$) by reaction (10).

The M^{3+} electron holes migrate from the oxide/gas interface to the oxide/metal interface by accepting electrons from neighboring M^{2+} cations:

$$M^{2+} = M^{3+} + e^-. \tag{14}$$

No ionic movements, only electron transfers, are required for electron hole migration. Upon reaching the oxide/metal interface, M^{3+} electron holes are converted to M^{2+} cations by accepting electrons generated by the anodic reaction (8). Electronic charge is, in effect, transferred from the oxide/metal interface to the oxide/gas interface without ionic migration, by the opposite movement of electron holes.

The absence of the cation creates a cation vacancy of charge −2, relative to the normal site occupancy. One or two electron holes may be attracted to a negatively charged cation vacancy, creating a vacancy with a charge of −1 or zero, respectively. The presence of these defects can be detected by their effects on electrical conductivity and oxidation rates, as a function of oxygen pressure (Section 12.3.2).

Li^+ additions, which reduce the cation vacancy concentration in a *p*-type oxide (Figure 12.10), decrease the oxidation rate of nickel. Cr^{3+} additions up to about 5% increase the oxidation rate, as shown in Figure 12.14,[16] because they increase the concentration of cation vacancies (Section 12.2.3). Chromium above 5% begins to form protective chromium oxides, which then decrease the oxidation rate of the alloy. The same additions to an *n*-type anion-vacancy oxide have opposite effects because Li^+ increases concentration of anion defects, while Cr^{3+} decreases them. These oxide doping effects are mainly of interest for understanding the processes that occur in the oxide film during the oxidation process. Doping effects are not strong enough to provide the basis for oxidation resistant alloys. From a practical standpoint, alloying elements such as chromium and aluminum are added in sufficient quantities to form their own highly protective oxides on corrosion-resistant alloys.

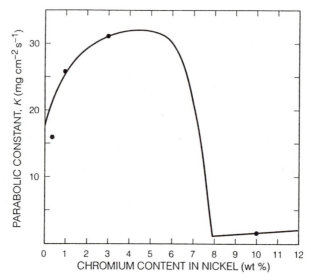

FIGURE 12.14   Effect of chromium on the parabolic rate constant, which is proportional to the oxidation rate (Section 12.3.1).   *(From data of Wagner and Zimens.[15])*

12.2.5 Oxide Properties

The oxidation rate of an alloy will be minimized if the oxide film has a combination of favorable properties which include:

1. The film should have good adherence, to prevent flaking and spalling.
2. The melting point of the oxide should be high.
3. The oxide should have low vapor pressure to resist evaporation.
4. The oxide film and metal should have close to the same thermal expansion coefficients.
5. The film should have high temperature plasticity to accommodate differences in specific volumes of oxide and parent metal and differences in thermal expansion.
6. The film should have low electrical conductivity and low diffusion coefficients for metal ions and oxygen.

The properties of relatively thin oxide films are difficult to measure or predict *in situ*. Therefore, the list above should be treated as qualitative. The actual behavior of metals and alloys in specific high-temperature oxidizing environments must be measured empirically and cannot be predicted from bulk oxide properties. And in fact, only chromium, aluminum, and silicon alloy additions have been found effective in forming protective high-temperature oxides. Nevertheless, an awareness of these factors leads to a better understanding of the processes involved in high-temperature oxidation.

12.2.6 Pilling-Bedworth Ratio

Like most ionically and covalently bonded chemical compounds, oxide scales are much stronger in compression than in tension. If the oxide scale has a greater specific volume than the parent metal, and if the oxide grows at the oxide/metal interface, the oxide as it forms will be in compression and will be more likely protective. Pilling and Bedworth[17] first proposed that the ratio of oxide to metal volume is a predictor of oxide protectiveness. The Pilling-Bedworth ratio is thus given by

$$\text{PB ratio} = \frac{\text{volume of oxide produced}}{\text{volume of metal consumed}} = \frac{Wd}{nDw},$$

where W is the molecular weight and D the density of the oxide, d is the density and w the atomic weight of the pure metal, and n is the number of metal atoms in the oxide molecule, e.g., 2 for Al_2O_3. It has been thought that the PB ratio should be slightly greater than 1, to foster a moderate compressive stress in the oxide and adherence to the substrate metal. Very high PB ratios may result in excessive compressive stresses, which buckle the film and destroy adherence.

PB ratios for selected metal/metal-oxides are given in Table 12.3, showing that most are greater than 1. However, contrary to the Pilling-Bedworth theory, most of the oxides listed are not protective, whether the ratio is near to or far from 1. For oxides which grow by oxygen migration to and reaction at the gas/oxide interface, compressive stresses are not necessarily formed in the oxide, invalidating the Pilling-Bedworth predictions. Specimen geometry may have an effect on the development of compressive stresses in the oxide film. Further, even metals with a PB ratio below 1 have continuous oxide films nearly up to the melting point of the

TABLE 12.3   Properties of Metal Oxides

Metal	Oxide	PB Ratio	Protec-tiveness[a]	Oxide Type[b]
Aluminum	Al_2O_3	1.28	P	n
Calcium	CaO	0.64	NP	n
Cadmium	CdO	1.42	NP	n
Cobalt	Co_2O_3	2.40	P	p
Copper	Cu_2O	1.67	P	p
Chromium	Cr_2O_3	2.02	P	p
Iron	FeO	1.78	P	p
Magnesium	MgO	0.81	P	n/p
Manganese	MnO_2	2.37	P?	n
Molybdenum	MoO_3	3.27	NP	n
Nickel	NiO	1.70	P	p
Lead	PbO	1.28	NP	p
Silicon	SiO_2	2.15	P	n
Tantalum	Ta_2O_5	2.47	NP	n
Titanium	Ti_2O_3	1.76	NP	n/p
Uranium	UO_2	1.97	NP	p
Tungsten	WO_3	1.87?	NP	n
Zinc	ZnO	1.58	NP	n
Zirconium	ZrO_2	1.57	P	n

[a] P: Protective; NP: Not Protective
[b] n: n-type semiconductor, metal excess oxide
   p: p-type semiconductor, metal deficit oxide

metal.[18] Thus, other properties, some listed in Section 12.2.5, must play a prominent role in most cases.

Furthermore, oxide films may be highly plastic and able to deform at high temperature, thus healing any porosity caused by a Pilling-Bedworth ratio < 1. Cracks and fissures may be filled by direct oxidation of metal at the base of such defects. Condensation of cation vacancies (Section 12.2.4) leads to oxide porosity which is not predicted from the PB ratio. Figure 12.9 shows that most oxides become less stable as the temperature increases, and many become volatile. For example, WO_3 has a PB ratio greater than 1 but is protective only up to 800°C, where it becomes volatile.

12.3 Oxidation Rate

Three kinetic laws—parabolic, linear, and logarithmic—describe the oxidation rates for many of the common metals and alloys. The oxygen reacted to form oxide on the metal surface, measured by weight gain, is shown as a function of time in Figure 12.15 for the three laws, which are discussed in more detail in the following three sections. The weight gain at any time, t, during oxidation is proportional to oxide thickness, x, and is commonly measured with a continuously recording microbalance, (e.g., Figure 12.16[19]).

Weight-gain measurements require careful interpretation, if the oxides evaporate or form nonadherent scales that continually spall off the surface. Certain metals, such as tungsten, tantalum, molybdenum, and columbium, desirable for high melting point and good high-temperature strength, must be coated for corrosion prevention in air, because of nonprotective, nonadherent oxides (Table 12.3) and otherwise high oxidation rates.

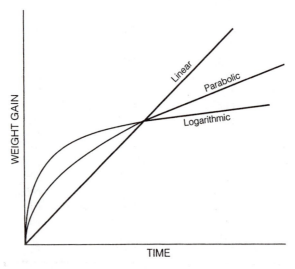

FIGURE 12.15 Weight gain versus time for the commonly observed kinetics laws for metal oxidation.

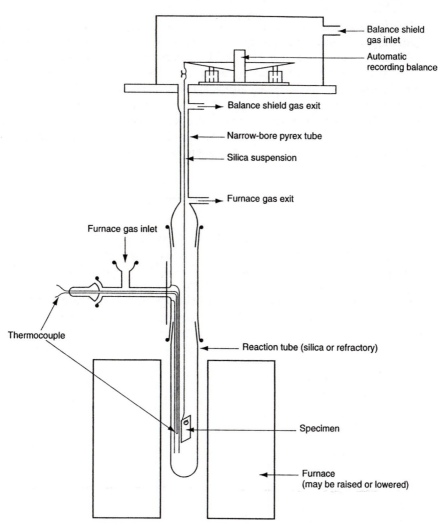

FIGURE 12.16   Schematic microbalance assembly for continuous recording of weight gain during high-temperature oxidation. *(From N. Birks and G. H. Meier, Introduction to High Temperature Oxidation of Metals, Arnold, London, p. 6, 1983. Reprinted by permission, Edward Arnold Ltd.)*

12.3.1 Parabolic Rate Law

The rate of high-temperature oxidation of many metals often follows the parabolic rate law, which requires that the square of film thickness, x, is proportional to time, t; that is,

$$x^2 = k_p t, \tag{14}$$

where k_p is known as the parabolic rate constant. The mechanism of parabolic growth has been explained by Wagner,[20] assuming that the oxide layer is compact and perfectly adherent and that migration of ions through the oxide layer is rate controlling. As indicated in Figure 12.12, either metal cations or oxygen anions may be the rate-controlling ionic species. Free electrons or electron holes are highly mobile compared to ions in the oxide lattice. For steady-state, reactant-ion concentrations (or more correctly, chemical potentials[15]) at the oxide-metal and oxide-gas interfaces and the ionic diffusivity, D, across the oxide scale may be assumed constant. Then, Fick's first law states that the steady-state flux of reacting ions is equal to $D(\Delta c/x)$ where Δc is the constant concentration difference across the oxide thickness, x. This flux is proportional to the rate of scale thickening, dx/dt, so that:

$$\frac{dx}{dt} = C D\left(\frac{\Delta c}{x}\right) \tag{15}$$

where C is a proportionality constant. Integration yields the parabolic relationship (14), where $k_p = C D\Delta c$ with $x = 0$ when $t = 0$. Thus, the parabolic rate constant, k_p, is directly proportional to the diffusivity of the rate-controlling ionic species for an electronically conducting oxide surface film.

Weight gain, W, is proportional to film thickness, x, and can be conveniently substituted for x in equation (14) with changes in the values of the constants k_p and C. Thus, a plot of x^2 (or W^2) versus time, t, will be linear if the parabolic rate law is obeyed. An example for pure cobalt is shown in Figure 12.17[21] at low oxygen pressures, where only the CoO oxide phase is present.

The parabolic rate law is the standard for analysis of high-temperature oxidation kinetics, in which diffusion through relatively thick films controls reaction

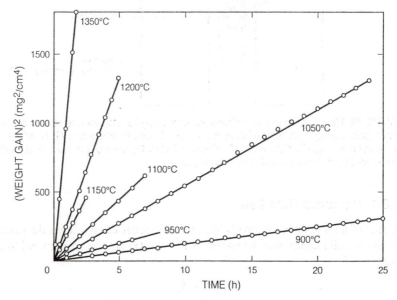

FIGURE 12.17 Parabolic plots of weight gain squared versus time for oxidation of cobalt at 900–1350°C. *(From F. R. Billman, J. Electrochem. Soc., Vol. 119, p. 1198, 1972. Reprinted by permission, The Electrochemical Society.)*

rates. Deviations are generally analyzed in terms of chemical and metallurgical effects on the relevant diffusion reactions. The parabolic law may not hold in early stages of oxidation before the film has developed sufficient continuity and thickness. Deviations for hafnium and zirconium, causing the exponent in equation (14) to be 3 (cubic law) rather than 2 (parabolic law), have been attributed to simultaneous dissolution of oxygen atoms into the substrate-metal lattice during oxidation.

Multiple-oxide phases on some metals further complicate the analysis of kinetics. Iron is a good example, shown in Figure 12.18,[22] forming a number of oxides because of its multivalent character. The oxide progresses from metal-rich to oxygen-rich; from predominantly Fe^{2+} in FeO at the oxide-metal interface to a mixture of Fe^{2+} and Fe^{3+} in the intermediate magnetite, Fe_3O_4 (FeO · Fe_2O_3), layer; and finally to Fe^{3+} in hematite, Fe_2O_3, at the outer surface exposed to the oxidizing environment. The numerous voids and fissures appearing in the oxide phases of Figure 12.18 are common in the oxide films of many other metals as well, and are often attributed to coalescence of cation vacancies (Figure 12.13c), migrating from the outer oxide surface toward the metal-oxide interface in the common p-type oxides. Such short-circuit diffusion paths as oxide grain boundaries, voids, fissures, precip-

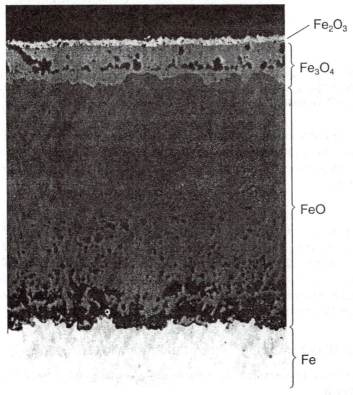

FIGURE 12.18   Microstructure of iron oxides formed on iron by high-temperature oxidation in air. *(From Fontana.[22])*

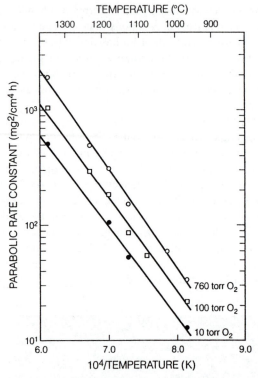

FIGURE 12.19 Arrhenius plot of parabolic rate constants for oxidation of cobalt, taken from Figure 12.16. *(From F. R. Billman, J. Electrochem. Soc., Vol. 119, p. 1198, 1972. Reprinted by permission, The Electrochemical Society.)*

itates, and brittle cracks in the thickened oxide film can cause deviations from the traditional parabolic rate law.

The temperature dependence of parabolic oxidation follows the usual Arrhenius relationship,

$$k_p = k_o e^{-Q/RT}, \tag{16}$$

where k_o is a constant, Q the activation energy, R the gas constant, and T absolute temperature. Thus, linearity is expected when logarithms of k_p, determined from the slopes in Figure 12.17, are plotted versus $1/T$ in Figure 12.19. The partial pressure of oxygen significantly affects the rate constant, but only weakly affects the activation energy. Oxygen pressure effects are discussed in further detail in Section 12.3.2, using the Wagner oxidation theory.

12.3.2 Effect of Oxygen Pressure on Parabolic Oxidation

The parabolic rate constant, k_p, is generally affected by oxygen partial pressure according to

$$k_p = C p_{O_2}^{1/n}, \tag{17}$$

where C is a proportionality constant. The oxidation of cobalt follows this relationship in Figure 12.18 with n = 3.1. It will be shown later in this section that n is negative for *n*-type oxides and positive for the *p*-type oxides.

The Wagner theory, incorporating the discussion of Section 12.2.4, provides insight into the effects of p_{O_2} on the rate constant, k_p. For the *n*-type interstitial oxides (Figure 12.12a), the summation of reactions (9) and (10) is

$$\tfrac{1}{2}O_2 + 2e^- + M_i^{2+} \rightarrow MO, \tag{18}$$

which is the overall reduction reaction affected by oxygen pressure. The equilibrium constant, K_{ZnO}, for reaction (18) is

$$K_{ZnO} = p_{O_2}^{-1/2}(e^-)^{-2}(M_i^{2+})^{-1},$$

where (e^-) and (M_i^{2+}) represent activities or concentrations of the indicated species in dilute solid solution, and the ZnO subscript represents any cation interstitial, *n*-type oxide. Because stoichiometrically, $2(e^-) = (M_i^{2+})$,

$$(M_i^{2+}) = K'_{ZnO}p_{O_2}^{-1/6}. \tag{19}$$

Equation (19) defines the effect of oxygen pressure, p_{O_2}, on the concentration of dominant defects in the *n*-type cation interstitial oxides. If an electron associates itself with an interstitial cation, a monovalent interstitial cation, M_i^+, would result, and reaction (18) would be modified as follows

$$\tfrac{1}{2}O_2 + e^- + M_i^+ \rightarrow MO.$$

Then equation (19) becomes

$$(M_i^{2+}) = K'_{ZnO}p_{O_2}^{-1/4}. \tag{20}$$

A determination of the effect of oxygen pressure on parabolic oxidation kinetics then proceeds as follows. From equation (15),

$$k_p = \frac{D_{M_i}[(M_i)_{M/Ox} - (M_i)_{Ox/g}]}{x}. \tag{21}$$

where D_{M_i} is the diffusivity for interstitial cations in the lattice. The M/Ox and Ox/g subscripts identify interstitial-cation concentrations at the metal-oxide and oxide-metal interfaces, respectively. Substituting (19) into (21) gives

$$k_p = \frac{D_{M_i}[p_{O_2,M/Ox}^{-1/6} - p_{O_2,Ox/g}^{-1/6}]}{x}. \tag{22}$$

Substituting (20) into (21) gives a similar result

$$k_p = \frac{D_{M_i}[p_{O_2,M/Ox}^{-1/4} - p_{O_2,Ox/g}^{-1/4}]}{x}. \tag{23}$$

The oxygen pressure at the oxide-gas interface is often much greater than that at the oxide/metal interface. Therefore, variations of $p_{O_{2,Ox/g}}$ have little effect on k_p in either of equation (22) or (23). Thus, oxygen pressure has no effect on k_p, irrespective of whether the interstitial cations are monovalent or divalent.

For the n-type, anion-vacancy oxide (Figure 12.13b), summation of (9) and (12) gives

$$\tfrac{1}{2}O^{2-} + 2e^- + \square_O^{2+} \rightarrow O_{ox}$$

for the overall reduction reaction involving oxygen. The equilibrium constant, K_{ZrO_2}, for this reaction is

$$K_{ZrO_2} = p_{O_2}^{-1/2}(e^-)^{-2}(\square_O^{2+})^{-1}.$$

The ZrO_2 subscript represents any anion vacancy, n-type oxide. In a manner similar to the other n-type oxide discussed previously, the effect of oxygen on the concentration of the dominant oxide defect is

$$(\square_O^{2+}) = K'_{ZrO_2}p_{O_2}^{-1/6}, \tag{24}$$

which is nearly identical to equation (19). The effect of oxygen pressure on the concentration of the dominant oxide defect is identical for both of the n-type oxides. Thus, according to equation (23), oxygen pressure has little or no effect on k_p for either of the n-type oxides, whether of the anion vacancy or the interstitial cation type.

For the p-type, cation vacancy oxide (Figure 12.13c), the summation of reactions (9), (11), (13), and (14) gives

$$\tfrac{1}{2}O_2 + 2M^{2+} \rightarrow O_{ox} + \square_M^{2-} + 2M^{3+}, \tag{25}$$

which is the overall reaction at the oxide-gas interface, affected by oxygen partial pressure. The equilibrium constant, K_{NiO}, for (25) is

$$K_{NiO} = (\square_M^{2-})(M^{3+})^2 p_{O_2}^{-1/2}. \tag{26}$$

The NiO subscript represents any cation-vacancy, p-type oxide. Activities of M^{2+} and O_{ox} are taken as unity in the oxide and do not contribute to (26). Therefore, the concentration of the dominant oxide defect is affected by oxygen pressure as follows;

$$(\square_M^{2-}) = K'_{NiO}p_{O_2}^{1/6}. \tag{27}$$

If an electron hole associates itself with a divalent-cation vacancy, (25) becomes

$$\tfrac{1}{2}O_2 + M^{2+} \rightarrow O_{ox} + \square_M^- + M^{3+}, \tag{28}$$

and (27) becomes

$$(\square_M^-) = K''_{NiO}p_{O_2}^{1/4} \tag{29}$$

for the monovalent-cation vacancy, $\square_M^-$, thus produced. If a pair of electron holes are associated with a divalent-cation vacancy, (25) becomes

$$\tfrac{1}{2}O_2 \rightarrow O_{ox} + \square_M, \tag{28}$$

and (27) becomes

$$(\square_M) = K'''_{NiO}p_{O_2}^{1/2} \tag{29}$$

for the neutral uncharged vacancy, $\square_M$.

The data for oxidation of cobalt (Figure 12.19) with n = 3.1 in equation (17) suggests that the oxide defects are primarily singly-charged cation vacancies, $\square_M^-$, with a significant fraction of neutral cation vacancies. The defect structure for zinc oxide (Figure 12.13a) is less easily verifiable experimentally, and although good evidence for zinc interstitial cations is available, details on relative numbers of, and charges on, the defects have not emerged.[15] Similarly, quantitative analysis of metal oxidation kinetics involving the *n*-type, anion-vacancy oxides (Figure 12.13b) has been difficult, because of oxygen solubility in such metals (e.g., zirconium).

12.3.3 Linear Rate Law and Breakaway

A linear law, as shown in Figure 12.15, may result, when a reaction at a phase boundary controls. Thus, any surface films or scales that may be present must be nonprotective. Strongly oxidizing conditions generally form thick, adherent scales, which result in parabolic kinetics. However, linear oxidation kinetics have been observed, when the oxidizing power of the environment is relatively low, such as for low oxygen pressures in a partial vacuum, low partial pressures of oxygen diluted with inert gas, and mixtures of CO and CO_2.[15] In such cases, oxidation reactions at the scale-metal interface have been found to control the linear oxidation kinetics. Control at the scale-gas interface has not been experimentally observed for any known oxidation process.[15]

In early stages of oxidation, film thickness may be sufficiently low to allow linear oxidation kinetics. As the film thickens, a transition to parabolic kinetics is often observed. Conversely, microcracking and porosity may develop as the film thickens, reducing the protectiveness of the oxide. The parabolic rate law may then fail and the kinetics approach linearity at some time after the start of reaction and scale growth. That is, linearity can develop with a constant oxidation rate after an initial period of parabolic behavior, as shown in Figure 12.20 for oxidation of tungsten at 700°C.[18] Some have suggested that the linear rate is governed by the steady-state

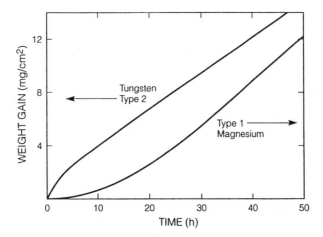

FIGURE 12.20 Linear oxidation of magnesium at 500°C and tungsten at 700°C showing linearity only after an initial nonlinear period. *(From J. E. Castle, Corrosion, Vol. 1, 2nd ed., L. L. Shreir, ed., Newnes-Butterworths, Sevenoaks, Kent, England, p. 1:241, 1976.)*

transport through an inner oxide, which remains constant in thickness with time. An outer layer, while thickening with time, remains porous and provides no additional barrier to the oxidation reaction. The controlling reaction then occurs at the boundary between the inner protective oxide layer and the outer nonprotective layer. This is the basis of so-called paralinear rate law,[15] which is often observed for such metals as niobium, tantalum, molybdenum, and tungsten. However, experimental observations have not always supported an inner layer of constant thickness. Instead, it is significant that all these metals have an appreciable solubility for oxygen, and the solid-state solution of oxygen may contribute to the observed oxidation kinetics.

A parabolic or cubic rate law may abruptly transform to linear nonprotective kinetics. This happens when the oxide film undergoes physical or chemical transformations that allow enhanced diffusion. Figure 12.21[18] shows transformation of parabolic to linear kinetics for magnesium at slightly lower temperature and oxygen pressure than shown in Figure 12.20.

If a protective surface film or scale breaks down at a continuously increasing number of sites, an increasing reaction rate, or breakaway, results. Uninhibited breakaway produces catastrophic reactions, as in the burning of magnesium in incendiary bombs.

Periodic oxide breakdown is well-known for zirconium alloys exposed to pressurized nuclear-reactor cooling water, as shown in Figure 12.22.[23] Weight-gain measurements are valid, because the oxide is adherent and insoluble. Consecutive parabolic or cubic increments have the same weight gain but shorter transition times. Thus, the weight gain curve evolves into linearity over long time periods, e.g., 5 years in Figure 12.22. Despite the eventual linearity, corrosion rates are low in the cooling water for nuclear reactors, where zirconium alloys, having high neutron transparency, are useful as cladding for uranium fuel elements. Such periodic transitions have been associated with recrystallization and resulting porosity in ZrO_2 surface films.[24]

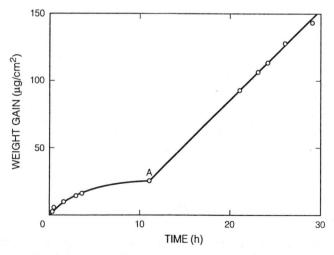

FIGURE 12.21   Transition of kinetics from parabolic to linear for oxidation of magnesium at 500°C. *(From J. E. Castle,* Corrosion, *Vol. 1, 2nd ed., L. L. Shreir, ed., Newnes-Butterworths.)*

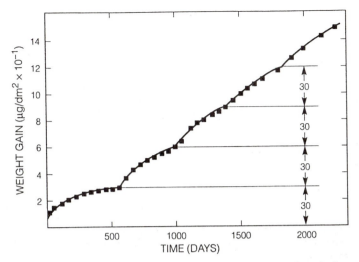

FIGURE 12.22 Period breakdown and deterioration to a linear law during oxidation of zirconium alloy in water at 288°C. *(From H. R. Peters, ASTM STP 824, ASTM, Philadelphia, p. 510, 1984. Reprinted by permission, American Society for Testing and Materials.)*

12.3.4 Logarithmic Rate Law

Often at lower temperatures, oxidation rate is inversely proportional to time, t:

$$\frac{dx}{dt} = \frac{k_e}{t}.$$

(13)

Integration results in the logarithmic rate law of oxidation,

$$x = k_e\log(at + 1),$$

(14)

where k_e and a are constants. Logarithmic oxidation is usually obeyed for relatively thin films at low temperatures. Thicker films and higher temperatures produce parabolic kinetics. The film thickness, measured by weight gain versus time behavior, shown in Figure 12.15, demonstrates a practical limiting oxide film thickness with time. Thin film reactions, in which the logarithmic rate law is significant, are seldom of concern for selection of materials for corrosion resistance. Therefore, the subject is treated only briefly here.

According to equation (14), a plot of film thickness versus logarithm of time should be linear if the logarithmic law is obeyed. An example is shown for nickel at relatively low temperatures in Figure 12.23.[25] Oxygen uptake was used as a sensitive measure of low film thickness. Logarithmic kinetics were obeyed faithfully at 200°C, but a changeover to parabolic kinetics occurred at 340°C above an oxygen uptake of 0.5 $\mu g/cm^2$, corresponding to a film thickness of about a 30-Å. This serves to emphasize that the logarithmic rate law is usually obeyed only for very thin oxide films.

The mechanism of logarithmic film growth is generally thought to involve electric fields across the film. Such fields are possible because of the much lower electronic oxide conductivity at the lower temperatures involved. Suggested controlling

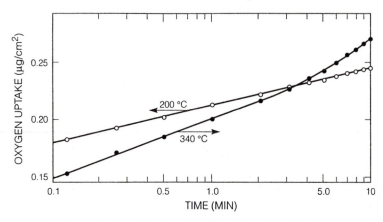

FIGURE 12.23   Logarithmic oxidation of nickel. *(From M. J. Graham and M. Cohen, J. Electrochem. Soc., Vol. 119, p. 879, 1972. Reprinted by permission, The Electrochemical Society.)*

processes include field-induced migration of cations across the film, quantum mechanical tunneling of electrons, and thermal emission of electrons from the metal-oxide interface.[26] None has been universally accepted, but the last by Uhlig[27] is of particular interest, because it accounts for observed effects of metallurgical and magnetic variables, whereas most others do not.

12.4   Alloying

The effects of individual alloying elements on high-temperature oxidation resistance[28,29] are discussed in the following section. Alloying effects on weldability and mechanical strength at high temperature also merit comment in many cases.

12.4.1   Effects of Individual Alloying Elements

Most alloys designed for high-temperature oxidation resistance contain *chromium*. Figure 12.24[30] shows the effects of increasing chromium content on the corrosion resistance and oxide morphology of Fe-Cr alloys oxidized at 1000°C. Oxidation rates decrease sharply as Cr is increased to 20%, and chromium oxides increase until they predominate in the oxide film. Similar effects are also present in nickel and cobalt alloys used in high-temperature service. Above 1000°C, Cr_2O_3 begins to vaporize, becoming critical above about 1200°C. If the oxide cannot be adequately replenished from the alloy, a type of breakaway (Section 12.3.3) corrosion will follow.

Simple iron-chromium alloys, the 400 series stainless steels, are acceptable in many applications. The major difficulty is the poor creep resistance of the body centered cubic ferritic structure at high temperature. The duplex, austenite-ferrite, stainless steels share this problem. Furthermore, these steels are difficult to weld,

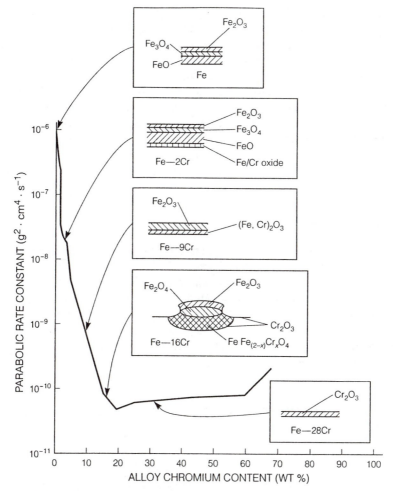

FIGURE 12.24 Effect of Chromium on the oxidation resistance and oxide morphology of iron-chromium alloys at 1000°C. *(From I. G. Wright,* Metals Handbook,
Vol. 13, Corrosion, *9th ed., ASM, Metals Park, OH, p. 97, 1987. Reprinted by permission, ASM International.)*

although many improvements in metallurgy and procedure have been implemented
recently (Section 9.4.4).

Nickel in conjunction with chromium improves high-temperature oxidation resistance of the stainless steels, just as it does aqueous corrosion resistance at lower temperatures. Alone, nickel has little effect on oxidation resistance of iron. Perhaps
more important, nickel stabilizes the austenite face centered cubic phase, which is
far more creep resistant than ferrite at high temperature. As will be shown later
(Section 12.5.1), nickel also gives improved resistance to thermal cycling during
high-temperature oxidation.

Silicon forms adherent resistant films alone and in conjunction with chromium. It
is added at 2 to 3% levels to many iron- and nickel-based chromium-bearing alloys

to improve corrosion resistance, especially at lower temperatures. However, its presence degrades creep resistance, thereby limiting high-temperature usefulness.

Aluminum forms protective oxides but at a lower rate than chromium. Also, aluminum forms brittle intermetallic phases with iron, restricting allowable alloy concentrations. Thus, efforts to replace strategic chromium with aluminum have been unsuccessful, because aluminum oxides do not form fast enough to repair mechanical damage during oxidation. Nevertheless, aluminum additions of a few percent are used frequently to enhance the oxidation resistance of iron- and nickel-based chromium-bearing alloys.

Molybdenum, tungsten, and niobium offer little improvement in oxidation resistance but are often added to enhance mechanical properties at high temperature.

12.4.2 Commercial Engineering Alloys

Chemical compositions of the common commercial heat-resistant alloys are given in the back flyleaf area of this book. Figure 12.25[31] summarizes oxidation in air of some common ferrous alloys as a function of temperature. Below 500°C, ordinary carbon steels perform suitably. Chromium-molybdenum steels may be required up to about 700°C, and about 1.5% Si additions improve performance considerably. Above 800°C, higher Cr levels are required in 400-grade ferritic stainless steels. Nickel additions in the austenitic stainless steels give improved resistance for equivalent chromium content.

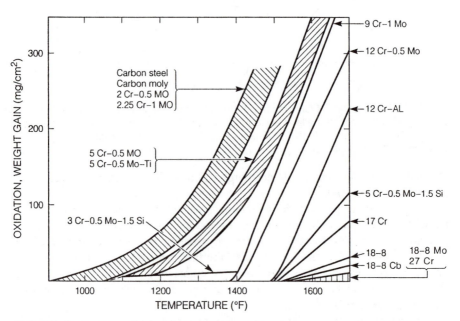

FIGURE 12.25 Weight gain of heat-resistant alloy compositions after 1000-hour exposure to air. *(From USX Corp.[29])*

Above 900°C, nickel-based chromium- and iron-bearing alloys are required in most applications. Additions of Cr and Al improve the oxidation resistance of nickel, much as they do for iron, but at higher temperatures for nickel. The same alloys used for service in oxidizing aqueous environments, such as Alloys 800 and 600, are also suitable for oxidation service in this temperature range. Additions of about 1.5% Al and increased Cr give superior resistance to cyclic oxidation and carburization in Alloy 601.[28] Additions of Mo give improved oxidation, as well as aqueous corrosion resistance, in Alloy 625. Further additions of Fe, Co, and W, as well as Mo, give metallurgically stable, oxidation- and creep-resistant Alloys 617, X, and S, which are useable in air up to about 1100°C.

Above about 1200°C, where Cr_2O_3 vaporizes, the refractory metals, molybdenum, tungsten, tantalum, and niobium, must be used for adequate mechanical strength. All form volatile or nonprotective oxides and must be coated for adequate oxidation resistance.

12.5 Oxidizing Service Environments

Isothermal testing in air or oxygen often provides the basis for selecting alloys for high-temperature service. In the absence of contaminants, oxygen normally forms resistant barrier scales on commercial heat-resistant alloys. However, many service environments incorporate thermal cycling and contain combustion products, such as water vapor, carbon monoxide and dioxide, and sulfur compounds. All have a detrimental influence on oxidation resistance and must be considered in alloy selection. Vapor contaminants establish their own equilibria with the various alloying elements in the metal, with the various oxides present, and with one another. Sulfur and carbon species can penetrate the oxide scales through pores and other defects and reduce resistance to oxidation. Many high-temperature gas streams in chemical and petroleum processing plants are intentionally oxygen deficient with variable sulfur, water, and carbon concentrations. The normally protective oxides are consequently less stable, and the alloys are further susceptible to enhanced scaling by less-resistant sulfides and embrittlement by carbides and hydrogen.

12.5.1 Thermal Cycling

Oxides that are ductile at high temperature are often brittle at lower temperatures. Most have different coefficients of thermal expansion than those of the metals from which they are formed, setting up thermal stresses when the temperature changes. Thus, an oxide formed at high service temperature may lose adherence to the substrate alloy, when cooled to lower temperatures, and become nonprotective when reheated. Alloys are generally less resistant to environments that include thermal cycling or periodic heating and cooling. Figure 12.26[32] shows the effects of thermal cycling at 982°C on weight losses for a number of common alloys. It is apparent that the higher nickel austenitic alloys are superior to those stainless steels of lower nickel but equal or nearly equal chromium content.

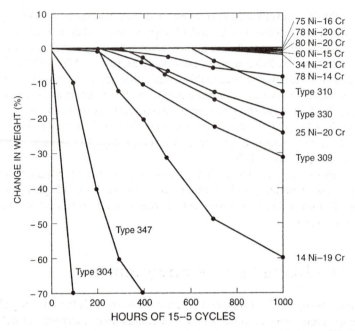

FIGURE 12.26 Weight loss in cyclic oxidation testing at 982°C (1800°F). Cycled 15 minutes at 982°C followed by 5 minutes of air cooling. *(From H. E. Eiselstein and E. N. Skinner,* ASTM STP 165, *ASTM, Philadelphia, p. 162, 1954. Reprinted by permission, American Society for Testing and Materials.)*

12.5.2 Sulfur

Sulfide scales and surface films are almost universally less protective than the corresponding oxides. Morphologies of the two are compared in Figure 12.27;[30] the sulfide scales are obviously more porous and less protective than the oxides. Oxygen and sulfur compete for reaction with the chemical elements present in the alloy. Oxygen is the stronger oxidizer, and protective oxides of aluminum and chromium normally form in preference to the sulfides. However, in many combustion-product gases, the oxygen content has been reduced, and a sulfur combustion product, such as SO_2 or H_2S, can more easily react with the alloy, especially at points of oxide breakdown. Sulfur then ties up the alloyed chromium and aluminum, allowing formation of base-metal (usually iron or nickel) sulfide scales, which are much less protective.[30]

Under certain conditions, sulfides formed at the metal-oxide interface can accelerate oxidation. Such sulfides are oxidized preferentially by the advancing oxide front, and the sulfides are continually displaced in "finger-like" protrusions of oxide and sulfide to greater depths in the metal.[30]

Alloys containing above 20% nickel may react with sulfur in oxygen-deficient environments to form the low-melting (645°C) $Ni-Ni_3S_2$ eutectic, which dissolves the nickel directly. It is notable that cobalt does not have such a low-temperature metal-sulfide eutectic, and cobalt alloys may be useful in some especially severe conditions. Similar "slagging" liquids, which dissolve the protective oxide film, may

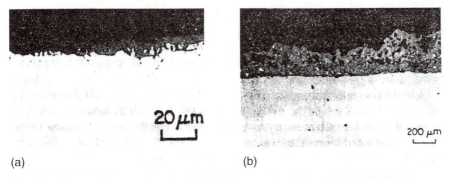

(a)   (b)

FIGURE 12.27   Oxide (a) and sulfide (b) film morphology on the Fe-Cr-Ni Alloy 800. *(From I. G. Wright, Metals Handbook, Vol. 13, Corrosion, 9th ed., ASM, Metals Park, OH, p. 97, 1987. Reprinted by permission, ASM International.)*

result from reactions between SO_2 and SO_3 and the chemical compounds in flue ash. Both iron and nickel alloys containing chromium and especially aluminum, preoxidized to form the protective alumina film, have been recommended for resistance to sulfur attack.[28]

12.5.3 Water and Carbon

Oxidizing environments containing water vapor are usually more aggressive than dry air or oxygen.[28] Thus, it has been a standard practice to include water vapor in engineering tests, because it is almost always present in service. Reaction (3) shows that oxidation by H_2O liberates hydrogen at the surface. Hydrogen, due to its small size and high mobility in solids can, in turn, cause any of the several hydrogen damage effects enumerated in Section 9.1. Hydrogen damage effects are, of course, aggravated by the presence of sulfides, especially H_2S. In the hydrogen attack mode (Section 9.1.6), hydrogen reacts with metallurgical carbon, decarburizing and blistering constructional steels. Carbon monoxide and dioxide can penetrate normally resistant oxides through pores, microcracks, and other defects much like SO_2 and SO_3. The metal can then be oxidized by reactions such as

$$M + CO_2 \rightarrow CO + MO$$

and

$$M + CO \rightarrow C + MO.$$

The carbon can then dissolve in the metal until it reaches sufficient activity to precipitate carbides. Carburization is again enhanced in high-temperature oxygen-deficient atmospheres containing water vapor, carbon dioxide, and/or carbon monoxide. Formation of a tenacious oxide film prevents carburization just as it does sulfur attack. Carburization is of great concern in hydrocarbon-gas streams of petrochemical and coking plants.

Carburization has several detrimental effects. Mechanical strength suffers; oxidation and sulfidation resistance is diminished, when protective chromium and nickel alloying elements are consumed as carbides; deeply embedded carbides can introduce stresses and distortion in parts of low cross section. "Green rot" is sometimes manifested in Fe-Cr-Ni alloys exposed to especially CO-rich environments, in which carburization is localized at grain boundaries. Bright-green nonprotective nickel oxides at the surface and on fracture surfaces are characteristic of the failures.

Carburization differs from the other oxidizing reactions in that the reduced carbon must diffuse through the metal before reacting. Thus, austenitic alloys with higher solubility for carbon are more susceptible, and the ferritic stainless steels are often recommended for resistance to carburization. Resistance of austenitic alloys to carburization, as well as sulfidization, is improved by additions of aluminum, which forms a resistant oxide film. Silicon additions are also effective, but only for relatively low-temperature service. Creep resistance suffers at higher temperature.

References

1. R. H. Heidersbach, CORROSION/89, Paper 129, NACE, Houston, 1989.
2. H. Kaesche, *Metallic Corrosion*, NACE, Houston, pp. 216–19, 1985.
3. P. W. Brown and L. W. Masters, *Atmospheric Corrosion*, W. J. Ailor, ed., Wiley, New York, p. 31, 1982.
4. *Metal Corrosion in the Atmosphere*, ASTM STP 435, ASTM, Philadelphia, p. 383, 1968.
5. C. P. Larabee and S. K. Coburn, *First Int. Cong. Metallic Corrosion*, Butterworth, Sevenoaks, Kent, England, p. 276, 1962.
6. W. P. Gallagher, *Application of the Scanning Electron Microscope to the Study of Rust Films*, presented at the 1970 Spring Meeting, Corrosion Research in Progress, Philadelphia, NACE, Houston, Mar. 1970.
7. S. K. Coburn and Y.-W. Kim, *Metals Handbook*, Vol. 13, *Corrosion*, 9th ed., ASM International, Metals Park, OH, p. 515, 1987.
8. C. P. Larabee, *Corrosion*, Vol. 9, p. 259, 1953.
9. U. R. Evans, *An Introduction to Metallic Corrosion*, 3rd ed., Arnold, London, p. 109, 1981.
10. E. Mattson, *Mater. Perform.*, Vol. 21, p. 9, July 1982.
11. N. Birks and G. H. Meier, *Introduction to High Temperature Oxidation of Metals*, Arnold, London, 1983.
12. P. Kofstad, *High Temperature Corrosion*, 2nd ed., Elsevier, New York, 1988.
13. P. Kofstad, *High Temperature Corrosion*, R. A. Rapp, ed., NACE, Houston, p. 123, 1983.
14. D. R. Gaskell, *Introduction to Metallurgical Thermodynamics*, 2nd ed., McGraw-Hill, New York, p. 287, 1983.
15. N. Birks and G. H. Meier, *Introduction to High Temperature Oxidation of Metals*, Arnold, London, pp. 32–62, 1983.
16. C. Wagner and K. E. Zimens, *Acta Chem. Scand.*, Vol. 1, p. 547, 1947.
17. N. B. Pilling and R. E. Bedworth, *J. Inst. Met.*, Vol. 29, p. 529, 1923.
18. J. E. Castle, *Corrosion*, Vol. 1, 2nd ed., L. L. Shreir, ed., Newnes-Butterworths, Sevenoaks, Kent, England, p. 1:241, 1976.
19. N. Birks and G. H. Meier, *Introduction to High Temperature Oxidation of Metals*, Arnold, London, p. 6, 1983.

20. C. Wagner, *Z. Physik. Chem.*, Vol. B21, p. 25, 1933.

21. F. R. Billman, *J. Electrochem. Soc.*, Vol. 119, p. 1198, 1972.

22. M. G. Fontana, *Corrosion Engineering*, 3rd ed., McGraw-Hill, New York, p. 509, 1986.

23. H. R. Peters, *ASTM STP 824*, ASTM, Philadelphia, p. 510, 1984.

24. B. Cox, *J. Nucl. Mater.*, Vol. 29, p. 50, 1969.

25. M. J. Graham and M. Cohen, *J. Electrochem.* Soc., Vol. 119, p. 879, 1972.

26. S. A. Bradford, *Metals Handbook*, Vol. 13, *Corrosion*, 9th ed., ASM International, Metals Park, OH, p. 61, 1987.

27. H. H. Uhlig, *Acta Metall.*, Vol. 4, p. 541, 1956.

28. R. H. Kane, *Process Industries Corrosion*, B. J. Moniz and W. I. Pollock, eds., NACE, Houston, p. 45, 1986.

29. G. Y. Lai, *High Temperature Corrosion of Engineering Alloys*, ASM International, Materials Park, OH, 1990.

30. I. G. Wright, *Metals Handbook*, Vol. 13, *Corrosion*, 9th ed., ASM International, Metals Park, OH, p. 97, 1987.

31. *Steels for Elevated Temperature Service*, USX Corp., Pittsburgh, PA, 1974.

32. H. E. Eiselstein and E. N. Skinner, *ASTM STP 165*, ASTM, Philadelphia, p. 162, 1954.

Exercises

12-1. Suggest possible causes of the following: **(a)** higher temperature decreases the atmospheric corrosion rate of steel, **(b)** the 25-m lot at Kure Beach is more corrosive than a 250-m lot, **(c)** architectural weathering steel panels corrode excessively against interior insulation, **(d)** a change in the fuel source increases atmospheric corrosion downstream from the exhaust stacks of a manufacturing plant.

12-2. Figure 12.2 shows linear corrosion kinetics and Figure 12.4 shows parabolic kinetics for atmospheric corrosion of carbon steel. Suggest a reason(s) for the difference.

12-3. Determine the oxygen partial pressure for decomposition of FeO at $1000°C$. Will a partial pressure of oxygen of 10^{-9} atm oxidize iron at this temperature?

12-4. The atmosphere in a processing furnace for silicon-integrated circuit boards has a measured ratio H_2/H_2O of 10^5. What temperature conditions must be maintained in the furnace to prevent decomposition of the SiO_2 insulating layers previously formed on the boards?

12-5. **(a)** Determine the free energy for oxidation of chromium at $300°C$ in an atmosphere with oxygen partial pressure of 10^{-4} atm. **(b)** Chromium plating in this environment is observed to remain bright with no apparent oxide tarnish. Does this observation violate thermodynamic principles? Explain.

12-6. Figure 12.13c indicates that electron charge migrates from the metal/oxide interface to the oxide/gas interface in *p*-type oxides by exchange between M^{2+} cations and neighboring M^{3+} electron holes. Would it be possible for conduction-band electrons in p-type oxides to migrate directly to the oxide/gas interface as they do in the n-type oxides? Explain.

12-7. Weight gain data for pure nickel at 900°C follows. Determine whether the data best follows parabolic or cubic oxidation kinetics.

Exposure time, min.	400	520	640	750	830	1000	1100	1200
Weight gain, mg/cm^2	1.511	1.73	1.87	2.09	2.21	2.37	2.51	2.79

12-8. Calculate the weight gain/cm^2 expected after 10 hours at 1000°C from the data of Problem 12-7, assuming an activation energy of 47.6 kcal/mole.

13

Cathodic Protection

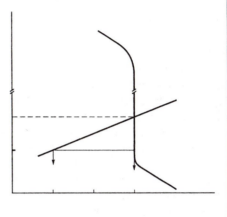

Cathodic protection is one of the most widely used methods of corrosion prevention. In principle, it can reduce or prevent the corrosion of any metal or alloy exposed to any aqueous electrolyte. Corrosion can be reduced to virtually zero, and a properly maintained system will provide protection indefinitely. The British first used cathodic protection on copper fittings of wooden sailing vessels in the 1820s. Unfortunately, the suppression of toxic corrosion products allowed growth of marine organisms, which impaired sailing speed. Cathodic protection was first used extensively in the 1920s for buried steel pipelines transporting petroleum products in the Gulf Coast oil fields of the United States. Cathodic protection of steel-hulled ships became prevalent in the 1950s, to supplement corrosion- and fouling-resistant coatings.

Cathodic protection finds its greatest use for coated carbon steels in many applications of intermediate corrosion rate. Thus, it extends the service life of thousands of miles of buried steel pipelines, oil and gas well casings, offshore oil-drilling structures, seagoing ship hulls, marine pilings, water tanks, and some chemical equipment.

13.1 Fundamentals

13.1.1 Principles and Applications

Cathodic protection reduces the corrosion rate by cathodic polarization of a corroding metal surface. Consider iron corroding in a dilute aerated neutral electrolyte solution. The respective anode and cathode reactions are

$$Fe \rightarrow Fe^{2+} + 2e^- \tag{1}$$
$$O_2 + 2H_2O + 4e^- \rightarrow 4OH^-. \tag{2}$$

Cathodic polarization reduces the rate of the half-cell reaction (1) with an excess of electrons, which also increases the rate of oxygen reduction and OH^- production by reaction (2).

A rectifier supplies impressed current for cathodic polarization by converting alternating current (ac) from the power lines to direct current (dc), as indicated in Figure 13.1.[1] In remote locations, where usual ac power lines are unavailable, external impressed current may be supplied by gas- or diesel-driven generators or solar power cells. Single or multiple anodes distribute the cathodic currents to the protected structure. For buried structures, the anodes are often inert graphite. For immersed seawater structures, they may be high-silicon cast iron or platinum-coated titanium.

A metal structure can be cathodically protected by connection to a second metal, called a sacrificial anode, which has a more active corrosion potential. The more noble (positive) structure in this galvanic couple is cathodically polarized, while the

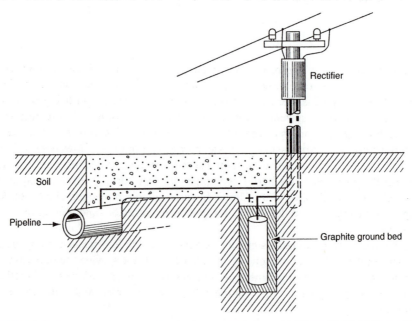

FIGURE 13.1   Cathodic protection by impressed current. *(From H. H. Uhlig and R. W. Revie,* Corrosion and Corrosion Control, *Wiley, New York p. 217, 1985. Reprinted by permission, John Wiley & Sons.)*

active metal is anodically dissolved, as discussed in Section 6.1.2. An example of buried steel protected with a sacrificial anode of magnesium is shown in Figure 13.2.[1] Magnesium, zinc, and aluminum sacrificial anodes (Section 13.2.8) of a few pounds to several hundred pounds, welded to buried and immersed structures, provide long-term cathodic protection. The sacrificial anodes must be replaced periodically as they are consumed by anodic dissolution. Figure 13.3 shows large aluminum sacrificial anodes installed on the legs of an offshore drilling platform.

Buried anodes, impressed current or sacrificial, are usually backfilled or prepackaged with conductive material, often granulated coke, to spread the current over a greater area and decrease the anode consumption.

Reaction (1) could be replaced by the anodic reaction for any metal, and the corrosion rate of any metal can be reduced by cathodic polarization. However, structures of carbon steel in mildly corrosive neutral aqueous environments, such as soil, water, and seawater are the greatest beneficiaries of cathodic protection. Some alloys, such as stainless steels and copper, are intrinsically corrosion resistant and have no need for additional protection. Others, such as aluminum and zinc, are subject to "cathodic corrosion" by the alkaline solutions formed by the cathodic reaction (Section 13.1.5), and the level of cathodic polarization must be carefully controlled. Cathodic protection is not economically feasible when corrosion rate is higher (e.g., steel in strong acid solutions), because adequate protection requires correspondingly high currents and large rectifiers. Copious evolution of flammable hydrogen by reaction (2) may be hazardous, and economic coatings to reduce the required currents generally fail in acid solutions.

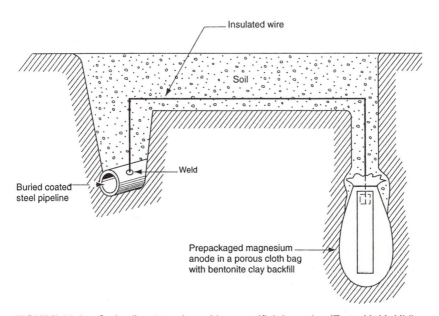

FIGURE 13.2 Cathodic protection with a sacrificial anode. *(From H. H. Uhlig and R. W. Revie,* Corrosion and Corrosion Control, *Wiley, New York p. 217, 1985. Reprinted by permission, John Wiley & Sons.)*

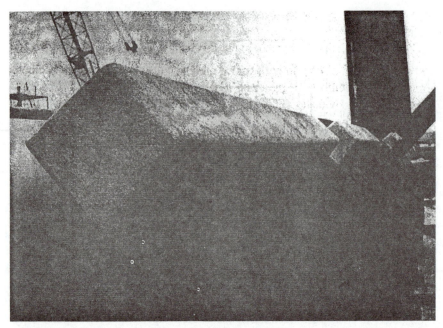

FIGURE 13.3   Aluminum sacrificial anodes installed on an offshore drilling platform. *(Photograph by courtesy of Kaiser Aluminum and Chemical Corp. Tulsa, OK.)*

The maximum corrosion rate of uncoated steel in aerated quiescent water is about 20 mpy (0.47 mm/y). This intermediate corrosion rate may be tolerable in many applications but not for permanent steel structures where long service life is expected. Thus, buried pipelines and offshore oil drilling platforms are the recipients of cathodic protection to reduce corrosion rate to negligible levels.

13.1.2 Impressed Current Cathodic Protection

The polarization diagrams introduced in Section 3.3.1[2] give a quantitative assessment of the decrease in corrosion rate caused by cathodic polarization. Figure 13.4 shows such a diagram for steel in acid solution, derived from the similar diagram of Figure 3.7 for zinc, with exchange current densities and Tafel constants in general agreement with those listed in Table 3.2. The unpolarized corrosion rate is about 10^3 $\mu A/cm^2$, or 460 mpy (Table 3.1). Cathodic polarization of about 120 mV will reduce corrosion rate by three orders of magnitude, to 1 $\mu A/cm^2$ (0.46 mpy), as indicated in Figure 13.4. The applied current, i_{app}, reduces the rate of the anodic reaction (1) to $i_a = 1$ $\mu A/cm^2$. The extent of reduction is governed by the value of the *anodic* Tafel constant, β_a. In this case, $\beta_a = 40$ mV, so that each 40 mV of cathodic polarization decreases corrosion rate by a factor of 10.

The cathodic polarization curve gives an indication of the impressed-current density, i_{app}, for cathodic protection to any required level of cathodic polarization. In the example of Figure 13.4, the required i_{app} of about 1.5×10^{-2} A/cm^2 is pro-

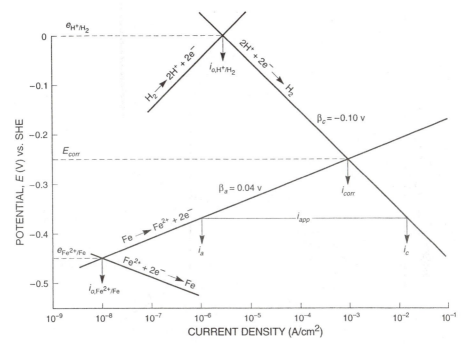

FIGURE 13.4 Cathodic protection with an impressed current density, i_{app}, for steel in an acid solution.

hibitively high. Only 1 m^2 of exposed area would require 150 A for protection! Unfortunately, the coatings usually used to decrease current requirements for cathodic protection are not resistant to strong acid solutions. Thus, cathodic protection is impractical for aggressive acid solutions with high corrosion rates, although Figure 13.4 clearly demonstrates that cathodic protection would be effective, if not practical, in acids.

Figure 13.5 illustrates the more usual case for cathodic protection of steel in neutral aerated water or seawater. Diffusion of dissolved oxygen to the corroding surface controls corrosion at about 100 $\mu A/cm^2$ (50 mpy). Corrosion rates in the nearly neutral pH range may be reduced to about 20 $\mu A/cm^2$ ($\approx$ 10 mpy) by accumulation of $Fe(OH)_2$ precipitated on the surface[3] in quiescent solutions (Section 11.1.1). Any degree of stirring or agitation restores the corrosion rate to the higher level. Assuming again that β_a = 0.040 V, as in Figure 13.4, a cathodic polarization of 120 mV reduces the corrosion rate now to 0.1 $\mu A/cm^2$. Because the applied current and corrosion rate are limited by i_L, the i_{app} is a maximum at 100 $\mu A/cm^2$. Although the required current is still sizable, it can be reduced in the presence of coatings, which in this case are resistant in neutral solutions. Also, in seawater containing magnesium and calcium salts, local alkaline conditions at cathodically protected surfaces precipitate insoluble protective coatings *in situ* (Section 13.1.5).

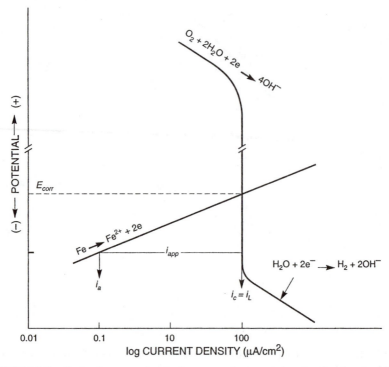

FIGURE 13.5 Cathodic protection by impressed current density, i_{app}, for steel in neutral aerated water.

If cathodic polarization becomes excessive in Figure 13.5, the direct reduction of water becomes thermodynamically possible:

$$2H_2O + 2e^- \rightarrow H_2 + 2OH^-. \tag{3}$$

The increase in current due to hydrogen evolution was shown earlier in polarization curves in Figure 3.16. Hydrogen evolution, even in neutral or alkaline solutions, (Section 14.1.4) destroys coatings and produces various forms of hydrogen damage (Section 10.1). Also, the new reduction reaction wastefully consumes additional cathodic current.

13.1.3 Sacrificial Anode Cathodic Protection

In a galvanic couple between dissimilar metals (Figure 6.1), the galvanic current cathodically protects the more noble metal and preferentially dissolves the more active metal (Section 6.1.2). Electrons flow from the active sacrificial anode to the noble cathode structure. The anodic reaction at the cathode structure, for example,

$$Fe \rightarrow Fe^{2+} + 2e^-, \tag{1}$$

is reduced by the surplus of electrons provided by the sacrificial anode. At the same time, the reduction of dissolved oxygen by reaction (2) or the evolution of hydrogen by (3) is accelerated. The cathode structure is cathodically protected, and the same electrochemical reactions are present at the cathode as when polarization is provided by impressed current.

In principle, when any two metals or alloys are galvanically coupled, the more active of the two in the galvanic series (Figure 6.1) becomes a sacrificial anode and cathodically protects the other. Thus, fasteners are cathodically protected when attached to alloys, which are active in the galvanic series. In a similar manner, a galvanized zinc coating serves as a distributed sacrificial anode, cathodically protecting the underlying steel at scratches and other breaks in the coating. The necessary galvanic currents flow in the electrolyte formed by thin aqueous films condensed from the atmosphere.

Anodic dissolution continually consumes sacrificial anodes which must be periodically replaced. Therefore, maximum anode life requires only minimum current for adequate protection. Only zinc, aluminum, and magnesium alloys have proven practical for sacrificial anodes. Properties of sacrificial anode alloys are discussed in Section 13.2.8.

Figure 13.6 shows a simplified polarization diagram for the galvanic couple (similar to Figure 6.7) between a sacrificial anode of $E_{corr(a)}$ and the cathodically protected structure of $E_{corr(c)}$. The two are polarized to the same "short-circuit" potential, E_{sc} (Section 6.3.1), with the galvanic current, $I_G(sc)$, flowing in the couple. The galvanic current provides cathodic protection in the same manner as an impressed current in Figures 13.3 and 13.4. At E_{sc} the corrosion rate of the structure has been reduced from I_{corr} to $I_{corr(sc)}$.

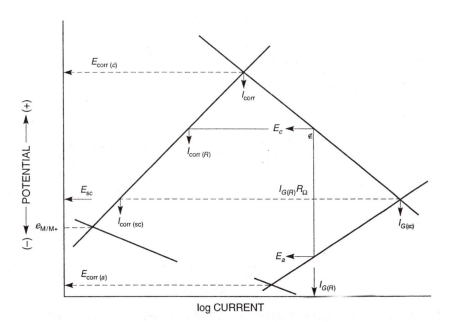

log CURRENT

FIGURE 13.6 Schematic polarization diagram for a sacrificial anode coupled to a cathodically protected metal structure. I_{corr} reduced to $I_{corr(sc)}$ by the galvanic current $I_{G(sc)}$. Presence of solution resistance R_Ω reduces galvanic current to $I_{G(R)}$, separates anode and cathode by a potential $I_{G(R)}R_\Omega$, and reduces I_{corr} to $I_{corr(R)}$.

The effect of solution resistance, R_Ω (Section 3.5.4), between a sacrificial anode and the cathode structure is also shown in Figure 13.6. With a resistance, R_Ω, a potential, $I_G(R)R_\Omega$ separates anode and cathode.[2] The corrosion current under protection decreases from $I_{corr(sc)}$ to $I_{corr(R)}$ by the presence of R_Ω, and the couple current is reduced to $I_G(R)$. $I_{corr(R)}$ may be adequately low, and anode consumption is reduced.

Figure 13.7 shows the polarization diagram for the galvanic couple formed by a zinc sacrificial anode coupled to a steel structure. The corrosive solution is assumed to be aerated water or seawater, and the polarization curves come from Figure 6.17. For equal 1-cm^2 areas of steel and zinc, the galvanic current I_G cathodically polarizes the steel surface (Section 6.3.1), and the corrosion current is reduced from $I_{corr} = I_L$ to some value below the lowest current, 1 μA, shown on the diagram. The cathode to anode area ratio is usually quite large, however, and a perhaps more realistic couple between 100 cm^2 of steel and 1 cm^2 of zinc is polarized by a current of I'_G. The corrosion current I_{Fe} is reduced to about 1 μA on 100 cm^2 for a current density of 10^{-8} A/cm^2 ($\approx$0.01 mpy), near the exchange current density (Table 3.2) for the anodic reaction, (1), Fe = Fe^{2+} + 2e$^-$.

A corrosion rate under cathodic protection of 0.01 mpy or less in Figure 13.7 is admirable but may be wasteful as well. The high current density on zinc, $\approx$20,000 μA/cm^2, produces rapid anodic consumption of the zinc anodes. This may be offset to some extent in seawater systems by using fewer anodes with greater solution resistance between the anode surface and more remote structure surfaces. The couple potential, E_G, is sufficiently active in Figure 13.7 to start hydrogen evolution on

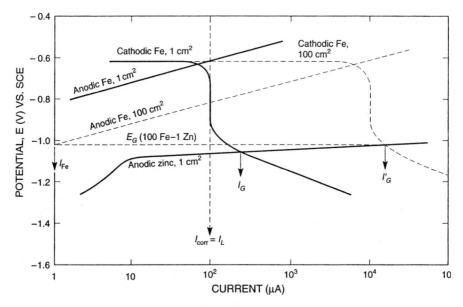

FIGURE 13.7 Polarization diagram for a zinc anode coupled to steel: polarization curves from Figure 6.17. I_G is the galvanic current for equal 1-cm^2 surfaces of steel and zinc. I'_G is the galvanic current for 1-cm^2 of zinc coupled to 100-cm^2 steel reducing the corrosion current to I_{Fe} = 1 μA.

the steel surface (Figure 13.4), as evidenced by a new cathodic Tafel slope at active potentials. Sacrificial anodes having somewhat less active potentials will relieve this situation, as described below.

Aluminum anodes have been developed for improved life in seawater in part because they polarize coupled steel surfaces less than zinc. Corrosion rates of the steel may be somewhat higher but are usually still acceptable. Other advantages of aluminum anodes are discussed in Section 13.2.8. As an additional example of desirable attenuation of anode activity, the life of 55% aluminum-45% zinc alloy coatings are two to four times greater than traditional zinc galvanized coatings, partly because the alloy polarizes the steel to a lesser degree, while still providing adequate protection[4] in thin film electrolytes condensed from the atmosphere (Section 14.3.1).

Magnesium sacrificial anodes are used frequently in soil, because the very active $E_{corr(a)}$ in Figure 13.6 is sufficient to overcome the large soil resistivities that are often present (Section 13.2.8). Magnesium consumption is modest, because the resultant high R_Ω limits the $I_{G(R)}$ that can flow between anode and cathode structure. Magnesium anodes are unsuitable in seawater, because the low solution resistivities allow high $I_{G(R)}$ and rapid consumption of the anodes. They are used frequently in glass- or enamel-lined (Section 14.2.2) domestic hot-water tanks, where the high electrolyte resistance again limits the magnesium consumption rates.

13.1.4 Coatings and Cathodic Protection

Cathodic protection provides an ideal supplement for coating systems. Imperfections in coatings are nearly always present in the form of pinholes, holidays, physical damage, and so on. Corrosion normally concentrates at the imperfections, undercuts the coating, and hastens failure. The coating is a high-resistance path for applied cathodic currents, which consequently concentrate at the coating imperfections. Thus, cathodic protection is effective at the weak points of the coating, which reduces the current requirements to a practical level for design.

Most immersed or buried steel structures use a combination of organic coatings with cathodic protection for corrosion prevention. Many such structures are coated during fabrication. However, some offshore structures may be left uncoated, because coatings are difficult to maintain in the hostile, deep, inaccessible ocean environment. The alkaline conditions promoted by cathodic protection in seawater deposit calcareous scales on the surface, which reduce corrosion rates and necessary cathodic protection currents, as discussed in the next section.

13.1.5 Cathodic Alkalinity

The reduction reactions,

$$O_2 + 2H_2O + 4e \rightarrow 4OH^- \tag{2}$$

and

$$2H_2O + 2e^- \rightarrow H_2 + 2OH^-, \tag{3}$$

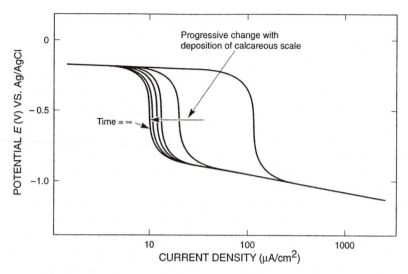

FIGURE 13.8 Effect of scale formation on cathodic polarization of steel in seawater. *(Adapted from Warne.[5])*

increase pH at the protected surface by generation of OH^- during cathodic polarization. To some degree this is beneficial, because the ferrous alloys are resistant to mildly alkaline solutions in which a protective oxide film is stable (Section 2.2.5). Also, generation of OH^- in seawater deposits calcareous scales by reaction with dissolved calcium and magnesium ions:

$$Ca^{2+} + HCO_3^- + OH^- \rightarrow H_2O + CaCO_3, \tag{4}$$

$$Mg^{2+} + 2OH^- \rightarrow Mg(OH)_2. \tag{5}$$

As shown in Figure 13.8,[5] scale deposition causes a continuous decrease in i_L for oxygen reduction, as scale thickness increases. The current necessary for cathodic protection decreases correspondingly. The reduced cathodic currents permit economic cathodic protection of immersed seawater structures without coatings, in some instances.

Alkaline solutions, however, degrade organic coatings. Resin coatings resistant to alkalinity are available, but overprotection is to be avoided. Excess cathodic polarization at exposed surfaces can also generate hydrogen by reaction (3), which can blister coatings. An "anode shield" (i.e., a supplemental coating of epoxy or other resistant material) near the impressed current anodes is often recommended[6] for seagoing ship hulls.

13.1.6 Controlled Potential Cathodic Protection

The cathodic protection industry has changed extensively in recent years from fixed-current to thyristor-controlled rectifiers, which automatically control current output to match corrosive conditions.[7] The advantages of a controlled-potential system[2] are shown in Figure 13.9. A steel surface exposed to seawater (seagoing ship hull) is

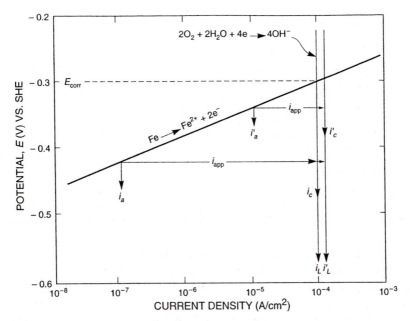

FIGURE 13.9 Controlled-potential cathodic protection compared with constant current cathodic protection in oxygenated salt water.

under diffusion control by dissolved oxygen. Variations in flow at the surface can change the limiting current density, i_L, by a factor of as much as 2. A 10% increase from 100 to 110 $\mu A/cm^2$ has a massive effect on the degree of protection when cathodic current is held constant. The corrosion rate increases 100-fold to 10 $\mu A/cm^2$ from only 0.1 $\mu A/cm^2$. However, corrosion rate stays constant under potential control at -0.42 V. The applied current increases from 100 $\mu A/cm^2$ to 110 $\mu A/cm^2$, to compensate for the greater i_L.

Controlled-potential cathodic protection is used extensively for ship hulls and many seawater installations in which flow velocity is variable. Figure 13.10 shows a seagoing vessel outfitted for controlled-potential cathodic protection.[8] A major design problem for such systems is the durability of reference electrodes.

The low polarization of the zinc anode in Figure 13.7 holds the couple potential, E_G, relatively constant, as the solution flow and dissolved-oxygen transport vary. Thus, a sacrificial anode of low polarization retains the advantages of controlled-potential cathodic protection. As solution flow rate varies, the rate of anode dissolution changes to maintain potential near E_G.

13.1.7 Comparison with Anodic Protection

Both anodic protection (Section 4.4) and cathodic protection utilize electrochemical polarization to reduce corrosion rates. Otherwise, their mechanisms are different, as well as the methods and equipment for implementation. Anodic protection uses an anodic current to polarize the corroding surface into a potential region where pas-

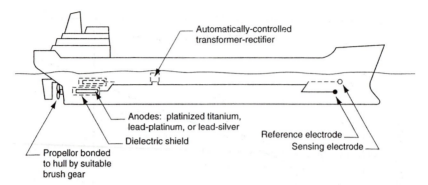

FIGURE 13.10 Controlled potential cathodic protection system on a ship hull. *(From J. S. Gerrard, Corrosion, Vol. 2, L. L. Shreir, ed., Newnes-Butterworths, Sevenoaks, Kent, England, p. 11:57, 1976. Reprinted by permission, Butterworths Publishers.)*

sivity is stable. Cathodic protection uses a cathodic current to polarize the surface to more active potentials that suppress the anodic dissolution rate. Anodic protection is effective only for active-passive metals or alloys that form a resistant passive film. Cathodic protection can be effective for any metal or alloy. The throwing power, or ability to be transferred over distance, is high for anodic protection because very low currents are applied, usually in low resistivity electrolytes. Throwing power is much lower for cathodic protection because higher currents are required, often in solutions of much higher resistivity. Other comparisons are summarized in Table 13.1.

13.2 Design Factors

The electrochemical principles of cathodic protection as outlined above are quite straightforward. However, their application in real systems is often sufficiently complex to preclude quantitative determinations. As a result, the design of cathodic protection systems is usually inexact and may depend largely on the experience of the cathodic protection practitioner. The purpose of this section is to list and discuss the factors that must be considered in evaluating proposed designs for cathodic protection systems.

13.2.1 Potential Gradients

In either an impressed-current or sacrificial-anode cathodic protection system with a current, I, the total potential change from anode to cathode, shown schematically in Figure 13.11,[9] includes the following (Section 3.5.2):

1. Cathodic overvoltage, $\varepsilon_{cathode}$, at the cathode
2. Ohmic potential change, IR_{Ω_c}, around the cathode structure
3. Ohmic potential change, $IR_{\Omega_{el}}$, through the intermediate electrolyte
4. Ohmic potential change, IR_{Ω_a}, around the anode
5. Anodic overvoltage, ε_{anode}, at the anode

TABLE 13.1 Comparison of Anodic and Cathodic Protection

	Anodic Protection	**Cathodic Protection**
Applicability		
Metals	Active-passive Metals and alloys only	All metals and alloys
Solution corrosivity	Moderate to aggressive	Weak to moderate for practical systems
Comparative cost		
Installation	High	Lower
Maintenance	High	Lower
Operation	Very low	Higher
Throwing power	Very high	Low
Rectifiers	Controlled potential	Constant current or controlled potential
Applied current	Very low Often a direct measure of corrosion rate during protection	Higher Depends on cathodic reduction current Is not an exact measure of corrosion rate but increases with corrosion rate
Operating conditions	Can be accurately determined by electrochemical measurements	Usually determined by empirical testing or experience

Overvoltages ε_{anode} and $\varepsilon_{cathode}$ are a few-hundred millivolts each, maximum. IR_{Ω_a} from C to D is on the order of a few volts in seawater systems up to 10 to 50 V for buried structures, depending on number and size of anodes and conductivity of seawater or soil. The change, $IR_{\Omega_{el}}$, in the electrolyte (e.g., soil or seawater) from B to C is quite small and not detectable in the figure because the current is distributed over a large electrolyte cross section. Finally, as the current carrying paths are channeled to the cathode surface, the change, IR_{Ω_c}, from A to B increases, but by a much lower amount than the corresponding change around the anode, due to the much larger cathode structure area.

It is apparent that the total ohmic, IR_Ω, potential (equal to the sum of IR_{Ω_a}, IR_{Ω_c}, and $IR_{\Omega_{el}}$) is orders of magnitude larger than the cathodic overvoltage, which controls the degree of cathodic protection. The major difficulty in design of cathodic protection systems is the correct measurement or calculation of small overvoltages at the cathode-structure surface, without interference from the accompanying large ohmic potential interferences.

The total potential change between anode and cathode must be overcome by the voltage output of the rectifier in an impressed current system or by the galvanic

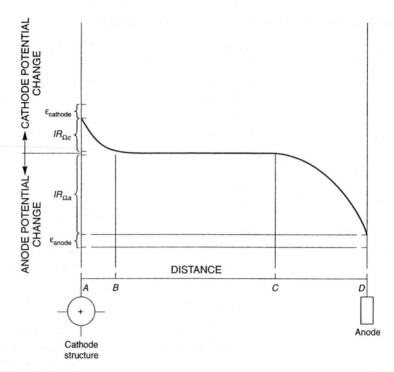

FIGURE 13.11 Schematic variation of potential with distance between cathode structure and anode during passage of current, *I*, in a cathodic protection system. *(Adapted from Jones.[9])*

potential difference between the two in a sacrificial anode system. Thus, for a buried structure, where resistances are high and large areas must be protected with remote anodes, impressed current systems are often necessary. When the expense and maintenance of an impressed current system are undesirable, sacrificial anodes of magnesium may be used, having a very active (negative) corrosion potential and a relatively large galvanic driving potential, when coupled to steel structures (Section 13.2.8).

13.2.2 Rectifiers

Electrical power is transmitted most efficiently as alternating current (ac), which must be converted by rectifiers to direct current (dc) for impressed current cathodic protection at the location of the protected structure (Figure 13.1). Rectifier circuits effectively "chop" one half of the ac cycle to produce a variable dc. Circuits and waveforms for half-wave and full-wave rectification are shown in Figure 13.12. Further filtering to obtain uniform dc is expensive and electrically inefficient. Thus, rectified cathodic-protection current invariably includes at least some degree of ac ripple. Figure 13.13[11] shows the effect on potential of full-wave rectified dc current. Although the polarized potential also exhibits a variation in concert

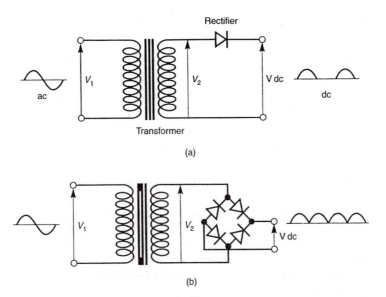

FIGURE 13.12 Half-wave (a) and full-wave (b) rectification of sinusoidal alternating current. *(From M. C. Bent,* Cathodic Protection Theory and Practice, *V. Ashworth and C. J. L. Booker, eds., Wiley (Horwood), Chichester, West Sussex, England, p. 271, 1986. Reprinted by permission, John Horwood Ltd.)*

with the variable current, protection is still effective at potentials negative to the criterion potential.

Thyristor-controlled rectifiers that control dc current to maintain relatively constant potential (still with dc ripple) are in common use for water and seawater systems, as discussed in Section 13.1.6.

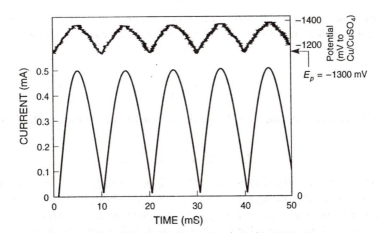

FIGURE 13.13 Polarization potential resulting from unfiltered full wave rectified dc. *(From B. A. Martin,* Cathodic Protection Theory and Practice, *V. Ashworth and C. J. L. Booker, eds., Wiley (Horwood), Chichester, West Sussex, England, p. 276, 1986. Reprinted by permission, John Horwood, Ltd.)*

13.2.3 Current and Potential Distributions on the Protected Structure

The ideal for cathodic protection system design is to distribute the necessary cathodic potential (Section 13.2.4) and current uniformly over the entire structure surface. Unfortunately, this standard is virtually impossible to attain in practice, due to certain physical restrictions, which include:

1. Current originates from finite anode locations. Thus, different points on the structure surface have variable access to current due to differing distances to the nearest anode surface.
2. Many solution environments appropriate for cathodic protection—soil, water, and dilute solutions, such as brackish and polluted waters—are of intermediate to low conductivity. Thus, structure potentials tend to be nonuniform because the low conductivity environment attenuates potential to different degrees, depending on the path length to a given point on the structure.
3. Protected structures are geometrically complex. Current from the anode(s) often has difficult access to remote or shielded surface locations on the cathode structure.

As a result, analytical approaches to determine current and potential distribution on real structures have been restricted to uncomplicated geometries with numerous simplifying assumptions. One such approach is described in the remainder of this section. Computer solutions to the problem are discussed in Section 13.3.5.

The potential, E_x, and current, I_x, as a function of distance, x, along a coated pipeline of infinite length are given by the attenuation equations[12]

$$E_x = E_o \exp(-\alpha x), \tag{6}$$

$$I_x = I_o \exp(-\alpha x) \tag{7}$$

where E_o and I_o are the potential and current at the point of connection to the anode (sometimes called the drainage point), and α is the attenuation coefficient, defined by

$$\alpha = \frac{R_S}{R_K}.$$

R_S is pipe resistance per unit length, and the "characteristic resistance" $R_K = (R_S R_L)^{\frac{1}{2}}$ where

$$R_L = (E_x - E_o)/(I_x - I_o). \tag{8}$$

R_L is known as the leakage resistance of the pipeline, and $E_x - E_o$ incorporates the ohmic potential drop through the coating and electrochemical polarization at the coating-metal interface. Note that E_x and I_x become zero at $x = $ infinity.

The attenuation equations between two drainage points separated by a distance $2d$ are

$$E_x = \frac{E_o \cosh \alpha(d - x)}{\cosh \alpha d} \tag{9}$$

and

$$I_x = \frac{I_o \sinh \alpha(d - x)}{\sinh \alpha d} \tag{10}$$

The variation of potential with distance between two drainage points, drawn from Equation (10), is shown in Figure 13.14.

Several simplifying assumptions are implicit in the attenuation equations:

1. The anodes are sufficiently removed or "remote" from the cathode structure, that the current has uniform access to all parts of the structure. This is often not valid even for impressed current anodes and virtually never for sacrificial anodes.
2. The electrolyte solution is uniform in the volume surrounding the anodes and the cathode structure. This is usually untrue for soil electrolytes over large distances with impressed current anodes.
3. The coating resistance is high, uniform, and ohmic. Coatings are inherently imperfect, and cathodic protection is used to supplement protection of coatings at the imperfections.
4. Polarization at the cathode is linear (ohmic) with increasing current. Instead, current at coating imperfections is either exponential (activation control) or invariant (concentration control) with current.

The theoretical attenuation curves are distorted in practice at or near anodes that are close enough to concentrate current, at breaks or imperfections in the coatings, and at soil inhomogeneities, as indicated by the dashed lines (---) in Figure 13.14. Nevertheless, the attenuation equations have been used frequently[13] as approximations for design of impressed current cathodic protection systems for pipelines.

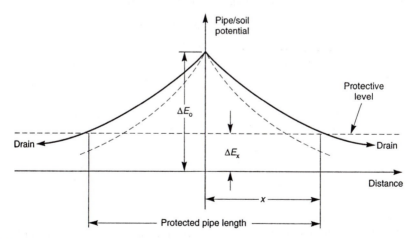

FIGURE 13.14   Theoretical attenuation of potential between two drainage points on a pipeline. More rapid attenuation of cathodic potential (---) is usual for nonideal conditions. *(From D. A. Jones,* Corrosion, *Vol. 2, L. L. Shreir, ed., Newnes-Butterworths, Sevenoaks, Kent, England, p. 11:3, 1976. Reprinted by permission, Butterworths Publishers.)*

The complexity of most corrosive systems defies analysis by conventional analytical methods. The many variables and difficult physical constraints require computer processing by numerical methods, which are discussed further in Section 13.3.5.

13.2.4 Cathodic Protection Criteria

Criteria for cathodic protection recommended[14] by the National Association of Corrosion Engineers (NACE) are listed in Table 13.2. Gummow[15] has given a recent review and criticism of the NACE-recommended criteria.

It may be concluded from the following discussion that all the listed criteria are deficient to some extent and therefore qualitative in practical application. Nevertheless, cathodic protection is effective, despite the uncertainties inherent in the various NACE criteria. Any level of cathodic polarization is beneficial, and a broad range of cathodically applied potentials will yield adequate protection. As a result, the use of any criterion listed in Table 13.2 will produce adequate cathodic protection if applied judiciously by experienced personnel. Although the exact cathodic potential is often not critical to reasonable performance of cathodic protection systems, the most quantitative possible design is desirable to optimize protection with minimum cost.

Probably the best-known criterion is number 1, polarization to -0.85 V versus Cu-saturated $CuSO_4$. It is strictly applicable only to steel in neutral environments, such as soil and seawater. The NACE-recommended practice does not preclude IR_Ω (Section 3.5.4), which is included in most practical measurements and is of uncertain value, depending on electrolyte conductivity and system geometry. Number 1 has some theoretical justification in neutral electrolytes, including soil, but overprotection is probable in the absence of IR_Ω (Section 13.2.5). Even without the uncertainties attributable

TABLE 13.2 Recommended Criteria for Cathodic Protection

Criterion	Measurement Condition	Comments
1. Potential less than -0.85 V versus Cu-saturated $CuSO_4$ for steel	Current on (IR_Ω present)	Meaningful in some environments Uncertain due to IR_Ω
2. Cathodic polarization more than 300 mV active to corrosion potential of structure	Current on (IR_Ω present)	Uncertain due to interferences from IR_Ω
3. Cathodic polarization more than 100 mV active to corrosion potential of structure	Current interrupted (IR_Ω absent)	Interruption techniques difficult to implement
4. Cathodic polarization to a potential where Tafel behavior achieved	Current variable (IR_Ω present)	Difficult to determine in presence of IR_Ω
5. Net protective current flows from electrolyte into the structure surface	Unspecified	Correct in theory Difficult to determine in practice

From NACE Standard RP-01-69.

to IR_Ω, the proper potential for cathodic protection depends on environmental chemistry of any metal or alloy. An earlier survey of practice[16] found no consensus on any potential for the cathodic protection criterion of steel in all aqueous media.

Number 2 specifies a potential change of 300 mV, which includes ohmic IR_Ω potential contributions. The IR_Ω is large and variable over the surface, as discussed in Section 13.2.3. A 100 mV cathodic potential change, free of IR_Ω (number 3), has been shown to be adequate in soils.[17] However, measurement of IR_Ω-free structure potentials is a difficult task (Sections 13.3.2 and 13.3.3).

Number 3 refers to a "break" or change of slope in the polarization curve, which has been shown to be without theoretical justification.[18] The assignment of breaks in a continuous curve is difficult at best and very much subject to the experience and preferences of the engineer or technician analyzing the data.[15]

Number 5 suggests that a cathodic current equal to the anodic corrosion current will stop corrosion. Examination of Figure 13.4 shows the corrosion rate is reduced, but not to zero, when the applied cathodic current is equal to I_{corr}. One cannot determine when all anodic corrosion currents have been suppressed, on the basis of current measurements alone.[15]

It will be shown in the next section that an IR_Ω-free cathodic polarization of about 100 mV (number 4 in Table 13.2) gives adequate protection for most applications. The inaccessibility of most cathodically protected structures makes the measurement of cathodic polarization on the structure without contributions from IR_Ω very difficult. Inadequate polarization and protection, or wasteful overprotection, may not become evident for years.

13.2.5 Electrochemical Basis for Cathodic Protection Criteria

When polarized to the half-cell potential, $e_{Fe/Fe^{2+}}$, the surface has a corrosion rate of zero, as originally postulated by Mears and Brown.[19] At the half-cell potential, the rate of the reverse (deposition) reaction, $Fe^{2+} + 2e^- \rightarrow Fe$, is the same as the forward (dissolution) reaction, $Fe \rightarrow Fe^{2+} + 2e^-$, (1), and the net reaction rate is zero. The half-cell potential for (1) in neutral electrolytes may be calculated as follows[20], from the Nernst equation (Section 2.1.3).

$$e_{Fe/Fe^{2+}} = -0.44 + \frac{0.059}{2} \log(Fe^{2+}),$$

where the activity of ferrous ions is (Fe^{2+}) = solubility product/$(OH^-)^2$, according to $Fe(OH)_2 = Fe^{2+} + 2OH^-$. The solubility product is 1.8×10^{-15} in neutral solutions,[20] and $(OH^-) = \frac{1}{2}(Fe^{2+})$. The calculated potential is -0.59 V versus SHE, which corresponds (Table 2.2) to -0.90 V versus Cu-saturated $CuSO_4$, not far from the empirical value of -0.85 V, routinely accepted as criterion number 1 in Table 13.2.

Although zero corrosion would seem desirable at the reversible half-cell potential for the anodic dissolution reaction, $Fe \rightarrow Fe^{2+} + 2e^-$, (1), corrosion rate may become adequately low at lesser cathodic polarization. For example, reducing corrosion below 0.1 $\mu A/cm^2$ in Figure 13.5 requires higher current and larger rectifiers, and generates alkalinity and hydrogen, which damage coatings and embrittle the protected steel. Furthermore, an accurate determination of the anodic half-cell

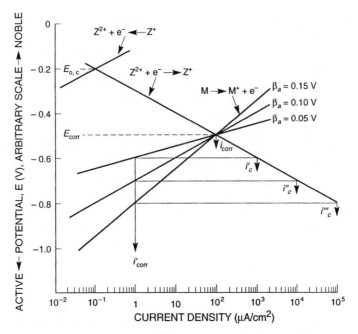

FIGURE 13.15 Effect of anodic Tafel constant on cathodic overvoltage necessary for reduction of corrosion rate by two orders of magnitude. *(From D. A. Jones, Corros. Sci., Vol. 11, p. 439, 1971. Reprinted by permission, Pergamon Press.)*

potential is not an easy task, since the activity of the dissolved metal must be assumed for substitution into the Nernst equation.

Inspection of Figure 13.4 reveals that the *anodic* process controls the degree of protection afforded by cathodic polarization. A corollary to the usual definition of the anodic Tafel constant (Section 3.2.1) is: β_a is the cathodic overvoltage required to reduce corrosion rate by one order of magnitude. Thus, in Figure 13.4, 40 mV is required to reduce corrosion from 1000 to 100 $\mu A/cm^2$ and 120 mV to reduce from 1000 to 1 $\mu A/cm^2$. To further illustrate, Figure 13.15[2] shows three hypothetical metals, with identical cathodic polarization and corrosion rates, but differing anodic Tafel constants. To reduce i_{corr} by two orders of magnitude to, 1 $\mu A/cm^2$, requires a potential change of 0.1 V for β_a = 0.05 V, 0.2 V for β_a = 0.1 V, and 0.3 V for β_a = 0.15 V.

It would be preferable to base cathodic protection criteria on the actual corrosion rate of the protected metal—that is, lower the corrosion rate using the anodic Tafel constant to some value that has been specified as adequate, as in the example above. However, this may be impractical because, in practice, the actual corrosion rate of the structure may not be known. A workable alternative would be to specify the potential change necessary to reduce corrosion by a given percentage. For example, in Figure 13.15, the corrosion rate has been reduced by 99% or two orders of magnitude from the original. This percentage should vary depending on the application and the criticality of a failure. Certainly, experience would play a part in determining the appropriate percentage for a given application. In any case, it is necessary to measure β_a for the protected alloy in the actual or simulated corrosive electrolyte.

Thus, the anodic Tafel constant provides a reasonable guide or criterion for cathodic protection to any selected rate of corrosion judged to be adequately low. The β_a criterion is rooted in accepted electrochemical theory and leads to a better understanding of how and why cathodic protection is effective. Even an estimate of β_a from laboratory soil or seawater measurements will give a more reliable and quantitative gauge of the cathodic polarization needed for adequate cathodic protection than any of the NACE criteria listed in Table 13.2. However, determination of an accurate anodic Tafel constant for the protected structure shares the same difficulties as others in Table 13.2, which require accurate measurement of surface potentials free of IR_Ω potentials (Section 3.5.4). Methods of measuring potentials free of IR_Ω are discussed for seawater and soil in Sections 13.3.2 and 13.3.3, respectively.

During long-term cathodic polarization, reactions (2) and (3) produce alkalinity (OH⁻) and consume water (Section 13.1.4). In soil where the mass transport and water content are limited, significant chemical changes can result in the near-surface environment surrounding the protected structure. Hours or even days may be required for surfaces to depolarize after cathodic protection currents are interrupted.[21] Time under protection will thus change the protection requirements initally considered appropriate for the surrounding bulk soil environment.

13.2.6 Current Requirements

Rectifiers are sized on the basis of protected surface area and required cathodic current density for a given environment (Section 13.3.4). Current requirements can be determined from laboratory- or small-scale field polarization curves, but scale-up to real structures is difficult. Field current-demand tests may also be conducted on the actual structure in service from installed anode groundbeds,[22] but delays in project completion are often unacceptable. Thus, current demands for rectifiers are usually estimated from experience. Approximate cathodic protection current requirements for various solutions are given in Table 13.3.[1] The high current requirement shows why cathodic protection in acid is rare. Only order-of-magnitude estimates are possible, usually. For coated structures an arbitrary estimate, usually 5 to 10%, must be made of the initial surface area exposed through the coating and of coating deterioration with time. More detailed tables of current requirements for soils,[23] seawater,[24] and a variety of other aqueous environments[8] are available.

TABLE 13.3 Estimated Current Density Requirements for Cathodically Protected Steel

Environment	(mA/m²)	(A/ft²)
Sulfuric acid (hot pickle)	400	35
Soil	0.01–0.5	0.001–0.05
Moving seawater, initial	0.15	0.015
final	0.03	0.003
Air-saturated water, hot	0.15	0.015
Moving fresh water	0.05	0.005

From H. H. Uhlig and R. W. Revie, Corrosion and Corrosion Control, *Wiley, New York p. 217, 1985. Reprinted by permission, John Wiley & Sons.*

13.2.7 Impressed-Current Anodes

The anodes are usually arranged in "groundbeds," either vertically or horizontally, depending on the excavation equipment available, nature of the soil strata, and proximity of nearby structures. Deep vertical groundbeds 15 to 30 cm (6 to 12 in.) in diameter and up to 100-m deep may be used for pipelines, where the surface soil is dry and nonconductive, and long distances must be protected. They can be installed often within the existing right-of-way for the pipeline. Deep groundbeds may be backfilled conventionally with coke breeze for improved electrical contact to the soil or with lubricated fluidizable coke, which can be pumped out later when the anodes need replacement.[25]

Selection of the best anode material must be considered as a part of system design. Low resistance to current flow, low rate of consumption, physical robustness, fabricability, and minimal cost are desirable properties of candidate anode materials. Although many electrically conductive solids can, in principle, be used for anodes, only a few have the necessary combination of properties to find extensive use as impressed-current cathodic-protection anodes. An extensive review of impressed-current anode-material properties, processing, installation, and operation is recommended to the interested reader.[26]

Resistance to current flow is determined primarily by electrochemical properties because charge transfer at the anode surface requires an anodic reaction (Section 3.5.2). Three such reactions are important:

$$M \rightarrow M^{n+} + ne^-, \tag{11}$$

$$2H_2O \rightarrow O_2 + 4H^+ + 4e^-, \tag{12}$$

$$2Cl^- \rightarrow Cl_2 + 2e^-. \tag{13}$$

Reaction (11) is to be avoided because it leads directly to anode dissolution and consumption. The other two prevail for the passive or inert materials that are usually specified for impressed current anodes. Methods to stabilize passivity and encourage oxygen (12) or chlorine (13) evolution will generally improve anode performance.

From Table 2.1, the half-cell potentials for reactions (12) and (13) are 0.82 and 1.36 V, respectively, in neutral, pH 7, solutions. Therefore, as anodic polarization increases, one would expect first oxygen evolution, followed by coevolution of chlorine. However, the kinetics for reaction (12), as measured by exchange current density, i_o, are so low on most surfaces that oxygen evolution is nearly undetectable. For example, the i_o for the oxygen evolution reaction (12) on platinum is about 10^{-12} A/cm^2 in solutions ranging from acid to alkaline (Table 3.2), whereas the i_o for the chlorine evolution reaction (13) is over 10^{-3} A/cm^2 on the same surface. Consequently, anodic current due to oxygen evolution, as shown in Figure 13.16, remains low up to the potential, e_{Cl^-/Cl_2}, for chlorine evolution, above which the applied anodic current (---) increases enormously (10,000-fold) at low overvoltage.

Scrap steel and cast iron have been used historically for impressed current anodes. Because the anodic reaction (11) is metal dissolution, the consumption rate is high, and contamination of the surrounding soil or water is also high. This may be

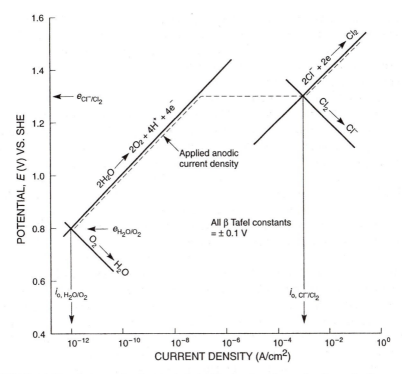

FIGURE 13.16 Schematic electrochemical behavior of a cathodic-protection anode evolving oxygen and chlorine under anodic polarization.

offset by low anode material cost in temporary or short-life systems. Anode life may be prolonged by designing with high surface area to minimize current density. However, such installations are currently quite rare.

In soil and water applications without Cl⁻, the oxygen-evolution reaction (12) predominates on passive and inert anode materials. The least expensive of the passive type are the chromium-bearing, *high-silicon cast irons* (Fe-0.95C-0.75Mn-14.5Si-4.5Cr). These find greatest use in soil, where cost is to be minimized, or in fresh water and seawater, when resistance to abrasion and rough handling is needed. Because metal dissolution by reaction (11), although low, is still significant, consumption rates are higher than for other commonly used anode materials. The common stainless steels are unsuitable, because the passive films break down near oxygen evolution potentials (Section 4.1.2), and localized pitting initiates at still lower potentials in the presence of chlorides (Chapter 7).

As shown in Figure 13.1, solid compacted *graphite* is commonly used for anodes in buried soil systems,[25] because of low cost and inertness. The term "inert" must be qualified, because, under high polarization with oxygen evolution, carbon is oxidized to CO_2. Thus, buried graphite anodes are limited in permissible current density. Chlorine is evolved efficiently at low polarization from graphite, which is consequently more resistant in seawater and brine. However, graphite is both brittle and friable and subject to damage and destruction in exposed locations.

Common practice is to backfill with granulated carbon (coke breeze) around buried anodes, both impressed current and sacrificial (Figures 13.1 and 13.2), to decrease the anode to electrolyte resistance. The porous coke-breeze backfill increases the anode surface area, partially displaces anodic reactions to the backfill carbon surfaces, and facilitates escape of evolved oxygen. The life of most anode materials is substantially extended by carbonaceous backfills.

Lead alloys (Pb-2Ag or Pb-1Ag-6Sb) resist anodic dissolution by formation of a conductive PbO_2 surface film in chloride solutions.[25] Whereas, Cl^- normally destroys passive surface films (Chapter 7), insoluble $PbCl_2$ is thought to heal breaks in a PbO_2 film, allowing polarization of lead surfaces to the chlorine and oxygen evolution potentials. Alloying additions of Ag and Sb improve performance and may facilitate conversion of nonconducting $PbCl_2$ to conducting PbO_2. The lead alloys owe their efficiency to low chlorine evolution overvoltages and are consequently limited to seawater applications. By the same token, they cannot be buried in the sea floor, where access by dissolved Cl^- is limited.

Some *conductive oxides* are useful as anode materials, because the metal is already in the oxidized state and cannot readily be consumed by anodic dissolution. All are nonstoichiometric oxides with sufficient electrical conductivity to carry the required anodic currents. Magnetite, Fe_3O_4, anodes are reportedly[25] produced in Europe for seawater use. Heidersbach[22] reports on dimensionally stable anodes (DSA), which consist of mixed ruthenium dioxide, RuO_2, and titanium oxide, TiO_2, coatings sintered onto titanium substrates. These or similar materials may be developed in the future for economic and relatively inert impressed current anodes.

Platinum is close to the ideal material for impressed-current anodes. The half-cell potential, $e_{Pt/Pt^{2+}}$, is more noble than the usual maximum reached by anodic polarization, so that anode consumption by reaction (11) is nearly nonexistent. Anodic polarization is low on platinum because the exchange current densities for most reactions are greater than those of any other natural or synthetic material. Only high cost has prevented Pt from displacing all competing materials for impressed-current anodes. Cost has been reduced substantially by coating (platinizing) the appropriate titanium shapes with platinum to a thickness of 1 to 5 μM. Titanium provides a strong, fabricable, corrosion-resistant current carrier for the electrochemically vital Pt coating. Platinized titanium anodes can provide more current for lower weight and volume than can any other anode material. Thus, platinized titanium has become the most widely used impressed current anode material for immersed water and seawater systems, despite the higher initial material cost. Consumption rates of the most frequently used, economically viable, impressed-current, anode materials are summarized in Table 13.4.[22]

Low-frequency dc ripple, <50 Hz, has been found to accelerate consumption of platinized titanium anodes in unfiltered half-wave rectifiers.[25] Therefore, single- or three-phase full-wave rectification is required. Apparently, high-silicon cast iron and lead alloy anodes are not affected by dc ripple.

13.2.8 Sacrificial Anodes

Sacrificial anodes depend on anodic dissolution by reaction (11) to supply electrons to the protected cathode by galvanic action. Therefore, anode consumption is

TABLE 13.4 Consumption Rates of Impressed-Current Anode Materials

Material	Typical Anode Current Density		Consumption rate per A-yr
	A/m^2	A/ft^2	
Pt on Ti	540–1080	50–100	6 mg
Pt wire or Pt clad Ti	1080–5400	100–500	10 mg
Pb-6Sb-1Ag	160–220	15–20	0.45–0.09 kg (0.1–0.2 lb)
Graphite	10.8–40	1–4	0.23–0.45 kg (0.5–1.0 lb)
Fe-14Si-4Cr	10.8–40	1–4	0.23–0.45 kg (0.5–1.0 lb)

From R. H. Heidersbach, Metals Handbook, Vol. 13, Corrosion, 9th ed., ASM, Metals Park, OH, p. 466, 1987. Reprinted by permission, ASM International.

required, and periodic replacement is expected for long-life systems. Anode alloys are characterized by certain desirable electrochemical properties:[2]

1. The corrosion potential of the alloy must be sufficiently active (negative) to drive protective current through the electrolyte. The higher the resistance of the electrolyte and the greater the separation of anode and structure, the more negative must be the potential of the anode.
2. Polarization at the sacrificial anode must be low enough to permit current flow. Passive alloys are precluded from use as anodes.
3. The electrochemical equivalent (output) of the anode alloy, which is the charge theoretically available to provide galvanic current per unit mass of the alloy, should be high.
4. The efficiency of the alloy, which is the percentage of the output delivered in practice, should be high.

Alloys of magnesium, zinc, and aluminum have the necessary combination of physical, chemical, and economic properties to produce feasible anode alloys. Selected properties are summarized in Table 13.5, which is compiled largely from Schreiber.[26]

Magnesium has a very active corrosion potential and low polarization. Thus, magnesium alloys are used primarily in soils or pure water (e.g., hot water tanks), where high electrolyte resistance is present. Magnesium alloys are usually not recommmended for seawater applications because of overprotection, inefficiency, and consequent low anode life. However, alloying elements have been added to somewhat deactivate magnesium so that it can be used in lower resistivity environments.

Zinc has the ideal combination of intermediate corrosion potential, low polarization, and high efficiency for sacrificial anodes in seawater applications. It is used in the pure form to minimize polarization.

Aluminum alloys have been developed in recent years for sacrificial anodes in seawater. Surface films usually cause noble (positive) corrosion potential and excessive polarization of pure aluminum and conventional alloys. New alloys with small (<1%) additions of zinc and mercury, indium, or tin remain active in the presence of

TABLE 13.5   Properties of Sacrificial Anode Materials in Neutral Water and Seawater

Property	Magnesium	Zinc	Aluminum
Corrosion Potential, volts, SCE	-1.68^a -1.48^b	-1.1	-1.05
Output: A-h/kg A-h/lb	2200 1000	810 370	2000^c 1250
Efficiency: percent	50–60	>90	>90
Density: gm/cm^3	1.7	7.1	2.7
1983 Cost: US$/A-yr	27	18	8

^{a} Mg-6Al-3Zn alloy

^{b} Mg-1Mn alloy

^{c} For Al-Zn-In alloy.[21] Al-Zn-Hg somewhat higher and Al-Zn-Sn somewhat lower.

chloride in seawater and have very high output because of aluminum's high valence and low density. The obvious low cost per unit output has favored aluminum anode alloys in recent years. Aluminum alloys are not usually recommended for use in soil or fresh water, where they passivate.

13.2.9 Stray-Current Corrosion

Stray currents in the past have resulted from dc-powered trolley systems, which have become obsolete. The major stray-current corrosion problems now result from interacting cathodic protection systems. If there is a nearby current path in earth provided by a low-resistance metallic object, such as a pipeline, current from an impressed-current cathodic protection system will pass through the metal for some distance before it returns to the protected structure, as shown in Figure 13.17.[27] Current to the protected structure is attenuated and protection made incomplete. Furthermore, although pipeline surfaces near the anode are cathodically protected, increased anodic corrosion occurs on the pipeline, where positive current leaves, returning to the protected structure (tank).

The best solution to stray-current problems is electrical bonding of nearby structures, as shown in Figure 13.18. Installation of additional anodes and possibly added rectifier capacity results in protection for both structures. However, if the two structures are separately owned, cooperation and cost sharing are necessary. An example of stray current detection on a pipeline is described in Section 13.3.3.

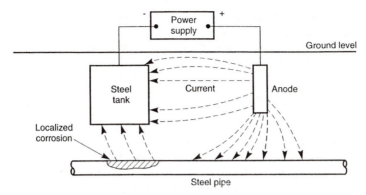

FIGURE 13.17   Stray currents resulting from cathodic protection. *(From M. G. Fontana, Corrosion Engineering, McGraw-Hill, New York, p. 299, 1986. Reprinted by permission, McGraw-Hill Book Company.)*

13.3 Monitoring and Design Procedures

13.3.1 Field-Potential Measurements

Monitoring cathodic protection system performance during design, installation, and operation requires potential measurements on the protected structure. $Cu\text{-}CuSO_4$ reference electrodes are used for buried structures, and $Ag\text{-}AgCl$ reference electrodes are used for seawater-immersed structures. Any measured potential must inevitably include a portion of the IR_{Ω_c} (Section 3.5.4) shown in Figure 13.11. Measurement of a polarized potential free of IR_{Ω_c} is difficult and sometimes impossible for inaccessible surfaces in complex structures. Nevertheless, potential measurements which include substantial IR_{Ω_c} components are common and still useful in

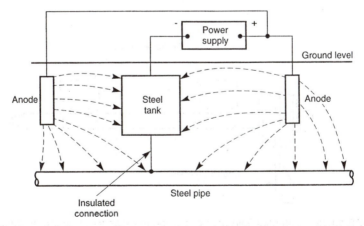

FIGURE 13.18   Prevention of stray-current corrosion by proper design. *(From M. G. Fontana, Corrosion Engineering, McGraw-Hill, New York, p. 299, 1986. Reprinted by permission, McGraw-Hill Book Company.)*

monitoring cathodic protection performance, especially for buried structures. As discussed below, variations in the IR_{Ω_c} itself are sensitive to coating defects or "hot spots." Also, the exact value of potential is often not absolutely critical to reasonable performance of cathodic protection, as discussed in Sections 13.1.2 and 13.2.4.

13.3.2 Monitoring in Seawater

Potential measurements on offshore structures and pipelines require the reference electrode to be placed as close as possible to the surface of the protected structure, to minimize IR_{Ω} errors.[28] Portable reference electrodes are available, which can be placed manually by divers in shallow locations. Reference electrodes may be installed permanently at selected locations by wiring through conduit or inside structure legs for damage protection. Ultrasonic transponders have been placed on immersed structures and pipelines to transmit potential data from permanent reference electrodes. A single interrogation unit can automatically log data from a number of transmitters at various locations on the protected structure(s). Deep, extensive cathodic protection surveys routinely use remote-controlled submersible vehicles with television monitoring for correct reference electrode positioning.

Current can be measured from gradients in the potential field around the protected structure. An immersion field-gradient current probe is shown in Figure 13.19.[28] Potential gradient or difference, E, is measured between the two balanced reference electrodes. Current, I, is calculated from Ohm's law, $E = IR_{\Omega}$, where R_{Ω} is the resistance between the two electrodes, calculated from analytical or finite-element methods and the measured or tabulated resistivity of the seawater. Structure potential can be measured between the probe tip contacting the structure surface and the nearer reference electrode. The probe can be manipulated again manually by a diver or automatically with a remote-controlled submersible.

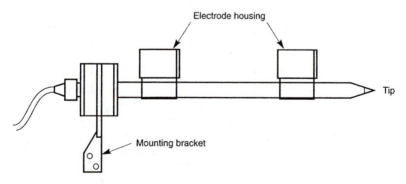

FIGURE 13.19   Potential-field gradient probe to measure structure potential and cathodic-protection currents in seawater. *(From G. H. Backhouse,* Cathodic Protection Theory and Practice, *V. Ashworth and C. J. L. Booker, eds., Wiley (Horwood), Chichester, West Sussex, England, p. 235, 1986. Reprinted by permission, John Horwood Ltd.)*

13.3.3 Monitoring in Soil

Correct potential, free of IR_Ω (Section 3.5.4) interference, is even more difficult to measure on buried structures. Routine placement of a reference electrode near the structure surface, as in offshore installations, is virtually impossible. Current and potential fields around a buried pipe are shown schematically in Figure 13.20.[13] A common and convenient compromise is to position the reference electrode at A on the earth surface above the protected pipe. While the potential at A still includes substantial IR_Ω, there is less than at a more remote surface position, such as B. Despite this inaccuracy, potential surveys along a pipeline will detect current concentrations at points of coating damage and deterioration or soil composition changes, because the IR_Ω field at the discontinuity is affected, as shown in Figure 13.21.[13] Stray current to or from nearby structures (Section 13.2.9) will also be detected, due to similar effects on potential field gradients, which are included in measured potentials.

Modern computer-logging procedures have facilitated continuous pipeline potential surveys.[29] The operator carries a battery-operated portable computer in a field backpack, while walking the line. Wire connected to a terminal point on the pipeline is reeled out through a distance counter. The potential of the buried surface is measured through a "walking stick" reference electrode thrust into the earth at about 3-foot intervals. Potential versus distance is automatically logged into the computer for later processing and graphic display. Figure 13.22 shows periodic variations due to bare joints in a coated line, and Figure 13.23 shows dramatic evidence of reduced cathodic protection at the crossing of a second pipeline due to stray currents (Section 13.2.9). Potential field gradient measurements, described in Section 13.3.2, are also routine with this equipment.[29]

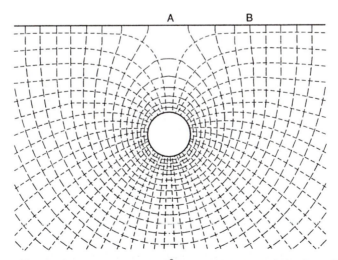

FIGURE 13.20   Current and potential profiles around a buried pipeline. *(From M. E. Parker, Pipeline Corrosion and Cathodic Protection, Gulf Publ. Co., Houston, 1962.)*

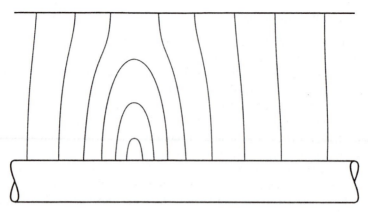

FIGURE 13.21 Potential field around a coating defect on a buried pipeline. *(From M. E. Parker,* Pipeline Corrosion and Cathodic Protection, *Gulf Publ. Co., Houston, 1962.)*

Various available methods for measuring IR_Ω-free polarization potentials on buried pipelines have been instructively reviewed by Martin.[11] Best known is the *"instant-off" method*, in which current is interrupted and potential response monitored, as in Figure 13.24.[11] The IR_Ω interference is distinguishable by its rapid initial decay (Section 3.5.4), followed by slower "depolarization" at the protected surface.

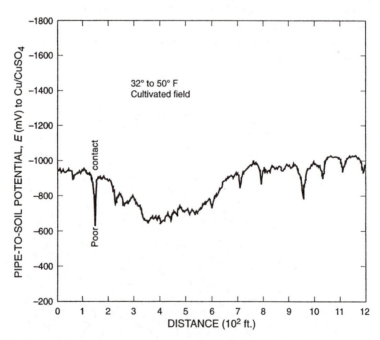

FIGURE 13.22 Computer-assisted potential survey. *(From J. Suprock and T. R. Wilken,* Cathodic Protection Theory and Practice, *V. Ashworth and C. J. L. Booker, eds., Wiley (Horwood), Chichester, West Sussex, England, p. 249, 1986. Reprinted by permission, John Horwood Ltd.)*

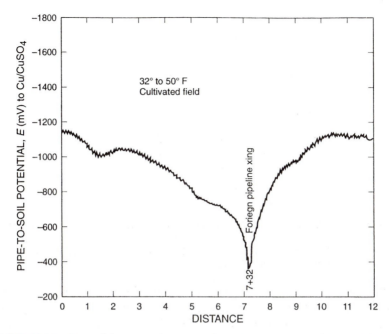

FIGURE 13.23 Potential survey showing effect of stray current. *(From J. Suprock and T. R. Wilken,* Cathodic Protection Theory and Practice, *V. Ashworth and C. J. Booker, eds., Wiley (Horwood), Chichester, West Sussex, England, p. 249, 1986. Reprinted by permission, John Horwood Ltd.)*

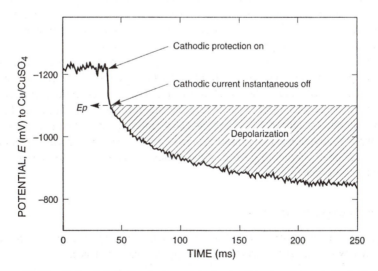

FIGURE 13.24 Potential decay during an instant-off measurement of ohmic interference. *(From B. A. Martin,* Cathodic Protection Theory and Practice, *V. Ashworth and C. J. L. Booker, eds., Wiley (Horwood), Chichester, West Sussex, England, p. 276, 1986. Reprinted by permission, John Horwood Ltd.)*

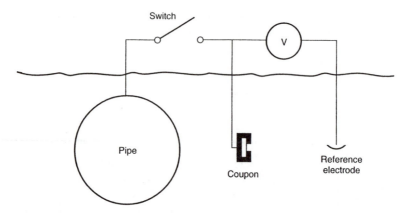

FIGURE 13.25   Coupon simulation circuit to determine ohmic interference *(From B. A. Martin,* Cathodic Protection Theory and Practice, *V. Ashworth and C. J. L. Booker, eds., Wiley (Horwood), Chichester, West Sussex, England, p. 276, 1986. Reprinted by permission, John Horwood Ltd.)*

Again, a simple measurement in principle is more complicated in practice. First, sophisticated instrumentation is required to detect sometimes very rapid potential changes. Valid measurements require that all cathodic protection systems, including those contributing stray currents, be interrupted *simultaneously*, no mean feat for numerous cathodic protection systems operated by different owners on the same and nearby structures. Furthermore, exposed surfaces at neighboring coating defects, polarized to different levels, will pass current between themselves when current is interrupted, developing a new IR_Ω, which decays slowly and conjointly with depolarization. Despite these difficulties, the instant-off potential can give at least an estimate of IR_Ω-free polarization potential for cathodic protection design and monitoring. Instant-off procedures have been incorporated into the computer-assisted cathodic protection surveys described in the preceding paragraph.[29]

A variation of the instant-off method[30,31] employs an insulated coupon, buried near enough to the protected structure to be characteristic of the chemical environment and electrical field at the structure surface. The exposed coupon surface then simulates a coating defect when shorted to the structure, as shown in Figure 13.25.[11] Cathodic protection current to the coupon can be interrupted by the indicated switch, and coupon potential decay monitored as in Figure 13.24, without any effect on the protected or nearby structures. Most of the difficulties cited above for the instant-off method are eliminated.

13.3.4 Design Procedures

Designs of both impressed-current and sacrificial-anode systems have some common steps, which are listed, as follows, in approximate sequence.

1. *Area to be protected.* The calculation of exposed structure areas is tedious and often not given necessary care. Areas may easily be underestimated in complex structures subject to a variety of environments, and the cathodic protection system(s) subsequently underdesigned. In coated systems the exposed

area at breaks in the coating is usually estimated as a percentage of the coated surface. An estimate of surface exposed by coating deterioration during the life of the system must also be included.

2. *Polarized potential.* The primary criteria in use for cathodic protection (Section 13.2.4) involve the appropriate polarized potential. The current density needed to achieve this potential is used in the design of the system.

3. *Current demand.* Required current is calculated simply from the product of area and the corresponding current-density requirement. The current-density demands on steel for various environments are known only approximately; some are tabulated in Table 13.3. Many structures have areas exposed to a number of environments. For example, a fixed offshore structure would have areas exposed to tidal splash, aerated near-surface seawater, seawater of progressively less aeration with depth, and saturated mud at the bottom. Separate calculations are required for each environment.

4. *Anode consumption.* Total required weight of anodes is determined from known consumption rates for the calculated current demand. For sacrificial anodes, consumption is necessary in order to supply current for protection. For impressed-current anodes, consumption is optimized to yield minimum cost for current provided from an external supply.

5. *Anode number and distribution.* The weight of anode material consumed must be divided into an appropriate number of anodes, which are distributed around the structure to give as close to uniform current distribution as possible. The choice of anode distribution is more art than science, and experience plays a major role because potential and current distributions in the electrolyte can only be approximated, even for the simplest system geometries (Section 13.2.3). For sacrificial anode systems, more anodes at closer spacing are necessary, because of the limited driving voltage. An estimate of current loss due to reduced anode surface during the life of the system must be included. Fewer anodes at higher driving voltage are possible with the impressed current systems. However, less design redundancy is present, and possibilities increase for underdesign at isolated locations on the structure.

6. *Anode resistance.* From the number and distribution of anodes, the anode (or groundbed) resistance can be calculated from one of the empirical formulas in Table 13.6. Other similar formulae in English units are given by Heidersbach et al.[22,32]

7. *Design output current.* From the anode resistance, R, and driving voltage, E, of the selected sacrificial anodes, the design output current can be calculated from Ohm's law, $E = IR$. In a proper design, the output current will at least match or exceed the required current calculated in Step 3. Impressed-current rectifiers must be sized to provide the design current from Step 3 at the required driving voltage, which is again calculated from Ohm's law. Cathode resistance is usually assumed negligible, compared to the anode resistance and neglected in the calculations.

Most design procedures ignore many of the fundamental factors discussed previously in this chapter. Nevertheless, successful cathodic protection systems have been

TABLE 13.6 Anode Resistance Formulae

Anode Configuration	Formula
Single horizontal	$R_h = \dfrac{P}{2\pi L}\left[\ln \dfrac{4L}{d} - 1\right]$
Single vertical	$R_v = \dfrac{P}{2\pi L}\left[\ln \dfrac{8L}{d} - 1\right]$
Single vertical (resistance to backfill)	$R_v = \dfrac{0.0171P}{L}\left[2.3\log \dfrac{8L}{d} - 1\right]$
Parallel connected group of n anodes	$R_v = \dfrac{0.0171P}{nL}\left[2.3\log \dfrac{8L}{d} - 1 + 2.3\log 0.656n\right]$
Plate or bracelet (flush mounted)	$R_p = \dfrac{0.315p}{\sqrt{A}}$

p: electrolyte (or soil) resistivity, ohm-cm
L: anode length, cm
n: number of anodes in group

P: backfill resistivity, ohm-cm
d: anode diameter, cm
A: anode area, cm^2

Compiled from Gerrard[8]

designed using workable procedures developed from experience. Detailed design examples in the literature are sparse, but some are given by Heidersbach et al,[22,32] with emphasis on seawater systems.

It may be concluded that the design and engineering of cathodic protection systems are inexact. Experience plays a predominant role, and it is not generally recommended that inexperienced personnel attempt design of complex cathodic protection systems. Numerous contractors specializing in design and/or installation of cathodic protection systems are in business. Careful attention should be given to experience and reputation during selection of a firm to design a cathodic protection project.

13.3.5 Computer Design

Efforts have increased in recent years to mathematically simulate or model cathodic protection systems. Modeling must include current and potential distributions in complex geometries as well as polarization at cathode-structure and anode surfaces. Earlier analytical estimates (Section 13.2.3) have been improved by numerical analysis techniques made possible by modern developments in computer hardware and programming software. Finite-element methods[33] divide the three-dimensional electrolyte volume into a network of finite "nodes" (Figure 13.26), whose electrical properties are connected to one another by linear equations. Simultaneous comput-

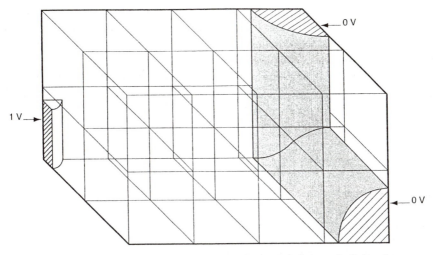

FIGURE 13.26 Simplified 3-dimensional mesh network for use in finite-element modeling of cathodic protection of a structure at 0-volt and an anode at −1.0 volt. *(From M. A. Warne, Cathodic Protection Theory and Practice, V. Ashworth and C. J. L. Booker, eds., Wiley (Horwood), Chichester, West Sussex, England, p. 44, 1986. Reprinted by permission, John Horwood Ltd.)*

erized solutions of these equations by numerical methods produce current and potential distributions within the electrolyte, which are useful for cathodic protection design.[34,35]

Finite-element methods yield potential and current distributions within the electrolyte volume, but incorporating polarization of anode and cathode surfaces is difficult at volume boundaries. Boundary-element methods have shown considerable promise in treating this problem.[36–39] The potential field through the electrolyte volume is determined traditionally by solving the LaPlace Equation,

$$\nabla^2 V = 0 = \frac{\delta^2 V}{\delta x^2} = \frac{\delta^2 V}{\delta y^2} = \frac{\delta^2 V}{\delta z^2} \qquad (14)$$

for the appropriate boundary values. The current-carrying electrode surfaces are divided into discrete boundary elements, which are treated numerically. Less computing power is required, since only the electrode boundaries, not the entire volume, need be divided into a network of nodes or elements. The mathematical functions describing the potential behavior at the boundary elements can be derived from appropriate experimental polarization curves.

Modeling has been applied primarily to cathodic protection systems in seawater of relatively high uniformity and conductivity. Buried structures have not been modeled readily, because of the added complexity of a nonuniform, low-conductivity electrolyte. However, modeling would appear to be a likely candidate for future development in impressed-current cathodic protection systems in view of the large current and potential fields that must be present and the inaccessibility of the protected surfaces in buried structures.

As computation methods develop, modeling will probably become further limited by a lack of experimental data to feed the models. Polarization data as a function of time, scale formation, and biological activity have not been determined for most practical environmental conditions. However, the ultimate goal of modeling, the computer design of cathodic protections systems—that is, specification of number, size, and placement of anodes, whether impressed current or sacrificial, together with the potential and current ratings achieved on the structure surface—appears to be attainable to some degree in the foreseeable future.

References

1. H. H. Uhlig and R. W. Revie, *Corrosion and Corrosion Control*, Wiley, New York p. 217, 1985.
2. D. A. Jones, *Corros. Sci.*, Vol. 11, p. 439, 1971.
3. H. H. Uhlig and R. W. Revie, *Corrosion and Corrosion Control*, Wiley, New York p. 96, 1985.
4. J. C. Zoccola, H. E. Townsend, A. R. Borzillo, and J. B. Horton, *ASTM STP 646*, ASTM, Philadelphia, p. 165, 1978.
5. M. A. Warne, *Cathodic Protection Theory and Practice*, V. Ashworth and C. J. L. Booker, eds., Wiley (Horwood), Chichester, West Sussex, England, p. 61, 1986.
6. J. Jensen, *Cathodic Protection Theory and Practice*, V. Ashworth and C. J. L. Booker, eds., Wiley (Horwood), Chichester, West Sussex, England, p. 128, 1986.
7. M. C. Bent, *Cathodic Protection Theory and Practice*, V. Ashworth and C. J. L. Booker, eds., Wiley (Horwood), Chichester, West Sussex, England, p. 271, 1986.
8. J. S. Gerrard, *Corrosion*, Vol. 2, L. L. Shreir, ed., Newnes-Butterworths, Sevenoaks, Kent, England, p. 11:57, 1976.
9. D. A. Jones, *Corrosion*, Vol. 2, L. L. Shreir, ed., Newnes-Butterworths, Sevenoaks, Kent, England, p. 11:3, 1976.
10. M. G. Fontana, *Corrosion Engineering*, McGraw-Hill, New York, p. 303, 1986.
11. B. A. Martin, *Cathodic Protection Theory and Practice*, V. Ashworth and C. J. L. Booker, eds., Wiley (Horwood), Chichester, West Sussex, England, p. 276, 1986.
12. H. H. Uhlig and R. W. Revie, *Corrosion and Corrosion Control*, Wiley, New York p. 223, 1985.
13. M. E. Parker, *Pipeline Corrosion and Cathodic Protection*, Gulf, Houston, 1962.
14. Control of External Corrosion on Underground or Submerged Metallic Piping Systems, *NACE Standard RP-01-69*, NACE, Houston, 1983.
15. R. A. Gummow, *Corrosion/86*, Preprint 343, NACE, Houston, 1986.
16. M. H. Peterson, *Corrosion*, Vol. 15, p. 483t, 1959.
17. T. J. Barlo and W. E. Berry, *Proc. 9th Int. Cong. Metallic Corrosion*, Vol. 4, National Research Council of Canada, Toronto, p. 86, June 7, 1984.
18. M. Stern and A. L. Geary, *J. Electrochem. Soc.*, Vol. 104, p. 56, 1957.
19. R. B. Mears and R. H. Brown, *Trans. Electrochem. Soc.*, Vol. 74, p. 519, 1938.
20. H. H. Uhlig and R. W. Revie, *Corrosion and Corrosion Control*, Wiley, New York, pp. 223–8, 1985.
21. N. G. Thompson, K. M. Lawson and J. A. Beavers, *Corrosion/94*, Preprint 580, NACE, Houston, 1994

22. R. H. Heidersbach, *Metals Handbook*, Vol. 13, *Corrosion*, 9th ed., ASM International, Metals Park, OH, p. 466, 1987.
23. F. W. Hewes, *Cathodic Protection Theory and Practice*, V. Ashworth and C. J. L. Booker, eds., Wiley (Horwood), Chichester, West Sussex, England, p. 226, 1986.
24. B. S. Wyatt, *Cathodic Protection Theory and Practice*, V. Ashworth and C. J. L. Booker, eds., Wiley (Horwood), Chichester, West Sussex, England, p. 143, 1986.
25. L. L. Shreir and P. C. S. Hayfield, *Cathodic Protection Theory and Practice*, V. Ashworth and C. J. L. Booker, eds., Wiley (Horwood), Chichester, West Sussex, England, p. 94, 1986.
26. C. F. Schreiber, *Cathodic Protection Theory and Practice*, V. Ashworth and C. J. L. Booker, eds., Wiley (Horwood), Chichester, West Sussex, England, p. 78, 1986.
27. M. G. Fontana, *Corrosion Engineering*, McGraw-Hill, New York, p. 299, 1986.
28. G. H. Backhouse, *Cathodic Protection Theory and Practice*, V. Ashworth and C. J. L. Booker, eds., Wiley (Horwood), Chichester, West Sussex, England, p. 235, 1986.
29. J. Suprock and T. R. Wilken, *Cathodic Protection Theory and Practice*, V. Ashworth and C. J. L. Booker, eds., Wiley (Horwood), Chichester, West Sussex, England, p. 249, 1986.
30. K. Kasahara, T. Sato, and H. Adachi, *Mater. Perform.*, Vol. 18, Number 3, p. 21, 1979.
31. K. Kasahara, T. Sato, and H. Adachi, *Mater. Perform.*, Vol. 19, No. 9, p. 45, 1980.
32. R. H. Heidersbach, R. Baxter, J. S. Smart, and M. Haroun, *Metals Handbook*, Vol. 13, *Corrosion*, 9th ed., ASM International, Metals Park, OH, p. 919, 1987.
33. M. A. Warne, *Cathodic Protection Theory and Practice*, V. Ashworth and C. J. L. Booker, eds., Wiley (Horwood), Chichester, West Sussex, England, p. 44, 1986.
34. J. W. Fu, *Corrosion*, Vol. 38, p. 296, 1982.
35. R. G. Kasper and M. G. April, *Corrosion*, Vol. 39, p. 181, 1983.
36. J. W. Fu and J. S. K. Chow, *Corrosion/82*, Preprint 163, NACE, Houston, 1982.
37. D. J. Danson and M. A. Warne, *Corrosion/83*, Preprint 211, NACE, Houston, 1983.
38. R. Stommen, W. Keim, J. Finnegan, and P. Mehdizadeh, *Mater. Perform.*, Vol. 26, p. 23, Feb., 1987.
39. S. Aoki, K. Kishimoto, and M. Miyasaka, *Corrosion*, Vol. 44, p. 926, 1988.

Exercises

13-1. Which of the following metals might be subject to cathodic corrosion during cathodic protection: **(a)** aluminum? **(b)** iron? **(c)** nickel? **(d)** zirconium?

13-2. Determine the cathodic overvoltage necessary to reduce the corrosion rate to 0.5 μA/cm^2 for a metal with an unpolarized corrosion rate of 300 μA/cm^2, having $\beta_a = 0.65$ V and $\beta_c = 0.12$ V at the corrosion potential.

13-3. Determine the applied current necessary in Problem 13-2 assuming **(a)** activation control and **(b)** concentration control of the cathodic process at the corrosion potential.

13-4. Taking the cathodic polarization curve for iron and the anodic polarization curve for aluminum from Figure 6.17, **(a)** determine the corrosion rate of the iron when coupled to an equal area of aluminum, assuming β_a for iron is 0.1 V. **(b)** Qualitatively, how would the cathodic protection of the iron be affected if aluminum of area one-tenth that of iron were coupled?

13-5. Taking the cathodic polarization curve for iron and the anodic polarization curve for magnesium from Figure 6.17, determine approximately the resistance needed between 1-cm^2 areas of iron and magnesium to obtain a corrosion rate of 1 μA/cm^2 for iron when the two are coupled. Why is such a resistance required for a sacrificial anode of magnesium? What would be the source of such a resistor?

13-6. In Figure 13-9, determine the corrosion rate (A/cm^2) if i_L were doubled due to increased water velocity.

13-7. Measure β_a in Figure 13-9. How would an increase or decrease in β_a affect i'_a, if all other parameters remained constant?

13-8. A large underground cylindrical steel storage tank (15 feet in diameter by 50 feet long) for aviation fuel is to be installed at a national airport where the soil is known to be corrosive with a resistivity of 10^4 Ω-cm. The Environmental Protection Agency (EPA) now requires that all such tanks be protected from soil corrosion with coatings and impressed current cathodic protection. Your task as an engineer is to design the corrosion prevention system. Consider all possible aspects of corrosion in this environment. Include in your design: **(a)** specifications for the coating system, including surface preparation, coating type and thickness (See Chapter 14) **(b)** current requirements and rectifier size **(c)** type, size, backfill, and distribution of anodes. Some of the information you need may not be available in this book, and may have to be derived from other references.

Coatings and Inhibitors

$$\begin{array}{c}
CH_3-CH_2 \\
O \qquad\qquad NH \\
CH_3-CH_2
\end{array}$$

In this chapter, coatings and inhibitors in corrosion prevention are described. Similar coatings on wood, concrete, and other materials are mentioned occasionally, but only in connection with the same or similar coatings on metals. Coatings and linings maintain the functional ability of structures and equipment to bear loads, maintain dimensions, store and transport liquids and gases, and so on. However, in addition, coatings can preserve cosmetic appearance, which is often a major concern for structures and equipment routinely exposed to public view. Organic coatings act primarily as a physical barrier between the substrate and the corrosive environment but may also serve as a reservoir for inhibiting compounds. Metallic coatings also act as a corrosion resistant physical barrier but may provide, in addition, sacrificial cathodic protection (Section 13.1.3) to exposed adjacent areas of a metal substrate of dissimilar composition.

Inhibitors are chemical compounds that deposit on exposed metal surfaces from the corrosive environment. The inhibitor may form a uniform film, which, like a coating, acts as a physical barrier. Often, however, a few monolayers or less are sufficient to alter the electrochemical reactivity of the surface to reduce the corrosion rate.

14.1 Liquid-Applied Organic Coatings

Organic coatings are conveniently applied as a liquid, primarily by brushing, rolling, and spraying. The liquid consists of solvent, resin, and pigment. The solvent carries the dissolved or suspended resin, which adhesively binds coating to substrate and inert pigment particles to one another after the solvent has evaporated and the coating has cured. The resin provides the chemical and corrosion resistance for the coating. The suspended pigment decreases permeability and provides opacity and color, thereby shielding the cured resin from degrading ultraviolet radiation and covering or "hiding" the underlying surface. Pigments may have corrosion inhibition functions as well. Other minor additives improve coating flow, emulsification, uniformity, and reduce pigment settling. The liquid-applied coatings cure while the solvent evaporates. Some undergo polymerization and cross-linking during the cure; others simply deposit as a consolidated layer as the solvent evaporates. Many industrial coatings are cured further by baking below about 300°C, to further cross-link the resin or expel residual solvent.

An industrial coating system may consist of several layers having different functions. A chemical-conversion coating of chromate or phosphate is often applied initially to provide a substrate of superior adherence. A primer coat of good surface adherence and inhibitive properties may be required, and one or more topcoats may provide a pleasing appearance and resistance to atmospheric weathering or a chemical environment. Thus, the total paint system is a composite designed for the particular service conditions.

The corrosion rate of the unprotected metal should not exceed about 1.3 mm/yr (50 mpy) for the usual liquid-applied coatings.[1] More aggressive environments require thicker coatings and liners (Section 14.2) or more corrosion-resistant alloys. Table 14.1[1] lists the types of paint that are generally suitable for various corrosive environments. It is apparent that liquid-applied coatings are especially suitable for atmospheric corrosion protection (Section 12.1.4) and relatively mild corrosive conditions, especially when cosmetic appearance is important.

The engineer should be generally acquainted with the properties of the common generic paint classifications described in the following sections. Space is not available to go further into the organic chemistry of coating formulation, which is available elsewhere.[1]

14.1.1 Classification and Uses

Organic coatings are usually classified according to the resin binder, which controls protectiveness and resistance to degradation. Coatings of any given classification may be used as primers or topcoats depending on the formulation. Characteristics of the various major types of organic coatings are summarized in Table 14.2[1] and amplified in this section. Some coating resins have good adhesion to the metal substrate and are readily amenable to further topcoating. These are often used as primer coats to improve the durability of the final topcoats, which have maximum resistance to weather and chemical environments. Many of the resin formulations cannot

TABLE 14.1 Paint Systems for Corrosive Conditions

Conditions	Suggested Painting System
Dry interiors where structural steel is embedded in concrete, encased in masonry, or protected by membrane or contact-type fireproofing	Leave unpainted
Interior: normally dry, very mild (oil-based paints last 10 years or more)	Latex or one coat
Exterior: normally dry (oil-based paints last 6 years or more)	Oil base
Frequently wet by fresh water: involves condensation, splash, spray, or frequent immersion (oil-based paints last 5 years or less)	Vinyl, coal-tar epoxy, epoxy, chlorinated rubber
Frequently wet by salt water: involves condensation, splash, spray, or frequent immersion (oil-based paints last 3 years or less)	Vinyl, coal-tar epoxy, epoxy, zinc-rich
Chemical exposure: acidic (pH 5 or lower)	Vinyl
Chemical exposure: neutral (pH 5 to 10)	Zinc-rich
Chemical exposure: mild organic solvents, intermittent contact with aliphatic hydrocarbons (mineral spirits, lower alcohols, glycols, etc.)	Epoxy
Chemical exposure: severe, includes oxidizing chemicals, strong organic solvents, or combinations of these at high temperature	May be none suitable

From K. B. Tator, Corrosion and Corrosion Protection Handbook, P. A. Schwietzer, ed., Dekker, New York, p. 355, 1983. Reprinted by permission, Marcel Dekker, Inc.

be easily coated over and thus are suitable only as topcoats. Only general guidelines for the various classifications are given in Table 14.2 and in subsequent discussion; more specialized treatments,[1,2,3] as well as manufacturers and distributors, should be consulted for further information and recommendations.

Alkyd resins provide the protectiveness of the well-known oil-based coatings used in household and industry. The oil-based coatings, often called enamels, polymerize and cross-link (cure) by oxidation from ambient air during drying and evaporation of the oil solvent used in manufacture of the resin. So-called long-chain drying oils allow easy application over poorly prepared surfaces but are less volatile, lengthen drying time, and decrease chemical and weathering resistance. Continuing oxidation with time reduces chemical stability of long-chain drying oils. Shorter oils give faster-drying, more-stable, weather-resistant coatings, which are relatively hard and brittle as well as more sensitive to surface preparation. Alkyd resins made from soybean oils yield coatings with optimal drying times and chemical resistance, and therefore constitute the majority of alkyd coating formulations.

TABLE 14.2 Organic Coating Classifications and Characteristics

Resin Type	Estimated Chemical and Weather Resistance[a]					Comments
	Acid	Alkali	Solvent	Water	Weather	
One Component Systems						
Alkyd (oil base)	F	P	P	F	VG	Excellent adhesion to poorly prepared surfaces. Adequate for mild chemical fumes. Not chemically resistant.
Modified alkyd silicon	F	F	F	F	VG	Improved durability, gloss retention, heat resistance.
amino	F	F	F	G	VG	High humidity resistance.
phenolic	F	F	F	VG	VG	Suitable for immersion service.
Epoxy ester	F	F	P	F–G	P–F	Similar to Alkyds. Somewhat better chemical resistance. More expensive.
Vinyl	VG	VG	P	VG	G	Widely used industrial coating. Easily recoated. Low toxicity, tasteless. High volatiles content— possible fire hazard.
Chlorinated rubber	G	G	P	G	VG	Good vapor barrier. Poor resistance to sunlight.
Acrylics (solvent base)	F	G	G	G	VG	Automotive topcoats.
Two Component Systems						
Phenolic	F	F	G	VG	VG	Immersion service.
Polyamide Cured epoxy	F	VG	VG	P	G	Tough, flexible, abrasion resistant. Difficult to topcoat.
Coal tar epoxy	VG	VG	G	VG	G	Can be applied without primer. Very adherent. Excellent chemical resistance.
Polyester urethane	F–G	G	VG	F–G	VG	Can be designed for wide range of service depending on modifying groups.

[a] VG:very good, G:good, F:fair, P:poor

From: H. G. Lasser, Metals Handbook, Vol. 5, Coating and Finishing, 9th ed., ASM International, Metals Park, OH, p. 498, 198.

It has been estimated that two-thirds of the coatings sold for corrosion protection are oil-based alkyds.[2] They are used especially for exterior and interior nonchemical atmospheric exposures. Thus, they are included as a comparison for the protectiveness of other paint systems in Table 14.2. Ease of application and low cost make the alkyds highly suitable for steel surfaces that must be protected from atmospheric rusting in mild atmospheres. They have no long-term resistance to more than occasional water condensation or immersion.

Modified alkyds incorporate additional organic groups having specialized properties. The coating takes on characteristics of the modifier, while retaining ease of application and adhesion from the alkyd. Silicon-modified alkyds have greatly improved durability, gloss retention, and moisture and heat resistance. They are used, for example, in marine maintenance coatings and for petroleum processing equipment. Amino-resin-modified aldehydes are used in high-humidity applications, such as within refrigerators and washing machines. Phenolic modifications are suitable for water immersion.

Epoxy esters, also prepared with drying oils, are similar to the alkyds. They have somewhat improved protectiveness but are correspondingly more expensive.

Vinyls do not cross-link during curing and soften at elevated temperature (thermoplastic). They consist generally of an 86% polyvinyl chloride (PVC), 14% polyvinyl acetate (PVA) copolymer. The PVA disrupts the inert PVC chain structure sufficiently to allow dissolution in strong organic solvents such as ketones, while retaining chemical resistance in the cured coating. Curing comes about by simple deposition of the uncross-linked long-chain polymer as the solvent evaporates into the surrounding atmosphere. The long-chain copolymer consolidates into a film, which is resistant to water, weathering, immersion, acids, alkalis, and many other chemicals. Light-reflective pigments such as rutile, TiO_2, are added to protect the copolymer from ultraviolet degradation. Coatings that cure by solvent evaporation without cross-linking are known as lacquers. Liberated volatile solvents are prohibited in many locations as a health hazard.

Vinyl coatings readily dissolve in many organic solvents. The coating formulations are low in dissolved solids and require multiple coats for coverage but are extremely durable. Later coatings or touch-ups adhere readily by partial dissolution of the old coating. Thus, vinyl coatings are useful as primers for more resistant or attractive topcoats for specialized service. Recoating intervals of over 20 years are reported[2] for locks and dam gates maintained by the U.S. Army Corps of Engineers, although more frequent maintenance painting of damaged or deteriorated areas may be needed.

Chlorinated-rubber coatings are also thermoplastic and cure by solvent evaporation. Thus, they share many of the properties of the vinyls. Chlorinated-rubber resins are prepared by reacting chlorine with natural and synthetic rubber. Formulation is complex, and a careful balance of plasticizers and stabilizers is necessary. The resultant coatings are especially known for their resistance to water-vapor passage. Chalking may occur in sunlight, and decorative applications are limited.

Acrylics may be dissolved in organic solvents or dispersed as emulsions in water. The resultant coatings are thermoplastic and uncross-linked in both cases. The solvent-based coatings are especially resistant to weathering and retain color and gloss to a high degree. Heating flows the coating and heals imperfections. Thus, their best known application is in automotive topcoats. The water-based acrylics retain many of the good properties of the solvent-based equivalents. Cleanup is more convenient, and the solvent is of course nonflammable and nontoxic. The water-based paints are somewhat more porous, due to the emulsifiers, and water vapor passes more readily, an advantage for concrete and wood surfaces to prevent water blistering. Thus, the water-based acrylics are widely used in household-maintenance paints.

The acrylic resins are also used as modifiers for alkyd and epoxy coatings for improved weathering and ultraviolet resistance. The epoxy modifications are still known as acrylics because of their reputation for outdoor durability.

Phenolics are a simple but important example of coatings cured by chemical reaction between two organic molecules. Such resins are called thermosets, because the reaction proceeds to completion with increasing temperature, and the cross-linked structure does not soften on reheating. Phenol and formaldehyde dissolved in alcohol are applied to the surface by usual methods of spraying, rolling, or dipping. Subsequent heating produces a reaction (Figure 14.1) that forms a cross-linked polymer at multiple sites on the ring structure of the phenol, with water as a by-product. The coating must be applied in thin coats, allowing the by-product water to escape between applications. Phenolic coatings are highly resistant to water immersion and are sufficiently inert for use as can linings for food products.

Polyurethanes are formed by reaction between isocyanates and polyhydroxyls (polyols). Multiple isocyanate ($-N=C=O$) locations on the resin molecule react with the hydroxyl:

$$R-N=C=O + R'-OH \rightarrow R-\underset{H}{\overset{\overset{\textstyle O}{\|}}{N}}-C-O-R'$$

isocyanate   polyol   urethane

The urethane product polymerizes and cross links by further reaction between additional isocyanate groups on the R radical of the isocyanate and hydroxyl groups on the R′ radical of the polyol. R and R′ radicals are many and varied, yielding correspondingly different properties in the resulting coatings. Earlier aromatic iso-

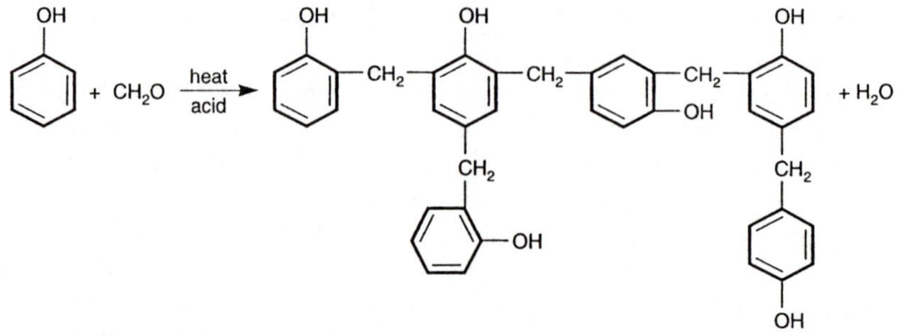

FIGURE 14.1   Reaction between phenol and formaldehyde in phenolic coatings.

cyanates, in which R includes the benzene ring, are being displaced by aliphatic, straight-chain or cyclic, varieties of R, which are more expensive but show superior light fastness and gloss retention. The polyol may have acrylic, epoxy, polyester, and vinyl structures, among others. The resulting urethane coatings have properties reflected by the groups included in the structure.

Epoxies are thermosetting coatings prepared by mixing the dissolved resin and curing agent just before application. Curing begins immediately, and application must be expedited. Low "pot-life" formulations are sometimes mixed during spraying. The resin package includes pigment, as well as the solvent. Most industrial resins are of the glycidyl ether type, made from reacting epichlorohydrin with bisphenol acetone (Figure 14.2). The curing agent is either a polyamine or a polyamide that copolymerizes with the epoxy resin. The polyamine is a smaller molecule and results in a tighter, more chemical-resistant coating. The polyamide, on the other hand, gives coatings that are more flexible but retain resistance to water and slightly reduced resistance to chemicals. Epoxy coatings are highly cross-linked and unreactive; they are, thus, difficult to topcoat or touch-up.

Further modified epoxies can give properties tailored for a particular application. Coal-tar epoxies are formulated as above with coal tar as a filler. The coal tar improves the moisture resistance, while retaining toughness, adhesion, ultraviolet resistance, and thermal stability of the epoxy. Phenolic cross-linked epoxies are highly resistant to alkalies, acids, and solvents and are used as coatings for tanks, cans, drums, and other process equipment. The phenolformaldehyde constituent of phenol, mixed with the epoxy resin, is applied to the surface and cured at elevated temperature as for normal phenolic coatings. The phenol serves as the curing agent to cross-link the epoxy resin, in place of the usual amine or amide curing agent. Whereas ordinary phenolic coatings mentioned earlier are attacked by alkalies, the epoxy-phenol coatings are resistant.

Zinc-bearing coatings have zinc powder as a filler. These fall into two categories.[4] Paints with *zinc pigments* contain about 80% by weight pigment, of which 20% is zinc oxide. *Zinc-rich* coatings contain 92–95% metallic zinc, by weight, in the dry coating with little or no oxide. The binder may be one of the organic resins discussed above or, in the case of the zinc-rich coatings, an inorganic silicate.

There is not sufficient metallic zinc in the zinc-pigmented paints to permit galvanic protection by zinc. However, the zinc/zinc-oxide combination serves as an

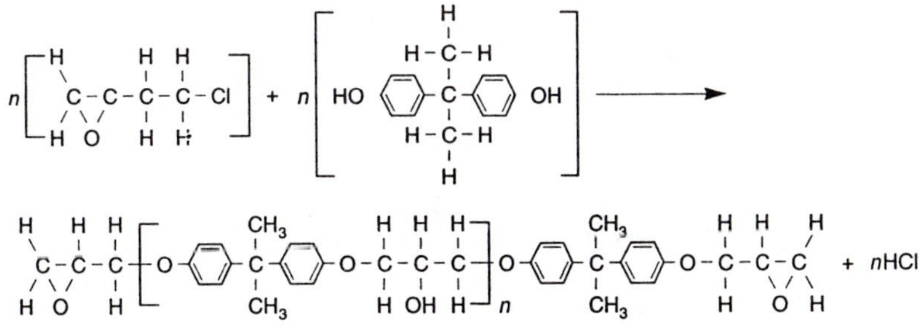

FIGURE 14.2 Reaction between epichlorohydrin and bisphenol acetone to produce an epoxy resin.

inhibitor for atmospheric rusting, in both primer and finish coats. Zinc-pigmented paints provide excellent coverage, adhesion, and abrasion resistance. They can be applied by brush, roller, or spraying and can be applied over lightly rusted surfaces in some formulations. Thus, surface preparation and application specifications are not nearly as stringent as those for the zinc-rich coatings. The zinc-pigmented coatings are generally effective for rural and mild industrial atmospheres, while the zinc-rich coatings are useful in more aggressive atmospheres, as discussed below.

Higher zinc content and absence of the oxide allows galvanic protection in the zinc-rich coatings. Zinc seals coating defects with zinc-corrosion products, notably zinc carbonates, zinc hydroxides, and complex zinc salts,[1] and confers cathodic protection on exposed steel at defects by galvanic action (Sections 6.2.4 and 13.2.8). Thus, the common mode of coating failure, undercutting at the coating-metal interface, is prevented or suppressed—a characteristic unique to the zinc-rich coatings.

Application of zinc-rich coatings requires considerable care[2] to ensure electrical contact between zinc particles and the coated surface. Surface cleaning before application is critical. Complete removal of all organics, oxides, and mill scale to expose bare metal is an absolute necessity to achieve the galvanic protection afforded by the zinc. The zinc is usually added in the spray container immediately before application and must be agitated to prevent settling. The suspension must be moved through the spray lines expeditiously to prevent settling and setup in the lines or a dry spray, in which insufficient liquid binder is present in the coating. Despite these difficulties, zinc-rich primers and coatings are widely used, due to the incremental improvement in performance over competitive systems.

Organic zinc-rich primers are often used under exterior topcoats for automotive finishes. Zinc-rich primers are used in many applications where resistance to undercutting is critical, but where decorative colors and finishes are provided by topcoats. Zinc-rich coatings are used to protect marine structures, ship hulls, ballast tanks, and ship superstructures. Highway bridges, chemical process plant equipment, sewage and water treatment plants, and many other applications requiring resistance to water immersion, salt, high humidity, and mild chemical fumes have employed zinc-rich coatings. However, they are not suitable for environments of pH<5 or >10, where the zinc is more readily dissolved.

The *inorganic* zinc-rich coatings are generally the most resistant, but also the most difficult to apply and the most sensitive to surface preparation. Post-cured water-based zinc-rich coatings are applied initially as a suspension of zinc dust in a sodium silicate solution. After evaporation of the water solvent, the hardened coating is washed with an acid phosphate curing solution, which converts the zinc to insoluble silicates and phosphates.[2] Excess acid must be thoroughly washed from the surface before topcoats can be applied. Self-cured water-based zinc-rich coatings are formulated such that formation of carbonic acid by dissolving CO_2 from air will cure the coatings by the same chemistry as for the postcured coatings. Rinsing and washing are not usually necessary for the self-cured coatings.

Organic zinc-rich coatings are less sensitive to surface preparation than the inorganic coatings but are protective for generally shorter times. The most suitable organic zinc-rich coatings utilize epoxy polyamide, urethane, vinyl, or chlorinated rubber as a binder.[2] As would be expected from previous discussion, the vinyl- and chlorinated-rubber-based organics are easier to topcoat as primers, whereas the thermoset epoxies and urethanes are more heat and abrasion resistant.

14.1.2 Surface Preparation

A key property of coatings is their ability to adhere to surfaces contaminated with oxides, scales, loose dirt, and organic matter, such as greases and oils. However, any coating system will perform better over a well-prepared surface with little or no foreign substances present. It has been said[1] that a poor coating applied to a well-prepared surface is better than a good coating applied to a poorly prepared surface. Therefore, careful selection and quality control of surface-preparation processes are usually well worth the extra time and expense.

The Steel Structures Painting Council (SSPC) has published a series of surface preparation standards widely used in North America. These are summarized very briefly in Table 14.3.[5] The National Association of Corrosion Engineers (NACE) also distributes visual standards for steel surfaces cleaned by either sand blasting or steel-shot blasting.[6] Before initiating any coating project, published SSPC and/or NACE standards should be consulted for complete procedural details, as well as illustrative photographs, when available, of cleaned surfaces meeting specifications.

After surface preparation, various surface prepainting treatments may be employed to enhance adhesion and resistance of the subsequent coatings. These include *phosphate and chromate conversion coatings*. Both may also be used without primers and topcoats but are usually applied in coordination with an integrated coating system. The composition of conversion coating baths and chemistry of the coating processes are not well understood and are usually proprietary. The general reactions involved in phosphate conversion coating are as follows.[7]

$$2H^+ + M \rightarrow H_2 + M^{2+} \tag{1}$$

$$2H_2PO_4^- + 3M^{2+} \rightarrow M_3(PO_4)_2 + 4H^+. \tag{2}$$

The summation gives

$$2H^+ + 3M + 2H_2PO_4^- \rightarrow H_2 + M_3(PO_4)_2. \tag{3}$$

M^{2+} represents any one of several possible metal cations, including iron, manganese, and zinc. The acid phosphate bath first dissolves M by (1), which then reprecipitates as the metal phosphate by (2). A metal other than the one to be coated may also be deposited by included cations of that metal in the phosphate bath. Chromate reactions are similar, involving chromium in the trivalent, Cr^{3+}, and hexavalent, Cr^{6+}, states. Iron phosphate is most conveniently applied to steel substrates by reactions (1) and (2), but zinc phosphates provide still more resistant prepaint treatments. Chromium oxide coatings result from chromate treatments and form resistant barriers for decorative applications and paint pretreatments.

The detailed technology of conversion coatings[8,9] is beyond the scope of this book. It may be noted, however, that conversion coatings are inexpensive compared to the total paint system cost, which includes surface preparation, primer coats, and

TABLE 14.3   Summary of Surface Preparation Standards Steel Structures Painting Council

Specification	Description
SP 1: Solvent cleaning	Removal of oil, grease, soil, salts, and other contaminants by cleaning with solvent, vapor, alkali, emulsion, or steam. Does not remove rust or mill scale.
SP 2: Hand-tool cleaning	Removal of loose rust, loose mill scale, loose paint, and other loose contaminants to degree specified by hand chipping, scraping, sanding and wire brushing. Does not require removal of intact rust or mill scale.
SP 3: Power-tool cleaning	Removal of loose rust, loose mill scale, loose paint, and other loose contaminants to degree specified by power-tool chipping, descaling, sanding, wire brushing, and grinding. Does not require removal of intact rust or mill scale.
SP 5: White-metal blast cleaning	Removal of all visible rust, mill scale, paint, and foreign matter by blast cleaning with wheel or nozzle (dry or wet) using sand, grit, or shot.
SP 6: Commercial blast cleaning	Removal by blast cleaning of surface contaminants, except slight streaks or discolorations caused by rust stains, mill scale oxides, or slight, tight residues of rust or old paint or coatings. Slight residues of rust or old paint may remain in the bottoms of pits. Discolorations limited to one-third of every square inch.
SP 7: Brush-off blast cleaning	Removal by blast cleaning of all loose deposits, leaving only tightly adherent mill scale, rust, and paint or coatings, as long as the entire surface has been exposed to abrasive blasting.
SP 8: Pickling	Removal of all mill scale, rust, and rust scale by chemical reaction, electrolysis, or both.
SP 10: Near-white blast cleaning	Removal of all oil grease, dirt, mill scale, rust, corrosion products, oxides, paint, or any other foreign matter. Very light shadows, very slight streaks and stains caused by residues up to 5% maximum of any square inch may remain.

topcoats. Thus, the improvements to a coating system added by conversion coatings have often been found to be highly cost effective.

14.1.3 Exposure Testing and Evaluation

Laboratory testing of coatings usually requires some accelerating agent to produce results in a reasonable testing time. Service exposures in the field require several years to show any degradation of properly designed coatings—much too long for development of new coating systems. Accelerated laboratory testing methods must shorten testing time but should induce the same mechanisms of degradation as in service. Some of the more common coating evaluation procedures are described briefly for illustration, but no attempt is made to cover all possible procedures that have been used.

The *salt-fog test*,[10] used as an accelerated atmospheric exposure test, supplies a continuous spray of salt-laden water in a closed, high-humidity chamber (Figure 1.23b). Conditions approach continuous immersion, saturated in dissolved oxygen, as the spray condenses on the specimen and runs down the surfaces. Coatings are usually scribed to create a well-defined defect, and resistance is judged by the degree of attack at the scribe, as shown in Figure 14.3. The attack is usually characterized by undercutting at the coating-metal interface. The organic polymer that constitutes the coating is relatively inert and shows little deterioration in the short testing time. Other tests involving ultraviolet exposure are used to determine color and gloss retention and degradation of the coating itself.

Alternate-immersion cycles have been developed to enhance undercutting at the coating-metal interface. Equivalent undercutting was observed in one-half to one-

FIGURE 14.3 Undercutting at a coated and scribed surface, caused by salt fog exposure. *(Photograph by courtesy of R. J. Neville, Dofasco, Inc.)*

fourth the time required in continuous salt-fog tests, using the following daily exposure cycle:[11]

- 0.25 hr immersion in 5% NaCl, air saturated
- 1.25 hr drying at room temperature, ambient humidity
- 22.5 hr exposure at 120F (48°C) and 90% relative humidity

Another procedure, consisting of atmospheric exposure with twice-weekly spraying of 5% NaCl, showed better correspondence with service testing in automotive applications.[12] Figure 14.4[13] shows the result of alternate-immersion exposure of a scribed specimen without the phosphate conversion coat. The oxide corrosion product accumulated beneath the coating at the scribe. In salt-fog or other immersion tests, corrosion products remain as suspended colloids and cannot deposit easily on exposed surfaces and under coatings.

Accelerated laboratory tests obviously may not duplicate conditions in the field, and failure mechanisms in the laboratory cannot always be expected to be the same. Thus, the protectiveness of coating systems can be compared in the conditions of an accelerated laboratory test, but laboratory rankings of different coating systems may not compare well with rankings from field tests. Field exposure is clearly the best method to determine the resistances of coated systems. However, the time taken for field exposures is often too lengthy for development programs or quality control, and the only alternative is an accelerated laboratory test.

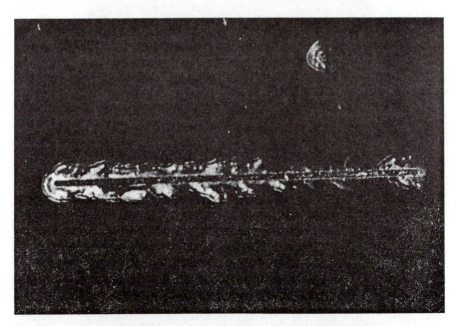

FIGURE 14.4 Undercutting at a 2.5-cm scribe in a coated steel surface caused by alternate immersion. *(From D. A. Jones, R. K. Blitz, and I. Hodjati, Corrosion, Vol. 42, p. 255, 1986. Reprinted by permission, National Association of Corrosion Engineers.)*

FIGURE 14.5   Samples suspended under a truck body for corrosion testing.
(Photograph by courtesy of R. J. Neville, Dofasco, Inc.)

Any coated product must eventually be qualified by *service testing* in the field. Samples in the form of panels or coupons are routinely exposed for atmospheric corrosion resistance on racks, such as the one shown in Figure 1.22. Coupon racks for exposure in process streams are shown in Figure 1.20. Exposure testing in the field can take many forms, depending on the nature of the equipment and environment and the ingenuity of the investigator. For example, sample coupons suspended beneath a truck body are shown in Figure 14.5. Coupons of various coated samples can be positioned around the test vehicle and removed for inspection periodically. Specimen configurations to capture mud and salt deposits on the surfaces are shown in Figure 14.6.[14] In automotive proving ground tests,[15] vehicles are operated continuously on a defined schedule that may include periodic water and salt spray, high humidity, freeze/thaw, high temperature, and gravel impact. The schedule is designed to simulate as closely as possible attack modes observed in field-tested vehicles. Thus, several years and many thousand miles of service are compressed into a few days or months on the proving ground. Coupons can be examined periodically, as in conventional laboratory testing, and the test vehicle itself can be disassembled and destructively examined at the conclusion of an exposure program.

14.1.4 Degradation and Failure Mechanisms

Many chemical mechanisms are possible for the degradation of coatings.[4] All require the penetration of the coating by reactant species, including water, oxygen, SO_2, and other electrolytes. Two possibilities of apparent technical importance, cathodic disbondment and oxide lifting, are discussed in this section to explain the

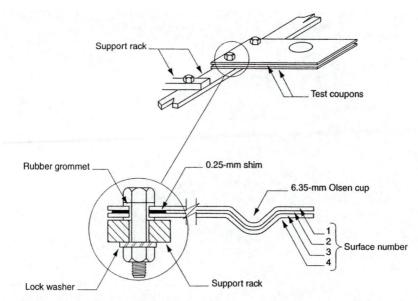

Support rack

Test coupons

Rubber grommet

0.25-mm shim

6.35-mm Olsen cup

1
2
3 } **Surface number**
4

Lock washer

Support rack

FIGURE 14.6   Test samples for undervehicle corrosion testing. *(From R. J. Neville,* SAE Technical Paper 800144, *Soc. Automotive Engineers, Warrendale, PA, 1980. Reprinted by permission, Society of Automotive Engineers.)*

test results reviewed in the preceding section. A schematic diagram describing the two appears in Figure 14.7.

Cathodic disbondment in the salt fog test has been attributed to formation of alkaline OH^- by cathodic reduction of dissolved oxygen,

$$O_2 + 2H_2O + 4e^- \rightarrow 4OH^-. \tag{4}$$

Macroscopic and microscopic defects, such as pinholes, voids, and mechanical scrapes and scratches, are inevitable in coatings, allowing access of the environment to the substrate metal. The anodic reaction

$$Fe \rightarrow Fe^{2+} + 2e^- \tag{5}$$

occurs at a coating defect that is coupled to the nearby cathode beneath the coating. Oxygen must migrate through the coating along with water, in order to support the cathodic reaction (4). Cathodically generated alkalinity can react with the organic polymer (saponification) to disbond the coating at the interface between coating and metal at a defect, as indicated in Figure 14.7.

Coating failures have been observed most often at macroscopic coating defects. However, cathodic disbondment may occur at apparently microscopic or smaller weak points in coatings to produce blisters, which require no physically obvious coating defect for initiation.

Oxide lifting occurs when anodic corrosion products accumulate under a coating, as illustrated in Figure 14.4. The lifting action of compacted oxides and resultant undercutting occurs only during alternate wetting and drying, not during continuous immersion. Although the mechanism is unclear, some observations summarized in

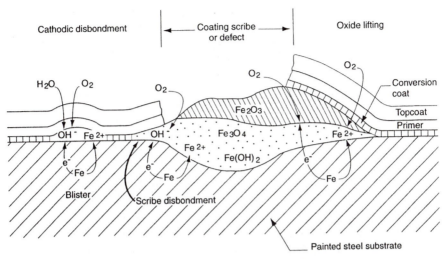

FIGURE 14.7 Schematic attack mechanisms at the scribe in a coating.

Figure 14.7 are possible, nevertheless. Flocculent oxide corrosion products in water are compacted by drying. The corrosion product (rust) scale forms an inner layer of dissolved ferrous ion, Fe^{2+} and precipitated $Fe(OH)_2$, which partially oxidizes to magnetite, Fe_3O_4 ($FeO \cdot Fe_2O_3$). An outer hydrated layer of fully oxidized $FeOOH$ completes the corrosion product deposit, as described in Section 12.1.3. Lamellar compact oxide corrosion-product layers are formed on uncoated steel by alternate wetting and drying cycles.[16]

The cathodic reaction during anodic lifting again may be

$$O_2 + 2H_2O + 4e^- \rightarrow 4OH^- \tag{4}$$

at the metal surface in the scribe or possibly

$$8FeOOH + Fe^{2+} + 2e^- \rightarrow 3Fe_3O_4 + 4H_2O \tag{5}$$

at the outer magnetite interface. Recall that magnetite is sufficiently conductive to act as an electrode for cathodic reaction and that Fe_3O_4 is reoxidized to $FeOOH$ by atmospheric oxygen (Section 12.1.3). SO_2, as an atmospheric pollutant, hydrates to sulfuric acid, increasing the scale porosity, the activity of the anodic reaction (5), the general atmospheric corrosion rate, and the breakdown of coatings at defects.

The phenomenon of oxide lifting, if not the exact mechanism, is similar to that occurring during filiform corrosion (Section 7.4.2) and distortion of bolted lap joints by packout (Section 7.4.3) during alternate wetting and drying in humid atmospheric exposures. All involve electrochemical corrosion in the condensed aqueous phase, during moist or "wet" exposures. Colloidal corrosion products deposit during subsequent dry periods and cannot be easily redissolved or dispersed on rewetting. The result is a buildup of compact corrosion products in constricted crevices or at the interface between coating and metal.

Oxygen is apparently more easily available at macroscopic coating defects, because disbondment occurs preferentially at such defects and at intentional scribes in coatings. Most efforts in development and application of coatings have been focused on improving the integrity of the bond between coating and metal at coating defects. For example, chromate and phosphate conversion coatings are generally thought to act by inhibiting disbondment at the coating-metal interface.

14.1.5 Electrochemical Testing

Much effort has been devoted to developing electrochemical methods to evaluate coatings. Cathodic polarization in chloride solutions has been found to produce rapid undercutting of the coating at a scribe.[17] Anodic polarization resulted in correspondingly little attack. From these results the *Paint Adhesion on a Scribed Surface* (PASS) test was developed.[18] As originally conceived, the test setup consists of a few drops of a 5% NaCl solution held on a scribed surface by an adhesively bonded O-ring gasket of 3/8-inch (0.95-cm) inside diameter. A platinum wire counter electrode is immersed in the upper surface of the electrolyte, and a current of 9 mA is passed for 15 minutes using a dc power supply and series-connected resistor (similar to Figure 3.17) to hold current at a constant level. Following this treatment, the loosened coating is stripped from the surface with adhesive tape, and the degree of undercutting is measured with a ruler or a low-power microscope. The PASS test was found to detect effects of process variables such as surface contamination, baking time and temperature, presence of phosphate conversion coat, and paint composition, in agreement with the salt-fog test.[10] However, the PASS test does not rank different coating systems in the same way as the salt-fog test or field tests and has not been used extensively.

Electrochemical impedance spectroscopy (EIS) is promising as an advanced research tool for coating development and evaluation. Theory and instrumentation for EIS were discussed in Section 3.5.6.; corrosion rate determinations from measurements of polarization resistance were described in Section 5.6.3.

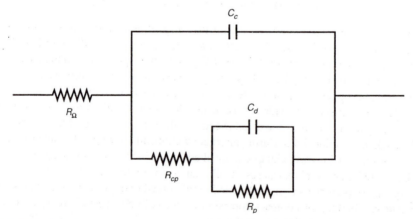

FIGURE 14.8 Electrical network analog for electrochemical impedance spectroscopy of a coated metal surface. R_Ω is solution resistance; R_{cp} is resistance of coating with pores; R_p is polarization resistance; C_c is capacitance of the coating; and C_d is the pseudo-double-layer capacitance.

The EIS technique requires development of an electrical analog network to simulate the electrochemical behavior of the metal-electrolyte interface. Figure 14.8 shows one such network that has been found generally suitable for modeling the behavior of coated-metal systems. Definitions of the network parameters are:

$R_\Omega \equiv$ solution electrolyte resistance,
$C_c \equiv$ coating capacitance,
$R_{cp} \equiv$ resistance of coating pores,
$R_p \equiv$ polarization resistance, which is inversely proportional to corrosion rate of exposed metal surface, and
$C_{dl} \equiv$ double-layer capacitance at metal surface.

The parallel connected Rp/Cdl circuit, simulating the metal-solution interface (Figure 3.25), is embedded in a second parallel circuit involving the coating parameters, Rcp and Cc, in Figure 14.8. A Warburg diffusion impedance (Section 3.5.6) may also be present at low frequencies, representing diffusion either in solution or in the coating pores and defects.

It may by possible to separate R_p and R_{cp} if their values are sufficiently different. Figure 14.9[19] shows the schematic Bode plot (Section 3.5.6) of the impedance spectrum expected from the network of Figure 14.8. At specific frequencies, the resistance plot shows a constant slope and the phase angle plot peaks, when the capacitances C_c and C_{dl} are present. The Nyquist format (Section 3.5.6) for data presentation has been found less discriminative between different processes, and the Bode format has been recommended.[19]

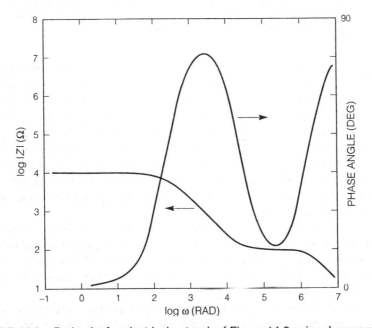

FIGURE 14.9   Bode plot for electrical network of Figure 14.8 using dummy values of R_Ω = 1 ohm, R_{cp} = 100 ohm, R_p = 10,000 ohm, C_C = 6.2 × 10^{-9} farad, C_d = 4 × 10 farad. *(From F. Mansfeld,* Corrosion, *Vol. 44, p. 558, 1988. Reprinted by permission, National Association of Corrosion Engineers.)*

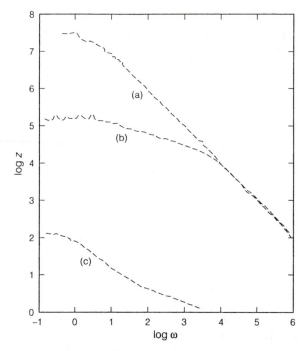

FIGURE 14.10   Bode plots for AISI 1010 carbon steel exposed 2 hours to aerated 0.5 *N* NaCl: Degreased, alkaline derusted, coated with 8μm polybutadiene; degreased, rust spots present, coated with 8 μm polybutadiene; uncoated. *(From F. Mansfeld, M. W. Kendig, and S. Tsai,* Corrosion, *Vol. 38, p. 478, 1982. Reprinted by permission, National Association of Corrosion Engineers.)*

Figure 14.10 shows typical examples of impedance spectra for coated steel after some period of exposure to an aqueous solution.[20] The curve, (c), for uncoated steel shows very low impedance at all frequencies, increasing to very high values for the cleaned (derusted) and coated steel, curve (a). The coating applied to a surface not completely derusted, curve (b), shows early deterioration of the low-frequency impedance, which measures the coating pore resistance, R_{cp}. It was concluded[20] that corrosion-product accumulation at the coating-metal interface induced coating defects, reducing R_{cp}. The coating capacitance, C_c, measured by the linear portion of curves (a) and (b) is identical for both. Therefore, this coating shows little or no water uptake, which would affect C_c.

Polarization resistance, R_p, for corrosion beneath the coating is apparently quite high and would require still-lower-frequency measurements, which are difficult and time consuming (Section 5.6.3). Without the coating, in the absence of R_{cp} and C_c, the impedance spectrum (c) measures the low value of R_p at low frequencies resulting from the comparatively high corrosion rate.

Thus, electrochemical impedance spectroscopy can give some direct indications of coating degradation by dielectric breakdown (R_{cp}) and water uptake (C_c). Other testing methods described earlier in this chapter, including conventional direct-current polarization techniques, measure attack only at an intentionally placed coating defect such as a scribe. Coating structural defects should also be detected easily by impedance methods, and work has begun[21] to confirm the effects.

Evaluation of coatings by impedance spectroscopy suffers from the same disadvantages mentioned in Section 5.6.3—complex, expensive instrumentation, difficulty in quantitative measurements at low frequency, and sometime problems with data interpretation. Nevertheless, EIS is expected to grow in usefulness for coating evaluation as instrument costs go down, more sophisticated analog networks are developed, and users become more familiar with the technique.

14.2 Thick Nonmetallic Coatings

In more aggressive chemical conditions and inaccessible locations, the corrosion rate is usually higher and appearance less important. The thickness of protective coatings must be increased to offset the inevitable defects in thinner coatings and improve protectiveness. Liquid-application methods may be inadequate, and adhesion to the metal surface is sometimes less critical.

14.2.1 Inert Liners

Many of the polymers used to formulate liquid-applied coatings also may be employed to form thick molded linings for tank, pipe, and other equipment. These include rubber, polyethylene, polyvinyl chloride, fluoroplastics, polyesters, and phenolics. The liners are molded in place and cured by conventional means without the solvent evaporation required by liquid-applied coatings. Reinforcement such as glass fiber and flake may be added to increase strength and impermeability. In the extreme, only the polymer itself is of importance; the metal shell provides strength only. Thus, many of these materials can be used to fabricate equipment directly, if strength can be increased to adequate levels. Nonmetallic plastic pipe, tanks, and pumps are widely used in the chemical process industry. Fabrication, design, and use of chemically resistant plastics are treated more fully elsewhere.[22]

Many resins may be formulated to the thicker consistency of *cement or mortar*. The same formulations are sometimes used as sealants and caulking compounds. Most contain an inert filler such as silica, quartz, clay, barites, or powdered carbon or coke mixed with a setting agent that must be added to the resin just prior to application. They are often proprietary, and the details of chemistry and methods of application are unavailable.

Chemical-setting cements based on sodium, and more recently potassium, silicates are cured with organic and inorganic setting agents.[23] They can be used as mortars for chemically resistant brick or as monolithic troweled coatings in the inside surfaces of equipment having complicated geometry, such as pipe, ductwork, chimneys, floors, trenches, and sumps. The potassium-silicate-based cements are generally preferable to the older sodium-silicate cements and are especially useful as coatings for resistance to acids from pH 1 to 7. Current technology for monolithic silicate liners utilizes intermediate membranes at the equipment surface.[24] The elastomeric membrane (resin, asphaltics, synthetic elastomers) cushions the cement lining from stresses caused by differential thermal expansion and provides a chemically resistant backup when the cement liner begins to crack and spall with age. At the same time, the cement liner protects the organic membrane from thermal and mechanical damage.

Chemically resistant brick linings are used in many chemical-process plants. The silicate and other resin-based cements and mortars for such linings are listed along with applications in Table 14.4, compiled from Read.[23] Modified silicate cements use proprietary setting agents for a one-component system, which requires only water and simplifies mixing. Molten sulfur is poured and solidified in the joints of some masonry walls. The remaining resin mortars listed in Table 14.4 are similar to

TABLE 14.4 Cement and Mortar Materials

Resin	Applications	Comments
Sodium silicate	Acids, pH 1–7	Long curing times, up to 96 hours
Potassium silicate	Acids, pH 1–7	Shorter curing times, easier workability
Modified silicate	Acids and Alkalies, pH 1–9; some alkalies to pH 14	One-part system needs no setting agent
Sulfur	Oxidizing acids and hydrofluoric acid below 88°C	Melted and poured into brick joints while hot, above 120C
Phenolic resin	Solution pH 0.7–9 up to 175°C; nonoxidizing acids	Not resistant to strong alkalies; poor storage life—must be refrigerated
Furan resin	Nonoxidizing acids, alkalies, salts, oils and detergents up to 175°C pH 1–13	Wide range of chemical resistance
Polyester resin	Most acids, mild oxidizing agents, pH 0.9–8 up to 120C	Not resistant to strong alkalies, most solvents; One- or two-component systems
Epoxy resin	Nonoxidizing acids and alkalies, organic solvents pH 2–14 up to 120°C	Good mechanical and physical properties

Derived from Read.[23]

the liquid-applied coatings but are formulated to high viscosity for troweling and cure at ambient temperature.

14.2.2 Porcelain Enamels

When a glass-powder overlay is fused (melted) on a metal surface and cooled, the result is a porcelain enamel coating. These vitreous-glass coatings are inert in water and weather, and special formulations are resistant to most chemicals, with the exception of hydrofluoric acid, which is well known to react with glasses. Excellent gloss and color combinations are attainable. Porcelain enamel coatings are routinely applied to steel, cast iron, and aluminum. The porcelain enamels are more expensive than the liquid-applied coatings discussed in Section 14.1 and can be applied only in the mill. Applications are limited to sizes and shapes that can be fired in a high-temperature furnace to fuse the coating. The brittleness of glass makes the porcelain enamels subject to chipping and cracking from erosion and impact.

The glass powder, or frit, may be applied to the surface either wet or dry. In the more-common wet process, a slip or suspension of the frit is applied to the surface by dipping, flowing, or spraying, followed by draining and drying. The residual frit remaining on the surface is then fired to fuse the glass coating. In the dry process, the frit is sprayed directly onto a hot surface, and the glass fuses on contact. In either method, clean surfaces prepared by chemical etching, mechanical abrasion, or shot blasting are critical to ensure good adhesion and absence of defects and discontinuities in the coatings.

The chemical resistance and appearance of porcelain enamel coatings are unmatched by any other coating system. Glass frit can be formulated to match the thermal-expansion properties of the substrate metal, and porcelain enamels routinely have excellent resistance to thermal shock over wide temperature ranges. The coatings are often applied in two layers. The ground coat, applied directly to the metal surface, has properties selected for adhesion and thermal expansion. Table 14.5[25] shows frit compositions for various applications. Acid and water resistance are obtained by increases in SiO_2 at the expense of B_2O_3 and BaO, with additions of TiO_2. ZrO_2 is desirable for both alkali and water resistance. Porcelain-enameled panels resist any evidences of deterioration for days or even weeks of salt-fog exposures. Color selection is limited in the ground-coat formulations. The ground-coat enamel is suitable for industrial and household applications, where appearance is not important. For example, porcelain-enamel coatings are used in chemical reactors, heat exchangers, food-processing vessels, mufflers, and transformer cases. The interiors of "glass-lined" household hot-water tanks are coated with porcelain enamels, specially formulated to resist hot water for up to 20 years.

A cover coat is often applied to the ground coat for optimal color, gloss, and appearance. The final porcelain-enamel products are noted for retention of color and gloss for decades in weather and sunlight. As a result, porcelain enamels are commonly seen on household appliances, signs, plumbing fixtures, exterior architectural panels and trim, and other products that require maximum appearance without maintenance for many years.

TABLE 14.5 Oxide Compositions for Ground Coat Enamel on Sheet Steel

Constituent	Composition, wt%			
	Regular Blue-Black Enamel	Alkali-Resistant Enamel	Acid-Resistant Enamel	Water-Resistant Enamel
SiO_2	33.74	36.34	56.44	48.00
B_2O_3	20.16	19.41	14.90	12.82
Na_2O	16.74	14.99	16.59	18.48
K_2O	0.90	1.47	0.51	...
Li_2O	...	0.89	0.72	1.14
CaO	8.48	4.08	3.06	2.90
BaO	9.24	8.59	...	...
ZnO	...	2.29	...	...
Al_2O_3	4.11	3.69	0.27	...
ZrO_2	...	2.29	...	8.52
TiO_2	...	...	3.10	3.46
CuO	...	...	0.39	...
MnO_2	1.43	1.49	1.12	0.52
NiO	1.25	1.14	0.03	1.21
Co_3O_4	0.59	1.00	1.24	0.81
P_2O_5	1.04	...	...	0.20
F_2	2.32	2.33	1.63	1.94

From Metals Handbook, Vol. 13, Corrosion, 9th ed., ASM International, p. 446, 1987. Reprinted by permission, ASM International

14.2.3 Pipeline Coatings

Thousands of miles of buried steel pipelines receive coatings in North America alone. Leaking pipelines not only waste valuable product, but also cause an environmental and safety hazard (Section 1.2). The economic investment is extraordinary in coatings to protect large surface areas for many years. Later costs of coating maintenance and equipment repair are still more extreme. A very recent, highly visible example is the failure of coatings to protect the trans-Alaska pipeline.[26] Initial investment and care in coating selection and application are well worthwhile in later savings in time, money, and safety.

Pipeline coatings first must provide a barrier against moisture reaching the steel surface. Thus, they are often applied to much greater thicknesses to minimize the holidays and defects present in conventional liquid-applied coatings for atmospheric and mild-chemical services. Thick coatings are also required to resist damage by handling during transport and installation, and later by soil and rock stresses applied during and after burial. High electrical resistivity retained over long periods is a special requirement, because cathodic protection is universally used in conjunction with coatings to prevent pipeline corrosion. Coatings must have good adhesion to the pipe surface and resist disbondment, especially from the alkaline products produced by the cathodic reactions during cathodic protection (Section 13.1.5). Finally, coatings must be resistant to degradation by biological organisms, which abound in most soils in the presence of moisture.

Trade-offs between coating protectiveness and cost are inevitable. No single coating is ideal for all applications. Careful selection and design are therefore required in view of the soil corrosivity and coating properties. The common coating systems used for buried pipelines[27] are discussed in the remainder of this section.

Bituminous coal tar enamels and *asphalt mastics* have been the workhorse materials for pipeline coatings. They are thermoplastic and are usually mill applied on the hot pipe surface as a hot melt or heated viscous paste, often with fillers and fiber tapes to increase thickness and resistance to mechanical and physical damage. The coal-tar enamels are also available dissolved in organic solvents and suspended in water-based emulsions for liquid application in the field. Melt-applied coatings generally have superior moisture and chemical resistance. These coatings become brittle below about 40°F (4°C), and installation at freezing temperatures requires special care. They also tend to degrade in ultraviolet light. This presents no difficulty after burial, but the surfaces must be shielded from sunlight during storage and transportation to the field installation site. Seamless extruded mastic coatings up to 1/2-inch thick are used in offshore pipelines.

Various *thermoplastics,* including polyethylene and polypropylene, have been successfully extruded onto pipe as coatings. *Epoxies and phenolics* have been applied to pipe surfaces as powder and liquid, where heat and chemical reaction initiate the polymerization reaction. The epoxies are especially known for resistance to the alkalies formed by cathodic protection.

Adhesive-backed tapes are applied in the mill, and sometimes in the field, to pipe surfaces that have been prime coated to enhance adhesion. The tape coats are generally applied in two layers, the inner for moisture impermeability and corrosion protection, and the outer for protection against physical damage. The tapes are applied cold, and thickness can easily be varied to allow for specialized service conditions.

14.3 Metallic Coatings

Metallic coatings usually serve two functions. First, the coating acts as a barrier to the environment, with greater corrosion resistance than the substrate metal. Second, the coating is usually selected with a more-active corrosion potential than the substrate and corrodes galvanically to provide cathodic protection at breaks in the coating. Metal coatings are applied by several methods, including immersion, plating, cladding, flame-spraying, arc-spraying, chemical deposition, and vapor deposition.[4,28] Space is not available to discuss them all in this book. The three most important from the standpoint of corrosion prevention, hot dipped, electroplated, and clad coatings, are discussed in the following sections.

14.3.1 Hot-Dipped Zinc and Aluminum

Galvanized zinc coatings applied by continuous or batch immersion (hot dipped) in a molten zinc bath have been used for many years to protect steel from atmospheric corrosion. The zinc serves as a barrier and provides galvanic protection. Nonprotective zinc hydroxide forms protective zinc carbonate by reaction with CO_2 and moisture in air. The zinc compounds are white or colorless and do not show obvious evidence of corrosion, as do the red rust products from iron and steel. Underlying steel exposed at scratches or cut edges are galvanically protected (Section 6.2.4) and do not show rusting when sufficient zinc is present. In exposure to weather, the time before appearance of rusting on panels is proportional to the weight or thickness of the zinc coating (Figure 14.11). Rusting appears sooner as the severity of the environment increases, in the order: rural, marine, industrial.[29]

Hot-dipped zinc coatings are very adherent, because a metallurgical bond forms between the steel and the coating at high temperature. Longer dipping times and slow cool-down promote excessive diffusion of iron into the coating, which impairs the appearance of the coating but not necessarily its corrosion resistance. Additions of 0.1 to 0.2% Al suppress iron transport into the coating but retain the metallurgical bond. Hot-dip galvanizing is used in all applications that require atmospheric corrosion protection of steel, including roofing and siding; automotive, truck, and bus parts; fencing; air conditioners; and farm implements.

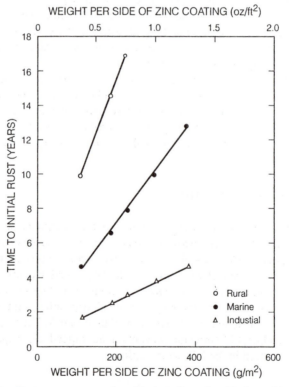

FIGURE 14.11 First appearance of rust in hot-dip galvanized steel in various atmospheres. *(From A. J. Stavros, Metals Handbook, Vol. 13, Corrosion, 9th ed., ASM International, Metals Park, OH, p. 432, 1987. Reprinted by permission, ASM International.)*

Galvanizing provides a good surface for conventional liquid-applied organic coatings after proper cleaning and priming. An organic coating extends the life of the galvanized zinc coating.[30] Corrosion products do not undercut coatings on zinc as they do on steel. The zinc still suppresses rusting of the steel exposed through the coatings.

Continuously applied *55% aluminum-zinc* has become the fastest-growing hot-dipped coating in recent years. This alloy coating exhibits significantly improved life over that of ordinary pure zinc coatings in atmospheric exposures, as indicated in Figure 14.12.[31] The alloy-coating microstructure consists of aluminum-rich dendrites with a zinc-rich interdendritic region. During corrosion, the zinc-rich areas corrode preferentially and are sealed into the exposed interdendritic spaces. There is still sufficient galvanic capacity to protect exposed steel at cut edges and other areas of coating damage, but the general-corrosion losses of the coating are limited by the resistance of the remaining aluminum-rich constituent.

Figure 14.12 also shows that weight losses of hot-dipped pure aluminum coatings are superior to those of both the pure-zinc and aluminum-zinc alloy coatings. However, aluminum often provides only barrier protection, especially in neutral solutions or in the atmosphere.[28] The corrosion potentials of the aluminum coating and the substrate steel are near one another and vary with solute and dissolved oxygen concentrations. Rust spots at coating defects and cut edges appear fairly soon, despite the low overall corrosion losses of the coating, indicated in Figure 14.12. Thus, thicker coatings are often required for equivalent protection against rusting. The higher-melting aluminum bath anneals the substrate steel and limits strength levels.

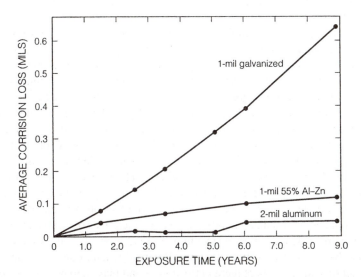

FIGURE 14.12   Losses in thickness with time for zinc, aluminum-zinc and aluminum hot-dip coatings on steel. *(From J. C. Zoccola, H. E. Townsend, A. R. Borzillo, and J. B. Horton, ASTM STP 646, ASTM, Philadelphia, p. 165, 1978. Reprinted by permission, American Society for Testing and Materials.)*

Flame-sprayed aluminum coatings have found use in naval-ship superstructures, where the dissolved sea salt in spray and condensate contributes to active aluminum potentials, which allow galvanic protection of the substrate steel. Aluminized coatings are also resistant in high-temperature applications such as automobile exhaust mufflers.

14.3.2 Electroplated Chromium and Other Metals

Electroplated metal coatings of any metal M are applied by cathodic polarization of the reaction,

$$M^{n+} + ne^- \rightarrow M, \tag{5}$$

which is the reverse of the usual anodic reaction for metal dissolution during corrosion. When the activity or concentration of dissolved M^{n+} is increased to high levels in an electroplating solution, the half-cell electrode potential of reaction (5) is made more noble, according to the Nernst relationship (Section 2.1.3). Polarization of an electrode to potentials active or negative to this redox potential reduces M^{n+} to M on the electrode surface. Various additives are usually present in electroplating baths to improve coating properties, including strength, uniformity, grain size, brightness, and so on.

Electrogalvanized-zinc coatings may be applied in preference to hot dipping for improved surface finish and finer control of dimensions. Galvanizing is one of the primary measures taken recently by automobile manufacturers to postpone corrosion by deicing salts from the interior of automotive body panels. The electroplated coating can be applied to both sides of sheet or to one side only, the "spangle" (surface crystals) in hot-dipped coatings is absent, and automotive finishes can be applied to electrogalvanized surfaces.

Chromium is a good example of a metal that finds considerable use in electroplated coatings on steel and other metals. It is perhaps a poor example from a corrosion viewpoint because its primary purposes are usually for decoration or wear resistance, although the hard chromium oxide on the surface does contribute to corrosion resistance. "Chrome-plated" coatings on automotive trim and bumpers are thin and porous; an intermediate electroplated layer of nickel provides corrosion protection for the underlying steel. Judicious additions of sulfur to the substrate nickel contributes to brightness and favorable galvanic interactions between the nickel and chromium layers.[28] Resistance to tarnishing, blue-white color, and high luster give chromium its decorative qualities. "Hard" chromium electrodeposits thicker than 1.2 μm (0.05 mil), applied directly to steel, provide abrasion and wear resistance; corrosion protection is useful but often of secondary importance.

Electroplated *tin* (tinplate) is well known as a protective coating for food and beverage containers. Tin is active to steel in most food products and provides galvanic as well as barrier protection in sealed cans. Relatively inert $FeSn_2$ forms during electrodeposition and subsequent thermal treatments and plays a major role in corrosion resistance of tinplate.[4] The dissolved corrosion products of tin are tasteless, colorless, and nontoxic. However, galvanic protection by tin is lost in the presence of dissolved oxygen; food should not be retained in tinplated cans after opening.

14.3.3 Cladding

There are many applications for cladding dissimilar metals on one another. The cladding process usually involves high-temperature roll bonding or coextrusion processes, which result in a continuous, pressure-welded, diffusion bond between the two alloys. The composite sheet, plate, or tubing retains a favorable combination of properties of the two alloys. For example, stainless steel may be clad onto copper for corrosion resistance, while retaining the thermal and/or electrical conductivity of the copper.

Another well-known example of cladding for corrosion protection is the roll bonding of a thin aluminum alloy layer, which is slightly active electrochemically, on the surface of relatively high-strength, aluminum-alloy plate. The cladding serves as a sacrificial anode, provides cathodic protection, and inhibits pitting, exfoliation, and stress corrosion cracking of the high-strength substrate alloy.

14.4 Inhibitors

In a sense, an inhibitor forms a protective coating *in situ* by reaction of the solution with the corroding surface. An inhibiting compound in small, but critical quantities reduces the corrosivity of the environment. Corrosion inhibition is reversible, and a minimum concentration of the inhibiting compound must be present to maintain the inhibiting surface film. Good circulation and the absence of any stagnant areas are necessary to maintain inhibitor concentration. A synergism, or cooperation, is often present between different inhibitors, and mixtures are the usual choice in commercial formulations. Specially designed mixtures are required when two or more alloys are present in a system. Extensive listings of inhibitor compounds[32] and commercial formulations[33] are available.

Certain compounds discussed earlier in this book to control water or solution chemistry are sometimes referred to as inhibitors. Sulfites and hydrazine, scavengers to control dissolved oxygen, were discussed in Section 11.1.4. Oxidizing ions, such as ferric salts and nitrates, will induce passivity in active-passive alloys as described in Section 4.1.3. Calcium, added to once-through water systems, fosters the formation of protective surface films by adjustment of the saturation index as discussed in Section 11.1.3. Certain weak bases, such as ammonia and a number of amine compounds (Section 11.1.4), are used to neutralize acidic boiler waters and thereby decrease corrosivity. Industrial applications of these and other inhibiting compounds are discussed in the next section. The chapter concludes with sections on inhibitor effects on electrochemical behavior and inhibitors that alter the properties of the metal-solution interface by surface adsorption and film formation.

14.4.1 Applications

Inhibitors find greatest use in recirculating systems. Once-through systems usually consume too much of the inhibiting chemical to be economically feasible. The effectiveness of inhibitors is diminished by increasing solution corrosivity, concentration,

TABLE 14.6   Effectiveness of Inhibitors in Near Neutral pH Water

Metal	Inhibitor*						
	Chromates	**Nitrites**	**Benzoates**	**Borates**	**Phosphates**	**Silicates**	**Tannins**
Mild steel	E	E	E	E	E	RE	RE
Cast iron	E	E	IE	V	E	RE	RE
Zinc	E	IE	IE	E		RE	RE
Copper	E	PE	PE	E	E	RE	RE
Aluminum	E	PE	PE	V	V	RE	RE
Pb-Sn solder	E	A	E			RE	RE

From A. D. Mercer, Corrosion, Vol. 2, 2nd ed., L. L. Schreir, ed., Newnes-Butterworths, p. 18:13, 1976. Reprinted by permission, Butterworths Publishers, Inc.

* E: Effective, IE: Ineffective, RE: Reasonably Effective, PE: Partially Effective, V: Variable, A: Aggressive

and temperature. Many inhibiting compounds are toxic (chromates, arsenic), and recent environmental regulations have limited their use. Nevertheless, inhibitors play a critical role in numerous corrosion-control strategies. Many are effective for more than one type of alloy (Table 14.6[34]), but pH, temperature, and other conditions are unique for each. An inhibitor for one metal may be corrosive to others.

Three types of environment find greatest use for inhibitors:[34]

1. Natural, supply, and industrial cooling waters in the nearly neutral (pH 5 to 9) range.
2. Acid solutions for pickling to remove rust and mill scale during the production and fabrication of metal parts or for postservice cleaning of such parts.
3. Primary and secondary production of crude oil and subsequent refining processes.

Inhibitors used commonly in these three classes of service are summarized in Table 14.7.

Inhibition in potable waters is limited by toxicity and cost. Only nontoxic chemicals are permitted, and since the systems are nonrecirculating, they must be inexpensive and can only be added in low quantities. Once-through cooling waters have the similar limitation of cost, and therefore only inexpensive chemicals similar to those above for potable waters are economically feasible, when added at low levels. More flexibility is available in recirculating waters. Often, combinations of inhibitors in commercial formulations give synergistic reductions and must be effective for several alloys in the system. Disposal of recirculating waters with toxic inhibitors must be handled carefully.

Corrosion in steam condensates is due to dissolved CO_2, which can be neutralized with the weak bases, including long-chain aliphatic amines listed in Table 14.7. Some amines apparently have inhibiting properties beyond simple acid neutralization, since several are effective even in strong acids, where their neutralization capabilities are inadequate.

TABLE 14.7   Inhibitors Used for Industrial Applications

Environment	Sample Inhibitors
Waters	
Potable water	$CaCO_3$ deposition, silicates, polyphosphates, zinc salts
Recirculating cooling water	Chromate, nitrate at 300–500 ppm calcium polyphosphates at 15–37 ppm silicates at 20–40 ppm
Automotive coolants	Nitrite, benzoate, borax, phosphate sodium mercaptobenzothiazole, benzotriazole
Steam condensates	Neutralizers: ammonia, morpholine, cyclohexamine, benzylamine long chain aliphatic amines such as octdecylamine at 1–3 ppm
Brines and seawater	Refrigeration brines: chromates, 2000–3300 ppm diluted seawater: sodium nitrite, 3–10% hot desalting brines: mixed chromate and phosphate, 50–100 ppm
Acid Pickling Solutions	
Sulfuric acid	Phenylthiourea, di-ortho-tolyl-thiourea, mercaptans, sulfides, 0.003–0.01%
Hydrochloric acid	Pyridine, quinoline, various amines, decylamine, phenylthiourea, dibenzylsulfoxides
Oil Production and Refining	
Primary and secondary recovery	Fatty imidazolines, various amines including primary amine, diamines, amido-amines, oxyethylated primary amines, alkyl pyridines, quaternized amines
Refining	Imidazoline and derivatives

Compiled from Mercer[34]

Chromate at high concentration must be used in high chloride systems. Nitrites at the percentage levels will inhibit ambient-temperature seawater. Chromate-phosphate mixtures exhibit a synergism and have been reported[34] effective at low levels of 5 ppm chromate and 30 to 45 ppm $NaHPO_4$.

In acid pickling solutions, sulfur-containing compounds have been found most useful for sulfuric acid, and nitrogen-containing compounds (amines) most useful for hydrochloric acid. A mixture may be better than either alone.

Corrosion in primary oil production derives from highly corrosive water extracted with the crude oil, containing H_2S, CO_2, and organic acids. Corrosion of well casings would be prohibitive without the development of effective inhibitors. The same water carries through to the refinery, causing corrosion in refinery equipment. Thus, the same inhibitors are often used throughout the production system. Again, amines are frequently useful in neutralizing weak acids from CO_2 and H_2S.

14.4.2 Electrochemical Behavior

Inhibitors may suppress the anodic, the cathodic, or both electrochemical reactions. Polarization measurements demonstrate which of these reactions are affected by an inhibitor, as illustrated in polarization studies of inhibition in brass-steel galvanic couples.[35] Figure 14.13 shows that nitrite is primarily an anodic inhibitor on steel, with little effect on the brass cathode. Conversely, chromate has the strongest effect on the cathodic polarization of the brass, with a weaker effect on anodic polarization of steel. Tolyltriazole affects both anodic and cathodic reactions equally, but not as strongly as the other two inorganic inhibitors.

Measurements by electrochemical impedance spectroscopy, described in Section 14.1.5 for coatings, also show considerable promise for evaluating mechanism and effectiveness of adsorbed and deposited inhibitor surface films.[36]

14.4.3 Fundamentals of Inhibitor Function

Although certain principles are common to the operation of many inhibiting compounds, there is no overall theory which permits prediction of the inhibiting performance of chemical compounds. Mechanisms are complex and often involve more than a simple barrier. Measured and calculated protective inhibitor deposits are often less than required for complete surface coverage. Nevertheless, some understanding of inhibiting mechanisms allows the engineer to evaluate proposed corrosion-control programs involving inhibitors more readily.

An inorganic anion, such as chromate or nitrite, forms an ionically bonded surface compound, which then serves as a barrier to corrosion reactions. Such inhibitors are more than simple oxidizers, like Fe^{3+}, that move potential into a noble region where the passive film is stable. They also form their own passive surface films or enhance the resistance of intrinsic passive films. This may be especially critical for carbon steels, in which the iron has a relatively high corrosion rate, even in the passive state.

The inorganic nitrite anion is known primarily as an anodic inhibitor. That is, nitrite has its primary effect on reducing the rate of the anodic dissolution reaction on steel, as shown in Figure 14.13. Other anodic inhibitors require oxygen to be effective, including molybdate, phosphate, silicate, and borate. They apparently enhance chemisorption of dissolved oxygen. Chromate affects both the anodic and the cathodic reactions for many alloys and is therefore known as a mixed inhibitor. Insufficient concentration of an anodic inhibitor can cause pitting. Therefore, special care is necessary to ensure that sufficient concentration is available to the inhibited surface.

Cathodic inhibitors result from surface deposits which impede the cathodic reaction, reduction of dissolved oxygen being the most important example. A natural inhibitor of this type is the calcium cation, which precipitates calcium carbonate in hard waters by reaction with the bicarbonate ion (Section 11.1.3);

$$Ca^{2+} + 2HCO_3^- \rightarrow Ca(HCO_3)_2 \rightarrow CaCO_3 + CO_2 + H_2O. \qquad (6)$$

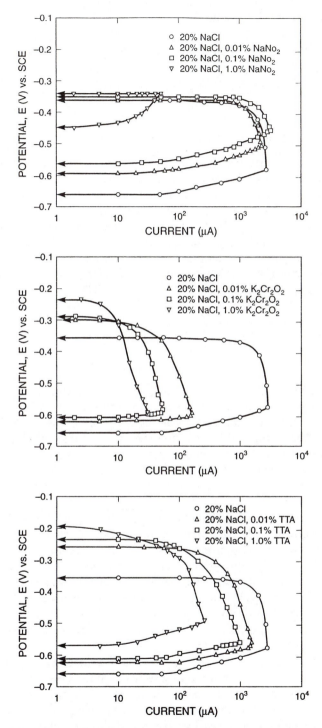

FIGURE 14.13 Effects of inhibitors on galvanic polarization diagrams for brass steel couples. *(From D. A. Jones,* Corrosion, *Vol. 40, p. 181, 1984. Reprinted by permission, National Association of Corrosion Engineers.)*

This reaction (6) is enhanced at cathodes where OH^- is produced by reduction of dissolved oxygen. That is,[37]

$$O_2 + 2H_2O + 4e^- \rightarrow 4OH^- \tag{4}$$

and

$$Ca(HCO_3)_2 + OH^- \rightarrow CaCO_3 + HCO_3^- + H_2O. \tag{7}$$

Zn^{2+} cations act similarly, reacting with the OH^- from reaction (4) to form $Zn(OH)_2$ deposits, which suppress dissolved oxygen reduction.

A deficiency in concentration of a cathodic inhibitor does not lead to excessive localized pitting, as is often the case with anodic inhibitors. Instead, the corrosion rate decreases uniformly over the surface with reduced inhibitor concentration. However, cathodic inhibitors are not generally as effective as the anodic ones.

The polyphosphates are inexpensive, effective, commonly used, cathodic inhibitors for aerated cooling-water systems.[37] The soluble sodium phosphates have the general form

$$NaO-\overset{\overset{\displaystyle O}{\|}}{\underset{\underset{\displaystyle ONa}{|}}{P}}-\left[-O-\overset{\overset{\displaystyle O}{\|}}{\underset{\underset{\displaystyle ONa}{|}}{P}}-\right]_x-ONa.$$

The noncrystalline (glassy) polyphosphate polymers with $x = 12$ or 14 can revert to lower-length crystalline polymers with $x < 3$ that are also effective cathodic inhibitors. However, the smaller phosphate molecules can form insoluble $Ca_3(PO_4)_2$ scales in recirculating hard-water cooling systems that are high in Ca^{2+}. Phosphonate compounds containing carbon are more stable and do not readily deposit phosphate scales. Both the polyphosphates and the phosphonates are also used as dispersants that inhibit crystal growth and thus deposition of $CaCO_3$ and other insoluble scales.[37]

Combinations of the above inhibiting elements can display beneficial synergism. For example, zinc salts added to chromates form $ZnCrO_4$, a cathodic inhibitor, which is more effective than the simple summation of the individual cathodic effects of Zn^{2+} and CrO_4^{2-} taken alone. However, both zinc and chromium (heavy metals) have an adverse environmental effect, and combinations of phosphates, polyphosphates, and phosphonates have partially or entirely replaced the two in many mixed inhibitor formulations.[36] Such nonheavy-metal inhibitors operate most efficiently in the slightly alkaline pH range, and good scale control with the phosphonates and other polymeric dispersants is a necessity.

Organic inhibitors are similar to the polyphosphates, inasmuch as they must be chemisorbed on the metal surface to be effective.[38] Since chemisorption processes are specific to the substrate metal, organic inhibitors vary in effectiveness for different alloy systems. Chemisorption of an organic compound is facilitated by the presence of polar groups in the molecular structure, which can readily attach to a metal surface. The most effective such polar groups include sulfur, nitrogen, hydroxyl, selenium, or phosphorous (e.g., phosphonate mentioned above). The effectiveness of an organic inhibiting molecule may be improved by larger size, asymmetry, greater molecular weight, and higher electron density. Molecular structures that include the benzene ring are especially prevalent. Schematic molecular structures[38,39] for some of the common organic inhibitors, listed in Table 14.7, and mentioned elsewhere in this chapter are shown in Figure 14.14.

The number and variety of organic inhibiting compounds are legion, structures are complex, and quantitative predictions of inhibiting capabilities are difficult. Thus, most organic inhibiting formulations are derived empirically and are often proprietary.

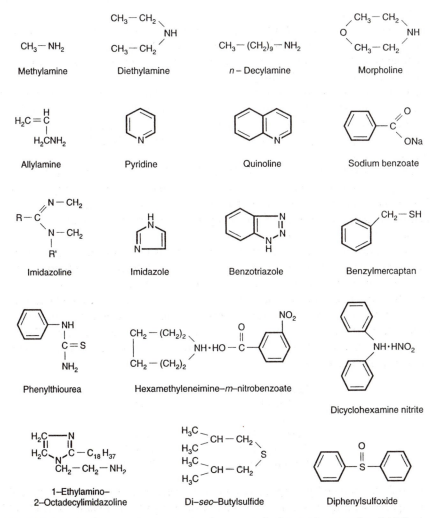

FIGURE 14.14 Schematic structures of some typical organic corrosion-inhibitor molecules.

14.4.4 Vapor-Phase Inhibitors

Some organic inhibitors with an appropriate vapor pressure (usually about 10^{-2} to 10^{-7} mm Hg[40]) will volatilize and condense on all surfaces in a closed volume. The inhibiting compound effectively increases the atmospheric-corrosion resistance of the exposed metal (usually steel) surfaces. A compound with a vapor pressure too low does not release sufficient vapor to effectively inhibit atmospheric corrosion. An inhibiting ion substituted into certain organic structures yields organic compounds having the necessary vapor pressure. This is particularly the case with nitrite (NO_2^-) and amine (NH_2^-), which are present in numerous organic inhibiting compounds, as may be seen in Figure 14.14. The volatility of the amines is also useful for carryover from boiler water into steam and condensate, where they continue to inhibit corrosion (Section 11.1.4).

A compound with a higher vapor pressure fills the enclosed space more rapidly, but the volatile solid or liquid inhibitor is consumed sooner, as the vapor escapes from nonairtight enclosures. It is impractical in normal service conditions to maintain seals which are totally airtight or to use vapor barriers which are absolutely impervious to vapor transport. Thus, the inhibitor of choice should have an optimum vapor pressure which fills the space to an adequate level before allowing significant corrosion, but which simultaneously yields adequate inhibitor life. A higher vapor-pressure inhibitor is appropriate for shorter storage times and for alloys of lower corrosion resistance.

Dicyclohexylamine nitrite (DAN) has been found especially effective as a vapor-phase inhibitor for steel surfaces but increases attack on copper alloys. Universal vapor-phase inhibitors, based on nitrobenzoate organic compounds that simultaneously protect ferrous, copper, and other alloy systems, have been reported.[41]

The major application of vapor-phase inhibitors is in packaging. In the simplest case, wrapping paper may be impregnated with the inhibitor for short-term protection. Frequently, the packaging material consists of an inner absorbent liner holding the inhibitor within an outer barrier membrane, which contains the inhibited vapor in proximity to the protected surfaces. The containment materials may be selected to maintain the inhibited environment for a few months to several years. The following containment materials yield the listed protection times (months), using DAN as the vapor-phase inhibitor:[41]

Kraft paper	3–15
Cardboard	9–14
Waxed kraft paper	24–54
Waxed cardboard	24–54
Plastic sheeting	60–120
Metal foil on paper	90–120

Vapor-phase inhibited packaging systems have been used to prevent temporarily atmospheric rusting of such items as small machine parts, up to large coils of cold rolled carbon steel, until the user is ready for them.

References

1. K. B. Tator, *Corrosion and Corrosion Protection Handbook*, P. A. Schwietzer, ed., Dekker, New York, p. 355, 1983.
2. K. B. Tator, *Metals Handbook*, Vol. 13, *Corrosion*, 9th ed., ASM International, Metals Park, OH, p. 399, 1987.
3. M. H. Sandler, *Metals Handbook*, Vol. 5, *Coating and Finishing*, 9th ed., ASM International, Metals Park, OH, p. 471, 1982.
4. H. Leidheiser, *Corrosion Mechanisms*, F. Mansfeld, ed., Dekker, New York, p. 165, 1987.
5. *SSPC Handbook*, Steel Structures Painting Council, Warrendale, PA.
6. Test Methods, TM-01-70 and TM-01-75, National Association of Corrosion Engineers, Houston.
7. T. W. Cape, *Metals Handbook*, Vol. 13, *Corrosion*, 9th ed., ASM International, Metals Park, OH, p. 383, 1987.
8. D. B. Freeman, *Phosphating and Metal Pretreatment*, Industrial Press Inc., New York, 1986.
9. K. A. Korinek, *Metals Handbook*, Vol. 13, *Corrosion*, 9th ed., ASM International, Metals Park, OH, p. 389, 1987.
10. ASTM Method B117, *Annual Book of ASTM Standards*, ASTM, Philadelphia.
11. V. Hospadaruk, J. Huff, R. W. Zurilla, and H. T. Greenwood, *Technical Paper 780186*, Society of Automotive Engineers, Warrendale, PA, 1978.
12. E. T. Nowack, L. L. Franks, and G. W. Froman, *Technical Paper 820427*, Society of Automotive Engineers, Warrendale, PA, 1982.
13. D. A. Jones, R. K. Blitz and I. Hodjati, *Corrosion*, Vol. 42, p. 255, 1986.
14. R. J. Neville, *Technical Paper 800144*, Soc. Automotive Engineers, Warrendale, PA, 1980.
15. G. Hook, The Historical Development of a Proving Ground Accelerated Corrosion Test, presented at CORROSION/80, Chicago, NACE, Houston, 1980.
16. A. J. Opinsky, R. F. Thomson, and A. L. Boegehold, *ASTM Bulletin 187*, ASTM, Philadelphia, p. 47, January 1953.
17. R. R. Wiggle, A. G. Smith, and J. V. Petrocelli, *J. Paint Technol.*, Vol. 40, p. 174, 1968.
18. J. Stone, *J. Paint Tech.*, Vol. 41, p. 661, 1969.
19. F. Mansfeld, *Corrosion*, Vol. 44, p. 558, 1988.
20. F. Mansfeld, M. W. Kendig, and S. Tsai, *Corrosion*, Vol. 38, p. 478, 1982.
21. J. N. Murray and P. J. Moran, *Corrosion*, Vol. 45, p. 34, 1989.
22. J. H. Mallinson, *Corrosion and Corrosion Protection Handbook*, P. A. Schweitzer, ed., Dekker, New York, p. 263, 1983.
23. G. W. Read, *Corrosion and Corrosion Protection Handbook*, P. A. Schweitzer, ed., Dekker, New York, p. 423, 1983.
24. G. D. Maloney, *Metals Handbook*, Vol. 13, *Corrosion*, 9th ed., ASM International, Metals Park, OH, p. 453, 1987.
25. *Metals Handbook*, Vol. 13, *Corrosion*, 9th ed., ASM International, Metals Park, OH, p. 446, 1987.
26. D. W. Freeman, *Popular Mechanics*, p. 68, August 1990.
27. C. G. Seigfreid, *Metals Handbook*, Vol. 13, *Corrosion*, 9th ed., ASM International, Metals Park, OH, p. 1288, 1987.
28. V. E. Carter, *Metallic Coatings for Corrosion Control*, Newnes-Butterworths, Sevenoaks, Kent, England, pp. 45–107, 1977.

29. A. J. Stavros, *Metals Handbook*, Vol. 13, *Corrosion,* 9th ed., ASM International, Metals Park, OH, p. 432, 1987.
30. J. W. Gambrell, *Metals Handbook*, Vol. 13, *Corrosion,* 9th ed., ASM International, Metals Park, OH, p. 436, 1987.
31. J. C. Zoccola, H. E. Townsend, A. R. Borzillo, and J. B. Horton, *ASTM STP 646*, ASTM, Philadelphia, p. 165, 1978.
32. M. G. Fontana, *Corrosion Engineering*, 3rd ed., McGraw-Hill, New York, p. 284, 1986.
33. E. W. Flick, *Corrosion Inhibitors*, Noyes, Park Ridge, NJ, 1987.
34. A. D. Mercer, *Corrosion*, Vol. 2, 2nd ed., L. L. Schreir, ed., Newnes-Butterworths, Sevenoaks, Kent, England, p. 18:13, 1976.
35. D. A. Jones, *Corrosion,* Vol. 40, p. 181, 1984.
36. W. J. Lorenz and F. Mansfeld, *Electrochim. Acta*, Vol. 31, p. 467, 1986.
37. B. P. Boffardi, *Metals Handbook*, Vol. 13, *Corrosion*, 9th ed., ASM International, Metals Park, OH, p. 487, 1987.
38. G. Trabanelli and V. Carassiti, *Adv. Corrosion Sci. Technol.*, Plenum, New York, p. 147, 1970.
39. O. L. Riggs, *Corrosion Inhibitors*, C. C. Nathan, ed., NACE, Houston, p. 7, 1973.
40. I. L. Rozenfeld, *Corrosion Inhibitors*, McGraw-Hill, New York, p. 133, 1981.
41. I. L. Rozenfeld, *Corrosion Inhibitors*, McGraw-Hill, New York, p. 297, 1981.

15

Materials Selection and Design

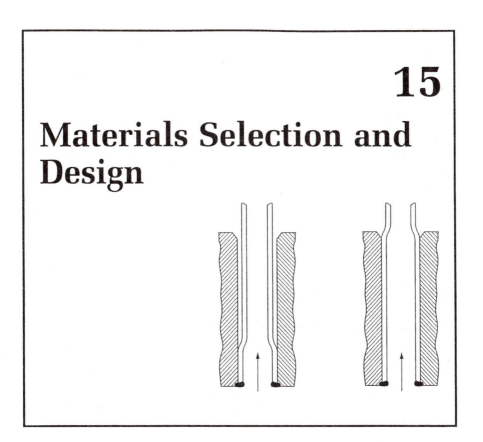

The most effective time to prevent corrosion is often during design, when specifications are determined for many factors affecting corrosion. These include chemical conditions of the environment, physical features of the designed structure/equipment, and methods of corrosion prevention. Conditions in a plant should be optimized for maximum production but must be adjusted so that equipment can perform without frequent and expensive interruptions and repairs. Environmental parameters affecting corrosion have been discussed extensively in this book and include chemical composition, temperature, and fluid velocity. Once conditions have been established for adequate production, the plant must be designed to avoid physical conditions that enhance corrosive effects. Again, physical parameters affecting corrosion have been discussed at length and include mechanical stress, wear, welding, crevices, and stagnant areas.

With preliminary design established, materials of construction must be selected, as well as appropriate prevention methods, such as coatings and cathodic protection. Often, the most effective method of corrosion prevention is proper selection of materials resistant to the specific corrosive environment. Yet if equipment design is poor, expensive corrosion-resistant alloys may be required and still may be inadequate in extreme cases. Thus, designers need a working knowledge of materials selection and corrosion prevention, and in some cases expert advice, to avoid costly, unexpected failures during startup and operation.

Prevention methods such as coatings, cathodic protection, and anodic protection have been discussed previously in this book, emphasizing various corrosion phenomena, and using particular alloys for illustration. In this concluding chapter, the important alloy systems are reviewed, emphasizing environments in which each finds extensive application. Design measures are discussed to prevent unpredictable conditions that enhance corrosion. The cost of corrosion prevention must be factored into the overall design scheme. Economic factors to be considered in selecting materials, determining prevention methods, and establishing design are described for corrosion prevention. Finally, economic calculations are sampled for design decisions on corrosion prevention.

15.1 Alloy Selection

The corrosion resistance and applications of the major constructional metal alloy systems are reviewed in this section. Nonmetallic materials are mentioned briefly in Section 14.2.1, with reference to more detailed treatments. The discussions in this section are confined largely to general corrosion, with comments on the localized forms of corrosion when appropriate. More detailed discussions of pitting and crevice corrosion are given in Chapter 7 and of stress corrosion cracking in Chapter 8. Alloy compositions are included on the back flyleaf.

15.1.1 Carbon and Low-Alloy Steels

Carbon steels are alloys of iron, with about 0.05 to 1% carbon. Low-alloy steels have other alloying elements, at levels usually below about 2%, mainly for improved mechanical properties. Carbon and low-alloy steels are relatively inexpensive, yet a wide range of strength and hardness can be obtained by simple variations in carbon content, alloying elements, and heat treatment. Carbon and low-alloy steels are mainly strengthened by the size and distribution of carbide phases and formation of hard, brittle martensite, when austenite is rapidly cooled from above 723°C to room temperature. The normally body-centered cubic structure of carbon and low-alloy steels leads to a ductile-to-brittle transition, which limits usefulness below about −20°C.

Carbon and low-alloy steels have relatively low corrosion resistance and often require protective coatings, even in relatively noncorrosive conditions. Coated carbon steels are used extensively for atmospheric exposures, as discussed in Chapter 14. Coatings can be omitted only when carbon-steel surfaces are sheltered and dry (Table 14.1). Carbon steel is the material of choice for most tanks, pipelines, and other buried structures. Coatings (Section 14.2.3) and cathodic protection (Section 13.2) are required for long-term service.

Carbon steels are used for ship hulls, piers, offshore drilling platforms, and other immersed marine applications (Section 11.1.5), but coatings and cathodic protection are required. The extent of corrosion in marine environments is generally dependent on access of dissolved oxygen. Uncoated carbon steel is used for boiler tubes, when the water is carefully deaerated and maintained slightly alkaline to prevent pitting.[1] However, boiling and evaporation in crevices, where heat transfer is low, will sometimes concentrate alkaline solutions and produce stress corrosion cracking (SCC), as described in Section 8.4.2.

Carbon steel is passive in alkaline solutions of high pH and is used to handle and store such solutions at ambient temperatures. Carbon steel is used for concrete reinforcement without coatings, because of the alkaline pH in concrete. However, coatings on reinforcement are sometimes necessary to control corrosion, when road deicing salts and marine salts are able to penetrate the protective concrete covering. Thus, design must provide for adequate thickness of the covering layer and proper proportions in the concrete mix to minimize porosity (Section 11.5).

Carbon steels will pit in neutral and alkaline salt solutions containing chloride (see the introductory comments in Chapter 7) and are subject to SCC in hot nitrate and caustic solutions (Section 8.4.2). Carbon-steel buried pipe has failed by SCC in bicarbonate solutions created by cathodic protection at coating breaks.

Carbon steel is used for tanks and pipe exposed to sulfuric acid at concentrations above 65% at ambient temperature and flow velocities below about 0.9 m/s.[2] However, use is limited above room temperature (Figure 11.20), and excessive flow or turbulence will disturb the protective sulfate surface film and cause localized erosion-corrosion (Section 10.2.1). Concentrations below 65% can be stored in carbon-steel tanks with anodic protection, avoiding the use of more expensive stainless steel.

Low-alloy steels behave much the same as carbon steels in aerated neutral water and solutions with corrosion controlled by dissolved oxygen. However, additions of Cu, Ni, Si, and Cr in the high-strength low-alloy weathering steels improve atmospheric corrosion significantly. These alloying elements facilitate the formation during alternate wetting and drying of an adherent, protective, surface film, which has an agreeable dark rust color. The weathering steels have thus been used successfully in the unpainted condition in applications where the surface is readily dried by sunlight after periodic wetting from dew and rain. Less success has been experienced in sheltered, shaded, or humid locations, where moisture can be retained on the surface without drying.[3] Crevices can also retain moisture and accumulate corrosion products that distort bolted lap joints fabricated from weathering steel (Section 7.4.3).

Both carbon and low-alloy steels are subject to hydrogen cracking and blistering in petroleum production and refining operations. Alloy steels are more susceptible, because they form susceptible martensitic structures more readily and are thus more difficult to weld. Susceptibility to hydrogen cracking increases as yield strength increases (Section 10.1.2).

15.1.2 Stainless Steels

Iron-based alloys containing at least 10.5% chromium are called stainless steels. They are classified as ferritic, austenitic, duplex, martensitic, and precipitation hardening. Each has its special characteristics,[4] as described in this section.

The *ferritic stainless steels* consist simply of chromium and sometimes molybdenum alloyed with iron. Chromium is a ferrite stabilizer and retains the alloys in the body-centered cubic crystal structure. Because of low carbon and nitrogen interstitial solubilities in ferrite, these alloys have had limited use because of easy sensitization to intergranular corrosion (IGC) as well as low ductile-to-brittle transition temperatures. Recent refinements in steel making, most notably argon-oxygen decarburization, have reduced interstitial impurity levels, lowered ductile-brittle transition temperatures, and improved the toughness of the ferritic stainless steels. However, they are generally confined to sheet and tubular products, because of inherent limitations in toughness of the ferritic structure, which is further reduced by increasing section thickness. A major advantage of the ferritic stainless steels is immunity to chloride stress corrosion cracking (SCC). Extreme care must be used during welding to avoid contamination with carbon and nitrogen. Nitrogen is obviously inappropriate as a cover gas. All oil (containing carbon) should be carefully removed from all surfaces before welding. Appropriate welding procedures for the ferritic stainless steels are discussed in Section 9.4.3.

The ferritic stainless steels are favored for use in thin-wall tubing for heat exchangers, where SCC is a concern, for example in the chemical, petroleum, and natural gas processing industries. Elevated service temperature further reduces the concern for low ductility and limited toughness. The lowest-alloyed ferritic stainless steel, Type 409, has been developed for use in automotive mufflers and catalytic converters. Types 405 and 409 have been used as a substitute for carbon steel in the tube-support plates of pressurized-water nuclear steam generators to prevent corrosion and denting (Section 7.4.4) in the crevices between the support plates and nickel Alloy 600 heat-exchanger tubes. Type 430 has been used in architectural and automotive trim for mild to moderate atmospheric exposure.

The *austenitic stainless steels* are widely used in industry and have been discussed extensively in this book. The addition of nickel to iron-chromium alloys stabilizes the face-centered cubic austenite phase and improves corrosion resistance synergistically with chromium. Corrosion resistance derives from a thin, hydrated, oxidized, chromium-rich, passive surface layer. Metallurgical segregation, causing localized depletion of chromium, can result in passive breakdown at grain boundaries and IGC (Section 9.1.2). Chlorides attack the passive layer and cause pitting, crevice corrosion (Section 7.1), and SCC (Section 8.1). These limitations can be avoided with the proper choice of alloy and heat treatment.

The austenitic stainless steels have been used for architectural trim in industrial and marine atmospheres, where the surfaces are not readily washed free of airborne deposits, but some surface discoloration of unprotected surfaces can be expected in such environments. Type 304 stainless steel has been used successfully for valve parts, pump shafts, fasteners, and screens in fresh water low in chloride. Immersion service in seawater is limited by crevice and pitting corrosion, especially if stagnant conditions occur. SCC is a possibility at any temperatures above ambient but is a real danger above 70°C, when chlorides are present. Addition of 3% molybdenum in Type 316 stainless steel improves resistance to chloride pitting and SCC. Higher alloyed stainless steels, such as 20Cb-3, are more resistant to pitting in chlorides and higher acid concentrations discussed below. Some nickel alloys are also resistant to chlorides (Section 15.1.3).

Austenitic stainless steels have been used widely to contain industrial chemicals. Nitric acid can be handled readily by the austenitic stainless steels in concentrations up to 65% and at temperatures up to boiling. The oxidizing nature of nitric acid promotes formation of the resistant passive film. Higher temperature and concentration require alloys of correspondingly higher chromium content. Hydrochloric acid attacks the passive film and cannot be used with any of the stainless steels. Austenitic stainless steels are resistant to very dilute and highly concentrated sulfuric acid but are attacked at intermediate concentrations. Aeration or other dissolved oxidizers will improve the resistance of austenitic stainless steels in sulfuric acid by stabilizing the passive film. Austenitic stainless steels are resistant to phosphoric acid over the full range of concentrations at temperatures up to about 65°C. However, impurities of chloride, sulfuric acid, and fluorides from phosphate rock in commercial plants reduce the resistance. The austenitic stainless steels are generally resistant to organic acids, but the stronger ones (e.g., formic acid) require special care (Section 11.6.5).

The austenitic stainless steels have been used extensively in the food-processing industry, where contamination affecting taste and color must be avoided. An added advantage is the easy cleanability of the austenitic stainless steels. Many food products contain chlorides and organic acids that require more highly alloyed austenitic stainless steels, especially at higher processing temperatures. Use in the pharmaceutical industry is similarly widespread, where contamination is even more critical, because dissolved corrosion products may catalyze unwanted biochemical reactions.

Corrosion-resistant alloys are required in the oil and gas industry to handle process streams containing water and H_2S. Austenitic stainless steels are used frequently but must be chosen judiciously to avoid pitting, crevice corrosion, and SCC, which are enhanced by H_2S. Thus, the higher-alloyed austenitic stainless steels and the duplex austenite-ferrite grades are often preferred.

The austenitic stainless steels are used extensively in electric power plants for turbine blades and vanes, heat exchangers, condenser tubing, and flue-gas desulfurization equipment. For the usual austenitic stainless steels, chloride and dissolved oxygen must be controlled to prevent SCC in boilers, as shown in Figure 8.17. More-resistant austenitic stainless steels, containing greater amounts of nickel, chromium, and molybdenum, must be used when chloride and oxygen cannot be removed. The modern low-interstitial ferritic stainless steels and duplex stainless steels have also been used for their resistance to SCC.

The pulp and paper industry makes extensive use of the austenitic stainless steels. Types 304, 316L, and 317 are used in process vessels and piping and for heat-exchanger tubing in evaporators. When chlorides are present, higher chromium and molybdenum grades are required.

The *duplex stainless steels* are chromium-molybdenum alloys of iron with sufficient austenite stabilizers, nickel and nitrogen, to achieve a balance of ferrite plus austenite. The result is a favorable combination of the two phases. The austenite confers ductility, and the ferrite, resistance to SCC. Molybdenum strengthens the passive film and improves pitting resistance. Carbides tend to precipitate at the disperse austenite-ferrite interfaces, preventing sensitization to intergranular corrosion by grain boundary precipitation (Section 9.2.1). The duplex stainless steels consequently find use in more severe conditions of temperature and chloride content, where the usual austenitic grades are susceptible to pitting, crevice corrosion, and SCC.

The *martensitic and precipitation hardening stainless steels* are chosen primarily for mechanical strength. Corrosion resistance is lower than the other grades of stainless steel, and applications are generally limited to mild environments. Because of the high-strength levels, both are susceptible to hydrogen induced cracking (Section 8.3.1).

15.1.3 Nickel Alloys

The face-centered cubic crystal structure of nickel has much higher solid solubility for alloying elements than that of body-centered cubic iron. Thus many nickel alloys have been custom designed for a wide variety of specific service conditions.[5] Further, the face-centered cubic structure makes the nickel alloys easily formable and not subject to ductile-to-brittle transitions with decreasing temperature. Nickel and its alloys are widely used in chemical process plants and oil refineries because there are corrosion-resistant alloys suitable for most of the corrosive environments present in these industries. The nickel alloys are used extensively for vessels, piping, and pumps in the production of most mineral and organic acids.

Nickel has intrinsically low corrosion rates in acid solutions in the active state. Alloying additions of copper, molybdenum, and tungsten lower the active rate even further. However, oxidizers increase the corrosion rates of these alloys considerably in the active state. Chromium additions induce passivity and improve corrosion resistance in the same oxidizing solutions. The electrochemical polarization curves shown in Figure 4.17 illustrate this behavior. The Ni-Mo, Alloy B, has resistance comparable to Alloys C and C-276 in sulfuric acid that is free of oxidizers. Chromium additions in Alloy C induce passivity, but control of Si and C stabilizes the passive current density (corrosion rate) over a much wider potential range in Alloy C-276. Figure 4.18 illustrates the superior resistance of Alloy C to chlorides in comparison to austenitic Type 304 stainless steel.

Commercially pure nickel, Alloy 200, is resistant to caustic solutions of NaOH and KOH at all concentrations and temperatures and is used in caustic evaporators to concentrate dilute caustic solutions up to 50%. Alloy 200 is not resistant to ammonium hydroxide, however. Applications requiring higher strength use Ni-Cu Alloy 400, Ni-Cr-Fe Alloy 600, and others in caustic service. Caustic concentrations above 70% at temperatures above 125°C can cause increased general attack of the alloys; Alloy 200 has the best resistance and is generally recommended. Stress corrosion cracking may be a danger in any of the alloys at higher temperatures and concentrations.

Nickel-copper alloys, such as Monel Alloy 400 (Ni-30% Cu), have improved resistance to nonoxidizing acids such as deaerated H_2SO_4, HCl, and H_3PO_4. Alloy

400 is particularly known for its resistance to all concentrations of deaerated hydrofluoric acid (HF). *Nickel-molybdenum alloys*, Alloys B and B-2, also have improved resistance to nonoxidizing acids. Molybdenum confers special resistance to deaerated hydrochloric acid (HCl). Alloy B-2 shows the highest resistance of all the nickel alloys in deaerated HCl. Oxidizers, as dissolved impurities or dissolved oxygen, increase attack on Ni-Mo and Ni-Cu alloys, and other nickel alloys containing chromium are usually specified.

Nickel-chromium-iron alloys passivate in the presence of oxidizers in acid due to chromium additions (Section 4.1.1). Chromium forms a chromium-rich, passive oxidized film that is highly resistant to acid attack. A few percent iron present in Alloy 600 decreases cost, without substantially impairing corrosion resistance. Alloy 800 is on the borderline between austenitic stainless steels and the nickel alloys. It contains about 44% Fe, 32% Ni, and 21% Cr; corrosion resistance and cost are correspondingly intermediate between those of the nickel-base alloys and stainless steels. All of these alloys are subject to increased corrosion, pitting, and crevice corrosion in the presence of chloride in oxidizing acids. However, all are more resistant to pitting, crevice corrosion, and chloride stress corrosion cracking (SCC) than the austenitic stainless steels.

Molybdenum added in the *nickel-chromium-iron-molybdenum alloys* increases resistance to chloride attack in the form of pitting, crevice corrosion, and SCC just as it does when alloyed with austenitic stainless steel. Further addition of about 3 to 4% tungsten with 13 to 16% molybdenum produces Alloy C-276, which has maximum resistance to localized pitting and crevice corrosion in chlorides, when silicon impurities are controlled.

The resistance of the nickel alloys is summarized in isocorrosion diagrams for sulfuric and hydrochloric acids in Figures 15.1 and 15.2,[5] respectively. The nickel alloys are resistant at intermediate and higher concentrations of sulfuric acid, where the austenitic stainless steels are unusable. Alloys containing both chromium and molybdenum (C-22 and C-276) are most resistant to hydrochloric acid. The nickel alloys with chromium are resistant to nitric acid, but the austenitic stainless steels are less costly and usually preferred. Nickel alloys containing substantial amounts of silicon, 9 to 11% (e.g. Alloy D), are used in castings, which are resistant to hot concentrated sulfuric acid.

Another class of nickel alloys containing chromium, aluminum, and molybdenum are designed for elevated temperature use. These alloys often require protective coatings for resistance to high-temperature oxidation and are described more fully in Section 12.4.2.

15.1.4 Copper Alloys

The thermodynamic tendency for copper corrosion is low, as measured by a low free energy of chemical reaction with aqueous solutions and relatively noble potentials in the emf series (Table 2.1) and galvanic series (Figure 6.1). Thus, copper and its alloys are quite corrosion resistant in many atmospheric and nonoxidizing aqueous environments. Strong-acid and oxidizing-acid solutions attack copper, but corrosion resistance is improved by appropriate alloying additions. Ammonia forms complex-

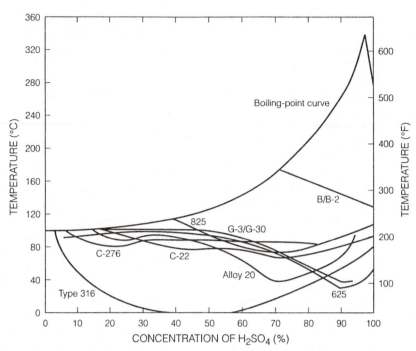

FIGURE 15.1 Isocorrosion diagram for nickel alloys in sulfuric acid at 0.5 mm/y (20 mpy). *(From N. Sridhar,* Metals Handbook, *Vol. 13,* Corrosion, *9th ed., ASM International, Metals Park, OH, p. 643, 1987. Reprinted by permission, ASM International.)*

es with copper and increases general corrosion and stress corrosion cracking (SCC) of copper alloys.

Commercially pure copper (>99% Cu) is used routinely in home plumbing systems to handle potable and waste waters and has excellent resistance to corrosion in soils. The tendency to pitting corrosion in some potable waters is described in Section 7.5.1. Atmospheric corrosion resistance is excellent; the characteristic, protective, green $CuSO_4$ patina forms with time. Copper-sheet roofing has endured for centuries on many historical buildings and is still used for roofing and architectural trim. Pure copper is also resistant to dilute nonoxidizing sulfuric, phosphoric, and acetic acids.

The major alloy of copper is *brass,* formed by additions of zinc (10 to 40%), which improves mechanical properties by solid solution strengthening and reduces cost of the alloy. Yellow brass (30% Zn) is the most common. Zinc additions reduce the inherent corrosion resistance of pure copper. Brasses are susceptible to SCC (Section 8.4.4), dealloying (Section 9.5.2), and impingement attack or erosion corrosion (Section 10.2.1). The lower-zinc brasses, around 10% Zn, are much more resistant to stress corrosion and dezincification, but strength levels are too low for many applications, and they are more susceptible to impingement attack. Additions of about 1% tin to 30% and 40% Zn brasses form alloys known as Admiralty metal and Naval Brass, respectively, which have increased resistance to dezincification. Additions of P, As, or Sb (0.02 to 0.1%) further increase the resistance to dezincification. Arsenical aluminum brass (76%Cu, 22%Zn, 2%Al, 0.02 to 0.10%As) is resistant to impingement attack and is used in seawater condensers and heat exchangers, where impingement attack is often a problem. Again, the As inhibits dezincification.

Bronzes are copper alloys containing 8 to 10% tin, often with small amounts of phosphorus (phosphor bronze) to improve impingement attack. The bronzes are more often used in the cast form, with further alloying additions of lead, tin, and nickel for seawater valves, pumps, gears, and bushings. Bronzes have somewhat better resistance to impingement attack and SCC than the brasses.

Aluminum bronzes are copper, 5 to 12% aluminum, alloys that are especially resistant to impingement attack. Above 8% Al, dealloying of aluminum-rich phases is a problem in some environments. The alloys are also susceptible to SCC in many of the same environments as the brasses, especially ammonia solutions. Aluminum bronzes are more corrosion resistant than the brasses and are resistant to nonoxidizing mineral and organic acids, ammonium, and alkali metal hydroxides, and various natural waters, including seawater, brackish water, and potable waters. Like other copper alloys, aluminum bronzes are not resistant to oxidizing salts such as chromates or ferric salts or to nitric acid.

Copper-nickel alloys are the most corrosion resistant of the commercial copper alloys. Alloy 400, with 30% Ni, has maximum resistance to impingement attack and

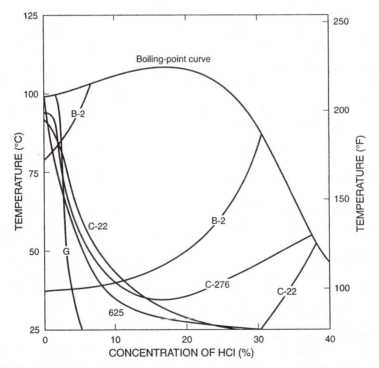

FIGURE 15.2 Isocorrosion diagram for nickel alloys in hydrochloric acid at 0.5 mm/y (20 mpy). *(From N. Sridhar, Metals Handbook, Vol. 13., Corrosion, 9th ed., ASM International, Metals Park, OH, p. 643 1987. Reprinted by permission, ASM International.)*

SCC, but the 10% alloy often has sufficient resistance at lower cost. Both have excellent corrosion resistance to nonoxidizing acids and find use in the chemical-process industries. Their greatest application, however, is in recirculating steam systems for condenser and heat-exchanger tubing. Amine inhibitors such as morpholine and hydrazine can decompose in steam systems to form ammonia, which may cause SCC in many copper alloys.

15.1.5 Aluminum Alloys

Aluminum is a reactive metal that owes its corrosion resistance to a thin, protective, barrier oxide surface layer, which is stable in air and neutral aqueous solutions from pH 4 to 8.5 (Figure 2.9). Thus, aluminum alloys are widely used in architectural trim, window and door frames, automotive trim, and cold- and hot-water storage vessels and piping. Aluminum alloys are resistant to food and pharmaceuticals in the neutral pH range, and the corrosion products are colorless, nontoxic, and relatively tasteless. Use in household cooking utensils and commercial food-handling equipment is traditional. Packaging of food and pharmaceuticals, especially beverage cans, has been a growing segment of the market for aluminum alloys. Because of its light weight, aluminum and its higher-strength alloys are used extensively for structural components and fuselage coverings in aircraft and aerospace vehicles. Lithium alloy additions have been the subject of recent development, because they further decrease density and increase modulus of elasticity.

In the neutral pH range, aluminum and most of its alloys may corrode by localized pitting in the presence of chlorides. The mechanisms of film breakdown and initiation and propagation of pitting corrosion are generally similar to those described for the stainless steels in Section 7.2. Nevertheless, aluminum alloys are resistant to seawater if surfaces are kept clean and solutions are flowing or regularly refreshed. Stagnation and accumulation of corrosion products may allow local pH to stray out of the neutral range in both seawater and fresh water, increasing general and pitting attack. Periodic cleaning of seawater-handling equipment is especially critical because nontoxic aluminum surfaces do not inhibit growth of marine fouling organisms, which promote pitting and crevice corrosion beneath the resulting deposits.

Aluminum alloys are active to most other metals in the galvanic series (Figure 6.1), with the exceptions of zinc, beryllium, and magnesium. Thus, galvanic corrosion may occur on aluminum, which acts as the anode when coupled to most other constructional alloys, such as steel, nickel and, copper alloys. Galvanic coupling often polarizes the aluminum anode above the pitting potential (Section 7.2.1) and results in failure by pitting in the presence of chlorides. Water with a few parts per million dissolved copper will deposit metallic copper at defect sites in the aluminum-oxide surface film, where the deposited copper will act as a local cathode and stimulate a pit-anode at the defect.

Corrosion resistance of aluminum alloys varies widely, depending on the alloying components that are present. Most of the alloying elements in aluminum decrease corrosion resistance and are added to improve mechanical properties by

TABLE 15.1   Wrought Aluminum Alloys and Their Corrosion Behavior

Alloy Class	Typical Temper*	Alloying Elements	Corrosion Resistance**			
			General	Pitting	Exfoliation	SCC
1xxx	All	Natural impurities in refinery Al	E	E	E	I
2xxx	T3,T4,	Cu	F	P	P	VS
	T8		F	P	F	R
3xxx	All	Mn, Mn + Mg	E	E	E	I
4xxx	All	Si	F	G	G	G
5xxx	Most	Mn, Mg, Cr	E	G	G	I-R
6xxx	All	Mg, Si	E	G	E	I
7xxx	T6,	Zn, Mg, Mn, Cu	F	F	F-P	S-VS
	T73		F	F	G	R

* T3, T4, T6: age hardened; T8, T73: overaged.

** E-excellent, G-good, F-fair, P-poor. I-immune, R-resistant, S-susceptible, VS-very susceptible.

solid-solution strengthening or age hardening. Heat treatment or tempering affects corrosion resistance and mechanical strength by controlling the distribution of alloying elements between solid solution and insoluble precipitates. Temper designations[6] are lengthy and specific to each alloy type. Heat-treatable, high-strength, age-hardened alloys are most susceptible to stress corrosion cracking (SCC), but may be made more resistant by slight overaging beyond maximum strength. Age-hardenable alloys are also susceptible to intergranular corrosion and exfoliation (Section 9.2.3). Table 15.1 summarizes the various alloy classes and their corrosion resistances. Each alloy class is specified by the first of the four-digit designation. Alloys within each class are specified by the other three, xxx, digits.

The 1xxx alloys consist of commercially pure aluminum with residual impurities controlled to maintain specified corrosion resistance in various applications. Minimal impurities and alloying elements give excellent corrosion resistance and immunity to SCC, but generally low strength. Thus, the 1xxx alloys have limited commercial use.

The 2xxx alloys contain copper as the primary alloying element, which produces age hardening, when metastable solid solutions are annealed at somewhat elevated temperature. These high-strength alloys have been used extensively in airframe components, because of the low density of aluminum. However, they are the least corrosion resistant of all aluminum alloys. High strength in the age-hardened condition leads to SCC susceptibility, which can be moderated by overaging. Copper corrosion products redeposit on the surface, forming galvanic cells that produce general, pitting, and exfoliation (Section 9.2.3) corrosion. Surfaces of 2xxx alloys are often protected by alcladding discussed later in this section.

The 3xxx alloys have low-level alloying additions of manganese and magnesium plus manganese. Manganese in solid solution and manganese-rich precipitates have nearly the same corrosion potential as aluminum. Thus, manganese has minimal effects on corrosion resistance of aluminum alloys, while allowing solid-solution strengthening and work hardening. The 3xxx alloys are used widely in household cooking utensils, food-processing equipment, and architectural products requiring high-corrosion resistance and modest strength.

The 4xxx alloys contain silicon as the primary alloying element. Silicon-rich precipitates, although noble to aluminum like copper, polarize strongly to limit any increased corrosion current. Thus, Al-Si alloys have comparable corrosion resistance to pure aluminum.

The 5xxx alloys contain somewhat higher levels of Mn and Mg than the 3xxx alloys, and some grades also contain chromium. As a result, the 5xxx alloys have higher strength but still good corrosion resistance. This combination of properties prompts use in food- and chemical-processing equipment, as well as applications involving seawater exposure. Alloys containing excess Mg with precipitates distributed exclusively at the grain boundaries may be susceptible to exfoliation and SCC. Tempers have been developed to alleviate the problem, however.

The 6xxx alloys are age hardenable by precipitation of Mg_2Si, which does not behave electrochemically much differently than the aluminum-rich matrix phase. Thus, these alloys retain good corrosion resistance with improved strength. Excess Mg over that required to form the precipitate increases susceptibility to intergranular attack.

The 7xxx alloys contain zinc, magnesium, manganese, and sometimes copper. Copper-containing alloys have maximum strength but least corrosion and SCC resistance. Alloys free of copper have moderate to high strength and corrosion resistance approaching that of the 3xxx, 5xxx, and 6xxx alloys. Judicious selection of alloy and temper, as well as good design practice, will minimize problems with SCC. Use of the copper-free 7xxx alloys has extended from traditional aircraft and aerospace to other transportation structural components, including military-armor plate and automotive bumpers.

15.1.6 Titanium Alloys

Titanium is highly reactive, forming a continuous, stable, protective, and adherent oxide film on the surface in the presence of oxygen and moisture. The TiO_2 oxide film is formed in all pH ranges, and titanium alloys are predictably resistant to a host of normally aggressive alkaline and acidic media. Corrosion resistance is generally unaffected by low levels of alloying additions in the simple, single-phase α (hexagonal close packed) alloys used in most industrial applications. Exceptions are certain precious metals, discussed later in this section. Because of the low alloy levels and absence of segregation on cooling, the single-phase titanium alloys can be welded readily, without losses in strength and corrosion resistance. More complex, high-strength, two-phase, $\alpha + \beta$, alloys have been developed for aircraft and aerospace applications, based on the very high strength-to-weight ratio of titanium, but these are not usually necessary for industrial applications, in which the inherent strength of unalloyed titanium is sufficient and corrosion resistance is the main requirement. Although the cost has come down in recent years, titanium is still used only in critical applications, where less expensive alloys cannot survive.

Titanium is often the material of choice for oxidizing nitric acid at all concentrations and temperatures up to about 80°C. Above this temperature, titanium may be subject to uniform or intergranular attack in the very pure acid. However, cationic oxidizing impurities, including Si^{4+}, Cr^{6+}, Fe^{3+}, and Ti^{4+}, effectively inhibit corrosion by raising potential into the passive range of the metal. Many such impurities accumulate during manufacture and handling of the acid, and titanium alloys are often usable at higher temperatures, up to and above boiling. This is a fortunate circumstance, because the stainless steels and nickel corrosion-resistant alloys fail by intergranular corrosion in boiling nitric acid. Titanium alloys are similarly useful in other oxidizing acids and salts, including chromic, perchloric, and hypochlorous acids, and chlorine and chloride solutions, in which the metal is highly resistant to pitting. The critical potential above which pitting occurs (Section 7.2.1) is found at very high values, up to about *10 volts* in chloride solutions.

Titanium has variable resistance to the strong reducing acids, hydrochloric and sulfuric. As acid concentration increases, the oxidizing power increases, as well, tending to passivate the metal. Oxidizing impurities inhibit corrosion by passivation, also. Temperature, on the other hand, increases anodic rates, diminishing the passivation tendency. Thus, resistance depends on purity, temperature, and concentration of the acid. Often, when feasible, oxidizing inhibitors are added to increase effectively the resistance of titanium to strong reducing acids. Titanium is also resistant to the less-reducing acids, sulfurous, boric, and carbonic, as well as H_2S solutions, without inhibition.

Titanium alloys are often unsuitable for hot, concentrated, reducing acids, in which the oxide film cannot form. However, certain "precious" alloying metals, such as platinum, palladium, and nickel, improve corrosion resistance dramatically by initial enrichment on the corroding surface. The enriched surface enhances passivation by catalytically increasing the rate of cathodic reduction of H^+ and galvanically polarizing the titanium into the passive potential region (Section 6.4.2). Molybdenum additions enhance passivation by reducing the critical current density needed for passivation (Section 4.1.5).

The fluoride ion complexes titanium, and titanium alloys are aggressively attacked by hydrofluoric acid and fluoride solutions. *Titanium can react explosively* with red fuming nitric acid, which readily breaks down the passive film and produces pyrophoric corrosion products.

Although boldly exposed titanium alloy surfaces are nearly immune to pitting corrosion, crevice corrosion is a possibility in hot (>70°C), oxidizing chloride solutions with tight crevices of the proper geometry. The mechanism is similar to that for the stainless steels (Section 7.3.2) involving concentration and hydrolysis of aggressive acid chloride solutions in the restricted crevice volume. Acid reduction provides hydrogen, which is damaging to reactive titanium, as described in the next paragraph. Conditions of temperature- and pH-promoting crevice corrosion and hydrogen pickup in three common alloys are summarized in Figure 15.3.[7]

Titanium alloys are subject to hydrogen-damage mechanisms, although the surface oxide film is a substantial barrier to hydrogen entry into the metal lattice. Titanium forms hydride platelets (Section 10.1.7), which decrease ductility and toughness. Impressed cathodic currents, galvanic coupling to active metals, surface abrasion and wear, hydrogen recombination poisons (e.g., sulfides, arsenic, antimo-

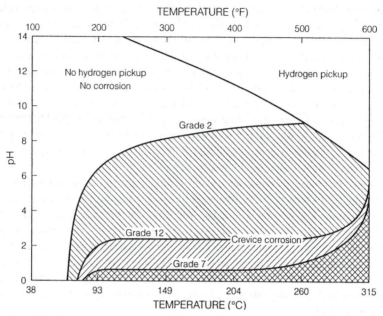

FIGURE 15.3 Crevice corrosion and hydrogen pickup in titanium alloys. *(From R. W. Schutz,* Metals Handbook, *Vol. 13,* Corrosion, *9th ed., ASM International, Metals Park, OH, p. 686, 1987. Reprinted by permission, ASM International.)*

ny), alkaline pH, and elevated temperature enhance hydrogen pickup in the metal and should be avoided. Figure 15.3 shows the accelerating influence of temperature and alkalinity.

15.1.7 Other Alloys and Materials

The alloy systems described previously in this section comprise the great majority of materials used in engineering applications. Others mentioned briefly in this section fill more limited, but often very important roles.

Cobalt alloys find greatest use for excellent wear resistance and high-temperature mechanical properties. Thus, they find considerable use in aircraft and rocket engines and other very high-temperature applications for handling hot flowing gases. Cobalt is metallurgically similar to nickel, and most constructional alloys contain substantial quantities of nickel to stabilize the face-centered cubic phase and increase ductility at high temperatures. As in nickel and iron, additions of chromium, nickel, molybdenum, and tungsten improve corrosion resistance.

The best-known ambient-temperature applications of cobalt alloys are for prosthetic devices in the human body. Figure 15.4[8] shows a cobalt hip joint for replacement of aged and diseased natural joints. The long, narrow, lower blade of the device is wedged and cemented into the central intramedullary canal at the upper end of the femur, and the ball joint is placed in the hip socket. The unique combination of wear and corrosion resistance have made cobalt-chromium alloys succesful in this application, where the stainless steels and corrosion-resistant nickel alloys have failed. Such devices are usually investment cast, to shape and ground to final dimensions. The wear resistance of the cobalt alloys makes them difficult to fabricate, as well. Cobalt is often used industrially as a hard facing material in critical wear regions, applied by weld overlay.

Zirconium has properties similar to titanium. It is resistant to all of the strong acids: sulfuric, nitric, and most notably, hydrochloric. Resistance to hydrochloric acid at all concentrations and temperatures above boiling makes zirconium alloys unique, along with tantalum. Like titanium, zirconium is attacked by hydrofluoric acid. However, zirconium is attacked by halide solutions containing oxidizing cations such as Fe^{3+} and Cu^{2+}.

The most common use of zirconium alloys is for cladding in nuclear fuel elements and other equipment requiring a low-capture cross section for neutrons. The combination of neutron transparency and corrosion resistance in high-temperature (315°C), high-pressure water makes zirconium dominant in this application. Zirconium alloys are subject to long-term oxidation and hydriding in pressurized-water systems, as described briefly in Section 12.3.3.

Tantalum is resistant to all the strong acids, such as hydrochloric, sulfuric, and nitric, at all concentrations and temperatures up to boiling. In contrast to titanium, tantalum is resistant to fuming nitric acid but is also susceptible to attack by hydrofluoric acid. It has a long history of use in specialized acid-handling applications in

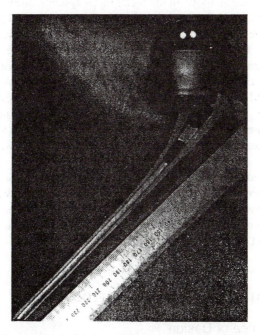

FIGURE 15.4 Hip prosthesis fabricated from cast cobalt chromium alloy. *(From P. J. Andersen, Metals Handbook, Vol. 13, Corrosion, 9th ed., ASM International, Metals Park, OH, p. 663, 1987. Reprinted by permission, ASM International.)*

the chemical-process industry. Tantalum's high cost has been a detriment to wide use, but its ductility has favored use as a cladding in recent years.

Magnesium alloys are relatively susceptible to corrosion. Nevertheless, extreme light weight makes magnesium alloys (often containing aluminum for solid solution strengthening) useful in aircraft, aerospace, and automotive components. The metal must be kept free of heavy-metal impurities—iron, nickel, and copper—to maximize corrosion resistance. The alloys are restricted primarily to mild atmospheric exposures, and various conversion and organic coatings are often required to achieve sufficient corrosion resistance.

Amorphous-metal systems (sometimes known as glassy metals) are formed by rapid solidification of very thin ribbon or wire to prevent nucleation and growth of the usual long-range crystalline structure. Some short-range order is often present, much as in normal liquids. The lack of grain boundaries, second phases, compositional segregation, and crystalline defects improves corrosion resistance in many cases.[9]

Amorphous structures are formed in two general systems. First, two or more transition metals such as copper-zirconium, copper-titanium, or nickel-niobium can form amorphous structures. Second, a combination of metals, including iron, chromium and nickel, with such metalloids as boron, phosphorous, silicon, or carbon, will also form amorphous structures. The corrosion behavior of each type is somewhat different.

In both types, the glassy or amorphous alloy has better corrosion resistance than the equivalent crystalline alloy. In the metal-metal systems, such as copper in titanium or zirconium, the improvement is relatively low (e.g. 20%), but the glassy alloy does not have better resistance than the corresponding titanium or zirconium. The improvement for the metal-metalloid combinations is more significant. The glassy state apparently improves the strength of the passive layer on Fe-Cr-Ni compositions. As chromium content increases, the corrosion rate drops to very low values, and pitting and crevice corrosion resistance increases dramatically at chromium levels much lower than the conventional stainless steels. For example, the pitting potential, a measure of pitting resistance (Section 7.2.1), increases to about 1-volt at only 7% Cr, as compared to 12% minimum Cr required in most stainless steels (Section 15.1.2).

Applications making use of the improved corrosion resistance of amorphous metals are limited at present. They are crystallized by only slightly elevated temperature and are currently available in only very thin cross section. However, deposition or *in situ* formation of corrosion-resistant surface coatings may become attractive, when the technology and cost of the coating processes can be reduced.

15.2 Designing to Prevent Corrosion

15.2.1 Importance of Design

Design is the first line of defense against corrosion. Here, the process conditions, materials of construction, and corrosion-prevention methods are specified for buildings, structures, and process equipment, all of which affect the corrosion subsequently experienced during service. Buildings, plant processes, and structures must be first designed to fulfill their intended function(s). However, inadequate consideration of corrosion at the design stage often results in time-consuming and expensive revisions during construction and operation. A British survey in the chemical-process industry[10] showed that a majority of corrosion failures were caused by inappropriate design.

The designer must *determine materials* of construction that are economical yet provide adequate resistance to the service conditions. This requires a choice of *operating conditions* and *physical design* to minimize corrosivity of the environment so that the least expensive material can be used to minimize the overall cost of the project. Overall costs include the initial capital costs, which are naturally emphasized during design and construction, but also subsequent operating costs, which often do not receive adequate consideration during design. Adequate design must include corrosion-prevention methods, such as coatings and cathodic protection to allow the use of less-expensive alloys. Initial design should also include operating procedures, which include continuing measures of corrosion monitoring and control.

Materials selection for corrosion resistance is, of course, but one part of the overall design process, which is shown schematically by the flowchart of Figure 15.5.[11] Corrosion-resistance and corrosion-prevention methods, as well as other factors, such as mechanical strength, must be satisfactory for an affirmative answer to the question, "Does a suitable material exist?" which appears in the flowchart. If analysis of an initial design concept reveals that a suitable material does not exist and that a new material cannot be developed, then redesign must be instituted and operating conditions altered to permit selection of satisfactory construction materials.

The owners and ultimate operators of a plant should be made aware of any calculated risks undertaken by equipment designers or vendors to meet plant production specifications on the basis of a low bid. The proper codes, both structural and corrosion, must be followed to ensure a safe, economical, efficient, and profitable project. Even then, judicious engineering analysis is necessary to ensure that such codes are adequate in any given situation.

The best of designs will be of no value unless properly implemented during equipment fabrication and plant construction. Therefore, careful quality-control *inspection schedules* during equipment fabrication, and routine inspections during plant construction are essential to ensure that contracted specifications of materials and design are met by equipment vendors and building contractors. In a complex project, seemingly identical components of different alloy composition, such as valves and piping, are often available at the construction site; interchanging these components during installation can lead to unexpected corrosion failures. Measures must be taken to retain equipment in operating condition during preconstruction storage. This may require at least prime-coating steel surfaces, packaging with vapor-phase inhibitors, or warehouse storage in a controlled environment, to prevent the accumulation of surface rust and scale, which may be difficult and expensive to remove later, before application of final protective coatings.

Once a plant or building is in service, *regular maintenance* is required to keep it operating safely and efficiently. Maintenance requirements vary with severity of

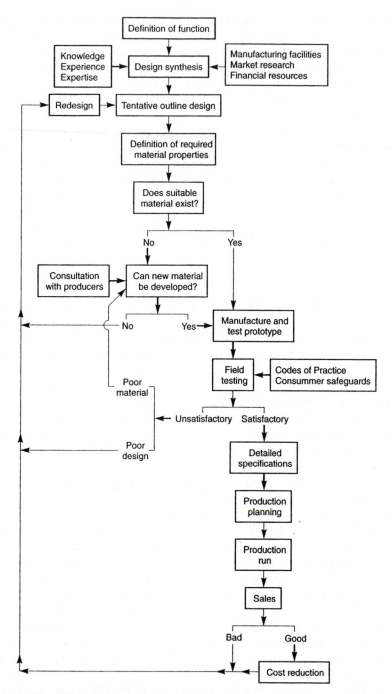

FIGURE 15.5 Flowchart illustrating the design process. *(From J. A. Charles and F. A. A. Crane,* Selection and Use of Engineering Materials, *2nd ed., Butterworths, Sevenoaks, Kent, England, p. 11, 1989. Reprinted by permission, Butterworths Publishers Ltd.)*

environment. Many buildings require only regular repainting and occasional inspection of electrical and plumbing lines, but chemical-process plant structures and marine equipment often need regular maintenance schedules, with periodic replacement of deteriorated coatings and plant components, to assure economical operation and avoid costly and unpredicted failures.

Even the best of designs cannot be expected to anticipate all conditions that may arise during plant operation. Thus, regular procedures of *corrosion monitoring* are usually desirable. This may simply require regular visual inspections, which are a part of the routine maintenance schedules. In a chemical-process plant, corrosion coupons may be installed in critical areas and retrieved for periodic assessment of corrosion rate. Special care should be exercised to assure that exposure conditions for the coupons duplicate critical conditions in the plant. Electrical (Section 1.6.6) and electrochemical (Section 5.4) instrumentation may be used to continuously monitor corrosion rates and solution corrosivity. Nondestructive-testing techniques, including ultrasonic and eddy current methods, may be useful to detect wall-thickness reductions by general corrosion and wear and cracks from fatigue, stress corrosion, and so on.

Unexpected failures during field testing (or during initial operation) in Figure 15.5 may require *remedial action* or redesign. While the choice of remedial action may appear obvious, experience has shown that considerable diagnostic effort is usually required before the proper or "final" choice is possible. The choices encompass all of the corrosion-control measures described in this book—substitution of a more resistant alloy or material, cathodic protection, coatings, modification of process or environmental conditions, and so on.

Unfortunately, corrosion problems do not necessarily go away, even when satisfactory materials have been selected for a profitable process or product. Ever-present intentional or unintentional material substitutions or process changes for operating convenience, increased productivity, or cost reduction (Figure 15.5) create corrosion problems, even in an established plant. Thus, redesign iterations are often continuous.

15.2.2 Considerations Affecting Design

Various design factors,[12] having indirect, others direct, influence on corrosion resistance and durability of buildings, structures, and plant components, are reviewed in this section. These considerations contribute to answering the question, "Does a suitable material exist?" in Figure 15.5.

Design philosophy is dictated by management, based on personal and company preference, usually influenced by previous experience. An optimum is usually chosen between the extremes of minimum-cost materials of low-corrosion resistance, requiring frequent maintenance and replacement, and high-cost materials of maximum corrosion resistance, requiring only infrequent maintenance and replacement. Corrosion prevention measures, such as coatings and cathodic protection, may be factored into the optimum economic design. Engineering economic calculations (Section 15.3) are often helpful to make more objective decisions on materials and design.

Process variables have an obvious effect on corrosion rates and materials of construction. Generally, increased temperature results in increased corrosion rates and requires more resistant materials of construction. For dilute corrodents, increased concentration often has a similar effect. However, for most strong acids, low-cost carbon steels can be used for the concentrated acids (e.g., 98% H_2SO_4) because they are less dissociated and therefore less corrosive. Carbon steels are also resistant in many alkaline solutions of higher concentration. However, combinations of high temperature and high alkalinity may result in stress corrosion and hydrogen induced cracking of many ferrous alloys.

Corrosion data may be derived from several sources: (a) previous design within the company or other companies from similar or identical plants or applications, (b) manufacturers' data, (c) published design data, and (d) specially developed (e.g., laboratory, pilot plant) corrosion data for the particular project. All corrosion data must be treated with care, recognizing the effects of all possible variables that may affect service performance, especially heat flux, localized concentrations, velocity, temperature, corrosive purity, weldments, crevices and other stagnant areas, and so on. Unfortunately, such variables are usually disregarded in simple corrosion tests used to generate corrosion-data compilations.

New materials and equipment designs are constantly being made available to deal with specific corrosion problems. The designer must keep abreast of any such developments, which may provide at least partial solutions to problems that have been intractable in the past. Marketing representatives and distributors of reputable material and equipment suppliers provide a valuable resource for such information.

Nature and composition of raw materials may be changeable. Thus, the designer must consider not only current feedstocks that may have less corrosivity, but also later extremes, which may include aggressive agents, such as chloride and sulfur, requiring additional control measures to offset increased corrosivity of the process solutions.

15.2.3 Corrosion-Prevention Measures

Prevention measures for the various forms of corrosion have been described through-out this book. These are reviewed briefly in this section, together with special considerations regarding design.

General corrosion is relatively predictable and can be included in design calculations by increasing the wall thickness. However, this may not always be feasible for thin-wall components subject to irregular or localized corrosion. Control of process-stream composition to promote general corrosion by minimizing agents causing localized corrosion (e.g., chlorides), is an obvious recommendation. An especially critical instance of general corrosion is at areas of acid-vapor condensation in oil-distillation units and acid-manufacturing and handling equipment. Design methods to control this type of attack include[4,12] (a) neutralization with ammonia or amines injected into the condensing vapor, (b) insertion of a more-resistant alloy at the critical area of condensation, (c) conformation of vessels to minimize collection of condensed vapor in corners (Figure 15.6), or (d) steam heating of vapor lines to

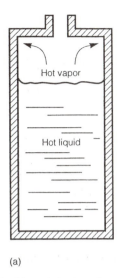
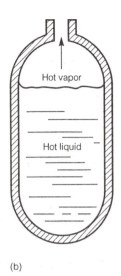

(a)   (b)

FIGURE 15.6   Vessel designs for containing hot liquids with corrosive vapors: (a) poor design allowing collection of vapor in corners; (b) good design preventing vapor collection in rounded corners. *(From R. M. Davison, T. DeBold, and M. J. Johnson,* Metals Handbook, *Vol. 13,* Corrosion, *9th ed., ASM, Metals Park, OH, p. 547, 1987. Reprinted by permission, ASM International.)*

prevent or postpone condensation. Chemical-plant design should also include proper valving to prevent leakage from corrosive to noncorrosive areas during shutdown or layup.

General corrosion, especially of carbon steel, in the atmosphere and in water, wastewater, seawater, and mildly corrosive solutions is most often controlled by organic coatings selected for resistance to the particular environment. Superstructures on seagoing vessels are often protected by flame-sprayed aluminum and zinc coatings, followed by further organic coatings. Cathodic protection combined with coatings is used to prevent general corrosion of pipelines and other buried structures, seagoing ship hulls, and other immersed marine structures. Alloy substitution is usually not an economically feasible option, because of the large tonnages of material involved in these applications.

General corrosion in the atmosphere is increased by any features that retain water. Channel and angle sections positioned to collect rainwater and debris may suffer increased corrosion. Proper design will position these parts to prevent water collection or will include holes for drainage, as shown in Figure 15.7. The underside of constantly shaded panels may have increased corrosion, because time-of-wetness is increased when dew and other condensation are not readily evaporated by sunlight. Insulation or lagging in contact with metal surfaces can also collect and retain condensation or rain water (Section 7.4.1). Structures designed for atmospheric corrosion resistance should always provide for easy drainage from all exposed surfaces.

Whenever possible, plant structures should be sited upwind from stacks emitting corrosive effluents and in geographic locations sheltered from sea salts or other pollutants carried by regular precipitation or prevailing winds (Figure 15.8[10]).

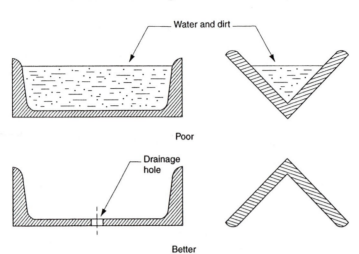

FIGURE 15.7   Positioning and draining angles and channel sections to minimize atmosphere corrosion.

Galvanized and painted constructional steel has excellent resistance to general atmospheric corrosion. Design should allow maximum access for maintenance and repair painting. Box sections (Figure 15.9[13]) have poor access to coatings, collect water and debris, and maximize possibilities of corrosion. Structures in critical chemical atmospheres often use cylindrical sections for ease and uniformity of paint application, as shown in Figure 15.10.[11] Edges and corners are difficult to coat uniformly, and thinly coated protrusions are susceptible to corrosion (Figure 15.11).

Stress corrosion cracking (SCC) can be prevented by substituting a more resistant alloy, removing the tensile stress, or making the environment less aggressive. For example, in SCC of austenitic stainless steel by chlorides, substitution of ferritic or duplex stainless steels will often (but not always) eliminate the problem. The ferritic stainless steels may be subject to pitting, but the duplex grades are more resistant. Stress-relief annealing will reduce tensile stresses to below threshold levels in many instances, but often will not prevent SCC in hot chlorides. A temperature of 70°C in chloride solutions is a rule-of-thumb minimum, below which SCC is infrequent.

Corrosion fatigue cracking (CFC) can be prevented by eliminating cyclic stress or the corrosive environment. Removal of the corrosive environment may be impossible, but vibrational stresses may be suppressed by a more rigid design. Removal of notches and other stress-concentrating features can be helpful, when feasible. Rounded filets and angles will also reduce stress concentrations. A more corrosion-resistant alloy will usually eliminate CFC in a particular environment. However, no alloy is totally immune to CFC.

Hydrogen damage effects are large in number, and prevention methods for most are discussed in Section 10.1. Briefly, it may be said here that hydrogen induced cracking may be reduced by choice of a lower-strength, less susceptible alloy.

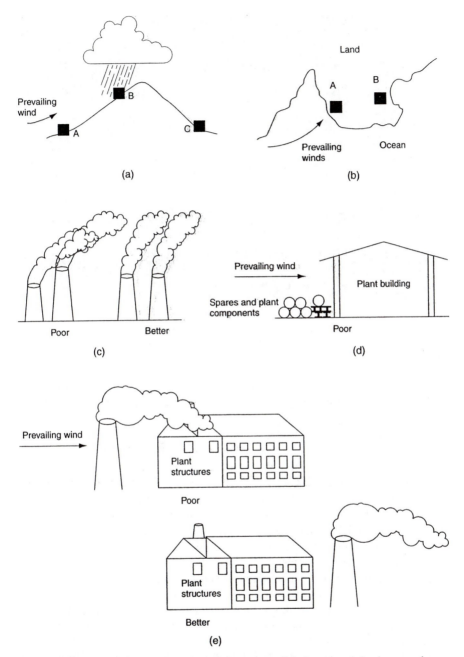

FIGURE 15.8 Geographic location of plants to minimize atmospheric corrosion. *(From P. Elliot,* Metals Handbook, *Vol. 13,* Corrosion, *9th ed., ASM, Metals Park, OH, p. 339, 1987. Reprinted by permission, ASM International.)*

FIGURE 15.9a Box section components representing poor design for corrosion resistance. *(From C. G. Munger,* Corrosion Prevention by Protective Coatings, *NACE, Houston, p. 174, 1984. Reprinted by permission, National Association of Corrosion Engineers.)*

Coatings will limit access of the corrosive or hydrogen bearing environment to the surface. Removal of cathodic-protection systems, galvanized coatings, or couples to more active alloys will reduce hydrogen at the surface.

Galvanic corrosion is controlled by selection of alloys that are near one another in the galvanic series (Figure 6.1) when they are in electrical contact with one another. In other unavoidable couples, the anode alloy should be large in area compared to the cathode. Both members of a galvanic couple should be coated, never the anode alone, to avoid any small anode area at coating defects and holidays. If feasible, dissimilar alloys should be electrically insulated from one another at their junction. A diagram of an insulated joint is shown in Figure 6.3. In atmospheric corrosion, continuous moisture drippage should not be allowed to carry corrosion products from a more noble to a more active metal to cause pitting of the latter. For example, drain waters from copper should not be allowed to contact zinc-galvanized rain gutters or aluminum siding. Copper deposits produce microgalvanic couples on the active surface. Titanium and zirconium alloys should not be in contact with a more noble alloy during corrosive exposures because they can be embrittled by hydriding in a galvanic couple.

Crevice corrosion can be eliminated by design of joints and junctions that minimize crevices. Welded joints are thus preferable to bolted and riveted joints, but the welds must be properly designed and constructed to eliminate crevices. In newer nuclear steam generators, the crevice between tubing and tube sheet is minimized by fully rolling the tubing into the tube sheet, as shown in Figure 15.12. Gaskets must be properly sized to minimize crevices exposed to the corrosive solu-

FIGURE 15.9b   (Cont.)

tion. Sealing compounds and inhibitive coatings on flange faces may also be helpful. Both crevice and *pitting corrosion* may be expected to increase in stagnant or slow-flowing solutions, where deposits and corrosion products can accumulate to create crevices. Periodic cleaning may be necessary to remove deposits. Tanks should be designed for complete drainage (Figure 15.13). Control or removal of dissolved species, such as chloride and sulfide, which cause crevice and pitting corrosion, will, of course, mitigate these forms of corrosion. Substitution of a more resistant alloy will also control localized forms of corrosion. Stainless alloys containing quantities of nickel, chromium, and molybdenum have increasing resistance to crevice and pitting corrosion. Titanium is usually immune to pitting corrosion but can experience attack in extreme conditions of temperature and chloride concentration in tight crevices. Substitution of a less-resistant alloy may even be a useful alternative. For example, a replaceable carbon-steel part corroding uniformly at a relatively high but predictable rate in the active state may be preferable to passive stainless steel that fails rapidly and unpredictably by localized stress corrosion, pitting, or crevice corrosion.

    Erosion-corrosion and *cavitation* are amenable to mitigation by design to reduce velocity and turbulence, as far as possible. Flow-channel dimensions and pumping

FIGURE 15.10 Cylindrical structural members preferable for uniform application of coatings. *(From C. G. Munger,* Corrosion Prevention by Protective Coatings, *NACE, Houston, p. 174, 1984. Reprinted by permission, National Association of Corrosion Engineers.)*

capacity can be adjusted to minimize velocity. Abrupt changes in flow direction should be avoided, using maximum-radius elbows, and eliminating right-angle T-junctions, when possible. Baffle plates or deflectors may be helpful in diverting or shielding surfaces from erosive streams. Abrupt changes in channel dimensions should be avoided with tapered reducers. Careful welding technique is needed to prevent weld-bead penetration into the flow path, creating turbulence and increased erosion-corrosion. A section of more-resistant material may be required downstream from any devices creating turbulence, such as valves, pumps, or flow measuring orifice plates. Similarly, valves and pumps themselves should be constructed from resistant material. Recall that erosion-corrosion often occurs by erosion of a friable corrosion product deposit on the surface. A more corrosion resistant alloy is thus usually more resistant.

15.3 Economics of Corrosion Prevention

15.3.1 Significance

Corrosion is often treated simply as a maintenance problem. The material having the least initial cost is chosen, and equipment repairs or replacements are made as required. Only when such corrosion-related costs become exorbitant does prevention become a consideration. The other extreme is to design for minimum corrosion, using materials of maximum resistance and all available prevention technology, regardless of expense. The resistant alloys are expensive, because they are usually more difficult to process and contain costly alloying additions such as chromium,

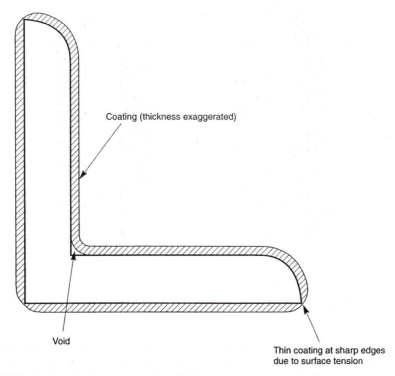

Coating (thickness exaggerated)

Void

Thin coating at sharp edges
due to surface tension

FIGURE 15.11   Inadequate coating thickness at corners and edges.

nickel, and molybdenum. Other prevention methods, including coatings and cathod-ic-protection systems, have their own maintenance and installation costs. In the next section, economic calculations are used to optimize the cost of corrosion prevention somewhere between the two extremes. Indeed, corrosion is essentially an economic problem and contributes to the overall considerations of engineering economics, a study of money as a resource, and as a measure of other resources.

The engineer concerned with corrosion needs an awareness of the important eco-nomic principles and some facility with economic calculations, in order to convince management that corrosion prevention measures can reduce production costs and thereby increase profits. For, indeed, companies are in business to generate profits, without which there is no justification for producing chemicals, automobiles, appli-ances, etc., *or for employing engineers.* Furthermore, today's engineers are often tomorrow's managers, who are responsible to the owners or stockholders for opera-tions that result in a profit.

The most expensive alloy to resist corrosion may not be the most cost-effective choice. The high initial cost of purchase and fabrication may override the periodic maintenance and replacement costs of a less-expensive alloy. A process may become technically obsolete in a relatively short time, and long-life equipment may be unjus-tifiable. The resources (money) tied up in equipment fabricated from an expensive alloy could be generating more money elsewhere, in an interest-bearing account or in other investments. Obviously, interest rates affect any decision on investment in

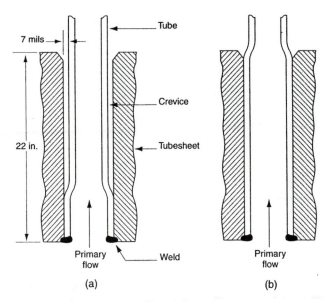

FIGURE 15.12   Design to minimize crevice between heat exchanger tubing and tubesheet.

corrosion prevention. Tax laws have a significant effect, as well. Tax rates and permissible depreciation schedules influence the recovery and reinvestment of money tied up in consumable equipment.

15.3.2 Discounted Cash Flow

Discounted cash flow (DCF) analyzes the value of money in time. Money available at some future date(s) has less value at the present time and must be "discounted" by the interest rate to a present value or *present worth*, PW. Terminology and notation

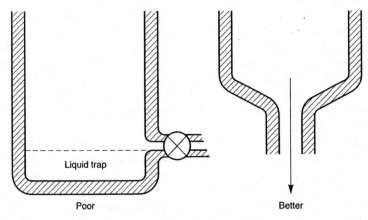

FIGURE 15.13   Tank design for complete drainage.

are major barriers to the understanding and application of DCF in engineering design. The definitions in this section are consistent with initial attempts at standardization[14] and have been adapted from presentations by Verink[15,16] and others.[17] The interested reader should pursue these references and the many others available for more detailed information.

A *present sum of money*, P, is a currently available amount of money or resources. *The major economic decision to be considered is how P should be utilized in the future.* Some options might be the purchase of new equipment, maintenance and repair or corrosion prevention of existing equipment, return to the owner or stockholders as profit, or investment in an interest-bearing account. Thus, any expenditure, including that for corrosion prevention, is justifiable only if it will produce more future income than competing forms of investment, especially the most conservative, i.e., ordinary interest-bearing deposits.

A *future sum of money*, F, is the amount of money or resources available at some time in the future after having invested P. For example, F, available as a single payment at the end of n yearly interest-bearing periods at an interest rate, i (expressed as a decimal), is given by

$$F = P(1 + i)^n. \tag{1}$$

The term $(1 + i)^n$, or (F/P), is called the single-payment compound-amount factor. A future amount of money, F, may be discounted to a present value, P, by the reciprocal multiplication factor (P/F), defined as the single-payment present-worth factor.

The *annual cash flow* in a uniform series of n annual receipts (positive cash flow) or disbursements (negative cash flow) is defined by the term, A. Such a series is commonly generated by regular receipts, as from service fees and sales, and regular expenditures such as rental or maintenance expenses. The present value of A, discounted by an interest rate, i, is given by

$$P = A\left[\frac{(1 + i)^n - 1}{i(1 + i)^n}\right]. \tag{2}$$

The term in brackets, (P/A), is called the uniform-series present-worth factor. Multiplication by (P/A) discounts any annual cash-flow term to present worth. Conversely, multiplication by (A/P) converts any present-value term to an equivalent annual cash flow.

The factors (P/F), (P/A), and (A/P) are commonly tabulated for various periods and interest rates, for example, as in Table 15.2. Present and discounted future values can be computed readily from (1) and (2) with hand-held calculators[18] or with the aid of the tabulated factors.

Depreciation, D, is an annual tax allowance for wear-and-tear, corrosion, chemical and biological deterioration, and functional obsolescence. Technically improved replacements can often perform more economically and efficiently, making old property "functionally obsolete." Over the depreciable life of the property, total depreciation amounts to the difference, $P - S$, between present and salvage cost, and the annual depreciation is

TABLE 15.2 Interest Factors in DCF Calculations

N	6.00% Interest Factors			10.00% Interest Factors			15.00% Interest Factors		
	(P/F)	(P/A)	(A/P)	(P/F)	(P/A)	(A/P)	(P/F)	(P/A)	(A/P)
1	0.9434	0.9434	1.0600	0.9091	0.9091	1.1000	0.8696	0.8696	1.1500
2	0.8900	1.8334	0.5454	0.8264	1.7355	0.5762	0.7561	1.6257	0.6151
3	0.8396	2.6730	0.3741	0.7513	2.4869	0.4021	0.6575	2.2832	0.4380
4	0.7921	3.4651	0.2886	0.6830	3.1699	0.3155	0.5718	2.8550	0.3503
5	0.7473	4.2124	0.2374	0.6209	3.7908	0.2638	0.4972	3.3522	0.2983
6	0.7050	4.9173	0.2034	0.5645	4.3553	0.2296	0.4323	3.7845	0.2642
7	0.6651	5.5824	0.1791	0.5132	4.8684	0.2054	0.3759	4.1604	0.2404
8	0.6274	6.2098	0.1610	0.4665	5.3349	0.1874	0.3269	4.4873	0.2229
9	0.5919	6.8017	0.1470	0.4241	5.7590	0.1736	0.2843	4.7716	0.2096
10	0.5584	7.3601	0.1359	0.3855	6.1446	0.1627	0.2472	5.0188	0.1993
11	0.5268	7.8869	0.1268	0.3505	6.4951	0.1540	0.2149	5.2337	0.1911
12	0.4970	8.3838	0.1193	0.3186	6.8137	0.1468	0.1869	5.4206	0.1845
13	0.4688	8.8527	0.1130	0.2897	7.1034	0.1408	0.1625	5.5831	0.1791
14	0.4423	9.2950	0.1076	0.2633	7.3667	0.1357	0.1413	5.7245	0.1747
15	0.4173	9.7122	0.1030	0.2394	7.6061	0.1315	0.1229	5.8474	0.1710
16	0.3936	10.1059	0.0990	0.2176	7.8237	0.1278	0.1069	5.9542	0.1679
17	0.3714	10.4773	0.0954	0.1978	8.0216	0.1247	0.0929	6.0472	0.1654
18	0.3503	10.8276	0.0924	0.1799	8.2014	0.1219	0.0808	6.1280	0.1632
19	0.3305	11.1581	0.0896	0.1635	8.3649	0.1195	0.0703	6.1982	0.1613
20	0.3118	11.4699	0.0872	0.1486	8.5136	0.1175	0.0611	6.2593	0.1598
21	0.2942	11.7641	0.0850	0.1351	8.6487	0.1156	0.0531	6.3125	0.1584
22	0.2775	12.0416	0.0830	0.1228	8.7715	0.1140	0.0462	6.3587	0.1573
23	0.2618	12.3034	0.0813	0.1117	8.8832	0.1126	0.0402	6.3988	0.1563
24	0.2470	12.5504	0.0797	0.1015	8.9847	0.1113	0.0349	6.4338	0.1554
25	0.2330	12.7834	0.0782	0.0923	9.0770	0.1102	0.0304	6.4641	0.1547
26	0.2198	13.0032	0.0769	0.0839	9.1609	0.1092	0.0264	6.4906	0.1541
27	0.2074	13.2105	0.0757	0.0763	9.2372	0.1083	0.0230	6.5135	0.1535
28	0.1956	13.4062	0.0746	0.0693	9.3066	0.1075	0.0200	6.5335	0.1531
29	0.1846	13.5907	0.0736	0.0630	9.3696	0.1067	0.0174	6.5509	0.1527
30	0.1741	13.7648	0.0726	0.0573	9.4269	0.1061	0.0151	6.5660	0.1523

TABLE 15.2 (Cont.)

n	8.00% Interest Factors			12.00% Interest Factors			20.00% Interest Factors		
1	0.9259	0.9259	1.0800	1.1200	0.8929	0.8929	1.2000	0.8333	0.8333
2	0.8573	1.7833	0.5608	0.5917	1.6901	0.7972	0.6545	1.5278	0.6944
3	0.7938	2.5771	0.3880	0.4163	2.4018	0.7118	0.4747	2.1065	0.5787
4	0.7350	3.3121	0.3019	0.3292	3.0373	0.6355	0.3863	2.5887	0.4823
5	0.6806	3.9927	0.2505	0.2774	3.6048	0.5674	0.3344	2.9906	0.4019
6	0.6302	4.6229	0.2163	0.2432	4.1114	0.5066	0.3007	3.3255	0.3349
7	0.5835	5.2064	0.1921	0.2191	4.5638	0.4523	0.2774	3.6046	0.2791
8	0.5403	5.7466	0.1740	0.2013	4.9676	0.4039	0.2606	3.8372	0.2326
9	0.5002	6.2469	0.1601	0.1877	5.3282	0.3606	0.2481	4.0310	0.1938
10	0.4632	6.7101	0.1490	0.1770	5.6502	0.3220	0.2385	4.1925	0.1615
11	0.4289	7.1390	0.1401	0.1684	5.9377	0.2875	0.2311	4.3271	0.1346
12	0.3971	7.5361	0.1327	0.1614	6.1944	0.2567	0.2253	4.4392	0.1122
13	0.3677	7.9038	0.1265	0.1557	6.4235	0.2292	0.2206	4.5327	0.0935
14	0.3405	8.2442	0.1213	0.1509	6.6282	0.2046	0.2169	4.6106	0.0779
15	0.3152	8.5595	0.1168	0.1468	6.8109	0.1827	0.2139	4.6755	0.0649
16	0.2919	8.8514	0.1130	0.1434	6.9740	0.1631	0.2114	4.7296	0.0541
17	0.2703	9.1216	0.1096	0.1405	7.1196	0.1456	0.2094	4.7746	0.0451
18	0.2502	9.3719	0.1067	0.1379	7.2497	0.1300	0.2078	4.8122	0.0376
19	0.2317	9.6036	0.1041	0.1358	7.3658	0.1161	0.2065	4.8435	0.0313
20	0.2145	9.8182	0.1019	0.1339	7.4694	0.1037	0.2054	4.8696	0.0261
21	0.1987	10.0168	0.0998	0.1322	7.5620	0.0926	0.2044	4.8913	0.0217
22	0.1839	10.2007	0.0980	0.1308	7.6446	0.0826	0.2037	4.9094	0.0181
23	0.1703	10.3711	0.0964	0.1296	7.7184	0.0738	0.2031	4.9245	0.0151
24	0.1577	10.5288	0.0950	0.1285	7.7843	0.0659	0.2025	4.9371	0.0126
25	0.1460	10.6748	0.0937	0.1275	7.8431	0.0588	0.2021	4.9476	0.0105
26	0.1352	10.8100	0.0925	0.1267	7.8957	0.0525	0.2018	4.9563	0.0087
27	0.1252	10.9352	0.0914	0.1259	7.9426	0.0469	0.2015	4.9636	0.0073
28	0.1159	11.0511	0.0905	0.1252	7.9844	0.0419	0.2012	4.9697	0.0061
29	0.1073	11.1584	0.0896	0.1247	8.0218	0.0374	0.2010	4.9747	0.0051
30	0.0994	11.2578	0.0888	0.1241	8.0552	0.0334	0.2008	4.9789	0.0042

$$D = \Phi(P - S),\tag{3}$$

where S is the salvage value after depreciation of N years (ideally the period of useful service) and Φ is the allowable fraction in any given year. For common, straight-line depreciation, $\Phi = 1/N$, and both Φ and D are the same in each year n from 1 to N. Simple multiplication of Φ by (P/A) gives a discounted value of fractional annual depreciation. Since depreciation is a tax deduction, tax laws often define the allowable depreciation period and depreciation method for various types of equipment and property.

Various methods have been devised to account for the fact that D is often greater in the early years of depreciation than in the later ones. Two such methods are sum of years digits and declining balance. To illustrate, the sum of years digits for five years is $5 + 4 + 3 + 2 + 1 = 15$. The depreciation factors, Φ, in Equation (3) for the first through fifth years are 5/15, 4/15, 3/15, 2/15, and 1/15, respectively. The numerator in each fraction is the years left in the depreciation period. The fraction, Φ, differs from year to year for SOYD, in contrast to straight-line and declining-balance methods in which Φ is constant.

In the declining-balance method, a specified percentage, Φ, is taken each year, n, on the "book" value, which is the original cost, P, less the total depreciation taken to date, ΣD_n. Depreciation, D_n, varies in each year of the depreciation period, N, so that

$$D_n = \Phi(P - \Sigma D_n),$$

where again n takes values from 1 to N. In the common double declining balance (DDB) method, $\Phi = 2/N$. The annual depreciation fraction is double that of straight-line depreciation, but the amount being depreciated, i.e., the book value, declines with increasing n. For double declining balance depreciation, the book value in any year, n, is given by[19]

$$P - \Sigma D_n = P\left[1 - \frac{2}{N}\right]^n.$$

U.S. tax laws allow property to be depreciated only down to the salvage value, S.

United States tax law changes in 1981, and especially in 1986, mandate a combination of declining-balance and straight-line depreciation methods for various property classes in the *accelerated cost recovery system* (ACRS). The most recent convention, the modified ACRS (MACRS), enacted in 1986, is discussed briefly here for illustration. The new laws ignore salvage value, S. MACRS property classifications and Internal Revenue Service (IRS) equipment examples in each classification are listed as follows:

Three-year property: Special handling devices for manufacture of food and beverages, rubber products, finished plastic products, fabricated metal products.

Five-year property: Computers, typewriters, copiers, duplicating equipment, heavy general purpose trucks, trailers, cargo containers, and trailer-mounted containers. Cars, light-duty trucks, computer-based telephone central office switching equipment, computer-related peripheral equipment, semiconductor manufacturing equipment, property used in research and experimentation.

Seven-year property: Office furniture and fixtures, railroad track, single-purpose agricultural and horticulture equipment.

Ten-year property: Vessels and water transportation equipment, assets used in petroleum refining or in manufacture of tobacco products and certain food products.

Fifteen-year property: Municipal sewage-treatment plants, telephone distribution plants, comparable equipment used by nontelephone companies for two-way exchange of voice and data communication.

Twenty-year property: In early years of the depreciation life, the three, five, seven, and ten year classifications use the double, or 200%, declining balance method; and the fifteen and twenty year properties use the 150% declining balance method. All transform to straight-line depreciation when this method gives a higher depreciation in later years of the depreciation life. Annual depreciation percentages of original cost for each property classification are shown in Table 15.3.[20]

Included in Table 15.3 is the IRS-prescribed half-year convention, in which equipment can be depreciated for only a half-year during the first year of service, irrespective of when it was placed in service during the year. The remaining half-year of depreciation is taken at the end of the depreciation period. In effect, the depreciation period extends through an extra tax reporting year (three-year property is depreciated during four years, five-year property during six years, and so forth), although the average depreciation period remains the same.

TABLE 15.3   Annual Depreciation Prescribed by Modified Accelerated Cost Recovery

Recovery Year	3-year Class (200%DB)	5-year Class (200%DB)	7-year Class (200%DB)	10-year Class (200%DB)	15-year Class (150%DB)	20-year Class (150%DB)
1	33.00	20.00	14.28	10.00	5.00	3.75
2	45.00	32.00	24.49	18.00	9.50	7.22
3	15.00*	19.20	17.49	14.40	8.55	6.68
4	7.00	11.52*	12.49	11.52	7.69	6.18
5		11.52	8.93*	9.22	6.93	5.71
6		5.76	8.93	7.37	6.23	5.28
7			8.93	6.55*	5.90*	4.89
8			4.46	6.55	5.90	4.52
9				6.55	5.90	4.46*
10				6.55	5.90	4.46
11				3.29	5.90	4.46
12					5.90	4.46
13					5.90	4.46
14					5.90	4.46
15					5.90	4.46
16					3.00	4.46
17						4.46
18						4.46
19						4.46
20						4.46
21						2.25

* Year of switch to straight-line depreciation to maximize deduction

15.3.3 Generalized Equations

Verink[15] has developed a generalized equation to determine present worth (PW) for various economic design situations using straight-line depreciation:

$$PW = -P + t\Phi(P - S)(P/A) - X(1 - t)(P/A) + S(P/F), \qquad (4)$$

where P is the present value of money when invested initially at time zero, t is the tax rate expressed as a decimal, S is salvage value, and X is the sum of expenses, excluding depreciation. (P/A) and (P/F) are functions of interest rate, i, and project life, N, in years and are tabulated in Table 15.2.

The first term, P, in (4) is the value of money at time zero and does not require further discounting to establish present worth. As an expenditure, P is given the negative sign by convention. The second term is the tax savings or income (therefore positive) derived from depreciation, which is discounted to a present worth when multiplying by (P/A). In the third term (really two terms combined), X is an expenditure and therefore is negative, while Xt is a tax savings and is therefore positive. Both are multiplied again by (P/A) to discount to a present worth at time zero. Finally, the fourth term discounts an estimated future salvage value to present worth. Since this is a one-time future payment rather than a uniform series of payments, the single-payment present-worth factor (P/F) is used. The present-worth values from equation (4) may be converted to annual cost at interest rate i over n years by use of equation (2):

$$A = (PW)(A/P).$$

Equation (4) is not readily applicable with MACRS depreciation, because the depreciation fraction is not uniform during the period of depreciation. The second depreciation term in (4) must be converted to the summation of annual incomes generated from depreciation. Total depreciation income through year N is given by the summation of products of tax rate, original cost, annual depreciation fraction, Φ_n, (from Table 15.3), and the single payment present worth factor $(P/F)_n$ for each year n from 1 to N. Equation (4) thus becomes

$$PW = -P + t(P)\Sigma[\Phi_n(P/F)_n] - X(1 - t)(P/A), \qquad (5)$$

where tax rate, t, is assumed uniform and S can be ignored for MACRS depreciation.

15.3.4 Examples of DCF Calculations

The following examples have been largely adapted from problems published previously by Professor Ellis Verink, University of Florida, with his kind cooperation.

Example 1: A new heat exchanger is required in conjunction with a rearrangement of existing facilities. Because of corrosion, the expected life of a carbon-steel exchanger is 5 years. The installed cost is $9500. An alternative to the carbon-steel heat exchanger is a unit fabricated of AISI type 316 stainless steel with an installed cost of $26,500 and an estimated life of 15 years to be written off in 11 years. The minimum acceptable rate of return is 10%, the tax rate is 48% (varies with current tax law), and the depreciation method is straight-line. It is necessary to determine which unit would be more economical based on annual costs.

Solution: Because the lives of the two heat exchangers are unequal, the economic choice cannot be based merely on the discounted cash flow over a single life of each alternative. Instead, comparison must be made on the basis of equivalent uniform annual costs, commonly referred to as annual cost, as mentioned above. In this example, data are given for only the first two terms of Equation (4). The third term, which involves maintenance expense, and the fourth term, involving salvage value, are both assumed to be zero.

We calculate the present worth over the life of each alternative and the annual cash flow for each:

$$PW = -P + \Phi\, t\, P(P/A)_{10\%,n}$$

For carbon steel: $P = \$9500$, $n = 5$, $\Phi = 1/5 = 0.2$.

$$PW_{cs} = -9500 + (0.2)(0.48)(9500)(3.791); \quad PW = -\$6043.$$
$$A_{cs} = (PW)(A/P)_{10\%,5y} = (-6043)(0.2638) = \underline{-\$1594}.$$

For stainless steel: $P = \$26500$, $n = 11$, $\Phi = 1/11 = 0.0909$.

$$PW_{ss} = -26500 + (0.0909)(0.48)(26500)(6.495); \quad PW = -\$18989$$
$$A_{ss} = (PW)(A/P)_{10\%,15y} = (-18989)(0.1315) = \underline{-\$2497}.$$

Therefore, the carbon-steel heat exchanger with lower annual cost is the more economical alternative under these conditions. This example illustrates the not unusual situation for stainless steel, in which the useful equipment life, 15 years, exceeds the depreciation period, 11 years.

Example 2: Further study shows that the carbon-steel heat exchanger of Example 1 will require $3000 in yearly maintenance (painting, inhibitors, cathodic protection, etc.). Is carbon steel still the preferred alternative?

Solution: Maintenance costs are treated as expense items with a 1-year life. The third expense term of Equation (4) thus becomes significant. Other terms are the same, so that the annual cost, including maintenance, can be obtained simply by subtracting the after-tax maintenance costs from the result of Example 1 for carbon steel:

$$A_{cs} = -1594 - 3000(1 - 0.48)(0.9091) = \underline{-\$3012}.$$

It is evident, by comparison to $A_{ss} = -\$2497$ from Example, 1 that the stainless-steel heat exchanger is favored if $3000 maintenance costs are included in the analysis. In this calculation, the $3000 is treated as a payment at the end of 1-year and is discounted by $(P/A)_{10\%,1y}$. No discount would be included if it is treated as an initial cost at time zero. The conclusion is the same for either assumption, however, that stainless steel is the favored alternative.

Example 3: Plant painting. A paint system originally cost $0.38 per square foot to apply and has totally failed after 4 years. (a) If the paint system is renewed twice at the same cost for a total life of 12 years, what is the annual cost, assuming the first application is capitalized and those in the 4th and 8th years are expense? Assume interest rate of 10%; tax rate of 48%; straight-line depreciation. (b) Total maintenance could be avoided by biennial touchup (wire brush, spot primer, and topcoat). What is the most that can be spent on this preventative maintenance?

Solution:

(a) $PW = -P + t(1/n)(P)(P/A)_{10\%,4y} - P(1 - t)(P/F)_{10\%,4y} - P(1 - t)(P/F)_{10\%,8y}$
$= -0.38[1 - 0.48(1/4)(3.1699) + 0.52(0.6830) + 0.52(0.4665)]$
$= -\$0.4626$ per square foot.

Therefore,

$A = -0.4626(A/P)_{10\%,12y} = -0.4626(0.1468) = \underline{-\$0.068}$ *per square foot*.

(b) Biennial touchup cost (X) at 2, 4, 6, 8, and 10 years can be expressed as an equivalent annual touchup cost, X':

$X' = X[(P/F)_{10\%,2y} + (P/F)_{10\%,4y} + (P/F)_{10\%,6y} + (P/F)_{10\%,8y}$
$+ (P/F)_{10\%,10y}](A/P)_{10\%,12y}$
$= X(0.8264 + 0.6830 + 0.5645 + 0.4665 + 0.3855)(0.1468) = 0.4295X$

Touchup avoids repainting in the 4th and 8th years, which has an annual cost of:

$A = -0.38[(P/F)_{10\%,4y} + (P/F)_{10\%,8}](A/P)_{10\%,12y}$
$= -0.38(0.6830 + 0.4665)0.1468 = -\$0.0641$ per square foot.

Equating the annual costs: $0.4295X = 0.0641$; $X = \underline{\$0.1492}$ per square foot. The biennial touchup costs cannot exceed $0.149 per square foot.

Example 4: A plant storage tank requires either (a) a replacement to last 4 years with no maintenance, at a cost of $6900, or (b) a coating system on the existing tank to achieve the same life. How much can be spent now for coating installation if annual coating maintenance costs are 10% of the installation costs, tax rate is 48%, and interest is 10% with straight-line depreciation?

Solution:

For (a) replacement:

$PW = -P + tP(1/n)(P/A)_{10\%,25y}$
$PW = -690,000 + (0.4800)(1/25)(690,000)(9.0770) = \$569,700.$

For (b) rectifier installation:

$PW = -P + tP(1/n)(P/A)_{10\%,25y} - X(1 - t)(P/A)_{10\%,25y}$
$= -P[1 - t(1/n)(P/A)_{10\%,25y} + 0.06(1 - t)(P/A)_{10\%,25y}]$
$= -P[1 - (0.48)(1/25)(9.0770) + (0.06)(0.052)(9.0770)]$
$= -P(1.109).$

We may compare the present worth of the two alternatives because the life of both is the same, 25 years. It is unnecessary to calculate annual cost. Equating present worths:

$PW = -569,700 = -1.109P; P = \underline{\$513,700}.$

Rectifier installation is the preferred option, if the initial investment is less than $513,700.

References

1. A. G. Preban, *Metals Handbook*, Vol. 13, *Corrosion*, 9th ed., ASM International, Metals Park, OH, p. 509, 1987.
2. S. K. Brubaker, *Process Industries Corrosion*, B. J. Moniz and W. I. Pollock, eds., NACE, Houston, p. 243, 1986.
3. S. K. Coburn and Y-W. Kim, *Metals Handbook*, Vol. 13, *Corrosion*, 9th ed., ASM International, Metals Park, OH, p. 515, 1987.
4. R. M. Davison, T. DeBold, and M. J. Johnson, *Metals Handbook*, Vol. 13, *Corrosion*, 9th ed., ASM International, Metals Park, OH, p. 547, 1987.
5. N. Sridhar, *Metals Handbook*, Vol. 13, *Corrosion*, 9th ed., ASM International, Metals Park, OH, p. 643, 1987.
6. C. R. Brooks, *Metals Handbook, Desk Edition*, ASM International, Metals Park, OH, pp. 28–79, 1985.
7. R. W. Schutz, *Metals Handbook*, Vol. 13, *Corrosion*, 9th ed., ASM International, Metals Park, OH, p. 686, 1987.
8. P. J. Andersen, *Metals Handbook*, Vol. 13, *Corrosion*, 9th ed., ASM International, Metals Park, OH, p. 663, 1987.
9. N. R. Sorenson and R. B. Diegle, *Metals Handbook*, Vol. 13, *Corrosion*, 9th ed., ASM International, Metals Park, OH, p. 864, 1987.
10. P. Elliot, *Metals Handbook*, Vol. 13, *Corrosion*, 9th ed., ASM International, Metals Park, OH, p. 339, 1987.
11. J. A. Charles and F. A. A. Crane, *Selection and Use of Engineering Materials*, 2nd ed., Butterworths, Sevenoaks, Kent, England, p. 11, 1989.
12. D. Fyfe, *Corrosion*, Vol. II, 2nd ed., L. L. Shrier, ed., Newnes-Butterworths, Sevenoaks, Kent, England, p. 10:12, 1976.
13. C. G. Munger, *Corrosion Prevention by Protective Coatings*, NACE, Houston, p. 174, 1984.
14. Engineering Economy, Z94.5, American National Standards Institute-Institute of Industrial Engineers.
15. E. D. Verink, *Metals Handbook*, Vol. 13, *Corrosion*, 9th ed., ASM International, Metals Park, OH, p. 369, 1987.
16. E. D. Verink, Corrosion/86, Paper 383, NACE, Houston, 1986.
17. E. L. Grant, W. G. Ireson, and R. S. Leavenworth, *Principles of Engineering Economy*, 8th ed., Wiley, New York, 1990.
18. J. M. Smith, *Financial Analysis and Business Decisions on the Pocket Calculator*, Wiley, New York, 1976.
19. D. G. Newnan, *Engineering Economic Analysis*, Engineering Press, San Jose, CA, p. 195, 1976.
20. *A Complete Guide to the Tax Reform Act of 1986*, Prentice-Hall, Paramus, NJ, p. 205, 1986.

Exercises

15-1. Suggest the most appropriate alloys/materials for the following applications. Include any recommended corrosion prevention measures:

a. knife blades which are resistant to rusting in household and commercial food processing.
b. buried storage tanks for automotive fuel storage.
c. concentrators containing boiling hydrochloric acid.
d. heat exchanger tubing exposed to water at 110°C with appreciable chlorides.
e. pipe to transport 50% nitric acid at 90°C in a chemical process plant.
f. truck tanks to transport 95% sulfuric acid.
g. seawater condensers in a naval ship power plant.
h. pipe to transport hot chloride bearing sulfuric acid in a chemical plant.
i. storage vessels for 60% caustic at 50°C and atmospheric pressure in a paper manufacturing plant.
j. mixing vessels to dilute concetrated pure hydrochloric acid at room temperature.
k. aircraft landing gear.
l. food-processing equipment of minimum cost.
m. concentrators containing boiling nitric acid.
n. architectural trim around windows in a high-rise office building.
o. large storage tanks dilute nonoxidizing sulfuric acid.

15-2. You have been recently hired as an engineer in operations of a new chemical process plant which is about to commence operation. One of your assignments is plant maintenance which includes corrosion. List the steps that you should take to anticipate any corrosion problems during startup and subsequent routine plant operation. Assume that you have some junior engineers or technicians reporting to you who will be responsible for conducting many of the measures which you recommend.

15-3. $100,000 has been set aside by management for repairs which will be required by corrosion. The funds will be available in 5 years. How much will be available if the principal ($100,000) has been deposited in an account paying 8%, compounded annually?

15-4. A new investor offers to pay the same interest on principal in Problem 15-3 but agrees to compound quarterly. Is this a better deal? How much will be gained (or lost) over the conditions of Problem 15-3?

15-5. Management agrees only to have $100,000 available for repairs required by corrosion at the end of 5 years in Problem 15-3. How much must they deposit now for $100,000 to be paid in 5 years for 8% interest compounded annually?

15-6. More careful evaluation leads to the conclusion that uniform annual payments for corrosion prevention are required over 5 years instead of a lump sum payment of $100,000 as suggested in Problem 15-3. How much money must be available at the beginning of the project to guarantee $20,000 per year for 5 years assuming 8% compounded annually?

15-7. A high-quality epoxy paint system for a large water system storage tank would cost $40,000 and would last for 12 years. A lower cost vinyl system would cost $30,000 to install initially and then after 8 years another $20,000 for renewal to extend the total life to 12 years. Which is the more economic alternative for an interest rate of 12% compounded annually?

15-8. A cast iron pump has an initial cost of $5000. What is the equivalent uniform annual cost over a life of 5 years for an interest rate of 10% compounded annually?

15-9. A cast stainless steel pump identical to the one in Problem 15-8 costs $10,000, has a life of 12 years and a salvage value of $2000. Which is more economical, cast iron or cast stainless steel, assuming the same interest rate?

Compositions given are approximate and are included for convenience in the discussion of this book. Authoritative references published by ASM-International and ASTM should be consulted for exact specifications of chemical composition and other alloy properties.

Composition Ranges for AISI-SAE Standard Carbon Steels—Structural Shapes, Plate, Strip, Sheet and Welded Tubing

AISI-SAE designation	UNS designation	C	Mn	AISI-SAE designation	UNS designation	C	Mn
1006	G10060 .... 0.08 max		0.25–0.45	1043	G10430 .... 0.39–0.47		0.70–1.00
1008	G10080 .... 0.10 max		0.25–0.50	1045	G10450 .... 0.42–0.50		0.60–0.90
1009	G10090 .... 0.15 max		0.60 max	1046	G10460 .... 0.42–0.50		0.70–1.00
1010	G10100 .... 0.08–0.13		0.30–0.60	1049	G10490 .... 0.45–0.53		0.60–0.90
1012	G10120 .... 0.10–0.15		0.30–0.60	1050	G10500 .... 0.47–0.55		0.60–0.90
1015	G10150 .... 0.12–0.18		0.30–0.60	1055	G10550 .... 0.52–0.60		0.60–0.90
1016	G10160 .... 0.12–0.18		0.60–0.90	1060	G10600 .... 0.55–0.66		0.60–0.90
1017	G10170 .... 0.14–0.20		0.30–0.60	1064	G10640 .... 0.59–0.70		0.50–0.80
1018	G10180 .... 0.14–0.20		0.60–0.90	1065	G10650 .... 0.59–0.70		0.60–0.90
1019	G10190 .... 0.14–0.20		0.70–1.00	1070	G10700 .... 0.65–0.76		0.60–0.90
1020	G10200 .... 0.17–0.23		0.30–0.60	1074	G10740 .... 0.69–0.80		0.50–0.80
1021	G10210 .... 0.17–0.23		0.60–0.90	1078	G10780 .... 0.72–0.86		0.30–0.60
1022	G10220 .... 0.17–0.23		0.70–1.00	1080	G10800 .... 0.74–0.88		0.60–0.90
1023	G10230 .... 0.19–0.25		0.30–0.60	1084	G10840 .... 0.80–0.94		0.60–0.90
1025	G10250 .... 0.22–0.28		0.30–0.60	1085	G10850 .... 0.80–0.94		0.70–1.00
1026	G10260 .... 0.22–0.28		0.60–0.90	1086	G10860 .... 0.80–0.94		0.30–0.50
1030	G10300 .... 0.27–0.34		0.60–0.90	1090	G10900 .... 0.84–0.98		0.60–0.90
1033	G10330 .... 0.29–0.36		0.70–1.00	1095	G10950 .... 0.90–1.04		0.30–0.50
1035	G10350 .... 0.31–0.38		0.60–0.90	1524(b)	G15240 .... 0.18–0.25		1.30–1.65
1037	G10370 .... 0.31–0.38		0.70–1.00	1527(b)	G15270 .... 0.22–0.29		1.20–1.55
1038	G10380 .... 0.34–0.42		0.60–0.90	1536(b)	G15360 .... 0.30–0.38		1.20–1.55
1039	G10390 .... 0.36–0.44		0.70–1.00	1541(b)	G15410 .... 0.36–0.45		1.30–1.65
1040	G10400 .... 0.36–0.44		0.60–0.90	1548(b)	G15480 .... 0.43–0.52		1.05–1.40
1042	G10420 .... 0.39–0.47		0.60–0.90	1552(b)	G15520 .... 0.46–0.55		1.20–1.55

Composition Ranges for AISI-SAE Standard Alloy Steels—Plates

AISI-SAE designation	UNS designation	C	Mn	Si	Cr	Ni	Mo
1330	G13300 .... 0.27–0.34		1.50–1.90	0.15–0.30	...	...	...
1335	G13350 .... 0.32–0.39		1.50–1.90	0.15–0.30	...	...	...
1340	G13400 .... 0.36–0.44		1.50–1.90	0.15–0.30	...	...	...
1345	G13450 .... 0.41–0.49		1.50–1.90	0.15–0.30	...	...	...
4118	G41180 .... 0.17–0.23		0.60–0.90	0.15–0.30	0.40–0.65	...	0.08–0.15
4130	G41300 .... 0.27–0.34		0.35–0.60	0.15–0.30	0.80–1.15	...	0.15–0.25
4135	G41350 .... 0.32–0.39		0.65–0.95	0.15–0.30	0.08–1.15	...	0.15–0.25
4137	G41370 .... 0.33–0.40		0.65–0.95	0.15–0.30	0.80–1.15	...	0.15–0.25
4140	G41400 .... 0.36–0.44		0.70–1.00	0.15–0.30	0.08–1.15	...	0.15–0.25
4142	G41420 .... 0.38–0.46		0.70–1.00	0.15–0.30	0.80–1.15	...	0.15–0.25
4145	G41450 .... 0.41–0.49		0.70–1.00	0.15–0.30	0.80–1.15	...	0.15–0.25
4340	G43400 .... 0.36–0.44		0.55–0.80	0.15–0.30	0.60–0.90	1.65–2.00	0.20–0.30
E4340	G43406 .... 0.37–0.44		0.60–0.85	0.15–0.30	0.65–0.90	1.65–2.00	0.20–0.30
4615	G46150 .... 0.12–0.18		0.40–0.65	0.15–0.30	...	1.65–2.00	0.20–0.30
4617	G46170 .... 0.15–0.21		0.40–0.65	0.15–0.30	...	1.65–2.00	0.20–0.30
4620	G46200 .... 0.16–0.22		0.40–0.65	0.15–0.30	...	1.65–2.00	0.20–0.30
5160	G51600 .... 0.54–0.65		0.70–1.00	0.15–0.30	0.60–0.90	...	...
6150	G61500 .... 0.46–0.54		0.60–0.90	0.15–0.30	0.80–1.15	...	...
8615	G86150 .... 0.12–0.18		0.60–0.90	0.15–0.30	0.35–0.60	0.40–0.70	0.15–0.25
8617	G86170 .... 0.15–0.21		0.60–0.90	0.15–0.30	0.35–0.60	0.40–0.70	0.15–0.25
8620	G86200 .... 0.17–0.23		0.60–0.90	0.15–0.30	0.35–0.60	0.40–0.70	0.15–0.25
8622	G86220 .... 0.19–0.25		0.60–0.90	0.15–0.30	0.35–0.60	0.40–0.70	0.15–0.25
8625	G86250 .... 0.22–0.29		0.60–0.90	0.15–0.30	0.35–0.60	0.40–0.70	0.15–0.25
8627	G86270 .... 0.24–0.31		0.60–0.90	0.15–0.30	0.35–0.60	0.40–0.70	0.15–0.25
8630	G86300 .... 0.27–0.34		0.60–0.90	0.15–0.30	0.35–0.60	0.40–0.70	0.15–0.25
8637	G86370 .... 0.33–0.40		0.70–1.00	0.15–0.30	0.35–0.60	0.40–0.70	0.15–0.25
8640	G86400 .... 0.36–0.44		0.70–1.00	0.15–0.30	0.35–0.60	0.40–0.70	0.15–0.25
8655	G86550 .... 0.49–0.60		0.70–1.00	0.15–0.30	0.35–0.60	0.40–0.70	0.15–0.25
8742	G87420 .... 0.38–0.46		0.70–1.00	0.15–0.30	0.35–0.60	0.40–0.70	0.20–0.30

Composition Ranges for Selected HSLA and Alloy Steel Plate (ASTM Specifications)

ASTM specifi-cation	Type or grade	UNS designation	C	Mn	P max	S max	Si	Cr	Ni	Mo	V	Other
A242	Type 1	K11510	0.15 max	1.00 max	0.45	0.05	...	...	...	...	...	0.20 min Cu
	Type 2	K12010	0.20 max	1.35 max	0.04	0.05	...	...	...	...	...	0.20 min Cu if both 0.5 Si and 0.5 Cr not present
A572	Grade 42	... ...	0.21 max	1.35 max	0.04	0.05	0.30 max	...	...	...	...	0.20 min Cu(a)
	Grade 45	... ...	0.22 max	1.35 max	0.04	0.05	0.30 max	...	...	...	...	0.20 min Cu(a)
	Grade 50	... ...	0.23 max	1.35 max	0.04	0.05	0.30 max	...	...	...	...	0.20 min Cu(a)
	Grade 55	... ...	0.25 max	1.35 max	0.04	0.05	0.30 max	...	...	...	...	0.20 min Cu(a)
	Grade 60	... ...	0.25 max	1.35 max	0.04	0.05	0.30 max	...	...	...	...	0.20 min Cu(a)
	Grade 65	... ...	0.26 max	1.65 max	0.04	0.05	0.30 max	...	...	...	...	0.20 min Cu(a)
A588	Grade A	K11430	0.10–0.19	0.90–1.25	0.04	0.05	0.15–0.30	0.40–0.65	...	...	0.02–0.10	0.25–0.40 Cu
	Grade B	K12043	0.20 max	0.75–1.25	0.04	0.05	0.15–0.30	0.40–0.70	0.25–0.50	...	0.01–0.10	0.20–0.40 Cu
	Grade C	K11538	0.15 max	0.80–1.35	0.04	0.05	0.15–0.30	0.30–0.50	0.25–0.50	...	0.01–0.10	0.20–0.50 Cu
	Grade D	K11552	0.10–0.20	0.75–1.25	0.04	0.05	0.50–0.90	0.50–0.90	...	...	...	0.30 max Cu; 0.05–0.15 Zr; 0.04 max Nb

Nominal Chemical Compositions of Some Typical Nickel-Base Alloys

Common alloy designation	UNS designation	C(a)	Nb	Cr	Cu	Fe	Mo	Ni	Si(a)	Ti	W	Other
Nickel												
200	N02200	0.1	...	...	0.25 max	0.4 max	...	99.2 min	0.15	0.1 max	...	...
201	N02201	0.02	...	...	0.25 max	0.4 max	...	99.0 min	0.15	0.1 max	...	...
Nickel-copper												
400	N04400	0.15	...	...	31.5	1.25	...	bal	0.5	...	...	...
R-405	N04405	0.15	...	...	31.5	1.25	...	bal	0.5	...	...	0.0435
Nickel-molybdenum												
B-2	N10665	0.01	...	1.0 max	...	2.0 max	28	bal	0.1	...	...	...
B	N10001	0.05	...	1.0 max	...	5.0	28	bal	1.0	...	...	...
Nickel-chromium-iron												
600	N06600	0.08	...	16.0	0.5 max	8.0	...	bal	0.5	0.3 max	...	...
601	N06601	...	...	23.0	...	14.1	...	bal	...	...	...	1.35Al
800	N08800	0.1	...	21.0	0.75 max	44.0	...	32.5	1.0	0.38	...	...
800H	N08810	0.08	...	21.0	0.75 max	44.0	...	32.5	1.0	0.38	...	...
Nickel-chromium-iron-molybdenum												
825	N08825	0.05	...	21.5	2.0	29.0	3.0	42	0.5	1.0	...	...
G	N06007	0.05	2.0	22.0	2.0	19.5	6.5	43	1.0	...	1.0 max	...
G-2/2550	N06975	0.03	...	24.5	1.0	20.0	6.0	48	1.0	1.0	...	...
G-3	N06985	0.015	0.8	22.0	2.0	19.5	7.0	44	1.0	...	1.5 max	...
H		0.03	...	22.0	...	19.0	9.0	42	1.0	...	2.0	...
G-30	N06030	0.03	0.8	29.5	2.0	15.0	5.5	43	1.0	...	2.5	...
Nickel-chromium-molybdenum-tungsten												
N	N10003	0.06	...	7.0	0.35 max	5.0 max	16.5	71	1.0	0.5 max	0.5 max	...
W	N10004	0.12	...	5.0	...	6.0	24.0	63	1.0	...	...	...
625	N06625	0.1	4.0	21.5	...	5.0 max	9.0	62	0.5	...	...	...
690	N06690	0.02	...	29.0	...	10.0	...	61	...	0.3	...	...
C-276	N10276	0.01	...	15.5	...	5.5	16.0	57	0.08	...	4.0	...
C-4	N06455	0.01	...	16.0	...	3.0 max	15.5	65	0.08	...	...	...
C-22	N06022	0.015	...	22.0	...	3.0 max	13.0	56	0.08	...	3.0	...
ALLCORR ..	N06110	0.15	2.0 max	30.0	...	...	10.0	53	...	1.5 max	4.0 max	...
Nickel-silicon												
D		0.12	...	1.0 max	3.0	2.0 max	...	86	9.5	...	...	...
Precipitation-hardening												
K-500	N05500	0.25	...	...	29.0	2.0 max	...	63	0.5 max	0.6	...	2.7Al
R-41	N07041	0.09	...	19.0	...	5.0 max	10.0	52	0.5 max	3.1	...	1.5Al
718	N07718	0.05	5.0	18.0	...	19	3.0	53	...	0.4 max	...	...
X-750	N07750	...	0.9	15.5	...	7.0	...	bal	...	2.5	...	...
925	N09925	0.02	...	21.0	2.0	28	3.0	43.0	...	2.1	...	...

(a) Maximum

Nominal Compositions of Titanium Alloys

Common alloy designation	UNS designation	Nominal composition, %	ASTM grade	Alloy type
Grade 1	R50250	Unalloyed titanium	1	α
Grade 2	R50400	Unalloyed titanium	2	α
Grade 3	R50550	Unalloyed titanium	3	α
Grade 4	R50700	Unalloyed titanium	4	α
Ti-Pd	R52400/R52250	Ti-0.15Pd	7/11	α
Grade 12	R53400	Ti-0.3Mo-0.8Ni	12	Near-α
Ti-3-2.5	…	Ti-3Al-2.5V	9	Near-α
Ti-6-4	R56400	Ti-6Al-4V	5	α-β
Ti-6-2-1-.8	…	Ti-6Al-2Nb-1Ta-0.8Mo	…	Near-α
Ti-5Ta	…	Ti-5Ta	…	Near-α
Ti-5-2.5	…	Ti-5Al-2.5Sn	…	α
Ti-8-1-1	…	Ti-8Al-1V-1Mo	…	Near-α
Ti-6-2-4-2	…	Ti-6Al-2Sn-4Zr-2Mo	…	Near-α
Ti-4-3-1	…	Ti-4Al-3Mo-1V	…	α-β
Ti-550	…	Ti-4Al-2Sn-4Mo-0.5Si	…	α-β
Ti-6-6-2	…	…	…	α-β
Corona 5	…	Ti-4.5Al-1.5Cr-5Mo	…	α-β
Ti-6-2-4-6	R56260	Ti-6Al-2Sn-4Zr-6Mo	…	α-β
Ti-10-2-3	…	Ti-10V-2Fe-3Al	…	Near-β
Transage 129	…	Ti-2Al-11.5V-2Sn-10Zr	…	Near-β
Transage 207	…	Ti-2.5Al-2Sn-9Zr-8Mo	…	Near-β
Ti-15-3-3-3	…	Ti-15V-3Sn-3Cr-3Al	…	β
Ti-3-8-6-4-4	R58640	Ti-3Al-8V-6Cr-4Zr-4Mo	…	β
Ti-13-11-3	…	Ti-3Al-13V-11Cr	…	β
Ti-8-8-2-3	…	Ti-8V-8Mo-3Al-2Fe	…	β
Ti-15-5	…	Ti-15Mo-5Zr	…	β

Compositions of Typical Wrought Coppers and Copper Alloys

Alloy number (and name)	Nominal composition, %
C10100 (oxygen-free electronic)	99.99 Cu
C11000 (electrolytic tough pitch copper)	99.90 Cu, 0.04 O
C22000 (commercial bronze, 90%)	90.0 Cu, 10.0 Zn
C26000 (cartridge brass, 70%)	70.0 Cu, 30.0 Zn
C26800, C27000 (yellow brass)	65.0 Cu, 35.0 Zn
C28000 (Muntz metal)	60.0 Cu, 40.0 Zn
C38500 (architectural bronze)	57.0 Cu, 3.0 Pb
C46400 to C46700 (naval brass)	60.0 Cu, 39.25 Zn, 0.75 Sn
C48200 (naval brass, medium-leaded)	60.5 Cu, 0.7 Pb, 0.8 Sn, 38.0 Zn
C48500 (leaded naval brass)	60.0 Cu, 1.75 Pb, 37.5 Zn, 0.75 Sn
C50500 (phosphor bronze, 1.25% E)	98.75 Cu, 1.25 Sn, trace P
C51000 (phosphor bronze, 5% A)	95.0 Cu, 5.0 Sn, trace P
C51100	95.6 Cu, 4.2 Sn, 0.2 P
C52100 (phosphor bronze, 8% C)	92.0 Cu, 8.0 Sn, trace P
C52400 (phosphor bronze, 10% D)	90.0 Cu, 10.0 Sn, trace P
C60800 (aluminum bronze, 5%)	95.0 Cu, 5.0 Al
C61000	92.0 Cu, 8.0 Al
C65100 (low-silicon bronze, B)	98.5 Cu, 1.5 Si
C65500 (high-silicon bronze, A)	97.0 Cu, 3.0 Si
C66700 (manganese brass)	70.0 Cu, 28.8 Zn, 1.2 Mn
C67400	58.5 Cu, 36.5 Zn, 1.2 Al, 2.8 Mn, 1.0 Sn
C67500 (manganese bronze, A)	58.5 Cu, 1.4 Fe, 39.0 Zn, 1.0 Sn, 0.1 Mn
C68700 (aluminum brass, arsenical)	77.5 Cu, 20.5 Zn, 2.0 Al, 0.1 As
C69400 (silicon red brass)	81.5 Cu, 14.5 Zn, 4.0 Si
C70600 (copper nickel, 10%)	88.7 Cu, 1.3 Fe, 10.0 Ni
C71000 (copper nickel, 20%)	79.0 Cu, 21.0 Ni
C71500 (copper nickel, 30%)	70.0 Cu, 30.0 Ni
C71700	67.8 Cu, 0.7 Fe, 31.0 Ni, 0.5 Be
C72500	88.2 Cu, 9.5 Ni, 2.3 Sn
C73500	72.0 Cu, 10.0 Zn, 18.0 Ni

AA No.	Si	Fe	Cu	Mn	Mg	Cr	Ni	Zn	Ti	Others Each	Others Total	Al
1100....	1.0 Si + Fe		0.05–0.20	0.05	...	...	...	0.10	...	0.05	0.15	99.00 min
2014....	0.50–1.2	0.7	3.9–5.0	0.40–1.2	0.20–0.8	0.10	...	0.25	0.15	0.05	0.15	Rem
2024....	0.50	0.50	3.8–4.9	0.30–0.9	1.2–1.8	0.10	...	0.25	0.15	0.05	0.15	Rem
3003....	0.6	0.7	0.05–0.20	1.0–1.5	...	...	...	0.10	...	0.05	0.15	Rem
4043....	4.5–6.0	0.8	0.30	0.05	0.05	...	...	0.10	0.20	0.05	0.15	Rem
5005....	0.30	0.7	0.20	0.20	0.50–1.1	0.10	...	0.25	...	0.05	0.15	Rem
5050....	0.40	0.7	0.20	0.10	1.1–1.8	0.10	...	0.25	...	0.05	0.15	Rem
5052....	0.25	0.40	0.10	0.10	2.2–2.8	0.15–0.35	...	0.10	...	0.05	0.15	Rem
5154....	0.25	0.40	0.10	0.10	3.1–3.9	0.15–0.35	...	0.20	0.20	0.05	0.15	Rem
5183....	0.40	0.40	0.10	0.50–1.0	4.3–5.2	0.05–0.25	...	0.25	0.15	0.05	0.15	Rem
5252....	0.08	0.10	0.10	0.10	2.2–2.8	...	...	0.05	...	0.03	0.10	Rem
5254....	0.45 Si + Fe		0.05	0.01	3.1–3.9	0.15–0.35	...	0.20	0.05	0.05	0.15	Rem
5356....	0.25	0.40	0.10	0.05–0.20	4.5–5.5	0.05–0.20	...	0.10	0.06–0.20	0.05	0.15	Rem
5454....	0.25	0.40	0.10	0.50–1.0	2.4–3.0	0.05–0.20	...	0.25	0.20	0.05	0.15	Rem
5456....	0.25	0.40	0.10	0.50–1.0	4.7–5.5	0.05–0.20	...	0.25	0.20	0.05	0.15	Rem
6061....	0.40–0.8	0.7	0.15–0.40	0.15	0.8–1.2	0.04–0.35	...	0.25	0.15	0.05	0.15	Rem
6063....	0.20–0.6	0.35	0.10	0.10	0.45–0.9	0.10	...	0.10	0.10	0.05	0.15	Rem
6066....	0.9–1.8	0.50	0.7–1.2	0.6–1.1	0.8–1.4	0.40	...	0.25	0.20	0.05	0.15	Rem
6151....	0.6–1.2	1.0	0.35	0.20	0.45–0.8	0.15–0.35	...	0.25	0.15	0.05	0.15	Rem
6162....	0.40–0.8	0.50	0.20	0.10	0.7–1.1	0.10	...	0.25	0.10	0.05	0.15	Rem
6201....	0.50–0.9	0.50	0.10	0.03	0.6–0.9	0.03	...	0.10	...	0.03	0.10	Rem
6262....	0.40–0.8	0.7	0.15–0.40	0.15	0.8–1.2	0.04–0.14	...	0.25	0.15	0.05	0.15	Rem
6351....	0.7–1.3	0.50	0.10	0.40–0.8	0.40–0.8	...	...	0.20	0.20	0.05	0.15	Rem
6463....	0.20–0.6	0.15	0.20	0.05	0.45–0.9	...	...	0.05	...	0.05	0.15	Rem
6951....	0.20–0.50	0.8	0.15–0.40	0.10	0.40–0.8	...	...	0.20	...	0.05	0.15	Rem
7008....	0.10	0.10	0.05	0.05	0.7–1.4	0.12–0.25	...	4.5–5.5	0.05	0.05	0.10	Rem
7016....	.010	0.12	0.45–1.0	0.03	0.8–1.4	...	...	4.0–5.0	0.03	0.03	0.10	Rem
7021....	0.25	0.40	0.25	0.10	1.2–1.8	0.05	...	5.0–6.0	0.10	0.05	0.15	Rem
7029....	0.10	0.12	0.50–0.9	0.03	1.3–2.0	...	...	4.2–5.2	0.05	0.03	0.10	Rem
7049....	0.25	0.35	1.2–1.9	0.20	2.0–2.9	0.10–0.22	...	7.2–8.2	0.10	0.05	0.15	Rem
7050....	0.12	0.15	2.0–2.6	0.10	1.9–2.6	0.04	...	5.7–6.7	0.06	0.05	0.15	Rem
7072....	0.7 Si + Fe		0.10	0.10	0.10	...	...	0.8–1.3	...	0.05	0.15	Rem
7075....	0.40	0.50	1.2–2.0	0.30	2.1–2.9	0.18–0.28	...	5.1–6.1	0.20	0.05	0.15	Rem
7175....	0.15	0.20	1.2–2.0	0.10	2.1–2.9	0.18–0.28	...	5.1–6.1	0.10	0.05	0.15	Rem
7178....	0.40	0.50	1.6–2.4	0.30	2.4–3.1	0.18–0.28	...	6.3–7.3	0.20	0.05	0.15	Rem
7475....	0.10	0.12	1.2–1.9	0.06	1.9–2.6	0.18–0.25	...	5.2–6.2	0.06	0.05	0.15	Rem

Compositions of Typical Stainless Steels

UNS designation	AISI type	C	Mn	P	S	Si	Cr	Ni	Mo	Others
Austenitic grades										
S30100	301	0.15	2.00	0.045	0.03	1.00	16.00-18.00	6.00-8.00	...	...
S30400	304	0.08	2.00	0.045	0.03	1.00	18.00-20.00	8.00-10.50	...	...
S30403	304L	0.03	2.00	0.045	0.03	1.00	18.00-20.00	8.00-12.00	...	...
S30430	S30430	0.08	2.00	0.045	0.03	1.00	17.00-19.00	8.00-10.00	...	3.00-4.00Cu
	304N	0.08	2.00	0.045	0.03	1.00	18.00-20.00	8.00-10.50	...	0.10-0.16N
S30500	305	0.12	2.00	0.045	0.03	1.00	17.00-19.00	10.50-13.00	...	...
S30800	308	0.08	2.00	0.045	0.03	1.00	19.00-21.00	10.00-12.00	...	...
S30900	309	0.2	2.00	0.045	0.03	1.00	22.00-24.00	12.00-15.00	...	...
	309S	0.08	2.00	0.045	0.03	1.00	22.00-24.00	12.00-15.00	...	...
S31000	310	0.25	2.00	0.045	0.03	1.50	24.00-26.00	19.00-22.00	...	...
	310S	0.08	2.00	0.045	0.03	1.50	24.00-26.00	19.00-22.00	...	...
S31400	314	0.25	2.00	0.045	0.03	1.50-3.00	23.00-26.00	19.00-22.00	...	...
S31600	316	0.08	2.00	0.045	0.03	1.00	16.00-18.00	10.00-14.00	2.00-3.00	...
	316F	0.08	2.00	0.2	0.10 min	1.00	16.00-18.00	10.00-14.00	1.75-2.50	...
	316L	0.03	2.00	0.045	0.03	1.00	16.00-18.00	10.00-14.00	2.00-3.00	...
	316N	0.08	2.00	0.045	0.03	1.00	16.00-18.00	10.00-14.00	2.00-3.00	0.10-0.16N
S31700	317	0.08	2.00	0.045	0.03	1.00	18.00-20.00	11.00-15.00	3.00-4.00	...
S31703	317L	0.03	2.00	0.045	0.03	1.00	18.00-20.00	11.00-15.00	3.00-4.00	...
S32100	321	0.08	2.00	0.045	0.03	1.00	17.00-19.00	9.00-12.00	...	Ti:5 × C min
	329	0.10	2.00	0.04	0.03	1.00	25.00-30.00	3.00-6.00	1.00-2.00	...
	330	0.08	2.00	0.04	0.03	0.75-1.50	17.00-20.00	34.00-37.00	...	0.10Ta, 0.20Nb
S34700	347	0.08	2.00	0.045	0.03	1.00	17.00-19.00	9.00-13.00	...	Nb:10 × C min
S34800	348	0.08	2.00	0.045	0.03	1.00	17.00-19.00	9.00-13.00	...	Nb:10 × C min
	384	0.08	2.00	0.045	0.03	1.00	15.00-17.00	17.00-19.00	...	...
Ferritic grades										
S40500	405	0.08	1.00	0.04	0.03	1.00	11.50-14.50	...	...	0.10-0.30Al
S40900	409	0.08	1.00	0.045	0.045	1.00	10.50-11.75	...	...	Ti:6 × C-0.75
S42900	429	0.12	1.00	0.04	0.03	1.00	14.00-16.00	...	...	...
S43000	430	0.12	1.00	0.04	0.03	1.00	16.00-18.00	...	...	...
	430F	0.12	1.25	0.06	0.15	1.00	16.00-18.00	...	0.60	...
	430FSe	0.12	1.25	0.06	0.06	1.00	16.00-18.00	...	...	0.15Si min
S43400	434	0.12	1.00	0.04	0.03	1.00	16.00-18.00	...	0.75-1.25	...
S43600	436	0.12	1.00	0.04	0.03	1.00	16.00-18.00	...	0.75-1.25	Nb:5 × C-0.70
S44200	442	0.20	1.00	0.04	0.03	1.00	18.00-23.00	...	...	...
S44400	444	0.25	1.00	0.04	0.03	1.00	17.50-19.50	...	...	(Ti+Nb): 0.2+4(C+N)-0.8
S44600	446	0.20	1.50	0.04	0.03	1.00	23.00-27.00	...	...	0.25N
Martensitic grades										
S40300	403	0.15	1.00	0.04	0.03	0.50	11.50-13.00	...	...	...
S41000	410	0.15	1.00	0.04	0.03	1.00	11.50-13.50	...	...	...

Compositions of Typical Stainless Steels (Cont.)

UNS designation	AISI type	Composition, %(a)								
		C	Mn	P	S	Si	Cr	Ni	Mo	Others
S41400	414	0.15	1.00	0.04	0.03	1.00	11.50–13.50	1.25–2.50	…	…
S41600	416	0.15	1.25	0.06	0.15 min	1.00	12.00–14.00	…	0.60	…
	416Se	0.15	1.25	0.06	0.06	1.00	12.00–14.00	…	…	0.15Se min
S42000	420	0.15 min	1.00	0.04	0.03	1.00	12.00–14.00	…	…	…
	420F	0.15 min	1.25	0.06	0.15 min	1.00	12.00–14.00	…	0.60	…
S42200	422	0.20–0.25	1.00	0.025	0.025	0.75	11.00–13.00	0.50–1.00	0.75–1.25	0.15–0.30V, 0.75–1.25W
S43100	431	0.20	1.00	0.04	0.03	1.00	15.00–17.00	1.25–2.50	…	…
	440A	0.60–0.75	1.00	0.04	0.03	1.00	16.00–18.00	…	0.75	…
	440B	0.75–0.95	1.00	0.04	0.03	1.00	16.00–18.00	…	0.75	…
	440C	0.95–1.20	1.00	0.04	0.03	1.00	16.00–18.00	…	0.75	…
Duplex grades										
S31803	2205	0.03	2.00	0.03	0.02	1.00	21.0–23.0	4.50–6.50	2.50–3.50	0.08–0.2N
S31200	44LN	0.03	2.00	0.045	0.03	1.00	24.0–26.0	5.50–6.50	1.20–2.00	0.14–0.2N
S31260	DP-3	0.03	1.00	0.03	0.03	0.75	24.0–26.0	5.50–7.50	2.50–3.50	0.2–0.8Cu, 0.1–0.3N
S31500	3RE60	0.03	1.20–2.00	0.03	0.03	1.4–2.0	18.0–19.0	4.25–5.25	2.50–3.00	…
S32550	Ferralium 255	0.04	1.50	0.04	0.03	1.00	24.0–27.0	4.50–6.50	2.00–4.00	1.5–2.5Cu, 0.1–0.25N
S32950	7Mo-PLUS	0.03	2.00	0.035	0.01	0.60	26.0–29.0	3.50–5.20	1.00–2.50	0.15–0.35N

(a) Maximum unless otherwise indicated; all compositions include balance of iron.

Index